Studies in Logic
Volume 31

Nonmonotonic Reasoning
Essays Celebrating its 30th Anniversary

Studies in Logic Series Editor
Dov Gabbay   dov.gabbay@kcl.ac.uk

Nonmonotonic Reasoning

Essays Celebrating its 30th Anniversary

Edited by

Gerhard Brewka,

Victor W. Marek,

and

Mirosław Truszczyński

ISBN 978-1-84890-042-4

College Publications
Scientific Director: Dov Gabbay
Managing Director: Jane Spurr
Department of Informatics
King's College London, Strand, London WC2R 2LS, UK

http://www.collegepublications.co.uk

Original cover design by orchid creative   www.orchidcreative.co.uk
Printed by Lightning Source, Milton Keynes, UK

To John McCarthy
In recognition of his pioneering role
in the development of nonmonotonic reasoning
and the field of artificial intelligence.

Contents

Preface

The publication of the seminal special issue on nonmonotonic logics by the Artificial Intelligence Journal in 1980 resulted in a new area of research in knowledge representation. This development changed the mainstream paradigm of logic that originated in antiquity. It established an important area of mathematical logic and resulted in exciting discoveries of logical techniques that created new bridges between logic, knowledge representation and computation. The research contributed to mathematical logic, computer science and philosophy. Importantly, it also changed the perspective on applications of logic.

To sum up the experience of the first 30 years of nonmonotonic logics and to map paths into the future we organized the conference NonMon@30 – Thirty Years of Nonmonotonic Reasoning. The conference was held in Lexington, KY, USA, October 22-25, 2010. It comprised eighteen invited talks and several technical presentations, and gathered about fifty participants. This volume consists of the texts based on twelve of the invited presentations. In addition, some of the conference technical papers will appear in a forthcoming special issue of the Journal of Artificial Intelligence Research.

The invited talks that form this volume were contributed by

- Mario Alviano, Francesco Calimeri, Wolfgang Faber, Giovambattista Ianni and Nicola Leone
- Alexander Bochman
- Minh Dao-Tran, Thomas Eiter, Michael Fink and Thomas Krennwallner
- James Delgrande
- Marc Denecker, Victor Marek and Mirosław Truszczyński
- Didier Dubois and Henri Prade
- Michael Gelfond
- Michael Kaminski
- Victor Marek, Ilkka Niemelä and Mirosław Truszczyński
- Victor Marek and Jeffrey Remmel
- Jack Minker
- David Pearce and Levan Uridia

Most of these talks offer surveys or personal accounts of major subfields of nonmonotonic reasoning. Some present original results concerning topics of current interest.

Alviano et al. consider several versions of answer-set programming with uninterpreted function symbols and survey complexity results concerning this general and practically important extension of the standard answer-set programming formalism. Bochman gives a perspective of a two-layerd nature of nonmonotonic reasoning, in which one layer is some standard monotonic logic and the second one specifies the way default assumptions are made. Dao-Tran et al. consider nonmonotonic multi-context systems that are designed to represent knowledge contained in interconnected heterogeneous knowledge bases. They study the case when the knowledge bases evolve over time. Delgrande examines the meaning and the role of a default in non-monotonic reasoning. Denecker et al. focus on defaults as understood by Reiter. They argue that Reiter's defaults are essentially autoepistemic propositions, discuss problems with using Reiter's defaults to represent informal defaults that arise in many knowledge representation applications and propose that Reiter's logic is in some ways a better autoepistemic logic than that by Moore. Dubois and Prade cover topics of nonmonotonic reasoning as they relate to logical treatments of uncertainty. Gelfond offers his personal perspective on knowledge representation languages designed on top of logic programming and discusses how it has been evolving over the past two decades. Kaminski presents an account of his long-term project to develop first-order extensions of major nonmonotonic logics. Marek et al. reminisce on the authors' intertwined paths that took them from research on nonmonotonic logics in late 1980s and early 1990s to the point where they were able to formulate explicitly the paradigm of answer-set programming. Next, Marek and Remmel discuss extensions of answer-set programming with arbitrary constraints, including constraints on sets, as well as an extension that allows algorithms to be embedded in rule bodies. Minker provides a broad overview of the early days of nonmonotonic reasoning focusing on key events and individuals that shaped the field of nonmonotonic reasoning. Pearce and Uridia discuss how some familiar embeddings of stable models into modal logics can be derived from some translations that are well known in non-classical logic.

These papers offer unique insights into many of the key questions that have been driving the development of nonmonotonic reasoning and suggest problems worthy of consideration in the future. They paint the picture of the field that now has a well-established tradition, and remains vibrant and relevant to long-term goals of artificial intelligence. We hope the volume will be a useful reference for everybody interested in the past, present and future of nonmonotonic reasoning.

Finally, it gives us a great deal of pleasure to note that the participants of the conference felt it most appropriate to dedicate this volume to John McCarthy, one of the founders of the field. While absent physically at the meeting, John's presence was felt in every presentation made.

Gerhard Brewka, Victor Marek and Mirek Truszczyński
July 2011

Function Symbols in ASP: Overview and Perspectives

Mario Alviano
Francesco Calimeri
Wolfgang Faber
Giovambattista Ianni
Nicola Leone
Department of Mathematics
University of Calabria
P.te P. Bucci, Cubo 30B
I-87036 Rende, Italy

Abstract: Answer Set Programming (ASP) is a highly expressive language that is widely used for knowledge representation and reasoning in many application scenarios. Thanks to disjunction and negation, the language allows the use of nondeterministic definitions for modeling complex problems in computer science, in particular in Artificial Intelligence. Traditionally, ASP has often been used as a first-order language without function symbols, similar to Datalog, in order to deal with finite structures only. More recently, also uninterpreted function symbols have been frequently considered in the setting of ASP, enabling a natural representation of recursive structures. Function symbols can be used for encoding strings, lists, stacks, trees and many other common data structures. However, the common reasoning tasks are undecidable for programs with no restrictions on the usage of function symbols. Therefore, identifying relevant classes of programs with decidable reasoning is important for practical applications, and many authors have addressed this issue in the past decade.

This article provides a survey of the decidability results for ASP programs with functions. We classify the decidable ASP programs in three main groups: programs allowing for finite bottom-up evaluations; programs suitable for finite top-down evaluations; programs characterized by finite representations of stable models. We focus on the decidability of ground reasoning and computability of non-ground reasoning. Moreover, we consider decidability of coherence checking and of class membership; expressiveness issues are briefly discussed as well.

1 Introduction

Logic Programming (LP) under the answer set or stable model semantics, often called Answer Set Programming (ASP), is a convenient and effective method for declarative knowledge representation and reasoning [Baral, 2003, Gelfond and Lifschitz, 1991]. The success of ASP in many practical applications has been encouraged by the availability of some efficient inference engines, such as DLV [Leone, Pfeifer, Faber, Eiter, Gottlob, Perri, and Scarcello, 2006], SMODELS [Simons, Niemelä, and Soininen, 2002], CMODELS [Lierler, 2005], and CLASP [Gebser, Kaufmann, Neumann, and Schaub, 2007a].

ASP allows for disjunction in rule heads and nonmonotonic negation in rule bodies. Over finite structures, the language allows for expressing all properties in the second level of the polynomial hierarchy. Restricting terms to constants and variables guarantees structures to be finite. ASP with this restriction has been successfully used for knowledge representation and reasoning in numerous applications. If uninterpreted function symbols of positive arities are permitted, instead, the expressive power of the language increases considerably, up to the first level of the analytical hierarchy if disjunction or recursive negation are allowed [Cadoli and Schaerf, 1993]. However, this high expressive power implies that the common reasoning tasks are undecidable for programs with function symbols. This is a consequence of the fact that even for Horn programs these tasks are undecidable [Tärnlund, 1977]. In the past decade, several efforts have been made to identify large classes of ASP programs with function symbols for which some important reasoning tasks are still decidable. A couple of interesting classes have already been discovered, and the research in this area is quite active; the reader may refer, for instance, to papers by Baselice, Bonatti, and Criscuolo [2009], Bonatti [2002, 2004], Cabalar [2008], Calimeri, Cozza, Ianni, and Leone [2008, 2009], Eiter and Simkus [2009b], Gebser, Schaub, and Thiele [2007b], Lierler and Lifschitz [2009], Lin and Wang [2008], Simkus and Eiter [2007] and Syrjänen [2001].

In this work, a survey of the main results achieved in this research area is given. In particular, three groups of ASP fragments with decidable reasoning are identified: *bottom-up computable*, *top-down computable* and *finitely representable stable models*.

- Programs in the first group are characterized by the existence of a finite ground program, which is equivalent to the (infinite) program instantiation and can be obtained by a bottom-up computation. This allows for stable model computation and query answering by means of the standard stable model search techniques over these finite ground programs. Therefore, all of the programs in this first group are characterized by decidable ground and non-ground reasoning as well as a decidable coherence check. Classes of programs in this group are ω-*restricted* [Syrjänen, 2001], λ-*restricted* [Gebser, Schaub, and Thiele, 2007b], *finite domain* [Calimeri et al., 2008], *argument restricted* [Lierler and Lifschitz, 2009] and *finitely ground programs* [Calimeri et al., 2008].

- Classes in the second group have been defined having query answering and top-down computation in mind. In contrast to the first group, programs in these classes are usually characterized by an infinite number of stable models, each one possibly comprising an infinite number of atoms. To guarantee decidability of reasoning, it is necessary that a finite number of atoms is sufficient for answering an input query. Moreover, these atoms need to be effectively identifiable. Classes in this group are *FP2* [Baselice and Bonatti, 2010], *positive/stratified finitely recursive* [Calimeri et al., 2009] and [Alviano, Faber, and Leone, 2010], and *finitary programs* [Bonatti, 2002, 2004].

- Programs in the third group are characterized by finitely representable sets of stable models, where a set can potentially comprise infinitely many stable models of possibly infinite size. Typically, these stable models have the shape of a forest of trees. Classes in this group are *FDNC* [Simkus and Eiter, 2007] and *bidirectional programs* [Eiter and Simkus, 2009a].

For each of these classes we highlight whether decidability of ground reasoning and/or computability of non-ground reasoning are guaranteed. *Ground reasoning* consists of checking the presence of a specific ground atom among the consequences of a program, while *non-ground reasoning* means computing all answers for a given non-ground query. Decidability of coherence checks and of class membership are also considered. A *coherence check* consists of establishing whether a given program has at least one stable model. By *class membership*, instead, we refer to establishing whether a given program (possibly associated with a query) belongs to the class of programs at hand. Expressiveness is briefly discussed as well, highlighting whether a class still allows for representing every recursive relation or not. Furthermore, we give a description of the efforts for endowing the ASP system DLV with function symbols. In fact, DLV (and its branches) can compute stable models, answers to queries and decide class membership for a number of the surveyed classes.

The remainder of this article is organized as follows. First, syntax and semantics of ASP are introduced in Section 2. After that, classes of ASP programs characterized by decidable reasoning tasks are presented in Section 3. Finally, DLV– a system supporting a powerful class of function symbols – is presented in Section 4 and few concluding comments are given in Section 5.

2 ASP with Function Symbols

In this section, we recall the basics of ASP with uninterpreted function symbols. We start by introducing syntax and semantics of the language. Then, we briefly discuss undecidability of reasoning for arbitrary programs with function symbols.

2.1 Syntax

A *term* is either a *variable* or a *functional term*. A functional term is of the form $f(t_1, \ldots, t_k)$, where f is a function symbol (*functor*) of arity $k \geq 0$, and $t_1, \ldots, t_k$ are terms. A functional term with arity 0 is a *constant*. If p is a *predicate* of arity $k \geq 0$, and $t_1, \ldots, t_k$ are terms, then $p(t_1, \ldots, t_k)$ is an *atom*. A *literal* is either an atom α (a positive literal), or an atom preceded by the *negation as failure* symbol not α (a negative literal). A *rule* r is of the form

$$\alpha_1 \vee \cdots \vee \alpha_k :- \beta_1, \ldots, \beta_n, \text{not } \beta_{n+1}, \ldots, \text{not } \beta_m.$$

where $\alpha_1, \ldots, \alpha_k, \beta_1, \ldots, \beta_m$ are atoms and $k \geq 1$, $m \geq n \geq 0$. The disjunction $\alpha_1 \vee \cdots \vee \alpha_k$ is the *head* of r and the conjunction $\beta_1, \ldots, \beta_n, \text{not } \beta_{n+1}, \ldots, \text{not } \beta_m$ is the *body* of r. The set of head atoms is denoted by $H(r)$, while $B(r)$ is used for denoting the set of body literals. We also use $B^+(r)$ and $B^-(r)$ for denoting the set of atoms appearing in positive and negative body literals, respectively. A rule r is *normal* (or disjunction-free) if $|H(r)| = 1$, *positive* (or negation-free) if $B^-(r) = \emptyset$, a *fact* if both $B(r) = \emptyset$, $|H(r)| = 1$ and no variable appears in $H(r)$.

A *program* $\mathcal{P}$ is a finite set of rules. If all the rules of a program $\mathcal{P}$ are positive (resp. normal), $\mathcal{P}$ is a *positive* (resp. *normal*) program. If all functional terms appearing in a program $\mathcal{P}$ are constants, $\mathcal{P}$ is *function-free*. Stratified and odd-cycle-free programs constitute other interesting classes of programs. Intuitively, for a program $\mathcal{P}$, a predicate p occurring in the head of a rule $r \in \mathcal{P}$ *depends* on each predicate q occurring in $B(r)$; if q occurs in $B^+(r)$, p depends on q positively, otherwise negatively.[1] If no cycle of dependencies involves negative dependencies (i.e., there are no "negative cycles"), the program is *stratified*. If no cycle of dependencies contains an odd number of negative dependencies (i.e., there are no "odd cycles"), the program is *odd-cycle-free*.

2.2 Semantics

The set of terms constructible by combining function symbols and constants[2] appearing in a program $\mathcal{P}$ is the *universe* of $\mathcal{P}$ and is denoted by $U_{\mathcal{P}}$, while the set of ground atoms constructable from predicates in $\mathcal{P}$ with elements of $U_{\mathcal{P}}$ is the *base* of $\mathcal{P}$, denoted by $B_{\mathcal{P}}$. We call a term (atom, rule, or program) *ground* if it does not contain any variable. A ground atom α' (resp. a ground rule r') is an instance of an atom α (resp. of a rule r) if there is a substitution ϑ from the variables in α (resp. in r) to $U_{\mathcal{P}}$ such that $\alpha' = \alpha\vartheta$ (resp. $r' = r\vartheta$). Given a program $\mathcal{P}$, let $grnd(\mathcal{P})$ denote the set of instances of all the rules in $\mathcal{P}$.

An *interpretation* I for a program $\mathcal{P}$ is a subset of $B_{\mathcal{P}}$. A positive ground literal α is true with respect to an interpretation I if $\alpha \in I$; otherwise, it is false. A negative ground literal not α is true with respect to I if and only if α is false with respect to I. The body of a ground rule r is true with respect to I if and only if all the body literals of r are true with respect to I, that is, if and only if $B^+(r) \subseteq I$ and $B^-(r) \cap I = \emptyset$. An interpretation I *satisfies* a ground rule $r \in grnd(\mathcal{P})$ if either (i) at least one atom in $H(r)$ is true with respect to I, or (ii) the body of r is false with respect to I. An interpretation I is a *model* of a program $\mathcal{P}$ if I satisfies all the rules in $grnd(\mathcal{P})$.

Given an interpretation I for a program $\mathcal{P}$, the reduct of $\mathcal{P}$ with respect to I, denoted by $grnd(\mathcal{P})^I$, is obtained by deleting from $grnd(\mathcal{P})$ all the rules r such that some body literal in $B(r)$ is false with respect to I. The semantics of a program $\mathcal{P}$ is then given by the set $\mathcal{SM}(\mathcal{P})$ of the stable models of $\mathcal{P}$, where an interpretation M is a stable model for $\mathcal{P}$ if and only if M is a subset-minimal model of $grnd(\mathcal{P})^M$.

Given a ground atom α and a program $\mathcal{P}$, α is a *cautious* (resp. *brave*) consequence of $\mathcal{P}$, denoted by $\mathcal{P} \models_c \alpha$ (resp. $\mathcal{P} \models_b \alpha$), if $\alpha \in M$ for each (resp. some) $M \in \mathcal{SM}(\mathcal{P})$. The cautious (resp. brave) semantics of a query $\mathcal{Q} = \alpha$ for a program $\mathcal{P}$, where α is an atom,[3] is given by the set $Ans_c(\mathcal{Q}, \mathcal{P})$ (resp. $Ans_b(\mathcal{Q}, \mathcal{P})$) of substitutions ϑ for the variables of α such that $\mathcal{P} \models_c \alpha\vartheta$ (resp. $\mathcal{P} \models_b \alpha\vartheta$) holds.

---

[1] A similar notion of *dependency* among ground atoms can be given for ground programs.

[2] If a program has no constants, an arbitrary constant symbol ξ is introduced.

[3] Note that more complex queries can still be expressed using appropriate rules. We assume that each functor appearing in $\mathcal{Q}$ also appears in $\mathcal{P}$.

2.3 Undecidability

A well-known result about logic programming with uninterpreted function symbols is the undecidability of the reasoning. Indeed, Horn clauses (under the classic first-order semantics) can represent any partial recursive function [Tärnlund, 1977], and this result can be adapted to ASP (even without using disjunction and negation) [Alviano et al., 2010]. However, in the past decade, relevant classes of ASP programs with function symbols guaranteeing decidability of the common reasoning tasks have been identified. Some of them are presented in the next section.

3 Decidable Reasoning: Class Overview

In this section the most important classes of ASP programs with uninterpreted function symbols that guarantee decidability of reasoning tasks are surveyed. More specifically, for each class we discuss decidability of ground reasoning and computability of non-ground reasoning. We recall that by ground reasoning we mean checking the presence of specific ground atoms among the consequences of a program, while by non-ground reasoning we mean computing all answers to non-ground queries. For each class we also consider decidability of coherence checks and of class membership. By coherence check we refer to verifying whether the set of stable models of a given program is empty or not. By class membership, instead, we refer to establishing whether a given program (possibly associated with a query) belongs to the class in question. Expressiveness is briefly discussed and, in particular, we highlight whether a class allows for representing every recursive relation or not.

3.1 Bottom-Up Computable Classes

Programs in this group can be finitely instantiated by means of bottom-up procedures. Typically, these procedures are obtained by extending existing ASP grounding techniques. Since all of these bottom-up procedures yield a finite ground program, the stable models can be computed from this ground program by adopting standard techniques developed for ASP without function symbols. Classes in this group, which include *ω-restricted*, *λ-restricted*, *finite domain*, *argument restricted* and *finitely-ground programs*, are represented in Figure 1; their properties and relationships are discussed throughout the current section.

3.1.1 *ω-Restricted Programs [Syrjänen, 2001]*

The definition of ω-restricted programs is based on the concept of ω-stratification, a stratification on predicates (see Section 2.1) in which an extra stratum, defined to be the uppermost stratum, is considered; this extra stratum is called ω-stratum, and comprises all predicates involved in cycles with negative dependencies. A normal program is ω-restricted if there exists an ω-stratification satisfying the following condition: for each rule r, and for each variable X appearing in r, there is a positive body literal $\beta \in B^+(r)$ such that X occurs in β, and the predicate of β belongs to a *strictly lower* stratum than the predicate in the head atom. Extending the concept of ω-restriction to

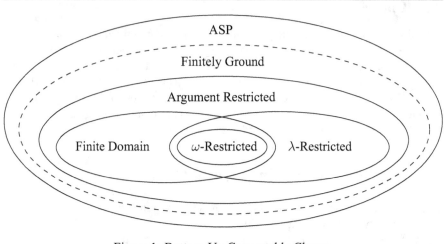

Figure 1: Bottom-Up Computable Classes

general, disjunctive ASP programs is straightforward: the predicate of β must belong to a strictly lower stratum than the predicates of all head atoms.

Example 1 Let us consider the program $\mathcal{P}_1$ consisting of

$$r_1 : \quad p(f(X)) :- q(X), \text{ not } p(X).$$

and some facts for predicate q (that are omitted for simplicity). It is easy to see that program $\mathcal{P}_1$ can be finitely instantiated. Intuitively, the only variable in r_1 is bound by q(X), which is only defined by facts. We will now show that $\mathcal{P}_1$ is ω-restricted. We start by observing that, in all ω-stratifications for $\mathcal{P}_1$, p must belong to the ω-stratum because it is involved in a negative cycle (actually, a self-loop). Moreover, q can be assigned to any stratum, in particular to a stratum different from the ω-stratum. Thus, $\mathcal{P}_1$ is ω-restricted because X is bound by q(X), and q belongs to a strictly lower stratum than p.   □

The above restriction ensures that reasoning on ω-restricted programs can be performed on a finite instantiation thereof, thus guaranteeing decidability of ground reasoning and computability of non-ground reasoning. For the same reason, the coherence check is decidable as well. Concerning class membership, it amounts to determining the existence of an ω-stratification for the program at hand, which is easily seen to be decidable as well. However, the restrictions imposed by ω-stratifications are fairly strong and cause a loss of expressive power: there are recursive relations that cannot be expressed by ω-restricted programs. Reasoning for the class of ω-restricted programs has been implemented in SMODELS [Simons et al., 2002] — one of the most popular ASP systems.

3.1.2 λ-Restricted Programs [Gebser et al., 2007b]

The notion of λ-restricted program is based on a level mapping that assigns an integer $\lambda(p)$ to each predicate p occurring in the program. A normal program is λ-restricted if for any rule r defining predicate p (i.e., where p occurs in the head), each variable occurring in r is bounded by means of an occurrence of a predicate q in $B^+(r)$ such that $\lambda(q) < \lambda(p)$. Intuitively, this means that the feasible ground instances of r are completely determined by predicates from levels lower than the one of p.

Example 2 Let us consider the following program $\mathcal{P}_2$:

$$r_2: \quad q(X) \; :\!- \; r(X), \, p(X).$$
$$r_3: \quad p(f(X)) \; :\!- \; q(X).$$

where r is assumed to be defined by facts. Program $\mathcal{P}_2$ above can be finitely instantiated. Intuitively, the instantiation of r_2 is finite because X is bound by r(X), which trivially has a finite extension. Thus, also the instantiation of r_3 is finite because of q(X) in the body. In fact, $\mathcal{P}_2$ is λ-restricted: a level mapping λ could be $\lambda(r) = 1$, $\lambda(q) = 2$, $\lambda(p) = 3$.

Program $\mathcal{P}_1$ in Example 1 is λ-restricted as well; a level mapping λ could be: $\lambda(q) = 1$, $\lambda(p) = 2$. On the other hand, $\mathcal{P}_2$ is not ω-restricted: in all possible ω-stratifications for $\mathcal{P}_2$, predicates p and q must belong to the same stratum (because of a cyclic, positive dependency). Hence, there is no ω-stratification for $\mathcal{P}_2$ such that variable X in r_3 satisfies the condition required by ω-restricted programs. □

The restriction above ensures a finite instantiation of λ-restricted programs, which in turn ensures decidability of ground reasoning, computability of non-ground reasoning, and decidability of the coherence check. For deciding class membership, the existence of a suitable λ has to be determined, which is easy to see to be decidable as well.

As in the case of ω-restricted programs, there are recursive relations that cannot be expressed by λ-restricted programs. Nevertheless, we note that the class of λ-restricted programs strictly contains the class of ω-restricted programs. Reasoning and a class membership test for λ-restricted programs have been implemented in the ASP grounder GRINGO [Gebser et al., 2007b].

3.1.3 Finite Domain Programs [Calimeri et al., 2008]

The notion of finite domain program has been introduced for general ASP and is based on syntactic restrictions on the arguments of head atoms. Basically, for an ASP program $\mathcal{P}$, a special dependency graph is defined such that there is a node for each argument of all predicates appearing in $\mathcal{P}$, and there are arcs according to dependencies as described in Section 2.1 — the only difference is that here arguments are considered instead of predicates, and a dependency is introduced only when arguments share a variable within a given rule. A program $\mathcal{P}$ is finite domain if, for each atom $p(t_1, \dots, t_n)$ in the head of a rule $r \in \mathcal{P}$, and for each argument p[i] of p, at least one of the following conditions is satisfied: (i) t_i is variable-free; (ii) t_i appears as a (sub)term of an atom in $B^+(r)$; (iii) all variables appearing in t_i are bound by argument terms in $B^+(r)$ which are not recursive with p[i].

Example 3 Let us consider the following program $\mathcal{P}_3$:

$$r_4 : \quad \mathbf{s}(\mathbf{X}) \; :- \; \mathbf{s}(f(\mathbf{X})).$$

Program $\mathcal{P}_3$ above can be finitely instantiated. Intuitively, for each instance of $\mathbf{s}(f(\mathbf{X}))$, only a finite number of instances of $\mathbf{s}(\mathbf{X})$ can be derived by means of r_4. In fact, $\mathcal{P}_3$ is a finite domain program: the term $\mathbf{X}$ in $\mathbf{s}[1]$ appears as a subterm of $\mathbf{s}(f(\mathbf{X}))$, which belongs to $B^+(r_4)$ (i.e., condition (ii) holds).

Program $\mathcal{P}_1$ in Example 1 is a finite domain program as well because variable $\mathbf{X}$ in $\mathbf{p}[1]$ is bound by a positive body literal, $\mathbf{q}(\mathbf{X})$, which is not recursive with $\mathbf{p}$. On the other hand, $\mathcal{P}_2$ in Example 2 is a λ-restricted program which is not finite domain. Indeed, $\mathbf{p}[1]$ in r_3 does not satisfy any of the required conditions: $f(\mathbf{X})$ contains variables, $f(\mathbf{X})$ does not occur as a (sub)term in $B^+(r_3)$, and $\mathbf{X}$ is only bound by $\mathbf{q}[1]$, which is however recursive with $\mathbf{p}[1]$. Finally, we observe that $\mathcal{P}_3$ is neither ω-restricted nor λ-restricted: $\mathbf{s}$ depends on itself because of r_4. □

Finite domain programs can be finitely instantiated. Therefore, for these programs, ground reasoning is decidable and non-ground reasoning is computable, and also the coherence check is decidable. Class membership can be done by checking conditions (i)–(iii) for each argument of head atoms and therefore is clearly decidable.

As for the classes discussed earlier, not all recursive relations can be expressed by finite domain programs. Comparing with the previously introduced classes, the class of finite-domain programs strictly contains the class of ω-restricted programs, while it is incomparable with the class of λ-restricted programs. Reasoning for finite-domain programs has been implemented in DLV [Calimeri et al., 2008], see also Section 4.

3.1.4 Argument Restricted Programs [Lierler and Lifschitz, 2009]

Similarly to λ-restricted programs, the definition of argument restricted programs relies on a level mapping γ, but defined for arguments rather than predicates. A program $\mathcal{P}$ is argument restricted if there is a γ such that, for each atom $\mathbf{p}(\mathbf{t}_1, \dots, \mathbf{t}_n)$ in the head of a rule $r \in \mathcal{P}$, and for each variable $\mathbf{X}$ occurring in some argument term $\mathbf{t}_i$, there is an atom $\mathbf{q}(\mathbf{s}_1, \dots, \mathbf{s}_m)$ in $B^+(r)$ such that $\mathbf{X}$ occurs in some argument term $\mathbf{s}_j$ satisfying the following inequality:

$$\gamma(\mathbf{p}[i]) - \gamma(\mathbf{q}[j]) \geq d(\mathbf{X}, \mathbf{t}_i) - d(\mathbf{X}, \mathbf{s}_j),$$

where $d(\mathbf{X}, \mathbf{t})$ is the maximum depth level of $\mathbf{X}$ in the term $\mathbf{t}$.

Example 4 Let us consider the following program $\mathcal{P}_4$:

$$\begin{aligned} r_2 : \quad & \mathbf{q}(\mathbf{X}) \; :- \; \mathbf{r}(\mathbf{X}), \mathbf{p}(\mathbf{X}). \\ r_3 : \quad & \mathbf{p}(f(\mathbf{X})) \; :- \; \mathbf{q}(\mathbf{X}). \\ r_4 : \quad & \mathbf{s}(\mathbf{X}) \; :- \; \mathbf{s}(f(\mathbf{X})). \end{aligned}$$

Note that $\mathcal{P}_4$ is the union of $\mathcal{P}_2$ and $\mathcal{P}_3$ from Examples 2 and 3, which have finite instantiations and disjoint predicate symbols. Thus, $\mathcal{P}_4$ also possesses a finite instantiation. In fact, we can show that $\mathcal{P}_4$ is argument restricted, for instance, by considering the level mapping $\gamma(\mathtt{q}[1]) = \gamma(\mathtt{r}[1]) = \gamma(\mathtt{s}[1]) = 1$, $\gamma(\mathtt{p}[1]) = 2$. Indeed, in this case, the following inequalities would be satisfied:

(r_2) $\quad 0 = 1 - 1 = \gamma(\mathtt{q}[1]) - \gamma(\mathtt{r}[1]) \geq d(\mathtt{X}, \mathtt{X}) - d(\mathtt{X}, \mathtt{X}) = 0 - 0 = 0$;

(r_3) $\quad 1 = 2 - 1 = \gamma(\mathtt{p}[1]) - \gamma(\mathtt{q}[1]) \geq d(\mathtt{X}, f(\mathtt{X})) - d(\mathtt{X}, \mathtt{X}) = 1 - 0 = 1$;

(r_4) $\quad 0 = 1 - 1 = \gamma(\mathtt{s}[1]) - \gamma(\mathtt{s}[1]) \geq d(\mathtt{X}, \mathtt{X}) - d(\mathtt{X}, f(\mathtt{X})) = 0 - 1 = -1$.

Clearly, γ and the above inequalities witness that $\mathcal{P}_2$ and $\mathcal{P}_3$ are argument restricted as well. Also $\mathcal{P}_1$ from Example 1 is argument restricted. For instance, for $\gamma(\mathtt{q}[1]) = 1$ and $\gamma(\mathtt{p}[1]) = 2$, we obtain

(r_1) $\quad 1 = 2 - 1 = \gamma(\mathtt{p}[1]) - \gamma(\mathtt{q}[1]) \geq d(\mathtt{X}, f(\mathtt{X})) - d(\mathtt{X}, \mathtt{X}) = 1 - 0 = 1$.

On the other hand, it can be shown that $\mathcal{P}_4$ is not ω-restricted, λ-restricted nor finite domain (by the observation made in Examples 2 and 3 for the rules of $\mathcal{P}_2$ and $\mathcal{P}_3$). □

As for the classes discussed before, argument restricted programs can be finitely instantiated, which implies that ground reasoning is decidable and non-ground reasoning is computable, and that also the coherence check is decidable. Class membership amounts to determining the existence of a suitable level mapping and is decidable as well. As in the previous cases, not all recursive relations can be expressed by programs in this class. However, the class of argument restricted programs include all ω-restricted, λ-restricted and finite domain programs. At the time of writing, no system implements a class membership check for argument restricted programs. Nonetheless, GRINGO [Gebser et al., 2007b] and DLV are able to compute the finite instantiation of an argument restricted program as a special case of finitely ground program, a broader class discussed next.

3.1.5 Finitely Ground Programs [Calimeri et al., 2008]

The definition of finitely ground program relies on the notion of "intelligent instantiation," obtained by means of an operator which is iteratively applied on program submodules. In order to properly split a given program $\mathcal{P}$ into modules, the *dependency graph* and the *component graph* are considered. The former connects predicate names, and represents dependencies in rules (see Section 2.1), the latter connects *strongly connected components*[4] (SCC) of the dependency graph. Each module is constituted by all rules defining predicates in a corresponding SCC. An ordering relation is then defined among modules/components: a *component ordering* γ for $\mathcal{P}$ is a total ordering such that the intelligent instantiation $\mathcal{P}^\gamma$, obtained by following the sequence given by γ, has the same stable models of $grnd(\mathcal{P})$. If $\mathcal{P}^\gamma$ is finite with respect to each possible component ordering γ, then $\mathcal{P}$ is finitely ground.

Example 5 Let us consider the following program $\mathcal{P}_5$:

$$r_5 : \quad \mathtt{p}(f(g(\mathtt{X}))) \; :\!- \; \mathtt{p}(g(\mathtt{X})).$$

---

[4]We recall here that a strongly connected component of a directed graph is a maximal subset C of its vertices, such that each vertex in C is reachable from all other vertices in C.

Program $\mathcal{P}_5$ is finitely ground and can be finitely instantiated. Intuitively, even if p is recursive in r_5, instances of $\mathrm{p}(f(g(\mathrm{X})))$ (the head atom) and instances of $\mathrm{p}(g(\mathrm{X}))$ (the body atom) are disjoint, so that the recursion is only apparent. Also $\mathcal{P}_4$ from Example 4 is finitely ground. In fact, $\mathcal{P}_4$ has two independent submodules, $\{r_2, r_3\}$ and $\{r_4\}$: the instantiation of the first submodule is limited by $\mathrm{r}(\mathrm{X})$, while for the second submodule a bound is provided by the depth nesting level of functions in $\mathrm{s}[1]$. By similar arguments, it can be shown that $\mathcal{P}_1$, $\mathcal{P}_2$ and $\mathcal{P}_3$ are finitely ground. On the other hand, we observe that $\mathcal{P}_5$ above is not argument restricted because X appears solely in $\mathrm{p}[1]$ and with a greater nesting depth in the head than in the body. Thus, for all possible level mappings γ, it holds that

$$(r_5) \quad 0 = \gamma(\mathrm{p}[1]) - \gamma(\mathrm{p}[1]) \quad \not\geq \quad d(\mathrm{X}, f(g(\mathrm{X}))) - d(\mathrm{X}, g(\mathrm{X})) = 2 - 1 = 1.$$

Since $\mathcal{P}_5$ is not argument restricted, it does not belong to any of the previously presented classes. In particular, $\mathcal{P}_5$ is neither ω-restricted nor λ-restricted: p depends on itself because of r_5; it is not finite domain because $\mathrm{p}[1]$ in r_5 does not satisfy any of the required conditions: $f(g(\mathrm{X}))$ contains variables, $f(g(\mathrm{X}))$ does not occur as a (sub)term in $B^+(r_5)$, and X is only bound by $\mathrm{p}[1]$ in $B^+(r_5)$, which however recursively depends on itself.   □

As Example 5 suggests, the class of finitely ground programs is the most general known class of ASP programs allowing for finite program instantiations, strictly including ω-restricted, λ-restricted, finite domain and argument restricted programs. Since the instantiation is guaranteed to be finite, ground reasoning is decidable and non-ground reasoning computable, and also the coherence check is decidable. In addition, it has been proved that all recursive relations can be expressed by finitely-ground programs [Calimeri et al., 2008]. However, this comes at a price: checking whether a program is finitely ground is semi-decidable in general. Finitely ground programs have been effectively implemented in DLV [Calimeri et al., 2008], which is now able to finitely compute all answer sets of any such program.[5]

3.2   Top-Down Computable Classes

The scientific community has also proposed some classes of programs specifically designed for query answering, and thus typically characterized by top-down computation schemata. Programs in these classes usually feature an infinite number of answer sets, each of which may contain an infinite number of atoms. In order to guarantee decidability of reasoning, a finite number of atoms must be sufficient for answering a query. Moreover, these atoms have to be effectively identified. Classes in this group, namely *FP2 programs*, positive and stratified *finitely recursive programs*, and *finitary programs*, are depicted in Figure 2; their properties and relationships are discussed throughout the current section.

3.2.1   FP2 Programs [Baselice and Bonatti, 2010]

The class of FP2 programs has been defined for normal programs only. The definition relies on two key concepts: recursion patterns and call-safeness. A recursion pattern π

---

[5] As a recent addition, GRINGO 3.0.X introduced a semi-decidable grounding procedure.

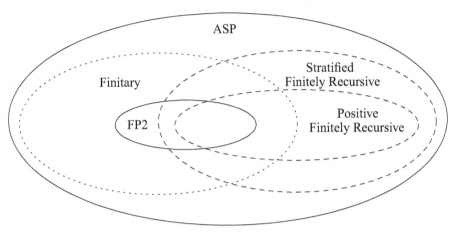

Figure 2: Top-Down Computable Classes

is a function mapping each predicate p to a subset of the arguments of p. Essentially, a recursion pattern for a program $\mathcal{P}$ ensures that the dependency graph of $grnd(\mathcal{P})$ is such that: (i) no cycle of dependencies is an odd cycle; and (ii) every path contains finitely many different atoms. A program is call-safe with respect to a recursion pattern π if, in a top-down computation, variables appearing in negative subgoals or in an argument of a subgoal selected by π are bound by previous resolved subgoals. Hence, a normal program $\mathcal{P}$ belongs to FP2 if there is a recursion pattern π such that $\mathcal{P}$ is call-safe with respect to π.

Example 6 Consider the following program $\mathcal{P}_6$:

$$
\begin{aligned}
r_6 &: \quad \text{p(X)} \ :- \ \text{not q(X)}. \\
r_7 &: \quad \text{q(X)} \ :- \ \text{not p(X)}. \\
r_8 &: \quad \text{p}(f(\text{X})) \ :- \ \text{p(X)}.
\end{aligned}
$$

The program is FP2. Indeed, a recursion pattern can be obtained by mapping p and q on their first, and unique, arguments; in addition, all rules are call-safe because, in each of them, the unique variable X appears in a selected argument of the head (see Definition 5.4 in the paper by Baselice and Bonatti [2010]). Hence, ground and non-ground reasoning over $\mathcal{P}_6$ is computable with respect to any query atom. □

FP2 programs have decidable ground reasoning and class membership, but cannot express all computable sets or relations. The coherence check is trivial (FP2 programs are odd-cycle-free). Non-ground reasoning is uncomputable.

3.2.2 Positive and Stratified Finitely Recursive Programs [Calimeri et al., 2009, Alviano et al., 2010]

Finitely recursive programs are based on a notion of *relevant atoms* (or subqueries). The relevant atoms for a ground query Q with respect to a program P are defined as follows: (i) Q itself is relevant; (ii) if a ground atom α is relevant and r is a rule in $grnd(P)$ having α in its head, all atoms appearing in r are relevant. A ground query Q is finitely recursive on a program P if the number of relevant atoms for Q with respect to P is finite. A program P is finitely recursive if every ground query Q is finitely recursive on P.

Example 7 Consider the query atom $p(f(f(0)))$ for the following program P_7:

$$r_8 : \quad p(f(X)) \; :- \; p(X).$$
$$r_9 : \quad p(X) \; \vee \; q(X).$$

Both the query and the program are finitely recursive. In fact, it can be observed that a bound to the depth nesting level of functors of all relevant atoms is implicitly provided by queries. The program is not FP2 because r_8 contains disjunction. On the other hand, P_6 from Example 6 contains recursive negation (see r_6 and r_7), and so it is not a positive or stratified finitely recursive program. $\qquad\square$

Decidability of ground reasoning has been proved for disjunctive positive programs and generalized to the case of disjunctive programs with stratified negation, while non-ground reasoning remains undecidable in general because an infinite number of queries would have to be considered in this case. In proving these results, an interesting link between top-down and bottom-up computable classes has been established: positive and stratified finitely recursive programs can be mapped to equivalent finitely ground programs by applying a Magic Set rewriting [Alviano et al., 2010].

Recognizing whether a query belongs to one of these classes is a semi-decidable task, while coherence checking is trivially decidable (positive and stratified programs are coherent). Concerning expressiveness, finitely recursive programs are sufficient to express all computable relations, even if disjunction and negation are forbidden. Positive finitely recursive programs are clearly contained in the class of stratified finitely recursive programs, while the class of FP2 programs is incomparable with respect to subset-inclusion.

3.2.3 Finitary Programs [Bonatti, 2002, 2004]

Finitary programs are a subset of finitely recursive programs. The class has been originally defined for normal programs and subsequently enlarged for allowing disjunctive heads. A program P is finitary if the following conditions are satisfied: (i) P is finitely recursive; (ii) only a finite number of odd cycles are present in the dependency graph of $grnd(P)$; (iii) only a finite number of head cycles are present in the dependency graph of $grnd(P)$.[6]

---

[6]A head cycle is a cycle of dependencies involving a pair of ground atoms occurring in the head of the same rule.

Example 8 Programs $\mathcal{P}_6$ and $\mathcal{P}_7$ from Examples 6 and 7 are finitary. Consider now the following program $\mathcal{P}_8$:

$$
\begin{aligned}
r_8 &: \quad p(f(X)) :- p(X). \\
r_9 &: \quad p(X) \vee q(X). \\
r_{10} &: \quad p(X) :- q(X). \\
r_{11} &: \quad q(X) :- p(X).
\end{aligned}
$$

Note that $\mathcal{P}_8$ is a positive finitely recursive program. Note also that the program is not finitary. Indeed, infinitely many head cycles are present in $grnd(\mathcal{P}_8)$ because of r_9, r_{10} and r_{11} (even if only a finite number of these cycles are relevant for answering a given ground query). □

Strictly speaking, knowing that a program is finitary does not guarantee decidability of ground reasoning. In fact, in order to ensure decidability, odd cycles and head cycles have to be provided as an additional input together with the program. Also class membership and the coherence check are undecidable for finitary programs, and non-ground reasoning is uncomputable. Finitary programs include FP2 programs, while they are incomparable with respect to positive and stratified finitely recursive programs.

3.3 Classes with Finitely Representable Stable Models

Classes in this group are FDNC and bidirectional programs, which are characterized by infinite stable models having a finite representation in the shape of a forest of trees.

3.3.1 FDNC Programs [Simkus and Eiter, 2007]

The class of FDNC programs allows for function symbols, disjunction, non-monotonic negation, and constraints. In order to retain the decidability of ground reasoning, rules have to be of the shape of one of seven predefined rule schemata. In particular, FDNC constraints function symbols to be unary (or constants), and predicates to be unary or binary. These syntactic restrictions ensure that programs have a forest-shaped model property, which means that stable models of FDNC programs are in general infinite, but have a finite representation that can be exploited for knowledge compilation and fast query answering.

Ground reasoning, coherence check and class membership are decidable for FDNC programs. Non-ground reasoning is also computable, but there are recursive relations that cannot be expressed by programs in this class. We also note that the restrictions imposed on the syntax of FDNC programs require that atoms occurring in rule heads have to be structurally simpler than atoms in rule bodies. These limitations have considerable impact on practical domains like reasoning about actions. In this context, indeed, rules of FDNC programs do not allow to refer naturally to the past.

3.3.2 Bidirectional Programs [Eiter and Simkus, 2009a]

Bidirectional programs are a close relative to FDNC programs, but allow for referring to both past and future events when reasoning about actions. The restrictions ensuring decidability of ground reasoning for bidirectional programs are based on the notion of t-atoms. An atom α is a t-atom if the first argument of α is t and each other argument of α is either a constant, or a variable not occurring in t. Hence, a safe program $\mathcal{P}$ is bidirectional if, for each $r \in \mathcal{P}$, there is a variable X such that each atom in r is either an X-atom, or an $f(X)$-atom, or a c-atom, where f is a unary functor and c a constant (fixed for all rules). Note that the definition of bidirectional programs does not limit the arity of predicates, while function symbols are constrained to be unary (or constants).

Ground reasoning, coherence check and class membership are decidable for bidirectional programs. Non-ground reasoning is also computable, but there are recursive relations that cannot be expressed by programs in this class. Note that FDNC and bidirectional programs are incomparable as there are FDNC programs that are not bidirectional and vice versa (see the example below).

Example 9 The following is an FDNC program that is not bidirectional:

$$p(X, f(X)) :- q(X).$$

Indeed, the structure of the rule above is allowed in FDNC, while bidirectional programs allow for using function symbols of positive arity only in the first argument of atoms. The following, instead, is a bidirectional program not belonging to FDNC:

$$p(X) :- q(f(X)).$$

Indeed, the rule above is allowed in bidirectional programs because $p(X)$ is an X-atom and $q(f(X))$ is an $f(X)$-atom, while it is forbidden in FDNC (in FDNC programs, depth levels of variables must be bound by body atoms). $\qquad\square$

3.4 Discussion

The main properties of each group of program classes are summarized in this section. Classes in the first group, graphically represented in Figure 1, are characterized by finite program instantiation, which in turn implies decidability of ground reasoning and coherence check, as well as computability of non-ground reasoning. These classes are particularly relevant for ASP because current systems can be easily adapted for computing stable models of programs in this group. Containment relationships between classes in this group are highlighted in Figure 1. The smallest class are ω-restricted programs, which is strictly contained in the intersection of λ-restricted and finite domain programs. The most general known class that can be syntactically recognized are argument restricted programs, a strict superset of λ-restricted and finite domain programs. Argument restricted programs are strictly contained in finitely ground programs, the most general known fragment of ASP in this group. A strength of finitely ground programs is its expressive power: all recursive relations can be expressed by finitely ground programs. However, this class cannot be recursively separated from the full ASP language. For this reason, the boundary between finitely ground and ASP programs is dashed in Figure 1.

Programs in the second group are suitable for finite top-down evaluations of queries and allow for reasoning about infinite stable models. Containment relationships between classes in this group are represented in Figure 2. We note that FP2 is strictly

Table 1: Classification of ASP fragments with decidable reasoning

Class	Ground Reason.	Non-gr. Reason.	Coher. Check	Recur. Compl.	Class Member.
ω-Restricted	decid.	comp.	decid.	no	decid.
λ-Restricted	decid.	comp.	decid.	no	decid.
Finite Domain	decid.	comp.	decid.	no	decid.
Argument Restr.	decid.	comp.	decid.	no	decid.
Finitely Ground	decid.	comp.	decid.	yes	semi-decid.
FP2	decid.	uncomp.	decid.	no	decid.
Pos. Fin. Rec.	decid.	uncomp.	decid.	yes	semi-decid.
Strat. Fin. Rec.	decid.	uncomp.	decid.	yes	semi-decid.
Finitary	decid.*	uncomp.	undecid.	yes	undecid.
FDNC	decid.	comp.	decid.	no	decid.
Bidirectional	decid.	comp.	decid.	no	decid.

* If odd cycles and head cycles are provided as input.

contained in finitary programs, while positive finitely recursive programs are clearly contained in stratified finitely recursive programs. Note also that FP2 is the only syntactically recognizable class in this group (represented by a solid boundary in Figure 2). The other classes, instead, allow for expressing all recursive relations, even if only positive and stratified finitely recursive programs guarantee decidability of ground reasoning. In fact, the evaluation of finitary programs requires knowing the set of all odd and head cycles, which is not known to be computable. Moreover, class membership is semi-decidable for positive and stratified finitely recursive programs (visualized by dashed boundaries in Figure 2), while it is undecidable for finitary programs (dotted boundary in Figure 2). We recall that an interesting link between top-down and bottom-up computable classes has been established in the literature [Calimeri et al., 2009, Alviano et al., 2010]: positive and stratified finitely recursive programs can be mapped to equivalent finitely ground programs by means of a Magic Set rewriting.

The third group contains FDNC and bidirectional programs that are characterized by forest-shaped stable models guaranteeing a finite representation. Ground reasoning, coherence check and class membership are decidable for these programs, and non-ground reasoning is computable. However, FDNC and bidirectional programs do not allow for representing all recursive relations.

Table 1 summarizes the most relevant properties discussed in this section. All classes guarantee decidability of reasoning for ground queries (ground reasoning), even if odd and head cycles have to be provided in input for ensuring decidability of ground reasoning for finitary programs. For non-ground queries (non-ground reasoning), instead, the set of answer substitutions can be effectively computed only by programs in the first and third group, which are also characterized by a decidable coherence check. Coherence is also decidable (actually, trivial) for FP2 and positive/stratified finitely recursive programs. The possibility of expressing all recursive relations (recursive completeness) is preserved by finitary, finitely ground and positive/stratified finitely recursive programs. However, while class membership (i.e., establishing whether a given program belongs to the class in question) is undecidable for finitary programs, it is semi-decidable for finitely ground and positive/stratified finitely recursive programs.

4   An ASP System Supporting Uninterpreted Function Symbols: DLV

The DLV system [Leone et al., 2006] is widely considered one of the state-of-the-art implementations of ASP. Since the first stable release, dated back in 1997, the DLV system has been significantly improved over and over, both thanks to the implementation of various optimization techniques and to the enrichment of its language that supports the most important ASP extensions such as aggregates and weak-constraints. DLV has been widely used in many practical application scenarios, including data integration [Leone, Gottlob, Rosati, Eiter, Faber, Fink, Greco, Ianni, Kałka, Lembo, Lenzerini, Lio, Nowicki, Ruzzi, Staniszkis, and Terracina, 2005], semantic-based information extraction [Manna, Ruffolo, Oro, Alviano, and Leone, 2011a, Manna, Scarcello, and Leone, 2011b], e-tourism [Ricca, Alviano, Dimasi, Grasso, Ielpa, Iiritano, Manna, and Leone, 2010], workforce management [Ricca, Grasso, Alviano, Manna, Lio, Iiritano, and Leone, 2011], and many more.

Beginning with version 2010–10–14, DLV supports a powerful (possibly recursive) use of function symbols, and list and set terms [Calimeri et al., 2008].[7] Within DLV, all recursive relations are expressible in a rich and fully declarative language. Termination is guaranteed on all finitely ground programs (which include all other program classes discussed in Section 3.1), or for all programs passing some syntactic check that can be performed on demand.

4.1   Language Overview

By supporting function symbols, DLV allows for aggregating atomic data, manipulating complex data structures and generating new symbols (*value invention* — see the paper by Calimeri, Cozza, and Ianni [2007] for a discussion). Strings, lists, sets, trees, and many other common data structures are representable by means of functions. In particular, for list and set terms, DLV offers an explicit notation. This and all the other features of DLV, such as aggregates and weak constraints, provide a solid basis for natural knowledge representation.

---

[7]Note that the notation for lists and sets are currently only available in the DLV-Complex branch. This functionality is scheduled to be taken over as soon as possible to the main distribution.

4.1.1 Function Symbols

Function symbols are supported by DLV with the syntax reported in this article. Strings, stacks, trees, and many other common data structures can be represented by means of function symbols. For instance, binary trees might be encoded by a function symbol bt of arity 3, where the first argument is associated with the root, and the other two stand for the left and the right subtrees. As an example, under such assumptions, the functional term $bt(\text{equal}, bt(\text{sum}, bt(\text{x}, \perp, \perp), bt(\text{y}, \perp, \perp)), bt(\text{sum}, bt(\text{y}, \perp, \perp), bt(\text{x}, \perp, \perp)))$ can be seen as a binary tree representing the structure of the formula $\text{x} + \text{y} = \text{y} + \text{x}$ (here, $\perp$ is a constant used for denoting the empty tree).

Example 10 Let F be a set of binary trees. Suppose we are interested in determining all subsets S of F such that S does not contain trees t_1, t_2 such that t_1 is a subtree of t_2. Assuming that the input set F is encoded by instances of a unary predicate tree, all such subsets can be determined by means of the following program:

$$
\begin{array}{ll}
r_{12}: & \text{in(T)} \lor \text{out(T)} := \text{tree(T)}. \\
r_{13}: & := \text{in}(T_1), \text{in}(T_2), \text{subtree}(T_1, T_2), T_1 \neq T_2. \\
r_{14}: & \text{subtree}(T, T) := \text{tree(T)}. \\
r_{15}: & \text{subtree}(L, T) := \text{subtree}(bt(X, L, R), T). \\
r_{16}: & \text{subtree}(R, T) := \text{subtree}(bt(X, L, R), T).
\end{array}
$$

In particular, possible subsets S of F are guessed by r_{12}, and those subsets not fulfilling the required condition are discarded by r_{13}.[8] Pairs of subtrees are determined by means of the rules r_{14}, r_{15} and r_{16}. Assuming that F contains two trees, $t_1 = bt(\text{a}, bt(\text{b}, \perp, \perp), bt(\text{c}, \perp, \perp))$ and $t_2 = bt(\text{b}, \perp, \perp)$, the following subsets are determined: $\emptyset$; $\{t_1\}$; $\{t_2\}$. Thus, the only subset of F not fulfilling the required condition is F itself. Indeed, t_2 is a subtree of t_1. □

4.1.2 Lists

Lists can be profitably applied in order to model collections of objects in which position matters and repetitions are allowed. Some examples of list terms are $[\text{jan}, \text{feb}, \text{mar}]$ and $[1, \text{i}, \text{s}, \text{t}]$. Lists can be nested arbitrarily, for instance, the following are valid list terms: $[[\text{jan}, 31], [\text{feb}, 28], [\text{mar}, 30]]$; $[[\text{t}, \text{h}, \text{i}, \text{s}], [\text{i}, \text{s}], [\text{a}], [1, \text{i}, \text{s}, \text{t}]]$. An element t can be appended to the front of a list 1 by using the "à la Prolog" syntax, i.e., $[\text{t}|\text{l}]$. For example, the list $[\text{jan}, \text{feb}, \text{mar}]$ can be equivalently written as $[\text{jan}|[\text{feb}|[\text{mar}|[\,]]]]$.

Example 11 A palindrome is a phrase that can be read the same way in either direction (ignoring punctuation and spaces). Examples of palindromes are the following phrases: "Was it a rat I saw?" and "Ai lati d'Italia." Assume that a set of phrases is given by means of a predicate phrase, where each phrase is encoded by a list of characters (for simplicity, all characters in these lists are lowercase letters); for example,

---

[8]Note that rules with empty heads (i.e., constraints) are not allowed by the syntax presented in Section 2.1. Intuitively, r_{13} is equivalent to the following rule: $\text{co} := \text{in}(T_1), \text{in}(T_2), \text{subtree}(T_1, T_2), T_1 \neq T_2, \text{not co}$, where co is a fresh symbol.

let us assume that the input comprises the following phrases:

```
phrase([w,a,s,i,t,a,r,a,t,i,s,a,w]),
phrase([a,i,l,a,t,i,d,i,t,a,l,i,a]) and
phrase([n,o,t,a,p,a,l,i,n,d,r,o,m,e]).
```

The following program determines palindromes among input phrases:

$$\text{palindrome}(X) :- \text{phrase}(X), \#\text{reverse}(X) = X.$$

Note that #reverse is an interpreted built-in function. Intuitively, for a list l, the term #reverse(l) is a list comprising the elements of l in reversed order. In our example, the first two phrases are palindromes, while the last one is not. Thus, the unique stable model of the program above contains

```
palindrome([w,a,s,i,t,a,r,a,t,i,s,a,w]) and
palindrome([a,i,l,a,t,i,d,i,t,a,l,i,a]).
```

□

Example 12 Let G be an undirected graph. A path is a sequence of nodes $x_1, \ldots, x_n$ such that for each $i = 1, \ldots, n-1$ there is an edge connecting x_i and x_{i+1} in G. A simple path is a path without repetition of nodes. Assuming that edges of G are represented by instances of edge, the following program derives all simple paths in G:

$$\text{path}([X, Y]) :- \text{edge}(X, Y).$$
$$\text{path}([X|[Y|W]]) :- \text{edge}(X, Y), \text{path}([Y|W]), \text{not } \#\text{member}(X, [Y|W]).$$

Note that #member is an interpreted built-in predicate. Intuitively, for a term t and a list l, the atom #member(t, l) is true if and only if t is an element in l. □

4.1.3 Sets

Set terms are used to model unordered collections of data, in which no duplicates are allowed. Examples of set terms are $\{a, b, c\}$ and $\{alice, bob, charlie\}$. Note that $\{a, b, c\}$ and $\{b, a, c\}$ are the same set term. Sets can be nested arbitrarily; for instance, the following are valid set terms: $\{\{1, 3, 5, 7\}, \{2, 4, 6, 8\}\}; \{0, \{0\}, \{\{0\}\}\}$. In addition, lists and sets can be combined: $\{[t, h, i, s], [i, s], [a], [s, e, t], [o, f], [1, i, s, t, s]\}; [\{t, h, i, s\}, \{i, s\}, \{a\}, \{1, i, s, t\}, \{o, f\}, \{s, e, ts\}].$

Example 13 Let $G = (V, E)$ be an undirected graph, where V is a set of vertices and E a set of edges. A dominating set for a graph G is a subset D of V such that every vertex in $V \setminus D$ is joined to a vertex of D by an edge in E. Assume that the input graph G is represented by instances of adjacent(v, a), where v is a vertex and a the set of its adjacent vertices; for example, let us assume that the following set of facts, representing the graph in Figure 3, is given:

$\{\text{adjacent}(a, \{b, c\}), \text{adjacent}(b, \{a, c, d\}), \text{adjacent}(c, \{a, b, e\}),$
$\text{adjacent}(d, \{b, e, f\}), \text{adjacent}(e, \{c, d\}), \text{adjacent}(f, \{d\})\}$

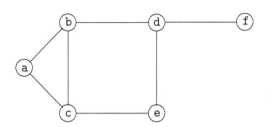

Figure 3: An undirected graph

All dominating sets of G can be determined by means of the following program:

$r_{17}:$ in(X) ∨ out(X) :− adjacent(X, A).
$r_{18}:$ dominated(X) :− out(X), in(Y), adjacent(Y, A), #member(X, A).
$r_{19}:$:− out(X), not dominated(X).

Intuitively, r_{17} guesses a subset D of vertices, r_{18} determines dominated nodes, and r_{19} discards D if not a dominating set. In our example, {a, e, f}, {c, d}, {b, d} and all their supersets are dominating sets.   □

4.2   Implementation Issues

4.2.1   Overview

The system architecture of DLV is shown in Figure 4. The *Rewriter* module is in charge of substituting all (non-constant) functional terms occurring in the input program, by introducing appropriate predefined built-in predicates (details are reported below in this section). The rewritten program is then passed to the *Finite Checker* module in order to verify the membership in a class for which termination is guaranteed (see Section 3.1). In fact, these two modules are essentially the only relevant ones for enabling DLV to deal with function symbols. For descriptions of the remaining modules we refer to the paper by Leone et al. [2006] for basic DLV and by Calimeri et al. [2007] for its extension by external built-ins.

Currently, main-line DLV checks for "strong safety," which is slightly more restrictive than the class of finite domain programs, while the DLV-Complex branch has a check for membership in the class of finite domain programs. It is planned to provide a check for argument restriction [Lierler and Lifschitz, 2009] in the near future. If the program is guaranteed to have a finite instantiation, answer sets can be computed as usual by processing the resulting ground program. The *Finite Checker* module can be skipped by specifying the command-line option -nofinitecheck. In this case, termination is only guaranteed for finitely ground programs (see Section 3.1.5), which however cannot be characterized in terms of a syntactic restriction. It is also possible to limit the nesting level of functional terms in the generated ground program, and in turn, ensure the termination of the instantiation process, by means of the command-line option -MAXNL=<N>.

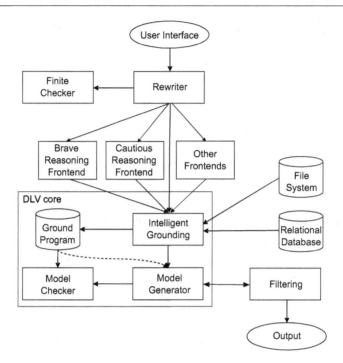

Figure 4: System Architecture

4.2.2 Rewriter

For each rule r in $\mathcal{P}$, each occurrence of a functional term of the form $f(X_1, \ldots, X_k)$ in r, where $X_1, \ldots, X_k$ are variables, is replaced by a fresh variable F. Moreover, depending on whether $f(X_1, \ldots, X_k)$ appears in the head or in the body of r, one of the following atoms is added to $B^+(r)$:

- #function_pack($f, X_1, \ldots, X_k, F$)   if $f(X_1, \ldots, X_k)$ appears in $H(r)$;

- #function_unpack($F, f, X_1, \ldots, X_k$)   if $f(X_1, \ldots, X_k)$ appears in $B(r)$.

This process is repeated until no functional terms of the form $f(X_1, \ldots, X_k)$ appear in r. Note that #function_pack and #function_unpack are built-in predicates [Calimeri and Ianni, 2005].[9]

Example 14 Let us consider the following rule:

$$p(f(f(X))) :\!- q(X, g(X, Y)).$$

---

[9]List and set terms are treated by means of *pack* and *unpack* built-in predicates, which act analogously to #function_pack and #function_unpack.

According to the process described earlier, the rule above is rewritten as follows:

$$p(F_2) :- q(X, F_1), \#function\_unpack(F_1, g, X, Y),$$
$$\#function\_pack(F_2, f, X), \#function\_pack(F_3, f, F_2).$$

Note that two $\#function\_pack$ atoms have been introduced in order to represent the nested functional term $f(f(X))$: $\#function\_pack(F_2, f, X)$ uses the variable F_2 to refer to the functional term $f(X)$; $\#function\_pack(F_3, f, F_2)$, instead, refers to $f(F_2) = f(f(X))$ by means of F_3. □

4.2.3 Intelligent Grounding

Rewritten programs are instantiated by means of standard techniques (the reader may refer, for instance, to the paper by Leone et al. [2006]). In this section, we only describe how the built-in predicates $\#function\_pack$ and $\#function\_unpack$, introduced by the *Rewriter* module, are interpreted during the instantiation phase.

Predicate $\#function\_pack$ is used to build ground functional terms. In particular, for a function symbol f of arity k, k ground functional terms $t_1, \ldots, t_k$ and a variable F, the built-in atom $\#function\_pack(f, t_1, \ldots, t_k, F)$ causes the creation of a ground functional term $f(t_1, \ldots, t_k)$, which is assigned to the variable F. The second predicate, instead, is used to disassemble functional terms. In particular, for a function symbol f of arity k, k ground functional terms $t_1, \ldots, t_k$ and k variables $X_1, \ldots, X_k$, the atom $\#function\_unpack(f(t_1, \ldots, t_k), f, X_1, \ldots, X_k)$ assigns the ground functional term t_i to the variable X_i, for each $i \in \{1, \ldots, k\}$. If some variable X_i has been previously bound to a ground functional term t, $\#function\_unpack$ simply checks that t is equal to t_i.

5 Conclusion

ASP is a highly expressive language that has found many application scenarios. However, until fairly recently, practical ASP was limited to express properties over finite structures. This major deficiency of the language prevented the representation of recursive structures, which are common in computer science. The introduction of uninterpreted function symbols allows for circumventing this problem. However, the common reasoning tasks become undecidable if functions are permitted without any restrictions. Many relevant classes of programs with function symbols for which decidability of reasoning is guaranteed were identified in the past decade. The most popular of these have been discussed in this paper. In particular, three groups of classes have been considered: programs allowing for finite bottom-up evaluations (ω-*restricted* [Syrjänen, 2001], λ-*restricted* [Gebser et al., 2007b], *finite domain* [Calimeri et al., 2008], *argument restricted* [Lierler and Lifschitz, 2009] and *finitely ground programs* [Calimeri et al., 2008]), programs suitable for finite top-down evaluations (*FP2* [Baselice and Bonatti, 2010], *positive/stratified finitely recursive* [Calimeri et al., 2009, Alviano et al., 2010] and *finitary programs* [Bonatti, 2002, 2004]) and programs characterized by finitely representable stable models (*FDNC* [Simkus and Eiter, 2007] and *bidirectional programs* [Eiter and Simkus, 2009a]). The main properties of these program classes and their interrelationships have been analyzed and summarized. Finally, a system that can deal with programs containing function symbols has been presented, which allows for terminating reasoning over programs of several of the presented program classes.

References

M. Alviano, W. Faber, and N. Leone. Disjunctive asp with functions: Decidable queries and effective computation. *Theory and Practice of Logic Programming, 26th Int'l. Conference on Logic Programming (ICLP'10) Special Issue*, 10(4–6):497–512, 2010. doi: 10.1017/S1471068410000244.

C. Baral. *Knowledge Representation, Reasoning and Declarative Problem Solving*. Cambridge University Press, 2003. ISBN 0-52181802-8.

S. Baselice and P. A. Bonatti. A decidable subclass of finitary programs. *Theory and Practice of Logic Programming, 26th Int'l. Conference on Logic Programming (ICLP'10) Special Issue*, 10((4–6)):481–496, July 2010.

S. Baselice, P. A. Bonatti, and G. Criscuolo. On Finitely Recursive Programs. *Theory and Practice of Logic Programming*, 9(2):213–238, 2009.

P. A. Bonatti. Reasoning with infinite stable models II: Disjunctive programs. In *Proceedings of the 18th International Conference on Logic Programming (ICLP 2002)*, volume 2401 of *LNCS*, pages 333–346. Springer, 2002.

P. A. Bonatti. Reasoning with infinite stable models. *Artificial Intelligence*, 156(1):75–111, 2004.

P. Cabalar. Partial Functions and Equality in Answer Set Programming. In *Proceedings of the 24th International Conference on Logic Programming (ICLP 2008)*, volume 5366 of *Lecture Notes in Computer Science*, pages 392–406, Udine, Italy, December 2008. Springer. ISBN 978-3-540-89981-5.

M. Cadoli and M. Schaerf. A survey on complexity results for non-monotonic logics. *Journal of Logic Programming*, 17:127–160, 1993.

F. Calimeri and G. Ianni. External sources of computation for Answer Set Solvers. In C. Baral, G. Greco, N. Leone, and G. Terracina, editors, *Logic Programming and Nonmonotonic Reasoning — 8th International Conference, LPNMR'05, Diamante, Italy, September 2005, Proceedings*, volume 3662 of *Lecture Notes in Computer Science*, pages 105–118. Springer Verlag, September 2005. ISBN 3-540-28538-5.

F. Calimeri, S. Cozza, and G. Ianni. External sources of knowledge and value invention in logic programming. *Annals of Mathematics and Artificial Intelligence*, 50(3–4):333–361, 2007.

F. Calimeri, S. Cozza, G. Ianni, and N. Leone. Computable Functions in ASP: Theory and Implementation. In *Proceedings of the 24th International Conference on Logic Programming (ICLP 2008)*, volume 5366 of *Lecture Notes in Computer Science*, pages 407–424, Udine, Italy, December 2008. Springer. ISBN 978-3-540-89981-5.

F. Calimeri, S. Cozza, G. Ianni, and N. Leone. Magic Sets for the Bottom-Up Evaluation of Finitely Recursive Programs. In E. Erdem, F. Lin, and T. Schaub, editors, *Logic Programming and Nonmonotonic Reasoning — 10th International Conference (LPNMR 2009)*, volume 5753 of *Lecture Notes in Computer Science*, pages 71–86. Springer Verlag, September 2009. ISBN 978-3-642-04237-9.

T. Eiter and M. Simkus. Bidirectional answer set programs with function symbols. In C. Boutilier, editor, *Proceedings of the 21st International Joint Conference on Artificial Intelligence (IJCAI-09)*. AAAI Press/IJCAI, 2009a.

T. Eiter and M. Simkus. Bidirectional Answer Set Programs with Function Symbols. In C. Boutilier, editor, *Proceedings of the 21st International Joint Conference on Artificial Intelligence (IJCAI-09)*, pages 765–771, Pasadena, CA, USA, July 2009b.

M. Gebser, B. Kaufmann, A. Neumann, and T. Schaub. Conflict-driven answer set solving. In *Twentieth International Joint Conference on Artificial Intelligence (IJCAI-07)*, pages 386–392. Morgan Kaufmann Publishers, January 2007a.

M. Gebser, T. Schaub, and S. Thiele. Gringo : A new grounder for answer set programming. In C. Baral, Gerhard B., and J. Schlipf, editors, *Logic Programming and Nonmonotonic Reasoning — 9th International Conference, LPNMR'07*, volume 4483 of *Lecture Notes in Computer Science*, pages 266–271, Tempe, Arizona, May 2007b. Springer Verlag. ISBN 978-3-540-72199-4.

M. Gelfond and V. Lifschitz. Classical Negation in Logic Programs and Disjunctive Databases. *New Generation Computing*, 9:365–385, 1991.

N. Leone, G. Gottlob, R. Rosati, T. Eiter, W. Faber, M. Fink, G. Greco, G. Ianni, E. Kałka, D. Lembo, M. Lenzerini, V. Lio, B. Nowicki, M. Ruzzi, W. Staniszkis, and G. Terracina. The INFOMIX System for Advanced Integration of Incomplete and Inconsistent Data. In *Proceedings of the 24th ACM SIGMOD International Conference on Management of Data (SIGMOD 2005)*, pages 915–917, Baltimore, Maryland, USA, June 2005. ACM Press.

N. Leone, G. Pfeifer, W. Faber, T. Eiter, G. Gottlob, S. Perri, and F. Scarcello. The DLV System for Knowledge Representation and Reasoning. *ACM Transactions on Computational Logic*, 7(3):499–562, July 2006.

Y. Lierler. Disjunctive Answer Set Programming via Satisfiability. In C. Baral, G. Greco, N. Leone, and G. Terracina, editors, *Logic Programming and Nonmonotonic Reasoning — 8th International Conference, LPNMR'05, Diamante, Italy, September 2005, Proceedings*, volume 3662 of *Lecture Notes in Computer Science*, pages 447–451. Springer Verlag, September 2005. ISBN 3-540-28538-5.

Y. Lierler and V. Lifschitz. One More Decidable Class of Finitely Ground Programs. In *Proceedings of the 25th International Conference on Logic Programming (ICLP 2009)*, volume 5649 of *Lecture Notes in Computer Science*, pages 489–493, Pasadena, CA, USA, July 2009. Springer. ISBN 978-3-642-02845-8.

F. Lin and Y. Wang. Answer Set Programming with Functions. In *Proceedings of Eleventh International Conference on Principles of Knowledge Representation and Reasoning (KR2008)*, pages 454–465, Sydney, Australia, September 2008. AAAI Press. ISBN 978-1-57735-384-3.

M. Manna, M. Ruffolo, E. Oro, M. Alviano, and N. Leone. The HiLeX System for Semantic Information Extraction. *Transactions on Large-Scale Data and Knowledge-Centered Systems*. Springer Berlin/Heidelberg, 2011a. To appear.

M. Manna, F. Scarcello, and N. Leone. On the complexity of regular-grammars with integer attributes. *Journal of Computer and System Sciences (JCSS)*. Elsevier, Academic Press, Inc., 77(2):393–421, 2011b. doi: doi:10.1016/j.jcss.2010.05.006.

F. Ricca, M. Alviano, A. Dimasi, G. Grasso, S. M. Ielpa, S. Iiritano, M. Manna, and N. Leone. A Logic–Based System for e–Tourism. *Fundamenta Informaticae, IOS Press*, 105(1–2): 35–35, 2010. Invited paper.

F. Ricca, G. Grasso, M. Alviano, M. Manna, V. Lio, S. Iiritano, and N. Leone. Team-building with Answer Set Programming in the Gioia-Tauro Seaport. *Theory and Practice of Logic Programming*. Cambridge University Press, 2011. To appear.

M. Simkus and T. Eiter. FDNC: Decidable Non-monotonic Disjunctive Logic Programs with Function Symbols. In *Proceedings of the 14th International Conference on Logic for Programming, Artificial Intelligence, and Reasoning (LPAR2007)*, volume 4790 of *Lecture Notes in Computer Science*, pages 514–530. Springer, 2007. ISBN 978-3-540-75558-6.

P. Simons, I. Niemelä, and T. Soininen. Extending and Implementing the Stable Model Semantics. *Artificial Intelligence*, 138:181–234, June 2002.

T. Syrjänen. Omega-Restricted Logic Programs. In *Proceedings of the 6th International Conference on Logic Programming and Nonmonotonic Reasoning*, pages 267–279, Vienna, Austria, September 2001. Springer-Verlag.

S.-Å Tärnlund. Horn clause computability. *BIT Numerical Mathematics*, 17(2):215–226, June 1977.

Logic in Nonmonotonic Reasoning

Alexander Bochman
Computer Science Department
Holon Institute of Technology (HIT)
Holon, Israel

Abstract: We present a conceptual description of nonmonotonic formalisms as essentially two-layered reasoning systems consisting of a monotonic logical system, coupled with a mechanism of a reasoned choice of default assumptions. On this 'surgery', the main features of a nonmonotonic formalism are determined by the interplay between its logical basis and the nonmonotonic overhead. Furthermore, it allows us to see the majority of existing nonmonotonic systems as implementations of just two basic nonmonotonic mechanisms in various logical formalisms, giving raise to two principal paradigms of nonmonotonic reasoning, preferential and explanatory reasoning. In addition, we discuss the requirements and constraints this view imposes on nonmonotonic systems, as well as some of the main problems such systems should try to resolve.

1 Nonmonotonic Reasoning versus Logic: Pre-History of Relations

Nonmonotonic reasoning is considered today an essential part of the logical approach to Artificial Intelligence. In fact, the birth of nonmonotonic reasoning can be loosely traced to the birth of Artificial Intelligence itself. This is due to the fact that John McCarthy, one of the fathers of AI, suggested [1959] and consistently developed a research methodology that used logic to formalize the reasoning problems in AI. McCarthy's objective was to formalize *common sense* reasoning about such problems. Despite this, the relationships between nonmonotonic reasoning and logic have always been controversial, part of a larger story of the relations between AI and logic in general [Thomason, 2003].

The origins of nonmonotonic reasoning proper lied in dissatisfaction with the traditional logical methods in representing and handling the problems posed by AI. Basically, the problem was that reasoning necessary for an intelligent behavior and decision making in realistic situations has turned out to be difficult, even impossible, to represent as deductive inferences in some logical system. The essence of this problem has been formulated by Marvin Minsky in his influential "frames paper" [1974], where he directly questioned the suitability of representing commonsense knowledge in a form of a deductive system:

There have been serious attempts, from as far back as Aristotle, to represent common sense reasoning by a "logistic" system .... No one has been able successfully to confront such a system with a realistically large set of propositions. I think such attempts will continue to fail, because of the character of logistic in general rather than from defects of particular formalisms.

According to Minsky, such a "logical" reasoning is not flexible enough to serve as a basis for thinking. Due to this deficiency, traditional logic cannot discuss what ought to be deduced under ordinary circumstances. Minsky was also one of the first who pointed to monotonicity of logical systems as a source of the problem:

Monotonicity: ... In any logistic system, all the axioms are necessarily "permissive" - they all help to permit new inferences to be drawn. Each added axiom means more theorems, none can disappear. There simply is no direct way to add information to tell such the system about kinds of conclusions that should not be drawn! To put it simply: if we adopt enough axioms to deduce what we need, we deduce far too many other things.

Last but not least, Minsky mentioned the requirement of consistency demanded by logic that makes the corresponding systems too weak:

I cannot state strongly enough my conviction that the preoccupation with Consistency, so valuable for Mathematical Logic, has been incredibly destructive to those working on models of mind. ... At the "logical" level it has blocked efforts to represent ordinary knowledge, by presenting an unreachable image of a corpus of context-free "truths" that can stand separately by themselves.

In some sense, nonmonotonic reasoning can be viewed as a rigorous solution to Minsky's challenges. It is important to note, however, that such philosophical deliberations were not the main source of the first nonmonotonic formalisms. Long before the emergence of formal nonmonotonic systems, there have been a number of problems and applications in AI that required some forms of nonmonotonic reasoning. Initial solutions to these problems *worked* (though in restricted applications), and this was an incentive for trying to provide them with a more systematic logical basis. It is these problems and their solutions that influenced the actual shape of subsequent nonmonotonic formalisms.

On a most general level, nonmonotonic reasoning is intimately related to the traditional philosophical problems of natural kinds and ceteris paribus laws. A most salient feature of these notions is that they resist precise logical definition, but involve a description of 'normal' or 'typical' cases. As a consequence, reasoning with such concepts is inevitably *defeasible*, that is, it may fail to 'preserve truth' under all circumstances, which has always been considered a standard requirement of logical reasoning.

The problem of representing and reasoning with natural kinds has reappeared in AI as a practical problem of building taxonomic hierarchies for large knowledge bases, when such hierarchies have been allowed to have exceptions. The theory of reasoning in such taxonomies has been called *nonmonotonic inheritance* (Horty [1994] provides an overview of the topic). The guiding principle in resolving potential conflicts in such hierarchies was a *specificity principle*: more specific information should override more generic information in cases of conflict. Nonmonotonic inheritance relied, however, more on graph-based representations than on traditional logical tools. Nevertheless, it has managed to provide a plausible analysis of reasoning in this restricted context.

Nonmonotonic reasoning of a different kind has been observed in the framework of already existing systems, such as databases, logic programming and planning algorithms. A common assumption in such systems has been that positive assertions that are not explicitly stated or derivable should be considered false. Accordingly, instead of explicitly representing such negative information, databases implicitly do so by appealing to the so-called *closed word assumption* (CWA), which states that if a positive fact is not derivable from the database, its negation is assumed to hold. The same principle has been employed in programming languages for AI such as Prolog and Planner. Thus, in Prolog, the goal **not** G succeeds if the attempt to find a proof of G using the program rules as axioms fails. Prolog's negation is a nonmonotonic operator: if G is not provable from some axioms, it need not remain nonprovable from an enlarged axiom set. It has turned out that this negation-as-failure can be used to implement important forms of commonsense reasoning, which eventually has led to developing modern declarative logic programming as a general representation formalism for nonmonotonic reasoning.

But first and foremost, nonmonotonicity has appeared in attempts to represent reasoning about actions and change. As was stated by McCarthy and Hayes [1969], a computer program capable of acting intelligently in the world must have a general representation of the world, and it should decide what to do by inferring in a formal language that a certain strategy will achieve its assigned goal. The main problem for an adequate formalization of this task is the *frame problem*: how *efficiently* determine which things remain the same in a changing world.

The frame problem arises in the context of *predictive* reasoning, a type of reasoning that is essential for planning and formalizing intelligent behavior, though relatively neglected in the traditional logical literature. Prediction involves the inference of later states from earlier ones. Changes in this setting do not merely occur, but occur for a reason. Furthermore, we usually assume that most things will be unchanged by the performance of an action. It is this *inertia assumption* that connects reasoning about action and change with nonmonotonic reasoning. In this reformulation, the frame problem is to determine what stays the same about the world as time passes and actions are performed (e.g., a red block remains red after we have put it on top of another block). In order to specify how propositions do not change as actions occur, McCarthy and Hayes [1969] suggested to write down special frame axioms. Of course, a huge number of things stay the same after a particular action, so we would have to add a very large number of frame axioms to the theory. This is precisely the frame problem. It becomes even more complicated due to the accompanying *ramification problem* that concerns the necessity of taking into account numerous derived effects (ramifications) of actions, effects created by logical and causal properties of a situation. Suppose we have a suitcase with two locks, and it is opened if both locks are open. Then the action of opening one lock produces an indirect effect of opening the suitcase if and only if the other lock is open. Such derived effects should be taken into account when combined with the inertia assumption, since they override the latter. The ramification problem has raised general questions on the nature of causation and its role in temporal reasoning, which had led, eventually, to the so-called causal approach to the frame problem.

Last but not least, there was the *qualification problem*, the problem of specifying what conditions must be true for a given action to have its intended effect. If I turn the ignition key in my car, I expect the car to start. However, many conditions have to be true in order for this statement to be true: the battery must be alive, the starter must work, there must be gas in the tank, there is no potato in the tailpipe, etc. — an open-ended list of qualifications. Still, without knowing for certain about most of these facts, we normally assume that turning the key will start the car. This problem is obviously a special instance of a general philosophical problem of ceteris paribus laws, laws or generalizations that are valid only under 'normal' circumstances which are usually difficult, if not impossible, to specify exactly. The qualification problem has turned out to be one of the most stubborn problems for the representation of action and change. The majority of initial nonmonotonic formalisms have failed to deliver a satisfactory solution for it.

The above problems and their first solutions provided both the starting point and basic objectives for the first nonmonotonic theories. These historical origins allows us to explain, in particular, an eventual significant discrepancy that has developed between nonmonotonic and commonsense reasoning. Ever since McCarthy's program, commonsense reasoning has appeared to be an attractive standard, and a promising way of solving AI problems. Nevertheless, the study of 'artificial reasoning' need not and actually has not been committed to the latter. As we will discuss later, the basic formalisms of nonmonotonic reasoning could hardly be called formalizations of commonsense reasoning. Still, in trying to cope with principal commonsense reasoning tasks, the suggested formalisms have succeeded in capturing important features of the latter and thereby have broken new territory for logical reasoning.

1.1   What Is Nonmonotonic Reasoning?

In our ordinary, commonsense reasoning, we usually have only partial information about a given situation, and we make a lot of assumptions about how things normally are in order to carry out further reasoning. For example, if we learn that Tweety is a bird, we usually assume that it can fly. Without such presumptions, it would be almost impossible to carry out the simplest human reasoning tasks.

Speaking generally, human reasoning is not reducible to collecting facts and deriving their consequences, but involves also making assumptions and wholesale theories about the world and acting in accordance with them. Both commonsense and nonmonotonic reasoning are just special forms of a general scientific methodology in this sense.

The way of thinking in partially known circumstances suggested by nonmonotonic reasoning consists in using reasonable assumptions that can guide us in our decisions. Accordingly, nonmonotonic reasoning can be described as a theory of making and selecting assumptions in a principled way. Of course, the latter are only assumptions (hence, beliefs), so they should be abandoned when we learn new facts about the circumstances that contradict them. However, nonmonotonic reasoning assigns a special status to such assumptions; it makes them *default* assumptions. Default assumptions (or simply defaults) are seen as acceptable in all circumstances unless they conflict with other defaults and current evidence. This presumptive reading has a semantic counterpart in the notion of *normality*; default assumptions are considered as holding for normal circumstances, and the nonmonotonic theory says us to always assume that the world is as normal as is compatible with known facts. This kind of belief commitment is a novel aspect contributed by nonmonotonic reasoning to a general theory of reasoning. It implies, in particular, that our assumptions are not abandoned, in general, in the light of evidence, but only postponed, and hence can still be used in other situations.

Already at this stage we can notice that this kind of reasoning conflicts with monotonicity of logical derivations. Monotonicity is a characteristic property of deductive inferences arising from the very notion of a proof: if C is provable from a set of propositions a, it is provable also from a larger set $a \cup \{A\}$. Nonmonotonic reasoning is of course non-monotonic in this sense, because adding new facts may invalidate some of the assumptions made earlier. In other words, nonmonotonicity is just a side effect, or a symptom, of such an assumption-based reasoning.

As was rightly noted once by David Poole, the default "Birds fly" is not a statement that is true or not of a world; some birds fly, some do not. Rather, it is an assumption used in building our theory of the world. Accordingly, nonmonotonic reasoning does not make any claims about the objective status of the default assumptions it uses. In particular, it does not depend on the objective confirmation of the latter. What it cares about, however, is an internal coherence of the choice of assumptions it makes in particular situations. Of course, if we use an entirely inappropriate claim as our default assumption, the latter will turn out to be either useless (inapplicable) or, worse, it may produce wrong conclusions. This makes nonmonotonic reasoning a risky business. Still, in most cases default assumptions we make are useful and give desired results, and hence they are worth the risk of making an error. But what is even more important, more often than not we simply have no 'safe' replacement for such a reasoning strategy. That is why it is worth to teach robots and computers to reason in this way.

2 Two Problems of Default Assumptions, and Two Theories thereof

Now it is time to turn to a more detailed analysis of the use of default assumptions in nonmonotonic reasoning.

An assumption like "Birds fly" has an unbounded list of exceptions – ostriches, penguins, Peking ducks, etc., etc. So, if we would try to use classical logic directly for representing "Birds fly", the first problem would be that it is practically impossible to enumerate all exceptions to flight with an implication of the form

$$(\forall x).Bird(x)\&\neg Penguin(x)\&\neg Emu(x)\&\neg Dead(x)\&... \supset Fly(x)$$

Furthermore, even if we could enumerate all such exceptions, we still could not derive $Fly(Tweety)$ from $Bird(Tweety)$ alone. This is so since then we are not given that Tweety is not a penguin, or dead, etc. The antecedent of the above implication cannot be derived, in which case there is no way of deriving the consequent. In a commonsense reasoning, however, we feel fully justified in assuming that Tweety can fly, if we know only that it is a bird.

Thus, the primary problem of a nonmonotonic reasoning theory boils down to how and when we can actually *use* such default assumptions. All three initial nonmonotonic formalisms, namely circumscription [McCarthy, 1980], default logic [Reiter, 1980] and modal nonmonotonic logic [McDermott and Doyle, 1980] have provided successful and rigorous answers to this primary problem. All these three formalisms have appeared thirty years ago in a celebrated issue of the Artificial Intelligence Journal. The approaches were formulated in three different languages – the classical first order language in the case of circumscription, a set of inference rules in default logic, and a modal language in modal nonmonotonic logic. Still, behind these differences there was a common idea of representing commonsense conditionals or rules as ordinary conditionals or rules with additional assumptions, assumptions that could readily be accepted in the absence of contrary information. In this respect, the differences between the three theories amounted to different mechanisms of making such default assumptions. In fact, subsequent studies have shown that default logic and modal nonmonotonic logics are largely interdefinable formalisms (see bellow), so they embody, in a sense, the same nonmonotonic mechanism. However, the differences between both these formalisms and circumscription have turned out to be more profound. In order to articulate these differences, we need to consider yet another, secondary problem of default assumptions, which, as we will argue, has become one of the main problems of nonmonotonic formalisms.

As we mentioned earlier, in order to preserve consistency of the resulting solutions, default assumptions should not be used when they contradict known facts and other defaults. Clearly, if a default plainly contradicts the facts, it should be 'canceled'. But if a number of defaults are jointly inconsistent with the facts, we have a *selection problem*: which of the defaults should be retained, and which abandoned in each particular case?

An apparently straightforward solution to the above selection problem is to choose all maximal consistent subsets of defaults. It is this solution that has been implicitly used, in effect, in the first variant of circumscription, described by McCarthy [1980].

Unfortunately, this solution turns out to be inadequate as a general solution to the problem. The main reason is that commonsense defaults are not born equal, and in most cases there exists an additional structure of dependence and priority between default assumptions themselves. As a result, not all maximal consistent combinations of defaults turn out to be adequate as options for choice. As a simplest example of this situation we can take already mentioned nonmonotonic inheritance, where the choice of defeasible rules in cases of conflicts is constrained by the specificity principle. We may safely assume both *Birds fly* and *Penguins don't fly*, but then, given that Tweety is a penguin (and hence a bird), we nevertheless univocally infer that it does not fly. This is because only the default *Birds fly* can be canceled in this conflict, since it is less specific than *Penguins don't fly*.

Speaking more generally, the selection problem can be viewed as a manifestation of the more profound fact that commonsense defaults have a structure and logic that involves much more than just a set of assumptions. This is the main reason why the primary problem of nonmonotonic reasoning, the problem how to make default assumptions, does not necessary provide an adequate solution to the selection problem. The latter presupposes a deeper understanding of the actual use of default assumptions in commonsense reasoning.

A general way of handling the selection problem in the framework of circumscription has been suggested by Vladimir Lifschitz and endorsed in the second version of circumscription [McCarthy, 1986]. The solution amounted to imposing priorities among minimized predicates and abnormalities. The corresponding variant of circumscription has been called *prioritized* circumscription.

The main idea behind prioritized circumscription can be described in our general terms as follows. Priorities among defaults can be naturally extended to preferences among sets of defaults. As a result, we acquire an opportunity to discriminate between maximal consistent sets of defaults, and to choose only preferred sets among them. Accordingly, the selection problem of default assumptions reduces on this approach to the problem of establishing an appropriate priority order on defaults. As a matter of fact, prioritized circumscription, combined with parallel developments in logic and belief revision theory, has evolved to a general *preferential approach* to nonmonotonic reasoning. We will describe the latter in more details later in this study.

Default and modal nonmonotonic logics suggested a different approach to nonmonotonic reasoning. They can be viewed as first systems of what we call an *explanatory approach* to default reasoning. From the outset, these formalisms were based on a more elaborate mechanism of choosing assumptions. In some sense, it could even be argued that this alternative approach has 'borrowed' and formalized a much larger piece of scientific and commonsense methodology than circumscription. In both scientific and commonsense discourse, a particular law may fail to explain the actual outcome due to interference with other mechanisms and laws that contribute to the combined result. In other words, violations of such laws are always *explainable* (at least in principle) by other laws that are active. It is precisely this justificational aspect of scientific reasoning that has been formalized in the notion of extension in default logic and corresponding models of modal nonmonotonic logic. An extension can be seen as a model generated by a set of defaults that is not only maximal consistent, but also, and most importantly, allows us to explain away, or refute, all other defaults that are left out. The latter requirement constitutes a very strong constraint on the coherence of potential solutions, which goes far beyond plain consistency.

Here it is important to note that the very notion of an extension, or, more precisely, its non-triviality, is based on the logical possibility of distinguishing between refutation and inconsistency, a possibility that does not exist in classical logic (and hence inaccessible for circumscription). Already this explains why these formalisms were based on more general logics than the classical one.

As an important by-product of the additional requirements on potential solutions imposed by the notion of extension, we acquire an opportunity to 'tune' this mechanism of choosing assumptions to intended combinations of defaults by supplying the underlying logic with appropriate refutation rules for default assumptions. In other words, explanatory formalisms are capable of selecting particular combinations of defaults by using tools that are available in the formalisms themselves[1]. In a hindsight, this might be seen one of the main reasons why these formalisms have been relatively slow in realizing the complexity of the selection problem.

At the time of working on his default logic, Ray Reiter believed that commonsense default claims *"A normally implies B"* are directly representable in his default logic by default rules of the form $A : B/B$, appropriately called *normal* default rules. He even remarked that he knows of no naturally occurring default which cannot be represented in this form. His views have changed quickly, however, and already Reiter and Criscuolo [1981] argued that in order to deal with default interactions, we need to represent such defaults using at least semi-normal rules $A : C \wedge B/B$, where C accumulates additional preconditions that can be refuted by other rules. Unfortunately, the problem of commonsense defaults and their interactions has 'survived' these initial attempts of formalization, and it has reappeared in a most dangerous form as a *Yale Shooting Anomaly* [Hanks and McDermott, 1987]. The latter paper has demonstrated that apparently plausible representations of defaults in default logic and other formalisms still do not provide an intended choice of assumptions required for the solution of the frame problem.

The anomaly has demonstrated, in effect, that also explanatory nonmonotonic formalisms, taken by themselves, do not provide a complete solution to the selection problem of default assumptions. Nevertheless, despite initial, radically anti-logicist reactions (cf. the paper by McDermott [1987]), subsequent studies have shown that the Yale Shooting problem can be resolved in the framework of these formalisms, though it may require, for example, the use of more general default rules than semi-normal ones. Yet even these solutions could have been taken as an indication of the basic fact that default logic, despite its name, is not primarily a theory of commonsense defaults. Still, they helped to preserve the hope that the latter theory can, in principle, be represented in the framework of default logic.

3 Logic in Nonmonotonic Reasoning

The advance of the first nonmonotonic formalisms has re-shaped the initial antagonistic relations between nonmonotonic reasoning and logic in general. First of all, these formalisms have demonstrated that nonmonotonic reasoning can be given a rigorous formal characterization. Furthermore, it has been shown that a nonmonotonic formalism can be defined, in principle, by supplying some background logical formalism with a new kind of semantics (a set of intended models), called a *nonmonotonic* semantics, which forms a distinguished subset of the corresponding *logical* semantics determined by the logical formalism itself. Thus, for circumscription, the underlying logical formalism is just the classical logic (and its semantics), while for modal nonmonotonic logics it is a particular modal logic with its associated possible worlds semantics. Though less immediate, a similar two-level structure can be provided also for default logic (see below).

---

[1]Thus, Lifschitz [1990] has shown that circumscription can be expressed in first-order default logic.

Unfortunately, it is this latter description that has also brought to life the problematic 'shortcut' notion of *nonmonotonic logic* as a formalism determined directly by syntax and associated nonmonotonic semantics. A seemingly advantageous feature of this notion was its obvious similarity with traditional understanding of logical formalisms as determined by syntax and corresponding semantics. On this view, a nonmonotonic logic has become just yet another logic determined by an unusual (nonmonotonic) semantics. However, we contend that this view has actually hindered in a number of ways an adequate understanding of nonmonotonic reasoning.

In ordinary logical systems, the semantics determines the set of logical consequences of a given theory, but also, and most importantly, it provides a semantic interpretation for the syntax itself. Namely, it provides propositions and rules of a syntactic formalism with *meaning*, and its theories with *informational content*. By its very design, however, the nonmonotonic semantics is defined as a certain *subset* of the set of logically possible models, and consequently it does not determine, in turn, the meaning of the propositions and rules of the syntax. As a result, two radically different theories may (accidentally) have the same nonmonotonic semantics. Furthermore, such a difference cannot be viewed as apparent, since it may well be that by adding further rules or facts to both these theories, we obtain new theories that already have different nonmonotonic models.

The above situation is remarkably similar to the distinction between meaning (intension) and extension of logical concepts, a distinction that is fundamental for modern logic. Nonmonotonic semantics provides, in a sense, the extensional content of a theory in a particular context of its use. In order to determine the meaning, or informational content, of a theory, we have to consider all potential contexts of its use, and hence 'retreat' to the underlying logic. Accordingly, the relevant distinction requires a clear *separation of the logical and nonmonotonic components of nonmonotonic reasoning*. This separation suggests the following more adequate understanding of nonmonotonic reasoning:

Nonmonotonic Reasoning = Logic + Nonmonotonic Semantics

Logic and its associated logical semantics are responsible for providing the meaning of the rules of the formalism, while the nonmonotonic semantics provides us with nonmonotonic consequences of a theory in particular situations.

In addition to a better understanding of the structure of nonmonotonic formalisms themselves, the above two-layered structure has important benefits in comparing different formalisms. Namely, it will allow us to see many such formalisms as instantiations of the same nonmonotonic mechanism in different underlying logics.

Fortunately, in many cases the underlying logic of a nonmonotonic formalism can actually be restored on the basis of the associated nonmonotonic semantics. A bit more precise description of such a 'reverse engineering' is as follows. Suppose we are given a nonmonotonic logic in its virgin sense, namely a syntactic formalism $\mathbb{F}$ coupled with a nonmonotonic semantics $\mathbb{S}$. The syntactic formalism determines the basic informational units that we will call default theories, for which the semantics supplies the nonmonotonic interpretation (i.e., a set of models). Default theories could be plain logical theories, as in circumscription or modal nonmonotonic logics, or sets of rules, as in default logic. For a default theory Δ, let $\mathbb{S}(\Delta)$ denote the nonmonotonic semantics of Δ. Then two default theories can be called *nonmonotonically equivalent* if they have the same nonmonotonic interpretation. This equivalence does not determine, however, the logical meaning of default theories. Note that precisely due to the nonmonotonicity of $\mathbb{S}$, we may have $\mathbb{S}(\Delta) = \mathbb{S}(\Gamma)$, though $\mathbb{S}(\Delta \cup \Phi) \neq \mathbb{S}(\Gamma \cup \Phi)$, for some default theory Φ.

Still, the above considerations suggest a simple solution. Recall that a standard definition of meaning in logic says that two notions have the same meaning if they determine the same extension in all contexts (e.g., in all possible worlds). In our case, a context can be seen simply as a larger theory including a given one, which leads us to the following notion:

Definition 1 *Default theories* Γ *and* Δ *are called* strongly $\mathbb{S}$-equivalent, *if* $\mathbb{S}(\Delta \cup \Phi) = \mathbb{S}(\Gamma \cup \Phi)$, *for any default theory* Φ.

This notion of strong equivalence has originally been suggested in logic programming [Lifschitz, Pearce, and Valverde, 2001], but it turns out to have general significance. Strong equivalence is already a logical notion, since strongly equivalent theories are interchangeable in any larger theory without changing the associated nonmonotonic semantics. This suggests that there should exist a logic $\mathbb{L}$ formulated in the syntax $\mathbb{F}$ such that default theories are strongly equivalent if and only if they are logically equivalent in $\mathbb{L}$. In this case, the logic $\mathbb{L}$ can be viewed as the underlying logic of the nonmonotonic formalism that will determine the logical meaning of default theories and, in particular, of the rules and propositions of the syntactic framework $\mathbb{F}$.

The attention to the underlying logics behind nonmonotonic reasoning is also rewarded with a better understanding of the range of such logics that are appropriate for particular nonmonotonic mechanisms. For the case of explanatory nonmonotonic formalisms, for instance, this approach reveals a whole range of logical possibilities for basically the same nonmonotonic semantics, starting from logic programming and ending with causal calculus and modal nonmonotonic logics.

To end this section, we should mention that John McCarthy was apparently the first to resist the understanding of nonmonotonic reasoning as a new logic. In concluding remarks in his 1980 paper, McCarthy mentioned that circumscription is not a "nonmonotonic logic"; it is a form of nonmonotonic reasoning augmenting ordinary first order logic. Actually, his main reason was that, due to the monotonicity property of logical derivations, we cannot get circumscriptive reasoning capability by adding sentences or ordinary rules of inference to an axiomatization. Unfortunately, the overall impact of McCarthy's arguments was somewhat weakened by his general resistance to all other logics beyond the classical one. Thus, we already mentioned that a change in the underlying logic was a necessary prerequisite for the explanatory nonmonotonic formalisms. Still, the basic 'monotonicity' argument remains intact also for the latter formalisms. Most importantly, it retains its force for subsequent 'nonmonotonic' inference systems such as preferential consequence relations [Kraus, Lehmann, and Magidor, 1990] suggested in the framework of preferential approach to nonmonotonic reasoning, an approach that can be seen as a systematic development of the very program started by McCarthy's circumscription. A somewhat complicated interplay of logic and nonmonotonicity in this latter approach will be our subject in the next section.

4 Preferential Nonmonotonic Reasoning

As we argued earlier, the main problem nonmonotonic reasoning has to deal with amounts to the selection problem, the problem of choosing appropriate sets of assumptions in each particular case. Now, the general preferential approach follows here the slogan *"Choice presupposes preference"*. According to this approach, the choice of assumptions should be made by establishing preference relations among them. This makes preferential approach a special case of a general methodology that is at least as old as the decision theory and theory of social choice.

Prioritized circumscription can be seen as the ultimate origin of the preferential approach. A generalization of this approach was initiated by Gabbay [1985] on the logical side, and Shoham [1988] on the AI side. Shoham argued that any form of nonmonotonicity involves minimality of one kind or another. He argued also for a shift in emphasis from syntactic characterizations in favor of semantic ones. Accordingly, he defined a model preference logic by using an arbitrary preference ordering of the interpretations of a language.

Definition 2 *An interpretation i is a* preferred model *of A if it satisfies A and there is no better interpretation $j > i$ satisfying A. A* preferentially entails *B (written $A \mid\sim B$) iff all preferred models of A satisfy B.*

Shoham's approach was very appealing, and apparently suggested a unifying perspective on nonmonotonic reasoning. A decisive further step has been made by Kraus et al. [1990] that described both semantics and axiomatizations for a range of such preferential inference relations. This has established logical foundations for a research program that has attracted many researchers, both in AI and in logic.

For our subsequent discussion, it is important to note that one of the strong advantages Kraus et al. have rightly seen in their approach was that preferential conditionals $A \mid\sim B$, the basic units of their consequence relations, provide a more natural, and more adequate, formalization of commonsense normality claims *"A normally implies B"* than, say, default logic. Accordingly, on their view preferential consequence relations stand closer to commonsense reasoning than previous nonmonotonic formalisms.

Epistemic States and Prioritization. The standard semantics of preferential inference relations is based on abstract possible worlds models in which worlds are ordered by a preference relation. However, a more specific semantic interpretation for such inference relations, suitable for real nonmonotonic reasoning, can be based on preference ordering on sets of default assumptions. Such a representation has been suggested by Bochman [2001], based on the notion of an epistemic state, defined as follows:

Definition 3 *An* epistemic state *is a triple $(\mathcal{S}, l, \prec)$, where $\mathcal{S}$ is a set of admissible belief states, $\prec$ a preference relation on $\mathcal{S}$, while l is a labeling function assigning a deductively closed belief set to every state from $\mathcal{S}$.*

On the intended interpretation, admissible belief states are generated as logical closures of allowable combinations of default assumptions. The preference relation on admissible belief states reflects the fact that not all combinations of defaults constitute equally preferred options for choice. For example, defaults are presumed to hold, so an admissible belief state generated by a larger set of defaults is normally preferred to an admissible state generated by a smaller set of defaults. In addition, defaults may have some priority structure that imposes, in turn, additional preferences among belief states.

Epistemic states guide our decisions what to believe in particular situations. They are epistemic, however, precisely because they say nothing directly about what is actually true, but only what is believed (or assumed) to hold. This makes epistemic states relatively stable entities; change in facts and situations will not necessary lead to change in epistemic states. The actual assumptions made in particular situations are obtained by choosing preferred admissible belief states that are consistent with the facts.

An explicit construction of epistemic states generated by default bases provides us with characteristic properties of epistemic states arising in particular reasoning contexts. An epistemic state is *base-generated* by a set Δ of propositions with respect to a classical Tarski consequence relation Th if

- the set of its admissible states is the set $\mathcal{P}(\Delta)$ of subsets of Δ;

- l is a function assigning each $\Gamma \subseteq \Delta$ a theory $\mathrm{Th}(\Gamma)$;

- the preference order is monotonic on $\mathcal{P}(\Delta)$: if $\Gamma \subset \Phi$, then $\Gamma \prec \Phi$.

The preference order on admissible belief states is usually derived in some way from priorities among individual defaults. This task turns out to be a special case of a general problem of combining a set of preference relations into a single 'consensus' preference order. As has been shown by Bochman [2001], a most natural and justified way of doing this is as follows.

Suppose that the set of defaults Δ is ordered by some *priority relation* $\lhd$ which will be assumed to be a strict partial order: $\alpha \lhd \beta$ will mean that α *is prior to* β. Then the resulting preference relation on sets of defaults can be defined in the following way:

$$\Gamma \preccurlyeq \Phi \;\equiv\; (\forall \alpha \in \Gamma \backslash \Phi)(\exists \beta \in \Phi \backslash \Gamma)(\beta \lhd \alpha)$$

$\Gamma \preccurlyeq \Phi$ holds when, for each default in $\Gamma \backslash \Phi$, there is a prior default in $\Phi \setminus \Gamma$. Lifschitz [1985] was apparently the first to use this construction in prioritized circumscription, while Geffner [1992] employed it for defining preference relations among sets of defaults.

Nonmonotonic Inference.   In particular situations, we restrict our attention to admissible belief sets that are consistent with the facts, and choose preferred among them. The latter are used to support the assumptions and conclusions we make about the situation at hand.

An admissible belief state $s \in \mathcal{S}$ will be said to be *compatible with* a proposition A, if $\neg A \notin l(s)$. The set of all admissible states that are compatible with A will be denoted by $\langle A \rangle$.

A *skeptical inference* with respect to an epistemic state is obtained when we infer only what is supported by each of the preferred states. In other words, B is a skeptical conclusion from the evidence A if each preferred admissible belief set compatible with A, taken together with A itself, implies B.

Definition 4 B *is a* skeptical consequence *of A in an epistemic state if $A \supset B$ belongs to belief sets of all preferred belief states in $\langle A \rangle$.*

Though apparently different from the original definition of Shoham, the above notion of nonmonotonic inference actually coincides with a principal system of preferential entailment [Kraus et al., 1990], which provided the latter with a complete, thoroughly logical axiomatization.

4.1   Logic in Preferential Reasoning

The above brief description of preferential approach to nonmonotonic reasoning forces us to return once more to the basic subject of this study, the relationships of nonmonotonic reasoning and logic. Indeed, on the face of it, preferential approach has apparently assimilated nonmonotonic reasoning to plain deductive reasoning in a certain 'nonmonotonic' logic. This cannot be right. And in fact, things turn out to be more complex than this picture suggests.

Preferential entailment is called nonmonotonic for the obvious reason that its rules or conditionals do not admit Strengthening the Antecedent: $A \hspace{-0.3em}\sim\hspace{-0.3em} B$ does not imply $A \wedge C \hspace{-0.3em}\sim\hspace{-0.3em} B$. However, it is a *monotonic*, logical system in another, far more important, sense that addition of new rules to a set of such rules preserves all the derivations that have been made beforehand. To put it in a different way, the inferences among conditionals sanctioned by preferential entailment are a priori restricted to derivations that will retain their force under any further additions of facts and rules. By our preceding discussion, this means that preferential entailment describes precisely the underlying *logic* of default conditionals, which inevitably implies, in turn, that it is too weak to capture the associated *nonmonotonic reasoning* with such defaults.

Preferential inference is severely sub-classical and does not allow us, for example, to infer "Red birds fly" from "Birds fly". Clearly, there are good reasons for not accepting such a derivation as a *logical* rule for preferential inference; otherwise "Birds fly" would imply also "Birds with broken wings fly" and even "Penguins fly". Still, this should not prevent us from accepting "Red birds fly" on the basis of "Birds fly" as a reasonable *default* conclusion, namely a conclusion made in the absence of information against it. By doing this, we would just follow the general strategy of nonmonotonic reasoning that involves making reasonable assumptions on the basis of available information. Thus, the logical core of skeptical inference, preferential inference relations, should be augmented with a mechanism of making nonmonotonic conclusions. This kind of reasoning will of course be defeasible, or *globally* nonmonotonic, since addition of new conditionals can block some of the conclusions made earlier.

On the semantic side, default conditionals are constraints on epistemic states in the sense that the latter should make them skeptically valid. Still, usually there is a huge number of epistemic states that satisfy a given set of conditionals, so we have both an opportunity and necessity to choose among them. Our guiding principle in this choice can be the same basic principle of nonmonotonic reasoning, namely that the *intended* epistemic states should be as normal as is permitted by the current constraints. By choosing particular such states, we will thereby adopt conditionals that would not be derivable from a given base by preferential inference alone.

The above considerations lead to the conclusion that preferential entailment assigns default conditionals a clear *logical* meaning and associated logical semantics, but this logic still should be extended to a truly nonmonotonic formalism of defeasible entailment by defining the associated *nonmonotonic semantics*. Actually, the literature on nonmonotonic reasoning is abundant with attempts to define such a theory of defeasible entailment. We will describe them briefly in the next section.

4.2 Theories of Defeasible Entailment

Daniel Lehmann, one of the authors of [Kraus et al., 1990], was also one of the first to realize that logical systems of preferential entailment are not sufficient, taken by themselves, for capturing reasoning with default conditionals and, moreover, that we cannot hope to overcome this by strengthening these systems with additional axioms or inference rules. Instead, he suggested a certain semantic construction, called *rational closure*, that allows us to make appropriate default conclusions from a given set of conditionals (see the papers by Lehmann [1989] and Lehmann and Magidor [1992]). An essentially equivalent, though formally very different, construction has been suggested by Pearl [1990] and called system Z. Unfortunately, both theories have turned out to be insufficient for representing defeasible entailment, since they have not allowed to make certain intended conclusions. Hence, they have been refined in a number of ways, giving such systems as lexicographic inference [Benferhat et al., 1993, Lehmann, 1995], and similar modifications of Pearl's system. Unfortunately, these refined systems have encountered an opposite problem, namely, together with some desirable properties, they invariably produced some unwanted conclusions. A more general, an apparently more justified, approach to this problem in the framework of preferential inference has been suggested by Geffner [1992].

Yet another, more syntactic, approach to defeasible entailment has been pursued, in effect, in the framework of an older theory of nonmonotonic inheritance (see the overview paper by Horty [1994]). Inheritance reasoning deals with a quite restricted class of conditionals constructed from literals only. Nevertheless, in this restricted domain it has achieved a remarkably close correspondence between what is derived and what is expected intuitively. Accordingly, inheritance reasoning has emerged as an important test bed for adjudicating proposed theories.

Despite the diversity, the systems of defeasible entailment have a lot in common, and take as a starting point a few basic principles. Thus, most of them presuppose that intended models should be described, ultimately, in terms of material implications corresponding to a given set of conditionals. More precisely, these classical implications should serve as defaults in the nonmonotonic reasoning sanctioned by a default base. This idea can be elaborated as follows: for a set $\mathcal{B}$ of default conditionals, let $\vec{\mathcal{B}}$ denote the the corresponding set of material implications $\{A \supset B \mid A \mathbin{\vert\!\sim} B \in \mathcal{B}\}$. The we can require

Default base-generation. The intended epistemic states for a default base $\mathfrak{B}$ should be base-generated by $\vec{\mathfrak{B}}$.

In other words, the admissible belief states of intended epistemic states for a default base $\mathfrak{B}$ should be formed by subsets of $\vec{\mathfrak{B}}$, and the preference order should be monotonic on these subsets (see the preceding section). In addition, it should be required that all the conditionals from $\mathfrak{B}$ should be skeptically valid in the resulting epistemic state.

Already these constraints on intended epistemic states allow us to derive "Red birds fly" from "Birds fly" for all default bases that do not contain conflicting information about redness. The constraints also sanction defeasible entailment across exception classes: if penguins are birds that normally do not fly, while birds normally fly and have wings, then we are able to conclude that penguins normally have wings, despite being abnormal birds. This excludes, in effect, Pearl's system Z and rational closure that cannot make such a derivation. Still, these requirements are quite weak and do not produce problematic conclusions that plagued some stronger systems suggested in the literature.

Though the above construction deals successfully with many examples of defeasible entailment, it is still insufficient for capturing some important reasoning patterns. Suppressing the details, what is missing in this representation is a principled way of constructing a preference order on default sets. Furthermore, here we may recall that establishing appropriate preferences and priorities among defaults is the main tool used by the preferential approach in general for resolving the selection problem of nonmonotonic reasoning. Accordingly, and as it could be expected, the problem of defeasible entailment boils down at this point to a general selection problem for defaults.

Unfortunately, the above problem has turned out to be far from being trivial, or even univocal. Till these days, we believe that Geffner's conditional entailment on the one hand, and inheritance reasoning, on the other hand, remain the most plausible solutions to this problem, suggested in the literature on preferential reasoning.

Conditional entailment of Geffner [1992] determines a prioritization of default bases by making use of the following relation among conditionals:

Definition 5 *A conditional α dominates a set of conditionals Γ if the set of implications $\{\vec{\Gamma}, \vec{\alpha}\}$ is incompatible with the antecedent of α.*

The origins of this relation can be found already in the work by Adams [1975], and it has been used in practically all studies of defeasible entailment, including the notion of preemption in inheritance reasoning. A suggestive reading of dominance says that if α dominates Γ, it should have priority over at least one conditional in Γ. Note that already this secures that α will be valid in the resulting epistemic state. Accordingly, a priority order on the default base is called *admissible* if it satisfies this condition. Then the intended models can be identified with epistemic states that are generated by all admissible priority orders on the default base (using the construction of the associated preference relation, sketched earlier).

Conditional entailment has shown itself as a serious candidate on the role of a general theory of defeasible entailment. Nevertheless, Geffner himself has shown that it still does not capture some desired conclusions, for which he suggested to augment it with an explicit representation of causal reasoning. The main point of the generalization was to introduce a distinction between propositions and facts that are plainly true versus facts that are *explainable* (caused) by other facts and rules. Actually, the causal generalization suggested by Geffner in the last chapters of his book has served, in part, as an inspiration for a causal theory of reasoning about actions and change [McCain and Turner, 1997], a formalism that belongs, however, to a rival, explanatory approach to nonmonotonic reasoning.

Yet another apparent problem with conditional entailment stems from the fact that it does not capture inheritance reasoning. The main difference between the two theories is that conditional entailment is based on absolute, invariable priorities among defaults, while nonmonotonic inheritance determine such priorities in a context-dependent way, namely in presence of other defaults that provide a (preemption) link between two defaults (see the paper by Dung and Son [2001]). In fact, in an attempt to resolve the problem of defeasible entailment, it has been shown by Bochman [2001] that nonmonotonic inheritance is representable, in principle, by epistemic states ordered by *conditional* (context-dependent) priority orders. Unfortunately, the emerged construction could hardly be called simple or natural. This also suggests (at least, it has suggested to the author) that preferential approach might not be a fully adequate framework for representing nonmonotonic inheritance, and thereby for a defeasible entailment in general.

Summing up our discussion of the preferential approach to nonmonotonic reasoning, we may conclude that the latter has suggested a natural and powerful research program that significantly advanced our understanding of nonmonotonic reasoning and even of commonsense reasoning in general. In fact, its most important achievement consisted in formalizing a plausible (monotonic) logic of default conditionals that could serve as a logical basis for a full, nonmonotonic theory of defeasible reasoning. Unfortunately, it has not succeeded in achieving this latter goal. Moreover, we have seen that a prospective solution to this larger problem was strongly related, in effect, to a still missing solution of the selection problem for default assumptions.

5   Explanatory Nonmonotonic Reasoning

In this section we return to an older, explanatory approach to nonmonotonic reasoning. In fact, this approach subsumes practically all nonmonotonic formalisms that are actively investigated in AI today. In addition to the initial formalisms of default and modal nonmonotonic logics, it encompasses now logic programming, argumentation theory and causal reasoning.

We called this approach *explanatory* nonmonotonic reasoning, since explanation can be seen as its basic ingredient. Propositions and facts may be not only true or false in a model of a problem situation, but some of them are explainable (justified) by other facts and rules that are accepted. In the epistemic setting, some propositions are *derivable* from others using admissible rules. In the objective setting, some of the facts are *caused* by other facts and causal rules acting in the domain. Furthermore, explanatory nonmonotonic reasoning is based on very strong principles of *Explanation Closure* or *Causal Completeness* [Reiter, 2001], according to which any fact holding in a model should be explained, or caused, by the rules of the domain. Incidentally, it is these principles that make explanatory reasoning nonmonotonic.

By the above description, abduction and causation are integral parts of explanatory nonmonotonic reasoning. In some domains, however, explanatory reasoning adopts simplifying assumptions that exempt, in effect, certain propositions from the burden of explanation. *Closed World Assumption* is the most important assumption of this kind. According to it, negative assertions do not require explanation. Nonmonotonic reasoning in databases and logic programming are domains for which such an assumption turns out to be most appropriate. It is important to note that, from the explanatory point of view, the *minimization principle* employed in McCarthy's circumscription can be seen as a by-product of the stipulation that negative assertions can be accepted without any further explanation, while positive assertions always require explanation.

5.1 Simple Default Theories

Let us recall first that a Tarski consequence relation can be defined as a set of rules of the form $a \vdash A$ (where A is a proposition, and a a set of propositions) that satisfy the usual postulates of Reflexivity, Monotonicity and Cut. A Tarski consequence relation can also be described using the associated provability operator Cn defined as follows: $Cn(u) = \{A \mid u \vdash A\}$. A theory of a Tarski consequence relation is a set u of propositions such that $u = Cn(u)$. In what follows, we will be mainly interested in *supraclassical* Tarski consequence relations that subsume classical entailment. Theories of a supraclassical consequence relation are deductively closed sets. Supraclassicality allows for replacement of classically equivalent formulas in premises and conclusions of rules, but the deduction theorem, contraposition, and disjunction in the antecedent are not in general valid for such consequence relations.

In what follows, for an arbitrary set of Tarski rules Δ, we will use Cn_Δ to denote the provability operator associated with the least supraclassical consequence relation containing Δ. It can be easily shown that a proposition A belongs to $Cn_\Delta(u)$ precisely when A is derivable from u using the rules from Δ and classical entailment.

Now, perhaps a simplest way of defining a nonmonotonic theory consists in augmenting a background logical theory, given by a set of (Tarski) rules, with a set of default assumptions:

Definition 6 *A* simple default theory *is a pair* $(\Delta, \mathcal{A})$, *where* Δ *is a set of rules, and* $\mathcal{A}$ *a distinguished set of propositions called* defaults.

Nonmonotonic reasoning in this setting amounts to deriving plausible conclusions of a default theory by using its rules and defaults. However, in the case when the set of all defaults $\mathcal{A}$ is jointly incompatible with the background theory, we must make a reasoned choice. Here explanatory reasoning requires that a reasonable set of defaults that can be actually used in this context should be not only consistent and maximal, but it should explain also why the rest of the assumptions should be rejected. The appropriate choices will determine the *nonmonotonic semantics* of a default theory.

Definition 7 *1. A set* $\mathcal{A}_0$ *of defaults will be called* stable *if it is consistent and refutes any default outside the set:*

$$\neg A \in Cn_\Delta(\mathcal{A}_0), \text{ for any } A \in \mathcal{A} \setminus \mathcal{A}_0.$$

2. *A set s of propositions is an* extension *of a simple default theory if* $s = \text{Cn}_\Delta(\mathcal{A}_0)$, *for some stable set of defaults* $\mathcal{A}_0$. *The set of extensions determines the non-monotonic semantics of a default theory.*

Combining the above two definitions, we obtain the following description of the nonmonotonic semantics:

Lemma 1 *A set s of propositions is an extension of a simple default theory* $(\Delta, \mathcal{A})$ *if and only if it satisfies the following two conditions:*

- *s is the closure of the set of its defaults:* $s = \text{Cn}_\Delta(\mathcal{A} \cap s)$;

- *s decides the default set: for any* $A \in \mathcal{A}$, *either* $A \in s$, *or* $\neg A \in s$;

Simple default theories and their nonmonotonic semantics provide a most transparent description of explanatory nonmonotonic reasoning. For our present purposes, it is important to note also that the logical basis and the nonmonotonic mechanism are clearly separated in this formalism.

There is an obvious correspondence between this formalism and two other general approaches to nonmonotonic reasoning, namely Poole's [1988] abductive theory and the assumption-based framework by Bondarenko, Dung, Kowalski, and Toni [1997]. In Poole's Theorist system a default theory is also described as a pair $(T, \mathcal{A})$, where T is a classical theory (a set of propositions), and $\mathcal{A}$ a set of assumptions. In this framework, an extension is defined as a logical closure of a maximal consistent set of assumptions. It should be clear that Poole's theories correspond precisely to simple default theories that contain only rules of the form $\vdash A$ in its logical basis. In the general setting, however, not every Poole's extension is an extension in our sense. The difference stems, ultimately, from a logical fact that, for supraclassical inference, the inconsistency $a, A \vdash \mathbf{f}$ does not imply refutation $a \vdash \neg A$.

From the point of view of Reiter's default logic, simple default theories correspond to a very simple kind of default theories, namely, to default theories that contain only plain inference rules $A:/B$ (without justifications) and 'super-normal' default rules of the form $:A/A$. However, despite its simplicity, it can be shown that the above formalism of simple default theories is equivalent to the full default logic. This can be done as follows (Bochman [2008b] provides further details).

First, let us extend the language with new propositional atoms A°, for any proposition A from the source language. For a set u of propositions, u° will denote the set of new atoms $\{A^\circ \mid A \in u\}$. Next, if D is a Reiter default theory, then D° will denote the simple default theory consisting of a set of defaults of the form A°, for any formula A that appears as a justification in the rules from D, plus the following plain inference rules in the extended language:

$$\{a, b^\circ \vdash C \mid a : b/C \in D\}$$
$$\neg A \vdash \neg A^\circ$$

It can be easily seen that the above translation is polynomial and modular. Moreover, the following theorem shows that this translation is also faithful, so it is actually a PFM translation in the sense of Janhunen [1999].

Theorem 1 *A set u is an extension of a Reiter default theory D iff there is a unique extension u_0 of the simple default theory $D°$ that coincides with u when restricted to the source language.*

5.2   Generalizing the Logic

Being a theory of a reasoned use of assumptions and beliefs, nonmonotonic reasoning is an inherently epistemic theory. Consequently, its logical basis should allow us to make more subtle distinctions than plain truth and falsity of propositions. Thus, we have seen already that the explanatory approach as such presupposes a distinction between logical inconsistency and refutation. However, in order to obtain adequate representation capabilities for actual reasoning tasks required by AI, this logical basis should be extended further. It has turned out that an appropriate generalization of the underlying logic can be achieved by adopting a well-known generalization of a theory of logical inference based on disjunctive rules $a \vdash b$, where a and b are sets of propositions. A generic informal interpretation of such an inference rule is

> If all propositions from a hold, then at least one proposition from b should hold.

Such a disjunctive generalization of inference rules is essential for adequate representation of many problem situations.

The theory of disjunctive inference is now a well-developed part of the general logical theory. The importance of disjunctive rules for describing logical inference has been realized already by Gerhard Gentzen [1934], the father of the sequent calculus. From a logical point of view, the main advantage of such rules consists in providing a more adequate encoding of semantic descriptions. Of course, in a classical setting such rules can be reduced to ordinary, Tarski inference rules using classical disjunction. Furthermore, in such a setting we basically do not need rules at all, since any rule is reducible to the corresponding material implication. Things become more complex, however, in epistemic contexts where the semantic objects (such as belief states) are inherently partial. In such contexts the information that A does not hold (i.e., is not believed) is weaker, in general, than the assertion that $\neg A$ holds (that is, $\neg A$ is believed), and consequently the inference rule $A \vdash B$ expresses a weaker claim than $A \supset B$. Similarly, $A \vee B$ provides less information in this epistemic setting than the claim that either A holds, or B holds (that is $\vdash A, B$). Here the full expressive capabilities of disjunctive inference rules become a necessity.

We will restrict our description of disjunctive consequence relations below to what will be strictly necessary for what follows.

A set of disjunctive rules forms a *Scott consequence relation* [Scott, 1974] if it satisfies the following 'symmetric' generalization of the postulates for Tarski consequence relations:

(Reflexivity) $A \vdash A$.

(Monotonicity) If $a \vdash b$ and $a \subseteq a'$, $b \subseteq b'$, then $a' \vdash b'$;

(Cut) If $a \vdash b, A$ and $a, A \vdash b$, then $a \vdash b$.

Let $\bar{u}$ denote the complement of the set u of propositions. Then a set u is a *theory* of a Scott consequence relation if $u \nvdash \bar{u}$. Theories can also be defined alternatively as sets of propositions that are closed with respect to the rules of a Scott consequence relation in the sense that, for any such rule $a \vdash b$, if $a \subseteq u$, then $u \cap b \neq \emptyset$.

Any family of sets of propositions $\mathcal{T}$ determines a Scott consequence relation $\vdash_{\mathcal{T}}$ defined as follows:

$$a \vdash_{\mathcal{T}} b \quad \text{if and only if} \quad \forall u \in \mathcal{T}, \text{ if } a \subseteq u, \text{ then } b \cap u \neq \emptyset. \qquad \text{(GS)}$$

Actually, any Scott consequence relation can be generated in this way by the set of its theories – this is precisely the basic result about Scott consequence relations, called Scott Completeness Theorem.

Theorem 2 (Scott Completeness Theorem) *If $\mathcal{T}_{\vdash}$ is the set of all theories of a Scott consequence relation $\vdash$, then $\vdash$ coincides with $\vdash_{\mathcal{T}_{\vdash}}$.*

As a consequence, Scott consequence relations are uniquely determined by their theories. Moreover, for more expressive languages the above representation serves as a basis of constructing full-fledged semantics, whereas the set of theories of a consequence relation constitutes, eventually, its canonical model.

Scott consequence relations provide an abstract description of disjunctive inference. As before, however, in this study we are mainly interested in consequence relations that subsume classical entailment. A Scott consequence relation in a classical language will be called *supraclassical*, if it satisfies:

Supraclassicality If $a \vDash A$, then $a \vdash A$.

Falsity $f \vdash$.

By the first condition, theories of a supraclassical Scott consequence relation are deductively closed sets. Falsity amounts, in addition, to exclusion of inconsistent theories. Accordingly, a Scott consequence relation is supraclassical if and only if all its theories are consistent deductively closed sets.

As before, supraclassicality allows for replacement of classically equivalent formulas in premises and conclusions of disjunctive rules, as well as replacement of premise sets by their classical conjunctions: $a \vdash b$ will be equivalent to $\bigwedge a \vdash b$. Disjunctive conclusions, however, cannot be replaced in this way by their classical disjunctions.

Finally, in the next section we will use the following notion. Given an arbitrary set Δ of disjunctive rules, we will say that a set of u of propositions is a Δ-*theory* if u is a deductively closed set that is closed also with respect to the rules of Δ. In fact, it can be shown that u is a Δ-theory if and only if it is a theory of the least supraclassical Scott consequence relation the contains Δ.

5.3   Simple Disjunctive Default Theories

The notion of a simple default theory can be naturally extended to disjunctive rules. Thus, a simple *disjunctive* default theory can be defined as a pair $(\Delta, \mathcal{A})$, where Δ is a set of disjunctive rules, and $\mathcal{A}$ a set of defaults. However, the disjunctive setting requires a somewhat more complex description of extensions.

Definition 8 *A set s of propositions is an* extension *of a disjunctive default theory* $(\Delta, \mathcal{A})$ *if it satisfies the following conditions:*

- *s is a minimal Δ-theory that contains $s \cap \mathcal{A}$;*
- *s decides the default set: for any $A \in \mathcal{A}$, either $A \in s$, or $\neg A \in s$;*

By analogy with the Tarski case, a set $\mathcal{A}_0$ of defaults can be called *stable* if it is consistent, and at least one minimal Δ-theory containing $\mathcal{A}_0$ includes also $\neg(\mathcal{A} \setminus \mathcal{A}_0)$. Then the following result shows that extensions are precisely theories that are generated by stable sets of defaults.

Lemma 2 *A set $\mathcal{A}_0$ of defaults is stable if and only if $\mathcal{A}_0 = u \cap \mathcal{A}$, for some extension* u.

As a partial converse of the above result, an extension can be characterized as a minimal theory s containing a stable set of defaults and such that it includes $\neg A$, for any default A not in s. It should be noted, however, that in the general disjunctive case extensions are not determined uniquely by their sets of defaults. Indeed, even in the extreme case when the set of defaults $\mathcal{A}$ is empty, a default theory may have multiple extensions (= minimal theories).

Using a translation sketched earlier for simple Tarski default theories, it can be shown that the formalism of simple disjunctive default theories is equivalent to a disjunctive generalization of default logic described by Gelfond, Lifschitz, Przymusińska, and Truszczyński [1991]. Moreover, it turns out to be equivalent even to powerful bimodal formalisms of nonmonotonic reasoning, suggested by Lin and Shoham [1992] and Lifschitz [1994] in an attempt to construct a unified formalism for nonmonotonic reasoning and logic programming.

Though really simple and transparent, the formalism of simple default theories still does not give us much insight into the characteristic properties of the associated nonmonotonic reasoning. In this sense it is similar to a formalization of classical logic in terms of, say, three axioms of Church. For a more detailed analysis of such a reasoning, as well as for comparing different nonmonotonic formalisms, we need a *structural* description of the corresponding underlying logic. More precisely, such a reformulation should provide us with a structural logical calculus for assumption-based reasoning. This is the subject of the next section.

5.4 Biconsequence Relations: A Logic of Assumption-Based Reasoning

A uniform account of various explanatory nonmonotonic formalisms can be given in the framework of biconsequence relations described below. Biconsequence relations are specialized consequence relations for reasoning with respect to a pair of contexts. On the interpretation suitable for nonmonotonic reasoning, one of these contexts is the main (objective) one, while the other context provides assumptions, or explanations, that justify inferences in the main context. This combination of inferences and their justifications creates a framework for an assumption-based reasoning.

A *bisequent* is an inference rule of the form $a : b \Vdash c : d$, where a, b, c, d are sets of propositions. According to the explanatory interpretation, it says

'If no proposition from b is assumed, and all propositions from d are assumed, then all propositions from a hold only if one of the propositions from c holds'.

A *biconsequence relation* is a set of bisequents satisfying the rules:

Monotonicity $\dfrac{a : b \Vdash c : d}{a' : b' \Vdash c' : d'}$, if $a \subseteq a', b \subseteq b', c \subseteq c', d \subseteq d'$;

Reflexivity $A : \Vdash A :$ and $: A \Vdash : A$;

Cut $\dfrac{a : b \Vdash A, c : d \quad A, a : b \Vdash c : d}{a : b \Vdash c : d} \qquad \dfrac{a : b \Vdash c : A, d \quad a : A, b \Vdash c : d}{a : b \Vdash c : d}.$

A biconsequence relation can be seen as a product of two Scott consequence relations. Accordingly, a pair (u, v) of sets of propositions will be called a *bitheory* of a biconsequence relation if $u : \bar{v} \nVdash \bar{u} : v$, where, as before, $\bar{u}$ denotes the complement of the set u of propositions. A set u of propositions is a (*propositional*) *theory* of $\Vdash$, if (u, u) is a bitheory of $\Vdash$.

Bitheories can be seen as pairs of sets that are closed with respect to the bisequents of a biconsequence relation. A bitheory (u, v) of $\Vdash$ is *positively minimal*, if there is no bitheory (u', v) of $\Vdash$ such that $u' \subset u$. Such bitheories play an important role in describing nonmonotonic semantics.

By a *bimodel* we will mean a pair of sets of propositions. A set of bimodels will be called a *binary semantics*.

Definition 9 *A bisequent $a : b \Vdash c : d$ is* valid *in a binary semantics $\mathcal{B}$, if, for any $(u, v) \in \mathcal{B}$, if $a \subseteq u$ and $b \subseteq \bar{v}$, then either $c \cap u \neq \emptyset$, or $d \cap \bar{v} \neq \emptyset$.*

The set of bisequents that are valid in a binary semantics forms a biconsequence relation. On the other hand, any biconsequence relation $\Vdash$ is determined in this sense by its canonical semantics defined as the set of bitheories of $\Vdash$. Consequently, the binary semantics provides an adequate interpretation of biconsequence relations.

According to a well-known idea of Belnap [1977], the four truth values $\top, \mathbf{t}, \mathbf{f}$, and $\perp$ of any four-valued interpretation can be identified with the four subsets of the set $\{t, f\}$ of classical truth-values, namely $\{t, f\}$, $\{t\}$, $\{f\}$ and $\emptyset$. Thus, $\top$ means that a proposition is both true and false (i.e., contradictory), $\mathbf{t}$ means that it is 'classically' true (that is, true without being false), $\mathbf{f}$ means that it is classically false, while $\perp$ means that it is neither true nor false (undetermined). This representation allows us to see any four-valued interpretation as a pair of ordinary interpretations corresponding, respectively, to independent assignments of truth and falsity to propositions. Accordingly, a bimodel (u, v) can be viewed as a four-valued interpretation, where u is the set of true propositions, while v is the set of assumed propositions. Biconsequence relations provide in this sense a syntactic formalism for four-valued reasoning [Bochman, 1998].

An important feature of biconsequence relations is a possibility of imposing structural constraints on the binary semantics by accepting additional structural rules. Some of them play an important role in nonmonotonic reasoning. Thus, a biconsequence relation is *consistent*, if it satisfies

Consistency $A : A \Vdash :$

Consistency says that no proposition can be taken to hold without assuming it. This amounts to restricting the binary semantics to *consistent* bimodels, that is, bimodels (u, v) such that $u \subseteq v$. For the four-valued representation, Consistency requires that no proposition can be both true and false, so it determines a semantic setting of three-valued logic that deals only with possible incompleteness of information.

A biconsequence relation is *regular* if it satisfies

Regularity
$$\frac{b : a \Vdash a : b}{: a \Vdash : b}$$

Regularity is a kind of an assumption coherence constraint. It says that a coherent set of assumptions should be such that it is compatible with taking these assumptions as actually holding. A semantic counterpart of Regularity is a *quasi-reflexive* binary semantics in which, for any bimodel (u, v), (v, v) is also a bimodel.

The descriptions of biconsequence relations so far have been purely structural, so it is not sufficient for capturing reasoning in default logic which is based on classical entailment. To this end, we should 'upgrade' biconsequence relations to a logical system that subsumes classical inference. This will naturally create a more elaborate, epistemic reading for the pair of associated contexts. More precisely, an epistemic understanding of biconsequence relations is naturally obtained by treating the objective and assumption contexts, respectively, as the contexts of knowledge and belief. Propositions that hold in the objective context can be viewed as known, while propositions belonging to the assumption context can be seen as forming the set of associated beliefs. Accordingly, both the objective and assumption contexts of bimodels will correspond in this case to deductively closed theories.

Supraclassical biconsequence relations are biconsequence relations in which both its component contexts respect the classical entailment:

Supraclassicality If $a \vDash A$, then $a : \Vdash A :$ and $: A \Vdash : a$.

Falsity $\mathbf{f} : \Vdash :$ and $: \Vdash : \mathbf{f}$.

A logical semantics of supraclassical biconsequence relations can be obtained from the general binary semantics by requiring that bimodels are pairs of consistent deductively closed sets. As before for Scott consequence relations, supraclassicality allows for replacement of classically equivalent formulas in both kinds of premises and conclusions of bisequents. In addition, it allows us to replace sets of objective premises and assumption sets in conclusions by their conjunctions; in other words, a bisequent $a{:}b \Vdash c{:}d$ is reducible to $\bigwedge a{:}b \Vdash c{:} \bigwedge d$. However, objective conclusion sets (c) and assumption sets in premises (b) are not replaceable in this way by their classical disjunctions.

A supraclassical biconsequence relation will be called a *default biconsequence relation* if it satisfies the above structural rules of Consistency and Regularity. As we will see in the next section, such a biconsequence relation will form an exact logical basis for the associated nonmonotonic formalism.

5.5   Nonmonotonic Semantics of Biconsequence Relations

Nonmonotonic semantics of a biconsequence relation is a certain set of its theories. Namely, such theories are *explanatory closed* in the sense that presence of propositions in the main context is explained (i.e., derived) using the rules of the biconsequence relation when the theory itself is taken as the assumption context.

Definition 10   *A set u is an* extension *of a biconsequence relation $\Vdash$, if (u, u) is a positively minimal bitheory of $\Vdash$. A default nonmonotonic semantics of a biconsequence relation is the set of its extensions.*

A direct correspondence between Reiter's default logic and biconsequence relations can be established by representing default rules $a : b/A$ as bisequents $a{:}\neg b \Vdash A{:}$. Then the above nonmonotonic semantics will correspond precisely to the semantics of extensions in default logic.

For a Reiter default theory D, let $\Vdash_D$ denote the least supraclassical biconsequence relation that includes bisequents corresponding to the rules of D. Then the following result shows that biconsequence relations provide an adequate logical framework for nonmonotonic reasoning in default logic. The proof of this result can be found in the monograph by Bochman [2005].

Theorem 3   *Extensions of a default theory D coincide with extensions of $\Vdash_D$.*

It should be noted that, though the formalism of biconsequence relations subsumes default logic, it is stronger than the latter. Taken in full generality, it corresponds to simple disjunctive default theories, described earlier, as well as to disjunctive default logic of Gelfond et al. [1991] and bi-modal nonmonotonic formalisms of Lin and Shoham [1992] and Lifschitz [1994].

The above representation theorem implies, in particular, that all the postulates of a supraclassical biconsequence relation are valid logical rules for default reasoning. These postulates still do not give us, however, a strongest possible logic underlying such a nonmonotonic reasoning. In order to find the latter, we can make use of the notion of a *strong equivalence* defined earlier for default theories (see Section 4). More precisely, the following result shows that default biconsequence relations, defined at the end of the preceding section, constitute a maximal logic suitable for the default nonmonotonic semantics.

Theorem 4   *Default theories are strongly equivalent if and only if they determine the same default biconsequence relation.*

In other words, default theories D_1 and D_2 are strongly equivalent if and only if each default rule of D_2 is derivable from D_1 using the postulates of default biconsequence relation, and vice versa.

It is also interesting to note that default biconsequence relations allow us to provide the following simplified description of extensions.

Proposition 1   *A set u is an extension of a default biconsequence relation $\Vdash$ if and only if*

$$u = \{A \mid : \overline{u} \Vdash A : u\}.$$

By the above description, an extension is precisely a set of formulas that are provable on the basis of taking itself as the set of assumptions. This description demonstrates, in particular, that the nonmonotonic reasoning in biconsequence relations is based essentially on the same idea as the original Reiter's logic.

In the next few sections we will briefly show how some important explanatory nonmonotonic formalisms, such as modal and autoepistemic logics, on the one hand, and the causal calculus, on the other hand, can be expressed in the framework of biconsequence relations. Actually, we will see that each of them can be obtained from the basic formalism just by varying its underlying logic.

5.6 Modal Nonmonotonic Logics

The paper by McDermott and Doyle [1980] marks the beginning of a modal approach to nonmonotonic reasoning. The modal nonmonotonic logic was formulated in a modal language containing a modal operator M with the intended meaning of MP "P is consistent with everything believed". The authors have suggested a fixed-point construction much similar to that of Reiter's extensions, but formulated this time entirely in the modal language.

For a set u of propositions, let Mu denote the set $\{MA \mid A \in u\}$, and similarly for $\neg u$, etc. As before, $\overline{u}$ will denote the complement of u with respect to a given modal language. Then a set s is a fixed point of a modal theory u, if it satisfies the equality

$$s = \mathrm{Th}(u \cup M\neg\overline{s}).$$

The initial formulation of modal nonmonotonic logic was unsatisfactory, however, since it secured no connection between a modal formula MC and its objective counterpart C. In response to this deficiency, McDermott [1982] developed a stronger version of the formalism based on the entailment relation of standard modal logics instead of first-order logic. The corresponding fixed points were defined now as sets satisfying the equality

$$s = \mathrm{CN}_{\mathcal{S}}(u \cup M\neg\overline{s}),$$

where $\mathrm{Cn}_{\mathcal{S}}$ is a provability operator of some modal logic $\mathcal{S}$ containing the necessitation rule. In what follows, we will call such fixed points $\mathcal{S}$-*extensions*.

The stronger is the underlying modal logic $\mathcal{S}$, the smaller is the set of extensions, and hence the larger is the set of nonmonotonic consequences of a modal theory. It has turned out, however, that the modal nonmonotonic logic based on the strongest modal logic S5 collapses to a monotonic system. So the resulting suggestion was somewhat indecisive, namely a range of possible modal nonmonotonic logics without clear criteria for evaluating the merits of the alternatives. However, this indecisiveness has turned out to be advantageous in the subsequent development of the modal approach. In particular, Schwarz [1990] has shown that autoepistemic logic of Moore [1985] is just one of the nonmonotonic logics in the general approach of McDermott [1982]. Namely, it is precisely the nonmonotonic logic based on KD45. This result revived interest in a systematic study of the whole range of modal nonmonotonic logics [Marek, Schwarz, and Truszczyński, 1993]. This study has shown the importance of many otherwise esoteric modal logics for nonmonotonic reasoning, such as S4F, SW5, and KD45.

It has been shown by Truszczyński [1991] that default logic can be translated into a range of modal nonmonotonic logics by using the following translation:

$$A : B_1, \ldots B_n/C \Rightarrow (LA \wedge LMB_1 \wedge \cdots \wedge LMB_n) \supset LC$$

In addition, the translation can be naturally extended to disjunctive default rules of Gelfond et al. [1991]. In this sense, modal nonmonotonic logics also subsumed disjunctive default logic.

A general representation of modal nonmonotonic reasoning can be given in the framework of modal biconsequence relations. The role of the modal operator L in this setting consists in reflecting assumptions (beliefs) as propositions in the main context.

Definition 11 *A supraclassical biconsequence relation in a modal language will be called* modal *if it satisfies the following postulates:*

Positive Reflection $A : \Vdash LA:,$

Negative Reflection $: LA \Vdash :A,$

Negative Introspection $: A \Vdash \neg LA:.$

For a modal logic $\mathcal{S}$, a modal biconsequence relation will be called an $\mathcal{S}$-*biconsequence relation*, if $\Vdash A$: holds for every modal axiom A of $\mathcal{S}$. A possible worlds semantics for K-biconsequence relations is obtained in terms of Kripke models having a last cluster, namely models of the form $M = (W, R, F, V)$, where (W, R, V) is an ordinary Kripke model, while $F \subseteq W$ is a non-empty *last cluster* of the model [Segerberg, 1971]. We will call such models *final Kripke models*.

For a final Kripke model $M = (W, R, F, V)$, let $|M|$ denote the set of modal propositions that are valid in M, while $\|M\|$ the set of propositions valid in the S5-submodel of M generated by F. Then the pair $(|M|, \|M\|)$ forms a bimodel, and it can be shown that the corresponding binary semantics is adequate for modal biconsequence relations.

A modal default biconsequence relation will be called an *F-biconsequence relation*, if it satisfies the following postulate:

F $\Vdash A, LA{\rightarrow}B : B.$

F-biconsequence relations provide a concise representation of the modal logic S4F obtained from S4 by adding the axiom $(A \wedge MLB){\rightarrow}L(MA \vee B)$. A semantics of S4F [Segerberg, 1971] is given in terms of final Kripke models (W, R, F, V), such that $\alpha R \beta$ iff either $\beta \in F$, or $\alpha \notin F$.

F-biconsequence relations play a central role in a modal representation of default nonmonotonic reasoning. Suppressing technical details, it can be shown that any ordinary, non-modal default biconsequence relation can be conservatively extended to a modal F-biconsequence relation. Now, a crucial fact about F-biconsequence relations is that any bisequent $a{:}b \Vdash c{:}d$ is already reducible in it to a plain modal formula

$$\bigwedge (La \cup L\neg Lb) \rightarrow \bigvee (Lc \cup L\neg Ld).$$

Consequently, any bisequent theory in such a logic is reducible to an ordinary modal theory. These logical facts provide a basic explanation for the possibility of expressing (rule-based) default nonmonotonic reasoning in a plain formula-based modal language.

Modal nonmonotonic semantics. We say that a set of modal propositions is an *S-extension* of a modal bisequent theory Δ, if it is an extension of the least S-biconsequence relation containing Δ. Then we obtain a whole range of *modal* nonmonotonic semantics by varying the underlying modal logic S. The stronger is the logic, the smaller is the set of extensions it determines.

Fortunately, it can be shown that any such modal nonmonotonic formalism is reducible back to a non-modal formalism of biconsequence relations. The crux lies in the fact that any modal extension is a modal *stable theory*, while the set of modal propositions belonging to the latter is uniquely determined by the set of its non-modal propositions. Consequently, modal extensions are uniquely determined by their objective, non-modal subsets. This creates a possibility of reducing modal nonmonotonic reasoning to a nonmodal one, and vice versa.

For any set u of propositions, let u_o denote the set of all non-modal propositions in u. Similarly, if $\Vdash$ is a modal biconsequence relation, $_o\Vdash$ will denote its restriction to the non-modal sub-language. Then we have

Theorem 5 *If $\Vdash$ is a modal biconsequence relation, and u a stable modal theory, then u is an extension of $\Vdash$ if and only if u_o is an extension of $_o\Vdash$.*

According to the above result, the non-modal biconsequence relation $_o\Vdash$ embodies all the information about the modal nonmonotonic semantics of $\Vdash$. Consequently, non-modal biconsequence relations turn out to be sufficiently expressive to capture modal nonmonotonic reasoning.

In the next section we will show how such a reduction works for a particular case of Moore's autoepistemic logic.

5.7 Saturation and Autoepistemic Logic

From a purely technical point of view, autoepistemic logic is a particular instance of a modal nonmonotonic formalism that is based on the modal logic KD45. Moreover, it has been shown by Bochman [2005] that any bisequent of a modal bisequent theory based on KD45 is directly reducible to a set of non-modal bisequents. As a result, autoepistemic logic is reducible to a non-modal formalism of biconsequence relations, though based on an apparently different notion of a nonmonotonic semantics, described below.

Note first that any deductively closed theory u always contains maximal deductive sub-theories; such sub-theories are sets $u \cap \alpha$, where α is a world (maximal deductive theory). Now, for a deductively closed set u, let $u \bot$ denote the set of all maximal sub-theories of u, plus u itself.

Definition 12 *A theory u of a supraclassical biconsequence relation $\Vdash$ is called an expansion of $\Vdash$, if, for every $v \in u\bot$ such that $v \neq u$, the pair (v, u) is not a bitheory of $\Vdash$. The set of expansions forms the* autoepistemic semantics *of $\Vdash$.*

By the above definition, u is an expansion if (u, u) is a positively 'pointwise-minimal' bitheory in the sense that (v, u) is not a bitheory for any maximal sub-theory v of u.

It follows directly from the respective definitions of extensions and expansions that any extension of a biconsequence relation will be an expansion, though not vice versa. It can be shown, however, that expansions are precisely extensions of biconsequence relations under a stronger underlying logic defined below.

A bimodel (u, v) will be called *saturated*, if $u \in v\bot$. A classical binary semantics $\mathcal{B}$ will be called *saturated* if its bimodels are saturated. Saturation corresponds to the following additional postulate for biconsequence relations:

Saturation $\Vdash A \vee B, \neg A \vee B : B.$

Now, it can be shown that saturated biconsequence relations preserve expansions, so they are admissible for the autoepistemic semantics. Moreover, the following theorem shows that, for such biconsequence relations, expansions actually collapse to extensions.

Theorem 6 *Expansions of a saturated biconsequence relation coincide with its extensions.*

Actually, saturated biconsequence relations constitute a maximal logic for the autoepistemic semantics [Bochman, 2005].

By the above results, autoepistemic logic can be viewed as a nonmonotonic formalism that employs the same nonmonotonic semantics as default logic, though it is based on a stronger underlying logic of saturated biconsequence relations.

To conclude this section, we should mention a peculiar fact that the above distinction between the underlying logics of default and autoepistemic semantics turns out to be inessential for certain restricted kinds of rules. For example, let as call a bisequent theory *positively simple*, if objective premises and conclusions of any its rule are sets of classical literals. Then we have[2]

Theorem 7 *Expansions of a positively simple bisequent theory coincide with its extensions.*

Bisequents $a{:}b \Vdash c{:}d$ such that a, b, c, d are sets of classical literals, are logical counterparts of program rules of extended logic programs with classical negation [Lifschitz and Woo, 1992]. The semantics of such programs is determined by *answer sets* that coincide with extensions of respective bisequent theories. Moreover, such bisequent theories are positively simple, so, by the above theorem, extended logic programs obliterate the distinction between extensions and expansions. This is why such programs are representable in autoepistemic logic as well.

5.8   Causal Logic

McCain and Turner [1997] have suggested to use *causal* rules for representing reasoning about action and change, in particular for a natural encoding of the commonsense principle of inertia. The logical basis of this formalism has been described by Bochman [2003] in the framework of causal inference relations that are based on conditionals of the form $A \Rightarrow B$ saying '*A causes B*', and required to satisfy the following postulates:

---

[2]The origins of this result are in the Main Lemma in the paper by Lifschitz and Schwarz [1993].

(Strengthening)   If $A \vDash B$ and $B \Rightarrow C$, then $A \Rightarrow C$;

(Weakening)   If $A \Rightarrow B$ and $B \vDash C$, then $A \Rightarrow C$;

(And)   If $A \Rightarrow B$ and $A \Rightarrow C$, then $A \Rightarrow B \wedge C$;

(Or)   If $A \Rightarrow C$ and $B \Rightarrow C$, then $A \vee B \Rightarrow C$.

(Cut)   If $A \Rightarrow B$ and $A \wedge B \Rightarrow C$, then $A \Rightarrow C$;

(Truth) $\mathrm{t} \Rightarrow \mathrm{t}$;

(Falsity) $\mathrm{f} \Rightarrow \mathrm{f}$.

From a logical point of view, the most significant 'omission' of the above set of postulates is the absence of reflexivity $A \Rightarrow A$. It is this feature that creates a possibility of nonmonotonic reasoning in such a system. A nonmonotonic semantics of causal inference is determined by the set of *exact worlds*, models that are not only closed with respect to the causal rules, but also such that every proposition in it is caused by other propositions accepted in the model.

Causal inference relations can be faithfully represented by a special class of *causal biconsequence relations* that satisfy the following postulate

(Negative Completeness) $: A, \neg A \Vdash.$

Negative Completeness restricts the assumption contexts to worlds, and hence a semantic representation of causal biconsequence relations can be given in terms of bimodels of the form (u, α), where α is a world. This implies, in particular, that any extension of a causal biconsequence relation should be a world.

The assumption context of bisequents in causal biconsequence relations is already 'fully classical'. In particular, the following classical reduction eliminates negative conclusions in bisequents:

$$a : b \Vdash c : d \equiv a : \bigwedge d \supset \bigvee b \Vdash c : .$$

Now, causal rules $A \Rightarrow B$ are representable in this formalism as bisequents of the form $: \neg A \Vdash B :$ or, equivalently, as $\Vdash B : A$ (due to the above reduction). This correspondence extends also to the associated nonmonotonic semantics. Namely, the nonmonotonic semantics of a causal inference relation coincides with the default semantics of the associated biconsequence relation. Moreover, taken in full generality, causal biconsequence relations constitute an exact non-modal counterpart of Turner's [1999] logic of universal causation (UCL).

6   Defeasible Entailment Revisited

Generally speaking, the problem of defeasible entailment is the problem of formalizing, or representing, commonsense rules of the form "*A normally implies B*". Since such rules play a key role in commonsense reasoning, they have been one of the main targets of the nonmonotonic reasoning theory from its very beginning. Moreover, we have argued earlier that this problem is intimately related to the central problem non-monotonic reasoning, the selection problem of default assumptions. Today, after more than thirty years of intensive studies, an impressive success has been achieved in our understanding of this problem, in realizing how complex it is, and, most importantly, how many different forms and manifestations it may have.

As we discussed earlier, the problem of defeasible entailment has been taken as the central problem in the preferential approach to nonmonotonic reasoning. Unfortunately, it has not succeeded in providing a commonly agreed solution to the problem. So it is only natural to inquire now whether the alternative, explanatory approach could do this job.

To begin with, however, we should acknowledge from the outset that default logic and related explanatory nonmonotonic formalisms are not primarily theories of defeasible entailment, or even of a commonsense notion of normality in general. Despite initial hopes, it was rightly argued already by Kraus et al. [1990] that there is an essential difference between a claim "*A normally implies B*" and, say, Reiter's normal default rule $A : B/B$. This could be seen already in the fact that "*A normally implies B*" is obviously incompatible with "*A normally implies ¬B*", while $A : B/B$ is perfectly compatible with $A : ¬B/¬B$ in default logic. Also, rules of default logic are fully monotonic in that they admit unrestricted strengthening of antecedents without changing the intended models. Finally, at least since the papers by Reiter and Criscuolo [1981] and Hanks and McDermott [1987] we know that the formalism of default logic does not provide by itself a solution to the problem of default interactions.

Nevertheless, the above discrepancy does not preclude a possibility that the tools provided by the explanatory theory still could be used for a rigorous (though indirect) *representation* of defeasible entailment. It is this possibility that we will briefly discuss below. Though our discussion in what follows will be obviously fragmentary and will not end up with an accomplished theory, it is intended to demonstrate that this possibility is worth to be explored in depth. Moreover, we will show that a plausible approach to this problem amounts once again not to an alternative understanding of nonmonotonicity, but rather to the use of already existing nonmonotonic tools in a suitably extended underlying logic.

6.1   The Language and Logic of Conditionals

On a commonsense understanding, a rule "*A normally implies B*" represents a claim that A implies B, given some (unmentioned and even not fully known) conditions that are presumed to hold in normal circumstances. Thus, a default rule $TurnKey \rightarrow CarStarts$ states that if I turn the key, the car will start given the normal conditions such as there is a fuel in the tank, the battery is ok, etc. etc.

McCarthy [1980] has suggested a purely classical translation of such normality rules as implications of the form

$$A \wedge ¬ab \supset B,$$

where ab is a new 'abnormality' proposition serving to accumulate the conditions for violation of the source rule. Thus, *Birds fly* was translated into something like "Birds fly if they are not abnormal".

In fact, viewed as a formalism for nonmonotonic reasoning, the central concept of McCarthy's circumscriptive method was not circumscription itself, but his notion of an abnormality theory - a set of classical conditionals containing the abnormality predicate ab. The default character of commonsense rules was captured in McCarthy's theory by a circumscription policy that minimized abnormality (and thereby maximized the acceptance of the corresponding normality claims $\neg ab$). Since then, this representation of normality rules using auxiliary (ab)normality propositions has been employed both in applications of circumscription, and in many other theories of nonmonotonic inference, sometimes in alternative logical frameworks.

Our approach below is based on the idea that the default assumptions of a defeasible rule jointly function as a conditional, that we will denote by A/B, that, once accepted, allows us to infer B from A. Accordingly, and similarly to McCarthy's representation, we may also represent such a rule as the classical implication

$$A \wedge (A/B) \supset B.$$

Now, the default character of the inference from A to B can be captured by requiring that A/B normally holds or, more cautiously, that A normally implies A/B. We will achieve this effect, however, by representing the latter rule simply as a normal default rule in default logic.

Our representation can be viewed as an 'unfolding' of the corresponding abnormality representation of McCarthy. Namely, in place of unstructured (ab)normality propositions, we suggest more articulated conditional propositions with a natural meaning. Note, for example, that this representation suggests an immediate explanation why the default assumptions required for "*A normally implies B*" and "*A normally implies C*" can be presumed to be independent, so violation of one of these rules does not imply rejection of another[3]. In fact, this feature goes also beyond the understanding of normality in preferential inference: according to the latter, violation of $A \hspace{-0.3em}\mid\hspace{-0.7em}\sim \hspace{-0.2em} B$ means that the situation at hand is abnormal with respect to A , so we are not entitled to infer anything that normally follows from A.

A most important presumption behind our representation, however, is that the normality defaults should have its own internal logic. In fact, we will stipulate below that the conditionals A/B should satisfy at least the usual rules of Tarski consequence relations. It is this internal logic that will allow us to formulate purely logical principles that will govern a proper interaction of defeasible rules in cases of conflict.

As a first step, we will augment our source, classical language L by adding new propositional atoms of the form A/B, where A and B are classical propositions of L. The conditionals A/B will be viewed as propositional atoms of a new type, so nesting of conditionals will not be allowed. Still, the new propositional atoms will be freely combined with ordinary ones using the classical propositional connectives. We will denote the resulting language by L_c.

As we mentioned, the essence and main functional role of our conditionals is expressed by the following 'modus ponens' axiom:

MP $A \wedge (A/B) \supset B.$

---

[3] This feature required introduction of *aspects* of abnormality in McCarthy's circumscription.

Now, conditionals can be viewed as ordinary inference rules that are 'reified' in the object language. Accordingly, we will require them to satisfy all the properties of a supraclassical consequence relation. The following postulates can be shown to be sufficient for this purpose:

$$\text{If } A \models B, \text{ then } A/B. \qquad \text{(Dominance)}$$

$$\frac{A/B \quad B/C}{A/C} \qquad \text{(Transitivity)}$$

$$\frac{A/B \quad A/C}{A/(B \wedge C)} \qquad \text{(And)}$$

Finally, we should formulate some principles of rejection for conditionals that will play a key role in our final representation. A number of choices are possible at this point. Thus, the following logical principle determines how the rejection is propagated along chains of conditionals. It turns out to be suitable for representing nonmonotonic inheritance.

$$\frac{A/B \quad \neg(A/C)}{\neg(B/C)} \qquad \text{(Forward Rejection)}$$

Forward Rejection can be viewed as a partial contraposition of the basic Transitivity rule for conditionals. Note however that, since Transitivity was formulated as an inference *rule*, Forward Rejection cannot be logically derived from the latter. In fact, Forward Rejection provides a natural formalization of the principle of *forward chaining* adopted in many versions of the nonmonotonic inheritance theory.

The above logic describes the logical properties of arbitrary conditionals, not only default ones. The difference between the two will be reflected in the representation of normality rules in the framework of default logic, described next.

6.2 Defeasible Inference in Default Logic

We will describe now a modular representation of defeasible rules in default logic. To begin with, we will assume that our default theory is defined in the conditional language L_c and respects the above logic of conditionals.

Now, for each accepted defeasible rule A *normally implies* B (that is, $A \hspace{-0.3em}\mid\hspace{-0.9em}\sim \hspace{0.1em} B$), we will introduce the following two rules of default logic:

$$A : A/B \vdash A/B$$
$$A : \vdash \neg(A/\neg B).$$

The first rule is a normal default rule. It secures that a corresponding conditional is acceptable whenever its antecedent holds and it is not rejected. The second rule is an ordinary inference rule that reflects the following *commitment principle*: if A is known to hold, then the opposite conditional $A/\neg B$ should be rejected (due to the commitment to $A \hspace{-0.3em}\mid\hspace{-0.9em}\sim \hspace{0.1em} B$). Note, however, that $A/\neg B$ may be a consequence of a chain of conditionals that starts with A and ends with $\neg B$. In this case Forward Rejection dictates, in effect, that the last conditional in any such chain should also be rejected.

Actually, it is this feature that allows us to reject a less specific conditional due to a commitment to a more specific one.

The following example will show how the resulting system works and, in particular, how it handles specificity.

Example 1 (A generalized Penguin-Bird story)

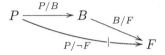

The above picture differs from the standard example about penguins and birds only in that $P{\rightarrow}B$ is not a strict but defeasible rule $P{\hspace{-0.3em}\sim}B$. Of course, the same results will be obtained also in the original case.

As could be expected, given the only fact B, the corresponding default theory has a unique extension that contains B and F. However, given the fact P instead, the resulting default theory also has a unique extension that includes this time P, B and $\neg F$. This happens because, given P, the commitment principle for $P{\hspace{-0.3em}\sim}\neg F$ implies $\neg(P/F)$. Taken together with P/B, this implies rejection of B/F by Forward Rejection. This is the way a more specific rule $P{\hspace{-0.3em}\sim}\neg F$ cancels a less specific rule $B{\hspace{-0.3em}\sim}F$. Note, however, that the situation is not symmetric, since the commitment to the less specific default $B{\hspace{-0.3em}\sim}F$ does not allow us to reject more specific rule $P{\hspace{-0.3em}\sim}\neg F$. That is why, for instance, we would still have a unique extension containing $\neg F$ even if both P and B were given as facts. □

The above formalism constitutes an exact formalization of the theory of nonmonotonic inheritance [Bochman, 2008a]. Moreover, relative simplicity and modularity provided by the above representation can be attributed almost exclusively to the use of an appropriate underlying logic behind default conditionals.

7 Conclusions

Today, after thirty years, we can confidently claim that nonmonotonic reasoning is not yet another application of logic, but a relatively independent field of logical research that has a great potential in informing, in turn, traditional logical theory as well as many areas of philosophical inquiry.

A nonmonotonic formalism is not just a syntax plus nonmonotonic semantics. An account of the underlying *logics* behind nonmonotonic reasoning systems allows us to see most of them as implementations of basically the same nonmonotonic mechanisms in different logics. It creates also a natural framework for comparing such formalisms, as well as for designing new ones for specific purposes. Last but not least, a conscious use of the tradeoff between logic and its nonmonotonic 'overhead' can provide an immense improvement in the quality of resulting combined representations.

General nonmonotonic formalisms do not resolve by themselves the actual reasoning problems in AI, just as the differential calculus does not resolve by itself the problems of physics. In this sense they are only logical frameworks for dealing with nonmonotonicity, in which real and useful kinds of nonmonotonic reasoning can hopefully be expressed and studied. Despite obvious successes, much work still should be done for a rigorous and adequate representation of the AI universum in all its actual scope and diversity. We contend, however, that an important part of this endeavor should amount to a further advance of the initial McCarthy's program, namely to representation of increasingly larger portions of our commonsense reasoning in nonmonotonic formalisms. Systematic use of defeasible rules, causation and abduction may serve as primary objectives of the nonmonotonic reasoning theory in this sense. Moreover, the latter theory presumably provides a most appropriate framework for analyzing these (essentially nonmonotonic) notions. In sharp contrast with the traditional logical analysis, nonmonotonic reasoning theory can achieve this aim by establishing a proper interplay of logical and nonmonotonic aspects of the corresponding kinds of commonsense reasoning.

References

E. W. Adams. *The Logic of Conditionals*. Reidel, Dordrecht, 1975.

N. D. Belnap, Jr. A useful four-valued logic. In M. Dunn and G. Epstein, editors, *Modern Uses of Multiple-Valued Logic*, pages 8–41. D. Reidel, 1977.

S. Benferhat, C. Cayrol, D. Dubois, J. Lang, and H. Prade. Inconsistency management and prioritized syntax-based entailment. In R. Bajcsy, editor, *Proceedings Int. Joint Conf. on Artificial Intelligence, IJCAI'93*, pages 640–645, Chambery, France, 1993. Morgan Kaufmann.

A. Bochman. Biconsequence relations: A four-valued formalism of reasoning with inconsistency and incompleteness. *Notre Dame Journal of Formal Logic*, 39(1):47–73, 1998.

A. Bochman. *A Logical Theory of Nonmonotonic Inference and Belief Change*. Springer, 2001.

A. Bochman. A logic for causal reasoning. In *Proceedings IJCAI'03*, Acapulco, 2003. Morgan Kaufmann.

A. Bochman. *Explanatory Nonmonotonic Reasoning*. World Scientific, 2005.

A. Bochman. Default theory of defeasible entailment. In *Principles of Knowledge Representation and Reasoning: Proceedings of the Eleventh International Conference, KR 2008, Sydney, Australia, September 16-19, 2008*, pages 466–475. AAAI Press, 2008a.

A. Bochman. Default logic generalized and simplified. *Ann. Math. Artif. Intell.*, 53:21–49, 2008b.

A. Bondarenko, P. M. Dung, R. A. Kowalski, and F. Toni. An abstract, argumentation-theoretic framework for default reasoning. *Artificial Intelligence*, 93:63–101, 1997.

P. M. Dung and T. C. Son. An argument-based approach to reasoning with specificity. *Artificial Intelligence*, 133:35–85, 2001.

D. M. Gabbay. Theoretical foundations for non-monotonic reasoning in expert systems. In K. R. Apt, editor, *Logics and Models of Concurrent Systems*. Springer, 1985.

H. Geffner. *Default Reasoning. Causal and Conditional Theories*. MIT Press, 1992.

M. Gelfond, V. Lifschitz, H. Przymusińska, and M. Truszczyński. Disjunctive defaults. In *Proc. KR'91*, pages 230–237, Cambridge, Mass., 1991.

G. Gentzen. Untersuchungen über das logische schließen. *Mathematische Zeitschrift*, 39:405–431, 1934.

S. Hanks and D. McDermott. Non-monotonic logics and temporal projection. *Artificial Intelligence*, 33:379–412, 1987.

J. F. Horty. Some direct theories of nonmonotonic inheritance. In D. M. Gabbay, C. J. Hogger, and J. A. Robinson, editors, *Handbook of Logic in Artificial Intelligence and Logic Programming 3: Nonmonotonic Reasoning and Uncertain Reasoning*. Oxford University Press, Oxford, 1994.

T. Janhunen. On the intertranslatability of non-monotonic logics. *Annals of Math. and Art. Intel.*, 27:791–828, 1999.

S. Kraus, D. Lehmann, and M. Magidor. Nonmonotonic reasoning, preferential models and cumulative logics. *Artificial Intelligence*, 44:167–207, 1990.

D. Lehmann. What does a conditional knowledge base entail? In R. Brachman and H. J. Levesque, editors, *Proc. KR'89*, pages 212–222. Morgan Kaufmann, 1989.

D. Lehmann. Another perspective on default reasoning. *Ann. Math. Artif. Intell.*, 15:61–82, 1995.

D. Lehmann and M. Magidor. What does a conditional knowledge base entail? *Artificial Intelligence*, 55:1–60, 1992.

V. Lifschitz. Computing circumscription. In *Proc. Int. Joint Conf. on Artificial Intelligence, IJCAI-85*, pages 121–127. Morgan Kaufmann, 1985.

V. Lifschitz. On open defaults. In J.W. Lloyd, editor, *Computational Logic: Symposium Proceedings*, pages 80–95. Springer, 1990.

V. Lifschitz. Minimal belief and negation as failure. *Artificial Intelligence*, 70:53–72, 1994.

V. Lifschitz and G. Schwarz. Extended logic programs as autoepistemic theories. In L. M. Pereira and A. Nerode, editors, *Proc. Second Int. Workshop on Logic Programming and Nonmonotonic Reasoning*, pages 101–114. MIT Press, 1993.

V. Lifschitz and T. Woo. Answer sets in general nonmonotonic reasoning (preliminary report). In *Proc. KR'92*, pages 603–614. Morgan Kaufman, 1992.

V. Lifschitz, D. Pearce, and A. Valverde. Strongly equivalent logic programs. *ACM Transactions on Computational Logic*, 2:526–541, 2001.

F. Lin and Y. Shoham. A logic of knowledge and justified assumptions. *Artificial Intelligence*, 57:271–289, 1992.

V. W. Marek, G. F. Schwarz, and M. Truszczyński. Modal nonmonotonic logics: ranges, characterization, computation. *Journal of ACM*, 40:963–990, 1993.

N. McCain and H. Turner. Causal theories of action and change. In *Proceedings AAAI-97*, pages 460–465, 1997.

J. McCarthy. Programs with common sense. In *Proceedings of the Teddington Conference on the Mechanization of Thought Processes*, pages 75–91, London, 1959. Her Majesty's Stationary Office.

J. McCarthy. Circumscription – a form of non-monotonic reasoning. *Artificial Intelligence*, 13: 27–39, 1980.

J. McCarthy. Applications of circumscription to formalizing common sense knowledge. *Artificial Intelligence*, 13:27–39, 1986.

J. McCarthy and P. Hayes. Some philosophical problems from the standpoint of artificial intelligence. In B. Meltzer and D. Michie, editors, *Machine Intelligence*, pages 463–502. Edinburgh University Press, Edinburgh, 1969.

D. McDermott. Nonmonotonic logic II: Nonmonotonic modal theories. *Journal of the ACM*, 29:33–57, 1982.

D. McDermott. Critique of pure reason. *Computational Intelligence*, 3(3):149–160, 1987.

D. McDermott and J. Doyle. Nonmonotonic logic. *Artificial Intelligence*, 13:41–72, 1980.

M. Minsky. A framework for representing knowledge. Tech. Report 306, Artificial Intelligence Laboratory, MIT, 1974.

R. C. Moore. Semantical considerations on non-monotonic logic. *Artificial Intelligence*, 25: 75–94, 1985.

J. Pearl. System Z: A natural ordering of defaults with tractable applications to default reasoning. In *Proceedings of the Third Conference on Theoretical Aspects of Reasoning About Knowledge (TARK'90)*, pages 121–135, San Mateo, CA, 1990. Morgan Kaufmann.

D. Poole. A logical framework for default reasoning. *Artificial Intelligence*, 36:27–47, 1988.

R. Reiter. A logic for default reasoning. *Artificial Intelligence*, 13:81–132, 1980.

R. Reiter. *Knowledge in Action: Logical Foundations for Specifying and Implementing Dynamic Systems*. MIT Press, 2001.

R. Reiter and G. Criscuolo. On interacting defaults. In *Proceedings of IJCAI-81*, pages 270–276, 1981.

G. Schwarz. Autoepistemic modal logics. In R. Parikh, editor, *Theoretical Aspects of Reasoning about Knowledge, TARK-90*, pages 97–109, San Mateo, CA, 1990. Morgan Kaufmann.

D. Scott. Completeness and axiomatizability in many-valued logic. In *Proc. Symp. In Pure Math., No. 25*, pages 431–435, 1974.

K. Segerberg. *An Essay in Classical Modal Logic*, volume 13 of *Filosofiska Studier*. Uppsala University, 1971.

Y. Shoham. *Reasoning about Change*. Cambridge University Press, 1988.

R. Thomason. Logic and artificial intelligence, 2003. URL `http://plato.stanford.edu/archives/fall2003/entries/logic-ai/`.

M. Truszczyński. Modal interpretations of default logic. In J. Myopoulos and R. Reiter, editors, *Proceedings IJCAI'91*, pages 393–398, San Mateo, Calif., 1991. Morgan Kaufmann.

H. Turner. A logic of universal causation. *Artificial Intelligence*, 113:87–123, 1999.

Dynamic Distributed Nonmonotonic Multi-Context Systems[1]

Minh Dao-Tran
Thomas Eiter
Michael Fink
Thomas Krennwallner
Institut für Informationssysteme
Technische Universität Wien
Favoritenstraße 9-11
A-1040 Vienna, Austria

Abstract: Nonmonotonic multi-context systems (MCS) provide a formalism to represent knowledge exchange between heterogeneous and possibly nonmonotonic knowledge bases (contexts). Recent advancements to evaluate MCS semantics (given in terms of so-called equilibria) enable their application to realistic and fully distributed scenarios of knowledge exchange. However, the current MCS formalism cannot handle open environments, i.e., when knowledge sources and their contents may change over time and are not known a priori. To improve on this aspect, we develop Dynamic Nonmonotonic Multi-Context Systems, which consist of schematic contexts that allow to leave part of the information interlinkage open at design time. A concrete interlinking is established by a configuration step at run time, where concrete contexts and information imports between them are fixed. We formally develop a corresponding extension and provide semantics by instantiation to ordinary MCS. Furthermore, we develop a basic distributed configuration algorithm and discuss several refinements that affect the resulting configurations, in particular by means of optimizations according to different quality criteria. This discussion is complemented with experimental results obtained with a corresponding prototype implementation.

1 Introduction

Developing modern knowledge-based information systems increasingly requires software engineers to handle knowledge integration tasks for accessing and aligning relevant information for particular application domains. As a result of recent developments of the World Wide Web, this information is in general distributed and comes with heterogeneous representation formalisms. Moreover, access to larger bodies of acquired knowledge is often organized via suitable interfaces, rather than providing

---
[1]This research has been supported by the Austrian Science Fund (FWF) project P20841 and the Vienna Science and Technology Fund (WWTF) project ICT08-020.

direct access to the respective knowledge base.

Nonmonotonic multi-context systems (MCS) of Brewka and Eiter [2007] provide a formalism to address such knowledge integration tasks in a principled way. They generalize seminal work by Giunchiglia and Serafini [1994] and Roelofsen and Serafini [2005] that served the purpose to integrate different monotonic inference systems, into a heterogeneous MCS. Intuitively, individual knowledge bases—for historic reasons called *contexts*—are represented in a logic formalism with associated *belief sets* as a high-level representation of the associated semantics. By reference to such beliefs, so-called *bridge rules* allow to model an information flow between contexts. Semantics is given to an MCS in terms of *equilibria*, i.e., belief sets—one for each context—such that each belief set is acceptable for the respective local knowledge base and the information flow is in equilibrium.

The initial MCS approach has been gradually extended, in particular allowing for nonmonotonicity, both at the level of bridge rules as well as in knowledge bases of homogeneous contexts [Roelofsen and Serafini, 2005, Brewka, Roelofsen, and Serafini, 2007], and more recently to accommodate heterogeneity of context logics also in the nonmonotonic case [Brewka and Eiter, 2007]. Furthermore, the practical importance of distributed settings has been perceived; they play e.g. an important role in ambient computing, where Antoniou, Papatheodorou, and Bikakis [2010] and Bikakis and Antoniou [2010] propose multi-context systems to integrate different entities, and develops a semantics for conflict resolution based on defeasible reasoning. In our work, the importance of distributed settings is reflected in the development of fully distributed evaluation algorithms for computing equilibria of nonmonotonic MCS [Dao-Tran, Eiter, Fink, and Krennwallner, 2010].

However, a characteristic that comes with many distributed application scenarios is that the environment is open, at least to some extent, meaning that participating knowledge sources and their contents may change over time and are not known a priori. This is in contrast with the static nature of current MCS in the sense that participating contexts and the corresponding information exchange need to be fixed completely at design time. Thus, atoms in bridge rules always point to a particular belief from a concrete context. This is prohibitive to formalizing systems where part of the behavior is instantiated at run-time only, as motivated by the following example.

Example 1 At the beginning of each semester, students in a group (including Alice, Bob, and Carol) need to choose courses from their curriculum. For each possible course, the students have three possible decisions, namely *select*, *hesitate*, and *eliminate* (in decreasing order). Intuitively, there is a potential for selecting a course if one finds it interesting. However, if the lecturer is known to be hard to please, they fear that it might be tough (or impossible) to get good marks and potentially eliminate the course. If there are reasons for both selecting and eliminating—or none—they are then in the state of hesitation, which dominates the other two potential decisions.

Moreover, the final decision of each student is supported by the decisions of their friends. If some friend gives a positive (resp., negative) opinion about a particular course, and no other friend shares an opposite opinion, then the group will adjust their final decision accordingly.

According to this strategy, the students do not specify in advance for a course which friends they will consult. This depends on the friends they will meet at the course orientation meeting. While attending the orientation meeting and exchanging opinions, every student in the group finally comes up with a list of courses that conforms with the choices of their colleagues.

For example, Alice may believe that if Bob hesitates or selects a course, then this is a positive sign, because he is very cautious; on the other hand, if Bob eliminates a course, then this is a negative sign. But she has a different opinion about Carol's choice, namely she is encouraged only when Carol selects the course and is discouraged otherwise. Carol, who is a bit more careful, might only accept that Bob's selection (resp., elimination) of the course as a positive (resp., negative) hint, i.e., she has no bias when Bob is hesitating. Finally, Bob may interpret the opinions of the two girls in the same way as Carol does with his, i.e., mapping selections to be positive, elimination to be negative, and having no preference w.r.t. hesitance.

When the three of them talk about the course on Answer Set Programming, which Bob finds interesting, but Alice has the impression that the professor is very demanding, they ask Carol, who has no additional opinion about it. One of the outcomes of the discussion is that Bob and Carol will select the course while Alice hesitates. □

The current MCS setting is sufficient to formalize the last part of the discussion between Alice, Bob, and Carol (see also Example 3), but lacks dynamicity to formalize the general setting of Example 1.

In this work, we address the above shortcoming of the MCS formalism concerning open environments of information exchange, that is when at design time the concrete knowledge sources participating in an information exchange are not known. Intuitively, what is needed to cope with such scenarios is a formalism for information exchange which is closer towards a peer-to-peer (P2P) approach, where so-called peers can at any time join or leave the system dynamically [Aberer, Punceva, Hauswirth, and Schmidt, 2002]. To this end, we present *Dynamic Nonmonotonic Multi-Context Systems*, which consist of schematic contexts that may leave some information interlinkage open at design time; this linkage is established by a configuration step at run time, in which concrete contexts and information imports between them are wired.

More specifically, our contributions are the following:

- We formalize dynamic multi-context systems, which extend the MCS formalism with so-called *schematic bridge rules*. Intuitively, schematic bridge rules may contain place holders that can range over both context identifiers and beliefs. Their semantics is defined via suitable notions of substitution and binding, where a *context substitution* maps context holders to concrete contexts and a binding maps schematic belief atoms to adequate concrete beliefs. To take into account that a perfectly matching belief might not exist, we use (unless exact substitution is forced) a 'similar'-based binding beliefs in which schematic beliefs are bound to 'similar' beliefs, which is assessed by a similarity function. To determine such beliefs, we foresee a matchmaking component as an oracle which returns on a call a list with similar beliefs. More precisely, it provides simple term substitutions according to an underlying similarity measure.

- We consider the problem of finding an instantiation of a dynamic multi-context system, starting from a specific context, i.e., a concrete "configuration" of the (open) system. To solve it, we first present a basic algorithm for computing configurations of dynamic MCS. As the number of configurations can be very large in general, we then consider different heuristics to generate 'good' ones, which take topological structure and/or different criteria of qualities of individual matches (bindings) into account. The algorithm is fully distributed, i.e., instances run at different contexts, and the configurations are found by local computations plus communication.

- Finally, we present some results of an experimental prototype implementation of the configuration algorithm under different heuristics for the configuration. The results show that the latter behave on a few considered topologies of potential interconnections as expected, and may also lead in particular cases to configurations which corresponds to natural social information interlinkage.

Using dynamic MCS, a broader range of application scenarios can be modeled which require the flexibility of taking changing context into account. In particular, group formation to satisfy information needs of heterogeneous components, with possible selection among different alternatives, can be readily expressed.

The remainder of this paper is structured as follows. The next section recalls basic concepts of answer set programming (ASP, the context logic of choice in our examples) and nonmonotonic MCS which are the starting point of this work. Dynamic MCS are subsequently introduced in Section 3, where we provide their formal definition and semantics in terms of instantiation to ordinary MCS. Section 4 contains then the description of our basic distributed configuration algorithm and discussions of refinements such as, e.g., different heuristics to drive the configuration. Furthermore, in Section 5 we report on a prototype implementation of the algorithm and some experimental results. Conclusions and issues for further work are eventually given in Section 6.

2   Preliminaries

We recall some basic notions of disjunctive logic programs under the answer set semantics [Gelfond and Lifschitz, 1991] and heterogeneous nonmonotonic multi-context systems [Brewka and Eiter, 2007].

Answer Set Programs. Let $\mathcal{A}$ be a finite alphabet of atomic propositions. A *disjunctive rule r* is of the form

$$a_1 \vee \cdots \vee a_k \leftarrow b_1, \ldots, b_m, \text{not } b_{m+1}, \ldots, \text{not } b_n \ , \tag{1}$$

$k + n > 0$, where all a_i and b_j are atoms from $\mathcal{A}$.[2] We let $H(r) = \{a_1, \ldots, a_k\}$, and $B(r) = B^+(r) \cup B^-(r)$, where $B^+(r) = \{b_1, \ldots, b_m\}$ and $B^-(r) = \{b_{m+1}, \ldots, b_n\}$. An *answer set program P* is a finite set of rules r of form (1).

An *interpretation* for P is any subset $I \subseteq \mathcal{A}$. It *satisfies* a rule r, if $H(r) \cap I \neq \emptyset$ whenever $B^+(r) \subseteq I$ and $B^-(r) \cap I = \emptyset$. I is a *model* of P, if it satisfies each $r \in P$.

---

[2]Gelfond and Lifschitz [1991] used classical literals as basic constituents rather than atoms. For simplicity, we disregard classical (also called strong) negation here; this does not affect the expressiveness of the formalism.

The *GL-reduct* [Gelfond and Lifschitz, 1991] P^I of P relative to I is the program obtained from P by deleting (i) every rule $r \in P$ such that $B^-(r) \cap I \neq \emptyset$, and (ii) all $not b_j$, where $b_j \in B^-(r)$, from every remaining rule r.

An interpretation I of a program P is called an *answer set* of P iff I is a $\subseteq$-minimal model of P^I.

Example 2 We will now model Example 1 as an answer set program. Let R_i be a set of the following rules:

$$s_i \leftarrow ps_i, \text{not } h_i, \text{not } e_i \qquad\qquad e_i \leftarrow pe_i, not h_i, not inc_i$$
$$s_i \leftarrow ph_i, \text{not } ps_i, \text{not } h_i, \text{not } e_i, inc_i \qquad e_i \leftarrow ph_i, dec_i$$
$$h_i \leftarrow ps_i, \text{not } e_i, dec_i \qquad\qquad ps_i \leftarrow inter_i$$
$$h_i \leftarrow ph_i, \text{not } inc_i, \text{not } dec_i \qquad pe_i \leftarrow hprof_i$$
$$h_i \leftarrow pe_i, \text{not } ph_i, inc_i \qquad\qquad ph_i \leftarrow ps_i, pe_i$$
$$ph_i \leftarrow \text{not } ps_i, \text{not } pe_i$$

The atoms have the following meaning: s_i, h_i, and e_i stand for the three decisions: select, hesitate, and eliminate, resp. Similarly, ps_i, ph_i, and pe_i stand for the potential to select, hesitate, and eliminate a course, resp. A course is interesting if $inter_i$ is true, and a professor is hard to please if $hprof_i$ is true. The atoms inc_i and dec_i mean that a student inclines and declines to select a course, resp.

The program $P_1 = R_1 \cup \{hprof_1\}$ has one answer set $\{e_1, pe_1, hprof_1\}$, $P_2 = R_2 \cup \{inter_2\}$ has the answer set $\{s_2, ps_2, inter_2\}$, while $P_3 = R_3$ has the answer set $\{h_3, ph_3\}$.

Intuitively, P_1 represents Alice's mind. She thinks that the professor is hard to please ($hprof_1$), hence she potentially eliminates (pe_1) the course and will eliminate it (e_1) if no more support information is provided. On the other hand, P_2 represents Bob's mind. He is really interested in the course ($inter_2$) and selects it (s_2) based on his potential of selecting the course (ps_2). Carol, modeled by P_3, adds no personal view about the course. She is currently hesitating (h_3, ph_3) in taking the course; her final decision can change depending on decisions of other friends. □

Multi-Context Systems. A *logic* is, viewed abstractly, a tuple $L = (\mathbf{KB}_L, \mathbf{BS}_L, \mathbf{ACC}_L)$, where

- $\mathbf{KB}_L$ is a set of well-formed knowledge bases, each being a set (of formulas),
- $\mathbf{BS}_L$ is a set of possible belief sets, each being a set (of formulas), and
- $\mathbf{ACC}_L: \mathbf{KB}_L \rightarrow 2^{\mathbf{BS}_L}$ assigns each $kb \in \mathbf{KB}_L$ a set of acceptable belief sets.

This covers many (non-)monotonic KR formalisms like description logics, default logic, answer set programs, etc.

For example, a (propositional) *ASP logic* L may be such that $\mathbf{KB}_L$ is the set of answer set programs over a (propositional) alphabet $\mathcal{A}$, $\mathbf{BS}_L = 2^{\mathcal{A}}$ contains all subsets of atoms, and $\mathbf{ACC}_L$ assigns each $kb \in \mathbf{KB}_L$ the set of all its answer sets.

Definition 1 *A multi-context system (MCS) is a set* $M = (C_1, \dots, C_n)$, *consisting of contexts* $C_i = (L_i, kb_i, br_i)$, *such that* $1 \leq i \leq n$, $L_i = (\mathbf{KB}_i, \mathbf{BS}_i, \mathbf{ACC}_i)$ *is a logic,* $kb_i \in \mathbf{KB}_i$ *is a knowledge base, and* br_i *is a set of* L_i*-bridge rules of the form*

$$s \leftarrow (c_1 : p_1), \dots, (c_j : p_j), \text{not } (c_{j+1} : p_{j+1}), \dots, \text{not } (c_m : p_m) \qquad (2)$$

where $1 \leq c_k \leq n$ *and* p_k *is an element of some belief set of* L_{c_k} *(i.e.,* $p_k \in \bigcup \mathbf{BS}_{L_{c_k}}$*),* $1 \leq k \leq m$, *and* $kb \cup \{s\} \in \mathbf{KB}_i$ *for each* $kb \in \mathbf{KB}_i$.

Informally, bridge rules allow to modify the knowledge base by adding s, depending on the beliefs in other contexts.

The semantics of an MCS M is defined in terms of particular *belief states*, which are sequences $S = (S_1, \ldots, S_n)$ of belief sets $S_i \in \mathbf{BS}_i$. Intuitively, S_i should be a belief set of the knowledge base kb_i; however, also the bridge rules br_i must be respected. To this end, kb_i is augmented with the conclusions of all $r \in br_{c_i}$ that are applicable.

Formally, r of form (2) is *applicable in* S, if $p_i \in S_i$, for $1 \leq i \leq j$, and $p_k \notin S_k$, for $j + 1 \leq k \leq m$. Let $app(R, S)$ denote the set of all bridge rules $r \in R$ that are applicable in S, and $head(r)$ the part s of any r of form (2).

Definition 2 *A belief state* $S = (S_1, \ldots, S_n)$ *of a multi-context system* M *is an equilibrium iff for all* $1 \leq i \leq n$, $S_i \in \mathbf{ACC}_i(kb_i \cup \{head(r) \mid r \in app(br_i, S)\})$.

Example 3 Let $M' = (C_1, C_2, C_3)$ be an MCS such that all L_i are ASP logics, with alphabets $\mathcal{A}_i = \{s_i, h_i, e_i, ps_i, ph_i, pe_i, inter_i, hprof_i, inc_i, dec_i\}$. Suppose $kb_i = P_i$, with P_i taken from Example 2, and

$$br_1 = \left\{ \begin{array}{rcl} inc_1 & \leftarrow & (2 : s_2), \text{not } (3 : h_3), \text{not } (3 : e_3), \text{not } (1 : dec_1) \\ inc_1 & \leftarrow & (2 : h_2), \text{not } (3 : h_3), \text{not } (3 : e_3), \text{not } (1 : dec_1) \\ dec_1 & \leftarrow & (2 : e_2), \text{not } (3 : s_3), \text{not } (1 : inc_1) \end{array} \right\},$$

$$br_2 = \left\{ \begin{array}{rcl} inc_2 & \leftarrow & (1 : s_1), \text{not } (3 : e_3), \text{not } (2 : dec_2) \\ dec_2 & \leftarrow & (3 : e_3), \text{not } (1 : s_1), \text{not } (2 : inc_2) \end{array} \right\}, \text{ and}$$

$$br_3 = \left\{ \begin{array}{rcl} inc_3 & \leftarrow & (2 : s_2), \text{not } (1 : e_1), \text{not } (3 : dec_3) \\ dec_3 & \leftarrow & (1 : e_1), \text{not } (2 : s_2), \text{not } (3 : inc_3) \end{array} \right\}$$

One can check that $S = (\{h_1, pe_1, hprof_1, inc_1\}, \{s_2, ps_2, inter_2\}, \{s_3, ph_3, inc_3\})$ is an equilibrium of M'. Intuitively, M' models the discussion between Alice (C_1), Bob (C_2), and Carol (C_3). Comparing this to Example 2, the decision of Bob influences those of Alice and Carol, as Alice now hesitates about the course even though having the potential of eliminating it, while Carol decided to select the course although she was hesitating about it before.   □

3   Dynamic Nonmonotonic MCS

In the following section, we will develop a framework that caters for dynamics in Multi-Context Systems using placeholders for contexts and beliefs in schematic bridge rules. The open parts of such rules can be made concrete by linking them to ordinary MCS.

3.1 Basic Notions

Let $\mathcal{V}_{ctx}$ be a vocabulary of *context holders*,[3] and let $\Sigma = \bigcup \Sigma_i$ a set of (possibly shared) *signatures*. Unless stated otherwise, elements from $\mathcal{V}_{ctx}$ (resp., Σ) are denoted with first letter in upper case (resp., lower case). Furthermore, we define the set $\Sigma_{@}$ (resp., $\Sigma_{\sim}$) of exact (resp., similar) *schematic beliefs* as the set of symbols $@[p]$ (resp., $[p]$) for all p in Σ. Let $bel(@[p]) = bel([p]) = p$ be a function for extracting the belief symbol from a schematic belief.

Definition 3 *A* dynamic multi-context system $M = \{C_1, \ldots, C_n\}$ *is a set of* schematic contexts $C_i = (L_i, kb_i, sbr_i)$, *where*

- $L_i = (\mathbf{KB}_i, \mathbf{BS}_i, \mathbf{ACC}_i)$ *is a logic based on a signature* Σ_i,

- $kb_i \in \mathbf{KB}_i$ *is a knowledge base, and*

- sbr_i *is a set of* L_i schematic-bridge rules *(s-bridge rules for short) of the form*

$$s \leftarrow B(r), \chi(r) \tag{3}$$

with $B(r) = (X_1 : P_1), \ldots, (X_j : P_j), \text{not}\,(X_{j+1} : P_{j+1}), \ldots, \text{not}\,(X_m : P_m)$, *where each* $sb_\ell = (X_\ell, P_\ell)$, $1 \leq \ell \leq m$, *is a* schematic bridge atom *(s-bridge atom for short) in which* $X_\ell \in M \cup \mathcal{V}_{ctx}$ *either refers to a context in* M *or is a context holder, and* $P_\ell \in \Sigma \cup \Sigma_@ \cup \Sigma_{\sim}$ *is either a belief (i.e.,* $P_\ell \in \Sigma$*) or a schematic belief (*$P_\ell \in \Sigma_@ \cup \Sigma_{\sim}$*); and* $\chi(r) = Y_{1_1} \neq Y_{1_2}, \ldots Y_{k_1} \neq Y_{k_2}$ *is a (possibly empty) list of inequality atoms* $Y_{i_1} \neq Y_{i_2}$ *(*$1 \leq i \leq k$*) where* Y_{i_1}, Y_{i_2} *are two different context holders from* $X_1, \ldots, X_m$.

For simplicity, we assume that context holders in rules are standardized apart, i.e., there exist no two context holders with the same name in two different rules, as they can be bound to different contexts.

Example 4 A group of n students in Example 1 can be modeled as a dynamic MCS $M = \{C_1, \ldots, C_n\}$, where, for each $C_i = (L_i, kb_i, sbr_i) \in M$, $kb_i = R \cup F_i$, with R from Example 2 and $F_i \subseteq \{inter, hprof\}$, and the following set of schematic bridge rules

$$sbr_i = \left\{ \begin{array}{ll} inc_i & \leftarrow \ (X_i : [pos_i]), \text{not}\,(Y_i : [neg_i]), \text{not}\,(i : dec_i), X_i \neq Y_i \\ dec_i & \leftarrow \ (Z_i : [neg_i]), \text{not}\,(T_i : [pos_i]), \text{not}\,(i : inc_i), Z_i \neq T_i \end{array} \right\}.$$

The first rule expresses that student i should be inclined to take a course, if some student in the group has a positive opinion and some student does not have a negative opinion, and student i herself is not declining to take the course. The second rule is similar, but for declining the course.

Here, the context holders set is $\mathcal{V}_{ctx} = \{X_i, Y_i, Z_i, T_i\}$, the local signature at each context C_i is $\Sigma_i = \mathcal{A}_i \cup \{pos_i, neg_i\}$ with $\mathcal{A}_i$ taken from Example 3. We use here only similar schematic beliefs, namely $[pos_i]$ and $[neg_i]$. □

---

[3]We use the term 'holder' rather than 'variable' to avoid confusion with variables as introduced for *relational MCS* [Fink, Ghionna, and Weinzierl, 2011].

Dynamic MCS differ from original MCS in the sense that s-bridge atoms in general are not specifically bound to some beliefs of other dynamic contexts in the system, but rather represent a collection of possibilities to point to different beliefs in other contexts. From a topological point of view, such a high-level representation incurs numerous dependencies between dynamic contexts in general. However, most of these dependencies are not reflected in intended instantiations, which provides evidence not to aim at defining equilibria of dynamic MCS in a direct way. Hence, for defining semantics one rather considers how to bind them to original MCS. We consider such bindings next, starting with the notion of binding a schematic bridge atom to ordinary bridge atoms based on potential matches.

A *context substitution* is a map $\sigma \colon (M \cup \mathcal{V}_{ctx}) \to M$ such that for every inequality atom $Y_{i_1} \neq Y_{i_2}$ occurring in bridge rules of a context $C \in M$, $\sigma(Y_{i_1}) \neq \sigma(Y_{i_2})$. For a context C_k, we denote by $\sigma|_{C_k}$ the *restriction* of σ to C_k, i.e., the subset of σ containing only maps from a context holder appearing in an s-bridge rule in C_k. Due to the assumption of standardization of context holders, the set of restrictions of σ to all individual contexts in M is a partitioning of σ.

The application of a context substitution σ to an s-bridge atom $sb = (X : P)$ is $\sigma(sb) = (C_j : P)$ where $P \in \Sigma_{@} \cup \Sigma_{\sim} \cup \Sigma_j$, and either $X = C_j$ or $X \in \mathcal{V}_{ctx}$ satisfying $(X \mapsto C_j) \in \sigma$. Intuitively, the application of a context substitution is responsible for instantiating a potential context holder of an s-bridge atom.

Example 5 Let $\sigma = \{X_1 \mapsto C_2, Y_1 \mapsto C_3\}$ be a context substitution, then the applications of σ to $sb_1 = (X_1 : [pos_1])$ and $sb_2 = (Y_1 : [neg_1])$ are $sb_1' = \sigma(sb_1) = (C_2 : [pos_1])$ and $sb_2' = \sigma(sb_2) = (C_3 : [neg_1])$. □

Let $f_M \colon \Sigma \times \Sigma \to [0, 1]$ be a function measuring the similarity between beliefs in an MCS M, where higher similarity of beliefs p and q is reflected by a larger value of $f_M(p, q)$. In particular, $f_M(p, q) = 1$ means that p and q are considered to have highest similarity (especially, if they are identical) and $f_M(p, q) = 0$ that p and q are completely dissimilar. We do not commit to a particular function f_M here, which may depend on the application; in what follows, we just assume that some such function f_M has been fixed and is available, for instance consider similarity of terms as defined by WordNet [Miller, 1995] or different types of matches on Larks specifications [Sycara, Widoff, Klusch, and Lu, 2002].

Definition 4 *Given an MCS M, a similarity function f_M, and a threshold t, a term substitution from C_i to C_j in M w.r.t. f_M and t, denoted by $\eta_M^t(C_i, C_j)$, is a relation $\eta_M^t(C_i, C_j) = \{(a, b) \mid a \in \Sigma_i, b \in \Sigma_j, f_M(a, b) > t\}$.*

By η_M^t we denote the collection of all pairwise term substitutions in M. The *density* of M w.r.t. η_M^t is $d_{\eta_M^t} = |\{(C_i, C_j) \mid \eta_M^t(C_i, C_j) \neq \emptyset\}|$. In the sequel, we pick a default value $t = 0$; furthermore, we use η instead of η_M^t when M is clear from the context.

The application of a term substitution η to an s-bridge atom $sb = (C_j : P)$ in context C_i, denoted by $\eta(sb)$, is defined by (i) if $P = a$, then $\eta(sb) = \{(C_j : a)\}$ if $a \in \Sigma_j$, and $\emptyset$ otherwise; (ii) if $P \in \Sigma_{@}$, then $\eta(sb) = \{(C_j : b) \mid (bel(P), b) \in \eta(C_i, C_j), f_M(bel(P), b) = 1\}$; (ii) if $P \in \Sigma_{\sim}$, then $\eta(sb) = \{(C_j : b) \mid (bel(P), b) \in \eta(C_i, C_j)\}$. Intuitively, the application of a term substitution to an s-bridge atom only

Table 1: Interesting part of similarity function

f_M	s_1	h_1	e_1	s_2	h_2	e_2	s_3	h_3	e_3
pos_1	0.0	0.0	0.0	**0.9**	**0.6**	0.0	**0.7**	0.0	0.0
neg_1	0.0	0.0	0.0	0.0	0.0	**0.8**	0.0	**0.5**	**0.7**
pos_2	**0.7**	0.0	0.0	0.0	0.0	0.0	**0.6**	0.0	0.0
neg_2	0.0	0.0	**0.7**	0.0	0.0	0.0	0.0	0.0	**0.6**
pos_3	**0.6**	0.0	0.0	**0.8**	0.0	0.0	0.0	0.0	0.0
neg_3	0.0	0.0	**0.7**	0.0	0.0	**0.8**	0.0	0.0	0.0

applies to s-bridge atoms with instantiated context holders, and then collects all possible substitutions for the schematic belief P.

Example 6 Continue with Example 5, suppose that we have a similarity function f_M whose interesting part is described in Table 1; and for the rest, f_M takes value 1 if the two parameters are identical and 0 otherwise.

The values of f_M are taken in conformity with the scenario in Example 1, e.g., Alice trusts the *select* and *hesitate* decisions of Bob as a positive sign at a measurement of 0.9 and 0.6, respectively. She considers Bob *eliminating* the course as a negative sign of 0.8. Hence, $f_M(pos_1, s_2) = 0.9$, $f_M(pos_1, h_2) = 0.6$, and $f_M(neg_1, e_2) = 0.8$. On the other hand, Alice is encouraged only when Carol selects the course, but with less confidence as $f_M(pos_1, s_3) = 0.7$; and she interprets other choices from Carol as discouragement with $f_M(neg_1, h_3) = 0.5$, and $f_M(neg_1, e_3) = 0.7$. The next rows in Table 1 show the opinions of Bob and Carol about the decisions of the others. Note that they do not take *hesitance* into account.

The term substitutions from C_1 to C_2 and C_3 with respect to f_M are

$$\eta(C_1, C_2) = \{(pos_1, s_2), (pos_1, h_2), (neg_1, e_2)\},$$

and

$$\eta(C_1, C_3) = \{(pos_1, s_3), (neg_1, h_3), (neg_1, e_3)\},$$

respectively. The applications of these substitution to sb_1' and sb_2' are $\eta(sb_1') = \{(C_2 : s_2), (C_2 : h_2)\}$ and $\eta(sb_2') = \{(C_3 : e_3)\}$. □

Based on σ and η, the notion of a bridge substitution is simply defined by their composition.

Definition 5 *Let σ be a context substitution of M. The* bridge substitution θ *for an s-bridge atom sb w.r.t. σ is $\theta(sb) = \eta(\sigma(sb))$.*

Thus, intuitively, the bridge substitution of an s-bridge atom is done in two steps. First, one uses σ to instantiate the context holders, and then η takes effect to instantiate the schematic beliefs.

Example 7 The bridge substitution of the s-bridge atom sb_1 from Example 5 w.r.t. σ from the same example and η from Example 6 is $\theta(sb_1) = \eta(\sigma(sb_1)) = \{(C_2 : s_2), (C_2 : h_2)\}$. Similarly, we have $\theta(sb_2) = \{(C_3 : e_3)\}$. □

Let us now turn to bridge rules. Given a schematic bridge rule r of form (3) in a context C_i, a context substitution σ is called a substitution of r iff there exist bridge substitutions θ_ℓ w.r.t. σ for all schematic bridge atoms sb_ℓ in $B(r)$, i.e., for $1 \le \ell \le m$, such that $\theta_\ell(sb_\ell) \ne \emptyset$. The bindings of r w.r.t. σ are defined as the set of bound rules $r\sigma$ where each bound rule is obtained by replacing $sb_\ell = (X : P)$ in $B(r)$ with

(i) some bridge atom $(C_i : b)$ such that $(C_i : b) \in \theta_\ell(sb_\ell)$, if $sb_\ell \in B^+(r)$; and

(ii) the sequence of negated bridge atoms not $(C_i : b_1), \ldots,$ not $(C_i : b_k)$ such that $\{(C_i : b_1), \ldots, (C_i : b_k)\} = \theta_\ell(sb_\ell)$; if $sb_\ell \in B^-(r)$.

The size m of $r\sigma$ is determined by $m = \Pi_{sb_\ell \in B^+(r)} |\theta_\ell(sb_\ell)|$.

Example 8 Continuing our example, pick r as the first s-bridge rule from Example 4 and consider it in context C_1 representing Alice's mind. Furthermore, regard the context substitution σ from Example 5. Taking the term substitutions of Example 6 into account, $r\sigma$ consists of two bound rules, namely the first two rules from br_1 in Example 3. □

Given a set of s-bridge rules R of a context C and a context substitution σ, the binding of R w.r.t. σ is defined as $R\sigma = \bigcup_{r \in R} r\sigma$. Then, the binding of a context C w.r.t. a context substitution σ is given by $C\sigma = (kb, sbr\sigma)$.

Definition 6 *Given a dynamic MCS M and a context substitution σ, the set $M\sigma = \{C_1\sigma, \ldots, C_\ell\sigma\}$, is a binding of M w.r.t. a context C_k iff*

1. $C_k \in \{C_1, \ldots, C_\ell\}$

2. $\{C_1, \ldots, C_\ell\} \subseteq M$

3. σ is a substitution for all s-bridge rules in all contexts $C_1, \ldots, C_\ell$, and

4. $\{C_1, \ldots, C_\ell\} = \bigcup_{C_j \in \{C_1 \ldots C_\ell\}} \{C \mid r \in sbr_j\sigma \wedge (C : a) \in B(r)\} \cup \{C_k\}.$

A *belief state* of $M\sigma$ is a sequence of belief sets $S = (S_1, \ldots, S_\ell)$, one S_i for each $C_i\sigma$. Such a belief state is an *equilibrium* of M w.r.t. C_k and σ iff for all $1 \le i \le \ell$, it holds that $S_i \in \mathbf{ACC}_i(kb_i \cup head(r) \mid r \in app(sbr_i\sigma, S))$.

The quality of a binding is

$$\frac{1}{|\mathcal{U}|} \cdot \sum_{(a,b) \in \mathcal{U}} f_M(a, b)$$

where $\mathcal{U}$ is the set of matches used in the binding, i.e., $\mathcal{U} = \{(a, b) \mid a = bel(P) \wedge sb = (X : P) \in C_i, 1 \le i \le \ell \wedge (C_j : b) \in \theta(sb), 1 \le j \le \ell\}$.

Intuitively, a binding of M w.r.t. a substitution σ and a context C_k consists of a subset of the contexts of M, which must contain C_k (hence conditions 1 and 2), which is properly instantiated by θ (condition 3) and, moreover, *closed* in the sense that every selected context, except for C_k, is used for instantiating bridge rules of other chosen contexts (condition 4). The notions of belief state and equilibrium are then inherited from ordinary MCS. The quality of a binding is simply the average of the similarities of all matches used in the binding.

3.2 From Dynamic to Ordinary Multi-Context Systems

Recapturing the idea of binding a dynamic MCS M to an original one, starting from a context C_{root}, one needs

1. to know all potential neighbors C_j for a context C_i and the term substitutions $\eta(C_i, C_j)$ between them;

2. a strategy to start from C_{root} and to expand the system by: first determining a context substitution σ for each context term in the s-bridge rules of C_{root}, and then to continue the process at each neighbor, until a closed system is obtained;

3. some decision criteria to guide the process to come up with a most suitable substitution to bind M.

Task (1) is in fact matching beliefs from different contexts. This problem shares similarities with the *matchmaking problem* in Multi-Agent Systems (MAS), which has been widely considered [Sycara et al., 2002, Ogston and Vassiliadis, 2001]. Our work in this paper is not doing matchmaking but rather using the matchmaker as a building block to configure the inter-linkage between contexts in a dynamic MCS to form ordinary ones. As such, we assume that there exists a *matchmaker* MatchMaker which, upon a call MatchMaker(P, C_i) from a context C_i, returns a set of potential neighbors such that

- if P is a schematic variable in $\Sigma_@$, then N is the set of context names C_j where the term substitution $\eta(C_i, C_j)$ contains at least one pair $(bel(P), a)$ with $f_M(bel(P), a) = 1$;

- if P is a schematic variable in $\Sigma_\sim$, then N is the set of context names C_j where the term substitution $\eta(C_i, C_j)$ is nonempty;

- if $P = p$ is an atom from Σ, then N is the set of all contexts C_j such that $p \in \Sigma_j$.

Further queries to the matchmaker such as MatchMaker(C_i, C_j) can give back $\eta(C_i, C_j)$ and/or the value of f_M for the pairs of atoms from this term substitution. This information is used for calculating the quality of the system after instantiating.

The main problems that we solve in this paper are those in (2) and (3). Concerning (2), we present a backtracking algorithm to enumerate all possible context substitutions σ, in a distributed, peer-to-peer like setting. This means that each context, knowing only its potential neighbors by asking the matchmaker, can only locally choose the matches for its own s-bridge atoms, which consequently decides its real neighbors in the resulting MCS, and then has to ask these neighbors to continue the configuration (hence our algorithm is called Iconfig).

The process starts at C_{root} and continues in a Depth-First Search (DFS) manner, carrying along the context substitution σ built up so far, until for all chosen contexts their s-bridge atoms are bound.

Regarding (3), we propose general methods to compare the outcome of different substitutions on two main aspects, namely (Q1) the matching quality of the bound rules and (Q2) the topological quality of the resulting MCS. These methods can be seen as heuristics that can be plugged into the basic version of lconfig to get the context substitutions returned ordered by quality.

For clarity and simplicity, in the sequel, we first present the very basic version of lconfig with generic possibilities for optimization. We then briefly go through such possibilities, where we choose some interesting ones to discuss in more detail and suggest potential realizations of them.

4 Multi-Context System Configuration

The question is now how one can actually compute substitutions as sketched above. We present a basic configuration algorithm which computes concrete bindings for a dynamic MCS. We start with a particular context in the system and gradually invoke some neighbors to get further solutions.[4]

4.1 Basic Algorithm

Given a dynamic MCS M and a starting context C_{root}, the algorithm lconfig presented in this section aims at enumerating all possible context substitutions that can lead to a binding for M, in a distributed way. It mutually calls an algorithm invoke_neighbors and makes use of the following primitives:

- a function get_contexts(σ), which takes a context substitution σ (containing substitutions of form $X \mapsto C$) as input and returns the set of contexts C used in σ.

- a DFS subroutine bind_rule, which given an s-bridge rule r as input consults the matchmaker MatchMaker and returns all context substitutions for the non-ordinary s-bridge atoms of r.

The algorithm lconfig has several parameters: the context C_{root} where the configuration started, the set R of s-bridge rules left to be bound, and the context substitution σ built up so far.

Intuitively, in a context C_k, lconfig first utilizes bind_rule in a DFS manner to enumerate all possible context substitutions for the s-bridge atoms in sbr_k (Step (b)). When this is done, in Step (a) it only refers to newly chosen contexts via a set $\mathcal{C}_{new}$ and calls invoke_neighbors to get the context substitutions of all members in $\mathcal{C}_{new}$.

The algorithm invoke_neighbors has the same parameters C_{root} and σ as lconfig, and carries in addition a set N of newly chosen neighbors of C_k where local configuration needs to be done. The algorithm first picks a neighbor C_j and calls lconfig at this context (Step (f)) to get all context substitutions updated with local substitutions for C_j, stored in $obuf_{C_j}$. Then, in Step (g), it picks each substitution from $obuf_{C_j}$ and continues invoking the remaining contexts in N. Note that in Step (h), the set of remaining neighbors to invoke is recomputed in N', as some of the contexts in N might already be chosen by the call to C_j and thus they are already invoked.

---

[4]In centralized settings, one might instead compute substitutions by making use of more standard declarative solvers, e.g., such as ASP solvers with external information access.

Algorithm 1: lconfig(C_{root}, R, σ) at C_k

Input: C_{root}: root context, R: set of s-bridge rules, σ: context substitution
Output: context substitution for C_k
Data: $obuf_r$ for every $r \in R$: substitutions for r
if $R = \emptyset$ **then**

(a) $\quad \mathcal{C}_{new} := \mathsf{get\_contexts}(\sigma|_{C_k}) \setminus (\mathsf{get\_contexts}(\sigma \setminus \sigma|_{C_k}) \cup \{C_{root}\})$
$\quad$ **if** $\mathcal{C}_{new} \neq \emptyset$ **then return** invoke_neighbors($C_{root}, \mathcal{C}_{new}, \sigma$)
$\quad$ **else return** $\{\sigma\}$

(b) **else**

(c) $\quad$ pick r from R, and $obuf_r := \mathsf{bind\_rule}(\chi(r), B(r), \sigma)$
$\quad$ $ctx\_sub := \emptyset$
$\quad$ **while** $obuf_r \neq \emptyset$ **do**

(d) $\quad\quad$ pick σ' from $obuf_r$ and $obuf_r := obuf_r \setminus \{\sigma'\}$
$\quad\quad$ $ctx\_sub := ctx\_sub \cup \mathsf{lconfig}(C_{root}, R \setminus \{r\}, \sigma')$

$\quad$ **return** ctx_sub

When all invocations of neighbors have finished, the substitution computed at this point is returned and is treated by lconfig either as an intermediate result for the context that invoked it, or as the final result for the user.

Example 9 Take the setting from Example 8 and run bind_rule over the rule body with a starting empty substitution. The call is bind_rule($B, \emptyset$) in which $B = \{(X_1 : [pos_1]), (Y_1 : [neg_1])\}$. Assume that the first s-bridge atom chosen at Step (i) is $sb = (X_1 : [pos])$. A call MatchMaker($[pos_1], C_1$) to the matchmaker returns $N = \{C_2, C_3\}$. The routine then tries all possibilities to bind sb and works recursively to bind the rest of the body. For example, if it chooses to bind X_1 to C_2, then the next call will be bind_rule($\{(Y_1 : [neg_1])\}, \{X_1 \mapsto C_2\}$) which returns $\{\{X_1 \mapsto C_2, Y_1 \mapsto C_3\}\}$ as the set of all context substitutions in which X_1 is mapped to C_2. The binding continues with $X_1 \mapsto C_3$ and in the end, we get two context substitutions, namely $\{X_1 \mapsto C_2, Y_1 \mapsto C_3\}$ and $\{X_1 \mapsto C_3, Y_1 \mapsto C_2\}$. $\quad\square$

Example 10 This example illustrates the run of lconfig and invoke_neighbors on a dynamic MCS from Example 4 with poolsize of contexts $n = 3$. Starting from C_1 we can pick one s-bridge rule from the non-empty set of s-bridge rules at (c), say the first one from Example 4. According to Example 9, the subroutine bind_rule returns a set of 2 possible context substitutions. Let us pick $\sigma = \{X_1 \mapsto C_2, Y_1 \mapsto C_3\}$ from this set and continue calling lconfig for the last rule. This gives 2 possible extensions of σ, one of which extends σ to $\{X_1 \mapsto C_2, Y_1 \mapsto C_3, Z_1 \mapsto C_2, T_1 \mapsto C_3\}$.

Having this context substitution carried to the next recursive call of lconfig, we reach the point where $R = \emptyset$, get $\mathcal{C}_{new} = \{C_2, C_3\}$, and continue calling lconfig at C_2 or C_3. The algorithm proceeds and in the end, we get a number of context substitutions, one is $\{X_1 \mapsto C_2, Y_1 \mapsto C_3, Z_1 \mapsto C_2, T_1 \mapsto C_3, X_2 \mapsto C_1, Y_2 \mapsto C_3, Z_2 \mapsto C_3, T_2 \mapsto C_1, X_3 \mapsto C_2, Y_3 \mapsto C_1, Z_3 \mapsto C_1, T_3 \mapsto C_2\}$. This substitution yields the MCS system in Example 3. $\quad\square$

Algorithm 2: invoke_neighbors(C_{root}, N, σ) at C_k

Input: C_{root}: root context, N: set of neighbors, σ: context substitution
Output: context substitutions for all neighbors of C_k

(e) **if** $N = \emptyset$ **then** **return** $\{\sigma\}$
else

(f)   pick C_j from N, and $obuf_{C_j} := C_j.\text{lconfig}(C_{root}, sbr_j, \sigma)$

     $ctx\_sub := \emptyset$

     while $obuf_{C_j} \neq \emptyset$ **do**

(g)   pick σ' from $obuf_{C_j}$ and $obuf_{C_j} := obuf_{C_j} \setminus \{\sigma'\}$

(h) $N' := N \setminus (\text{get\_contexts}(\sigma') \cup \{C_j\})$

         $ctx\_sub := ctx\_sub \cup \text{invoke\_neighbors}(C_{root}, N', \sigma')$

     return ctx_sub

When the pool size gets large, enumerating all bindings for each bridge rule and all bridge substitutions becomes infeasible. A practical approach would be to compute only a small number of bindings for each rule, and also just a few substitutions at each context. For the remainder of this section, let us use b and n to denote corresponding limits.

We have presented the basic algorithm for enumerating all possible context substitutions of a dynamic MCS M w.r.t. a context C_k in M with which M can be bound to original MCS. To keep it simple, in steps (c), (d), (f), (g), (i), and (j), we nondeterministically pick either a rule, a context substitution, or a context as no supporting information is provided. This leaves a lot of room for optimization. Furthermore, we did not mention how to deal with irregular cases such as when the matchmaker returns no potential neighbor, or the size of the partial MCS has passed some boundary; furthermore, no caching has been foreseen.

In the following subsection, we discuss different heuristics to enhance the search process when more support information is available, so that the context substitutions will be returned in some quality driven order. After that, we briefly describe a strategy for cutting off when reaching a size boundary, hence a possibility to tolerate partial bindings.

4.2   Quality-Driven Local Configuration

Quality for the topology (Q_T). Our experimental results reveal that evaluating equilibria of MCS in general does not scale up to very large systems, and Dao-Tran et al. [2010] and Bairakdar, Dao-Tran, Eiter, Fink, and Krennwallner [2010] showed that limitations on some specific topologies such as the diamond topology exist. Hence, one of the purposes for configuration is to restrain the size/topology of the resulting system to some boundary, e.g., by trying to reuse as many contexts from the local configuration of the parent as possible; or by trying to avoid troublesome topological properties, such as ones having *join contexts*, i.e., contexts C_i which are accessed from different contexts C_j and $C_{j'}$ which in turn are accessed (possibly by intermedi-

---

Algorithm 3: bind_rule(I, B, σ) at C_k

Input: I: set of inequality atoms, B: set of s-bridge atoms, σ: context substitution

Output: substitutions for B

(i) **if** $\exists a = (X : P)$ *non-ordinary in* B **then**

     $N := \mathsf{MatchMaker}(P, C_k)$ // N: set of potential neighbors

     if $\nexists (X \mapsto C)$ *in* σ **then**

         $dup := \{C_i \in N \mid (Y \mapsto C_i) \in \sigma \wedge (X \neq Y) \in I\}$

         $N := N \setminus dup$

         $ctx\_sub := \emptyset$

         while $N \neq \emptyset$ **do**

(j)   choose a context C_j from N, and $N := N \setminus \{C_j\}$

            $ctx\_sub := ctx\_sub \cup \mathsf{bind\_rule}(I, B \setminus \{a\}, \sigma \cup \{X \mapsto C_j\})$

         return ctx_sub

     else if $\sigma(X) \in N$ **then return** $\mathsf{bind\_rule}(I, B \setminus \{a\}, \sigma)$

     else return $\emptyset$

else return $\{\sigma\}$

---

ate contexts) from a single context C_k, or cycles.

For this purpose, the selection of neighbors (Step (j)) is crucial. To support a context C_k with more information for this task, we do a *one-step look-ahead* at all of its potential neighbors. In general, looking ahead into a potential neighbor C_i can give back any information that C_i is able to infer from its own knowledge and information provided by the matchmaker. In this paper, our setting allows the look-ahead to return the number of s-bridge atoms in C_i, denoted by nba_i.

We define in the following different heuristic possibilities of the topological quality function to reflect attempts to have the resulting MCS in some restricted shape (the smaller the value of the function, the better is the quality).

Assume that having started the configuration from context C_{root}, we are now doing local configuration at context C_k, choosing a binding for a schematic bridge belief $[p]$, considering the possible match (p, q) to a context C_i. As the context substitution σ is carried along, one can easily extract the set of chosen contexts so far, which is denoted here by $\mathcal{C}$. Consider the following topological quality functions:

(H1) $quality_{i,k} = nba_i$: with this function, we prefer potential neighbors with fewer s-bridge atoms.

(H2) $quality_{i,k} = \begin{cases} 0 & \text{if } C_i \in \mathcal{C} \\ 1 & \text{otherwise.} \end{cases}$

     This function gives priority to contexts which are already chosen, hence to keep the size of the resulting system small. On the other hand, this tends to introduce cycles.

One can imagine more complicated quality functions; for instance, to take the topology of the system built up so far into account in order to avoid cycles or join contexts. To this end, one must transfer not only the substitution σ between contexts, but also the system topology (i.e., the respective graph).

Along the same lines, for Step (c) (resp., (i)) one can define quality functions, for instance based on some syntactic criteria combined with some history information, to provide an heuristic ranking for choosing the next rule (resp., the next non-ordinary s-bridge atom).

Quality for bindings of a schematic rule (Q_S).   This type of quality measures the closeness between the bindings and the intended meaning of the schematic rule based on the matching quality of each single s-bridge atom. Notice that after getting the schematic substitution η from the matchmaker, θ is determined by the context substitution σ. Each realization of σ gives us a possibility to bind the schematic rules. What we need is a means to compare these possibilities. To make it generic, we define the quality function ρ of a bound rule $r' \in r\sigma$ of a schematic rule r of form (3) as follows:

$$\rho(r') = \text{op} \left\{ \begin{array}{l} \alpha \mid sb = (X : P) \in B(r) \wedge \\ (C_j : b) \in \theta(sb) \wedge (bel(P), b) \in \eta(C_i, C_j) \wedge \\ (C_j : b) \in B(r') \wedge f_M(bel(P), b) = \alpha \end{array} \right\}.$$

Basically, we take the measure of similarity of all bindings used to construct r' and apply an operator op on top. Here, op is generic and can be instantiated to any operator for a specific use. For example, two plausible options are (i) op $=$ min, and (ii) op $=$ avg.

Roughly, in case (i), following an overly cautious approach, the quality of the whole binding is determined as the minimal quality of all matches of the s-bridge atoms in its body. In a different approach, case (ii) takes all matches into account and respects a contribution of each match in the overall quality of the rule. Depending on different philosophies to establish the overall quality of a binding that is based on bindings of each single schematic bridge atom, one can provide more complicated operators and plug them into this scheme.

To benefit from this quality, one can sort the output buffer $obuf_r$ (resp., $obuf_{C_j}$) according to Q_S in Step (d) (resp., (g)), and then pick the best context substitution up to this point to continue with.

Combined quality (Q_C).   The two types of quality functions above look into two aspects of the resulting system, namely the (Q_T) topology and the (Q_S) similarity in meaning of the bindings. To exploit the latter, one needs to enumerate all possible bindings for a rule.

When we have a limit on the number of solutions (see discussion above), it is very important to approximate Q_T when choosing a binding, since then one cannot compute all bindings of a rule, and then sort them.

To this end, we modify Q_T in a way that it also takes care of the quality of the match. Intuitively, when two potential contexts are equally ranked by Q_T, one looks at the quality of the match, i.e., approximating Q_S, to rank them. More specifically, the heuristic functions H1 and H2 are changed to:

(H3) $quality_{i,k} = nba_i - \alpha$, and

(H4) $\quad quality_{i,k} = \begin{cases} -\alpha & \text{if } C_i \in \mathcal{C} \\ 2 - \alpha & \text{otherwise,} \end{cases}$

where α is the quality of the match being considered to bind the current s-bridge atom.

4.3 Dealing with Irregular Cases

In practice, it is convenient for the user to have the possibility to specify an upper bound for the size of the resulting system. However, a full substitution respecting the given limit may not always exist. A flexible approach to deal with such a situation, rather than to increase the bound, is to *cut off* when reaching the boundary and to tolerate partial answers.

By cutting off, more precisely we mean to remove all unbound negative s-bridge atoms from s-bridge rules, and remove all s-bridge rules with unbound positive bridge atoms. Intuitively, this amounts to consider a system where any further contexts that would exist for binding are considered to return empty belief sets (and thus the respective bridge atoms are pre-evaluated accordingly). Note that one also needs to undo the on-going substitutions for such s-bridge rules, and this might trigger the cancellation of substitutions in a backward manner: since once a context is not used anymore for instantiating other contexts, it is not needed and the part of the substitution w.r.t. this context should be removed from the final result.

Another case in which cutting off might be used is when substitutions do not exist due to non-existent matchings, i.e., when the matchmaker does not return any match for a schematic constant. We can apply the same strategy as above, i.e., remove the corresponding s-bridge atom if it appears in the negative body of an s-bridge rule, or remove the whole s-bridge rule if the s-bridge atom is in its positive body.

The cutoff is in fact easy to implement. One can create a dummy context C_0 which has only a single belief *dumb* with the unique acceptable belief set $\emptyset$ (i.e., intuitively *dumb* is *false*, and all beliefs in every other context are matchable to *dumb*. Then cutting off as discussed above is simply achieved by matching unbound s-bridge atoms to $(C_0 : dumb)$.

5 Implementation and Experimental Results

In this section, we report on some initial experiments with a prototype implementation of our MCS configuration algorithm. Full details of the implementation and the experiments are available on the web.[5]

Our implementation is written in C++ and uses a simple realization of MatchMaker, in which matchmaking results are statically given rather than dynamically computed. The respective matching information is stored in a text file, which is loaded into each context at start up time. During configuration, a call to the matchmaker is then simply performed by posing a query to the respective storage in the context.

For our experiments, we used a host system having an Intel Core 2 Duo 2.4GHz processor with 4GB RAM, running Mac OS X. We have tested three types of topologies, called random, rake, and grid, respectively, which model different interlinkage

---

[5]http://www.kr.tuwien.ac.at/research/systems/dmcs/.

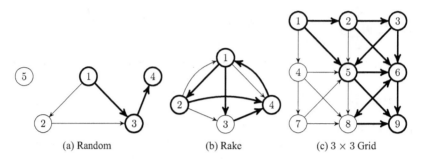

| (a) Random | (b) Rake | (c) 3 × 3 Grid |

Figure 1: Benchmark topologies and possible configurations (emphasized)

patterns (see Fig. 1 for exemplary instances). The *random* topology is generated by deciding whether there exist potential matches from context C_i to context C_j with a probability of 0.5; this leads to a rather dense system as random topologies have approximately half of the edges of a complete graph. Fig. 1a shows an example with five contexts, for which C_1, C_3, and C_4 has been chosen in the configuration. The *rake* topology requires that every context C_i has potential matches to all contexts C_j where $j > i$; furthermore, for each context C_i with $i > 1$, we randomly pick $\lceil i/3 \rceil$ distinct context(s) C_k where $k < i$, and with a probability of 0.7, place potential matches from C_i to C_k. The example instance in Fig. 1b has a configuration using all contexts C_1 to C_4, which forms a system with two cycles. And finally, in instances of the *grid* topology having size $m \times n$, each context has potential matches to its adjacent contexts on the next row/column, a diagonally adjacent context on the next row and next column; and for contexts on the last row/column, diagonally adjacent backward/forward contexts (see Fig. 1c).

For each topology, we experimented with dynamic MCS having a pool of 100 or 200 contexts. Each context contains a local knowledge base as in Example 2 and a set of bridge rules which is a subset of $sbr_i \cup \{r_i\}$ where sbr_i is from Example 4 and $r_i = inc_i \vee dec_i \leftarrow (X_i : [pos]), (X_i : [neg])$, in order to have a varying number of s-bridge atoms in contexts.Furthermore, the limits on the number of substitutions and bindings were set to $n = 3$ and $b = 8$, respectively.

The results of running our basic algorithm (no heuristics) and the four heuristics H1-H4 are shown in Tables 2-4. There, the test names are of the form $Name(ps, d)$ with poolsize ps and density d is the number of possible connections between contexts of the dynamic MCS instance. The measurements of the output include the running time of the algorithm, the size and the density (i.e., the number of connections between contexts) of the resulting MCS, and the average similarity of all matches used for instantiation.

As a general observation, we can see that most of the time the heuristics improved the result in different aspects, either the size respectively density, or the average quality, or both. The heuristics H1 and H2 concentrate on minimizing the size of the system, which was strongly confirmed in case of the rake topology. Here, the basic algorithm yielded large configurations: the default ordering for the next neighbor does not take advantage of back edges, because it always picks first a context with a higher index. Hence, for some first results, the configuration has to wait until the last context

Table 2: Experimental results for random topology

Test name	Heuristics	Running time (secs)	Size	Density	Quality
	None	0.22	5	10	0.604
	H1	6.16	9	18	0.450
Random (100, 4974)	H2	0.18	4	8	0.605
	H3	2.39	26	54	0.975
	H4	1.42	8	16	0.859
	None	0.18	7	14	0.643
	H1	1.01	9	18	0.603
Random (100, 4886)	H2	0.19	6	12	0.584
	H3	6.65	17	34	0.975
	H4	0.36	8	16	0.703
	None	0.62	6	12	0.584
	H1	2.23	7	14	0.480
Random (200, 19953)	H2	0.40	6	12	0.588
	H3	8.92	44	91	0.996
	H4	0.82	7	14	0.852
	None	1.02	6	12	0.510
	H1	0.84	6	12	0.605
Random (200, 19896)	H2	1.12	6	12	0.510
	H3	18.88	37	76	0.999
	H4	1.29	7	14	0.794

Table 3: Experimental results for rake topology

Test name	Heuristics	Running time (secs)	Size	Density	Quality
	None	10.61	99	198	0.529
	H1	1.00	35	70	0.530
Rake (100, 5575)	H2	1.01	10	20	0.500
	H3	7.65	28	57	0.983
	H4	0.48	10	20	0.908
	None	22.76	100	200	0.557
	H1	1.01	26	52	0.514
Rake (100, 5597)	H2	1.29	10	20	0.498
	H3	0.96	22	45	0.937
	H4	1.51	9	18	0.867
	None	170.10	197	394	0.541
	H1	17.11	73	146	0.507
Rake (200, 22579)	H2	0.82	9	18	0.518
	H3	64.24	59	119	0.999
	H4	0.88	5	11	0.790
	None	277.04	199	398	0.550
	H1	4.84	54	108	0.579
Rake (200, 22510)	H2	1.65	9	18	0.520
	H3	4.40	36	72	0.994
	H4	0.74	9	18	0.746

Table 4: Experimental results for grid topology

Test name	Heuristics	Running time (secs)	Size	Density	Quality
Grid (10×10, 279)	None	1.82	100	198	0.540
	H1	1.60	66	130	0.568
	H2	4.02	100	198	0.545
	H3	2.87	74	154	0.630
	H4	2.02	64	128	0.736
Grid (20×5, 274)	None	2.52	100	198	0.562
	H1	1.84	57	112	0.565
	H2	5.05	82	162	0.557
	H3	2.38	67	134	0.623
	H4	2.04	52	104	0.724
Grid (25×4, 270)	None	3.44	100	198	0.519
	H1	1.81	50	98	0.534
	H2	8.64	100	198	0.527
	H3	1.46	50	98	0.591
	H4	1.94	55	109	0.623
Grid (20×10, 569)	None	5.32	200	398	0.544
	H1	6.90	130	258	0.541
	H2	25.62	200	398	0.551
	H3	9.10	152	323	0.631
	H4	9.86	104	206	0.677
Grid (25×8, 566)	None	6.78	200	398	0.563
	H1	10.60	142	282	0.563
	H2	31.54	200	398	0.567
	H3	2.61	87	178	0.623
	H4	10.15	83	164	0.705
Grid (40×5, 554)	None	8.44	200	398	0.557
	H1	5.10	102	202	0.577
	H2	29.46	162	322	0.570
	H3	5.35	103	215	0.629
	H4	5.88	91	184	0.686

is bound; only then it can close the system.

The same behavior of the basic algorithm could be observed in the case of the grid topology: it always took the whole set of contexts for instantiation. On the other hand, when a heuristics was applied, the system size was often reduced by almost 50%, except for H2; this is because in this restricted structure, there are no backward edges at intermediate contexts, hence the effort to close the system can only take effect when contexts at the last row or column are involved in the instantiation.

However, since the random topology has no structure, the heuristics H1 and H2 behave similarly to running the algorithm without heuristics. The former even returned a slightly bigger system and many times one of worse quality.

On the other hand, the heuristics H3 and H4 combine the topology and quality criteria in order to intuitively gain an improvement; this indeed shows up in the experimental results. For the random topology, H3 dominated the result in quality; this was achieved by a considerably larger configuration, which is explained by the fact that heuristics H3 branches out to promising contexts for quality improvements, rather than looking at already chosen neighbors. In the other cases, either H3 or H4 gave the best quality in all the tests.

Moreover, the trade-off between the number of contexts and the quality of the configuration can be seen in comparing heuristics H3 and H4. While the former approximates the system size based on the number of bridge atoms, the latter has a more direct approach by looking at the chosen contexts. Hence H4 usually ended up with a smaller system, but not always with one of better quality. On the other hand, H3 always returned a bigger system compared to standard results, but interestingly it could not beat H4 in the case of grid topology, because this rigid structure gives H3 no chance to trade system size for quality.

On average, H4 appeared to have the best balance of all result aspects together: it has fast running time, almost the best quality, and acceptable system size/density.

To see the effects of inequality atoms in s-bridge rules, we have also run our algorithm on the same test cases with these special atoms removed. In general, the trends and the relationships between the basic algorithm and its heuristic extensions do not change (see the appendix). The only difference now is that the size/density of the resulting systems gets smaller, as a single context can be used to bind two different context holders in one rule, and the algorithm tries to exploit this property to come to the answers as soon as possible.

We have carried out further experiments with a social-groups topology that are not reported here. This topology forms groups of contexts with full potential connections, and there exist some rare potential connections from one group to another; in some sense this models loosely interconnected communities. The outcome of our configuration resembled typical human behavior of group formation in such a setting quite well, as the resulting MCS tended to contain contexts from a single group only.

6 Related Work and Conclusion

We have presented the framework of dynamic multi-context systems, which extend ordinary multi-context systems (MCS) with so-called schematic bridge rules in order to allow for open environments, in which concrete contexts are taken at run time to form an ordinary MCS by instantiating the schematic bridge rules. We have developed a distributed algorithm for instantiating such dynamic MCS to ordinary MCS, implemented this algorithm in a prototype system, and have shown some benchmark results that compare different heuristics for configuring dynamic MCS.

6.1 Related Work

The problem that we considered in this paper shares some similarities with configuration in Multi-Agent Systems using matchmaking [Sycara et al., 2002]. In this setting, provider agents advertise their capabilities to middle agents; requester agents do not directly go to a provider but first ask some middle agent whether it knows of providers with desired capabilities; the middle agent matches the request against the stored advertisements and returns the information about appropriate providers to the requester. In our setting, the matchmaker plays the role of the middle agent. A context has both roles, it is seen as a requester when being instantiated, and as a provider when being used to instantiate s-bridge atoms from other contexts.

A configuration problem for multi-agent system that is in a sense orthogonal respectively complementary to matchmaking is coalition formation. Here, the problem is the assembly of a group of agents for cooperation in order to get some task down (assuming that the agents already know that cooperation is possible) [Sandholm, 1999]. However, this problem is only remotely related to our configuration problem. Agents have goals and intentions, and decide their participation in a coalition based on utility and reward in a rational manner. This leads in interaction with other agents to complex behaviors, which may be studied using game-theoretic methods and tools. Contexts instead lack such goal and reward orientation, and offer in an altruistic manner information exchange in order to enable the assembly of an MCS. Thus, from a coalition formation point of view, the MCS configuration problem is trivial. The problems gets more complicated if constraints are imposed (e.g., on the solution size or quality), but there is still a difference: at no point, some context may decide not to participate in an MCS as it concludes its payoff is insufficient, or it is being cheated.

Naturally related to dynamic MCS are peer-to-peer (P2P) systems. However, in typical models such as the Peer-Grid [Aberer et al., 2002], a global system semantics does not play a role: peers are strictly localized and can join/leave the system at anytime. Our approach, on the other hand, aims at global model building for an ordinary MCS that is dynamically constructed, where the first step is instantiation, and then the (distributed) evaluation kicks in [Dao-Tran et al., 2010]; this tacitly assumes that no relevant contexts disappear during configuration and evaluation of the configured system. We note that also Calvanese, Giacomo, Lenzerini, and Rosati [2004] proposed a global model semantics for P2P systems, which is based on epistemic logic, and presented a distributed algorithm for query answering. This algorithm evaluates P2P mappings dynamically, but no system configuration like in our approach is performed. Similarly, Bikakis and Antoniou [2010] considered distributed query answering in a given P2P system of contexts, but under preferences using an argumentation based approach; however, no dynamic configuration in a potentially open environment is performed.

6.2   Issues for future research

While we have introduced in this paper dynamic MCS to accommodate open environments and we provided an algorithm for run time configuration, several issues remain for future work.

On the foundational side, a study of the computational complexity of dynamic MCS, and in particular of the configuration problem, could reveal important insight into computational resources needed to solve this problem, and may help to identify classes of systems for which it is efficiently solvable; here, the distribution and possible parallelism are interesting aspects. Furthermore, an improvement of the configuration algorithm, and in particular a deep investigation into heuristics would be an interesting task. Another aspect related to this is linkage cost. The size of a configuration is a crude measure of such cost, which clearly can be refined, taking, e.g., besides the topology also the cost (or value) of accessing particular beliefs into account.

On the implementation side, an obvious task is the implementation of a full-fledged configuration system that includes rich matchmaking, e.g., as in Larks [Sycara et al., 2002] instead of just hard-coded matches. Finally, another issue are applications of dynamic MCS. The student example in the Introduction suggests to consider possible applications in social group formation, complementing e.g., recent work in social Answer Set Programming [Buccafurri, Caminiti, and Laurendi, 2008, Buccafurri and Caminiti, 2008]. Another, less mundane area is configuration of small heterogeneous information systems, in which generic components (e.g., some domain ontologies, some decision component, and some fact base) must be suitably instantiated, given various possibilities. Here matchmaking may play an important role, e.g., if aspects like different levels of abstraction in the context knowledge bases should be handled. The usage of logic-based matchmaking approaches (cf. the work of Noia, Sciascio, and Donini [2007]), in combination with other techniques, might here be worthwhile to consider. In particular, configuration of small systems in mobile environments, where openness is a natural requirement, to further the use of multi-context systems in ambient intelligence [Antoniou et al., 2010, Bikakis and Antoniou, 2010] would be interesting.

References

K. Aberer, M. Punceva, M. Hauswirth, and R. Schmidt. Improving Data Access in P2P Systems. *IEEE Internet Computing*, 6(1):58–67, 2002. doi: 10.1109/4236.978370.

G. Antoniou, C. Papatheodorou, and A. Bikakis. Reasoning about Context in Ambient Intelligence Environments: A Report from the Field. In Lin and Sattler [2010], pages 557–559. URL http://aaai.org/ocs/index.php/KR/KR2010/paper/view/1209.

S. E.-D. Bairakdar, M. Dao-Tran, T. Eiter, M. Fink, and T. Krennwallner. Decomposition of Distributed Nonmonotonic Multi-Context Systems. In T. Janhunen and I. Niemelä, editors, *12th European Conference on Logics in Artificial Intelligence (JELIA 2010), Helsinki, Finland, September 13-15, 2010*, volume 6341 of *LNAI*, pages 24–37. Springer, September 2010. doi: 10.1007/978-3-642-15675-5_5. URL http://www.kr.tuwien.ac.at/staff/tkren/pub/2010/jelia2010-decompmcs.pdf.

A. Bikakis and G. Antoniou. Defeasible Contextual Reasoning with Arguments in Ambient Intelligence. *IEEE Transactions on Knowledge and Data Engineering*, 22(11):1492–1506, 2010. doi: 10.1109/TKDE.2010.37.

G. Brewka and T. Eiter. Equilibria in Heterogeneous Nonmonotonic Multi-Context Systems. In R. C. Holte and A. Howe, editors, *22nd AAAI Conference on Artificial Intelligence (AAAI'07)*, pages 385–390. AAAI Press, 2007. URL http://www.aaai.org/Papers/AAAI/2007/AAAI07-060.pdf.

G. Brewka, F. Roelofsen, and L. Serafini. Contextual default reasoning. In M. M. Veloso, editor, *International Joint Conference on Artificial Intelligence (IJCAI'07)*, pages 268–273. AAAI Press, 2007. URL http://ijcai.org/Past%20Proceedings/IJCAI-2007/PDF/IJCAI07-041.pdf.

F. Buccafurri and G. Caminiti. Logic programming with social features. *Theory and Practice of Logic Programming*, 8(5-6):643–690, 2008. doi: 10.1017/S1471068408003463.

F. Buccafurri, G. Caminiti, and R. Laurendi. A logic language with stable model semantics for social reasoning. In M. Garcia de la Banda and E. Pontelli, editors, *24th International Conference on Logic Programming (ICLP'08), Udine, Italy, December 9-13 2008*, volume 5366 of *LNCS*, pages 718–723. Springer, 2008. doi: 10.1007/978-3-540-89982-2_64.

D. Calvanese, G. De Giacomo, M. Lenzerini, and R. Rosati. Logical Foundations of Peer-To-Peer Data Integration. In *23rd ACM Symposium on Principles of Database Systems (PODS'04)*, pages 241–251. ACM, 2004. doi: 10.1145/1055558.1055593.

M. Dao-Tran, T. Eiter, M. Fink, and T. Krennwallner. Distributed Nonmonotonic Multi-Context Systems. In Lin and Sattler [2010], pages 60–70. URL http://aaai.org/ocs/index.php/KR/KR2010/paper/view/1249.

M. Fink, L. Ghionna, and A. Weinzierl. Relational Information Exchange and Aggregation in Multi-Context Systems. In J. Delgrande and W. Faber, editors, *11th International Conference on Logic Programming and Nonmonotonic Reasoning (LPNMR 2011), Vancouver, BC, Canada, 16-19 May, 2011)*, volume 6645 of *LNCS*, pages 120–133. Springer, 2011. doi: 10.1007/978-3-642-20895-9_12.

M. Gelfond and V. Lifschitz. Classical negation in logic programs and disjunctive databases. *New Generation Computing*, 9(3–4):365–385, 1991. doi: 10.1007/BF03037169.

F. Giunchiglia and L. Serafini. Multilanguage Hierarchical Logics or: How we can do Without Modal Logics. *Artificial Intelligence*, 65(1):29–70, 1994. doi: 10.1016/0004-3702(94)90037-X.

F. Lin, U. Sattler, and M. Truszczynski, editors. *12th International Conference on the Principles of Knowledge Representation and Reasoning (KR 2010), Toronto, Canada, May 9-13, 2010*, May 2010. AAAI Press. URL http://www.aaai.org/Library/KR/kr10contents.php.

G. A. Miller. Wordnet: A lexical database for english. *Commun. ACM*, 38(11):39–41, 1995.

T. Di Noia, E. Di Sciascio, and F. M. Donini. Semantic Matchmaking as Non-Monotonic Reasoning: A Description Logic Approach. *J. Artif. Intell. Res.*, 29:269–307, 2007. doi: 10.1613/jair.2153.

E. Ogston and S. Vassiliadis. Local Distributed Agent Matchmaking. In C. Batini, F. Giunchiglia, P. Giorgini, and M. Mecella, editors, *9th International Conference on Cooperative Information Systems (CoopIS'01)*, volume 2172 of *LNCS*, pages 67–79. Springer, 2001. doi: 10.1007/3-540-44751-2_7.

F. Roelofsen and L. Serafini. Minimal and Absent Information in Contexts. In L. Pack Kaelbling and A. Saffiotti, editors, *19th International Joint Conference on Artificial Intelligence (IJCAI'05)*, pages 558–563. Morgan Kaufmann, 2005. URL http://www.ijcai.org/papers/1045.pdf.

T. Sandholm. Distributed rational decision making. In G. Weiss, editor, *Multiagent Systems – A Modern Approach to Distributed Artificial Intelligence*, chapter 5, pages 201 – 258. MIT Press, 1999.

K. P. Sycara, S. Widoff, M. Klusch, and J. Lu. Larks: Dynamic Matchmaking Among Heterogeneous Software Agents in Cyberspace. *Autonomous Agents and Multi-Agent Systems*, 5 (2):173–203, 2002. doi: 10.1023/A:1014897210525.

Appendix: Experimental results (no inequalities)

Table 5: Experimental results for grid topology, without inequalities

Test name	Heuristics	Running time (secs)	Size	Density	Quality
Grid (10×10, 279)	None	0.13	19	18	0.558
	H1	0.13	16	16	0.553
	H2	0.12	19	18	0.558
	H3	0.73	20	25	0.589
	H4	0.10	17	17	0.592
Grid (20×5, 274)	None	0.17	24	23	0.597
	H1	0.19	28	28	0.547
	H2	0.16	24	23	0.597
	H3	1.10	30	37	0.598
	H4	0.16	24	23	0.658
Grid (25×4, 270)	None	0.22	28	27	0.529
	H1	0.22	31	31	0.579
	H2	0.20	28	27	0.529
	H3	1.42	37	50	0.623
	H4	0.25	34	34	0.616
Grid (20×10, 569)	None	0.20	29	28	0.480
	H1	0.21	30	30	0.459
	H2	0.20	29	28	0.480
	H3	1.54	49	61	0.605
	H4	0.20	26	25	0.712
Grid (25×8, 566)	None	0.25	32	31	0.566
	H1	0.26	32	32	0.592
	H2	0.25	32	31	0.566
	H3	1.73	35	41	0.638
	H4	0.91	28	28	0.617
Grid (40×5, 554)	None	0.49	44	43	0.538
	H1	0.85	46	46	0.535
	H2	0.37	44	43	0.538
	H3	2.02	53	65	0.628
	H4	0.53	53	52	0.630

M. Dao-Tran, T. Eiter, M. Fink and T. Krennwallner

Table 6: Experimental results for random topology, without inequalities

Test name	Heuristics	Running time (secs)	Size	Density	Quality
	None	0.39	2	2	0.567
	H1	0.46	3	3	0.638
Random (100, 4974)	H2	0.37	2	2	0.567
	H3	2.61	23	44	0.981
	H4	0.37	4	4	0.764
	None	0.43	3	3	0.573
	H1	0.47	3	3	0.573
Random (100, 4886)	H2	0.15	3	3	0.573
	H3	9.74	17	33	0.977
	H4	0.46	4	4	0.833
	None	0.38	2	2	0.711
	H1	0.78	5	5	0.415
Random (200, 19953)	H2	0.10	2	2	0.711
	H3	8.13	42	80	0.996
	H4	0.13	3	3	0.631
	None	0.39	2	2	0.483
	H1	0.46	2	2	0.467
Random (200, 19896)	H2	0.43	2	2	0.483
	H3	36.58	36	70	0.999
	H4	0.20	3	3	0.686

Table 7: Experimental results for rake topology, without inequalities

Test name	Heuristics	Running time (secs)	Size	Density	Quality
	None	3.08	91	91	0.521
	H1	0.52	32	32	0.533
Rake (100, 5575)	H2	0.34	6	6	0.367
	H3	8.46	27	49	0.983
	H4	0.08	5	5	0.744
	None	3.15	90	90	0.568
	H1	0.25	22	22	0.508
Rake (100, 5597)	H2	0.69	5	5	0.445
	H3	0.82	20	38	0.951
	H4	0.16	3	3	0.682
	None	11.70	174	174	0.551
	H1	1.09	66	66	0.491
Rake (200, 22579)	H2	0.42	4	4	0.494
	H3	39.95	57	111	0.999
	H4	0.64	4	4	0.815
	None	6.04	173	173	0.541
	H1	0.81	48	48	0.593
Rake (200, 22510)	H2	2.91	4	4	0.481
	H3	4.47	36	72	0.993
	H4	0.55	3	3	0.667

What's in a Default?
Thoughts on the Nature and Role of Defaults
in Nonmonotonic Reasoning

James Delgrande
School of Computing Science
Simon Fraser University
Burnaby, B.C., Canada V5A 1S6

Abstract: This paper examines the role and meaning of defaults in nonmonotonic reasoning (NMR). Defaults, that is, statements that express a condition of normally such as "adults are normally employed", are crucial in commonsense reasoning and in artificial intelligence in general. The majority of research concerning defaults has focussed on (default) inference mechanisms, rather than representational issues involving the meaning of a default. I suggest that, despite the very impressive formal work in the area, it would be useful to (re)consider defaults with respect to the phenomena that they are intended to model.

To start, I briefly consider how defaults have been represented in NMR, along with informal interpretations of defaults. Two major distinctions are explored. The first considers the view of a default as an assertion about some domain, as opposed to an inferential procedure for deriving properties of individuals. The second distinction considers the manner in which default application is informally treated, whether as a weak "rule" or essentially as a weak material implication. I suggest that the "weak material conditional" interpretation is not suitable in the case of defaults; this is problematic since most existing approaches take this latter interpretation.

Subsequently, I argue that defaults of normality are best regarded as statements in a naïve scientific theory. A theory of the meaning of such defaults can be given by a logic of weak conditionals, in which a default is treated as a counterfactual normative statement. From this vantage, nonmonotonic reasoning with such conditionals can be re-examined. To this end, the notion of relevant properties emerges as a key factor in drawing default conclusions about an individual. As well, other phenomena, such as reasoning about norms, or deontic assertions, or counterfactuals may be addressed in a similar fashion.

1 Introduction

Classical reasoning is *monotonic*, which is to say it adheres to a principle of monotonicity:

Monotonicity: If $\Gamma \vdash \phi$ then $\Gamma, \Delta \vdash \phi$

Thus, having proven a result in geometry, say, it is absurd to suggest that learning more information about the problem would invalidate the conclusion. On the other hand, our commonsense, everyday knowledge is for the most part *nonmonotonic*, in that it fails to satisfy monotonicity. Hence, for example, on being told that an individual is a bird, one will conclude that it flies, while on being later informed that it is a penguin or is a nestling, one will conclude that it does not fly.

The area of nonmonotonic reasoning (NMR) in artificial intelligence (AI) studies such reasoning. NMR then is a central and crucial area of AI, and is fundamental to commonsense reasoning. The past 30 years have seen much impressive and important work in NMR, beginning with the seminal Artificial Intelligence Journal issue on the topic [AIJ, 1980]. At this stage, 30 years on, we now have a good understanding of principles underlying NMR, and it would appear that the major approaches to NMR have been identified and are well explored.

In NMR, a fundamental notion is that of a *default*, where a default can be thought of as a weak, or defeasible, conditional. The principal use of defaults in NMR is to ascribe default properties to individuals. A default is generally expressed in English in the form "X's are (normally) Y's".[1] Examples include the hackneyed "birds (normally) fly", as well as "adults are employed" or "snow is white".[2] So a primary task in NMR is to come up with a principled means of dealing with such statements.

To this end, it can be observed that much research in NMR has dealt with the development of formal approaches for reasoning with defaults. Paradigmatically,[3] particularly during the early days of the 1980's and 1990's, a research program would involve proposing a formal mechanism, examining its suitability with respect to dealing with defaults, locating glitches in the representation, modifying the approach, and so proceed. The approaches developed, whether default logic, circumscription, non-monotonic inference relations, conditional logics, or others,[4] are arguably among the most impressive and important formalisms developed in AI.

On the other hand, part of the task of AI researchers is to apply such approaches to real-world problems. That is, these formalisms are intended to be used, and resulting knowledge bases will be used to encode information about and to reason about the world. The main thesis of this paper is that in some important cases default formalisms don't capture the phenomenon that they're intended to model. Arguably, a large part of the problem is that representational issues have received insufficient attention.

In this paper, I first review the notion of a (normality) *default*, specifically how defaults have been represented in NMR, along with informal interpretations of defaults. Subsequently, two major distinctions are explored. The first contrasts, on the one hand, the notion of a default as a general assertion about some domain with, on the other hand, "applying" a default to an individual to derive a property of that individual. That is, one can consider a default such as "birds fly" as asserting something about a domain; on the other hand, one can use a default to obtain a default conclusion, such

---

[1] In linguistics, such sentences are examples of the broader class of *generics* [Carlson and Pelletier, 1995]. Insofar as possible, linguistic issues and, particularly, linguistic conventions or communication conventions are avoided here. Rather, the focus is on purely representational and reasoning issues.

[2] This last is the well-known example attached to Tarski's theory of truth. But of course it isn't the case that snow is (unreservedly) white, but rather that snow is normally white.

[3] and stereotypically

[4] See the next section for references and brief descriptions.

as that a given bird flies. It can be observed that most approaches ignore this first aspect, or conflate these aspects. The second distinction considers the manner in which default application is informally regarded, whether as a weak "rule" or whether as a weak material implication. I suggest that the "weak material conditional" interpretation is not suitable in the case of normality defaults. This, insofar as reasoning with defaults goes, is problematic since most existing approaches take the latter interpretation. The overall conclusion is that, despite the very impressive formal apparatuses developed and despite the remarkable success in applying these approaches in areas such as reasoning about action and planning, diagnosis, database systems, and logic programming, nonetheless there remains a general problem with defaults of normality with "getting the inferences right".

I suggest, toward a direction for a solution, that defaults are best regarded as naïve scientific statements. After developing this argument, I also suggest that this view will lead to a better understanding of reasoning with defaults and that perhaps it will also allow a wider application of default reasoning to other types of weak conditionals, including counterfactuals, deontics, and statements of causality.

2 Defaults

To begin, it seems fair to ask, *What is a default?* This question will be addressed in part throughout this section. Commonly, defaults are expressed in the form "X's are Y's" or "If X then normally Y". As a starting point, a default can be taken as an assertion about the world. Thus "birds fly" says something about the class of birds. We can also ask *How are defaults used?*; and here it can be noted that the standard use of defaults is to draw plausible conclusions, or conclusions in situations where we have only incomplete information.

Various approaches have been proposed for inference involving defaults, notably, Default Logic [Reiter, 1980] (and encompassing, for our purposes, the stable models semantics and answer set programming [Gelfond and Lifschitz, 1988, 1991, Baral, 2003], as well as autoepistemic logic [Moore, 1985, Denecker, Marek, and Truszczynski, 2003]), circumscription [McCarthy, 1980, Lifschitz, 1985, McCarthy, 1986], nonmonotonic inference relations (and associated closure operations) [Kraus, Lehmann, and Magidor, 1990, Lehmann and Magidor, 1992], and conditional logics [Delgrande, 1988, Lamarre, 1991, Boutilier, 1994].[5] As described, a general problem (then and now) is getting the inferences right: obtaining plausible, commonsense conclusions given a set of defaults and general assertions about a domain.

2.1 Default Inference: Encodings

We give the briefest of introductions to approaches to nonmonotonic reasoning here; for details the reader should consult the aforecited references, or general accounts [Brewka, 1991b, Antoniou, 1997, Brewka, Niemelä, and Truszczynski, 2007].

A default can be encoded according to several quite different schemes:

---

[5]Comments on these approaches will be seen to apply to other approaches, including conditional entailment [Geffner and Pearl, 1992] and abductive approaches such as Theorist [Poole, 1988], as well as inheritance networks [Horty, 1994].

"Rule of Inference": This is the approach of Default Logic. "Birds fly" can be encoded either in propositional or first order logic as follows:

$$\frac{Bird \,:\, Fly}{Fly} \qquad \text{or} \qquad \frac{Bird(x) \,:\, Fly(x)}{Fly(x)}.$$

In the first case, if $Bird$ is true and Fly is consistent with what is believed, then Fly is concluded. The notion of "is consistent with" is subtle, and leads to an intricate, elegant fixed point definition. In the second case, the same intuitive account can be given, except that x is instantiated to some ground term. That is, a default rule with free variables can be regarded as a schema, standing for the set of its ground instantiations.

Several points can be noted:

1. Despite its name, Default Logic is not a logic of defaults per se, since it doesn't give an account of a notion of truth of a default. Instead what is provided is a means of drawing default conclusions in the absence of information.

2. Since Default Logic isn't a logic of defaults, but rather is a general and powerful mechanism, one must "program" desirable properties for defaults. For example, given the additional default that penguins don't fly, along with the information that birds are penguins, one has to stipulate explicitly that the more specific penguins-don't-fly default takes priority over the less specific birds-fly default.

3. Quite frequently in the literature a default is expressed propositionally. However it is not clear what a default such as $\frac{Bird:Fly}{Fly}$ means; specifically, it is not clear what the propositions $Bird$ and Fly refer to. Probably the most intelligible gloss is that, in default reasoning one most often is reasoning about an individual, say x, and the defaults are phrased with reference to this individual. Hence the default may be more mnemonically encoded propositionally as $\frac{x\text{-}is\text{-}a\text{-}Bird\,:\,x\text{-}Flys}{x\text{-}Flys}$.

Via Classical Logic: This is the approach taken by circumscription. The "birds fly" example can be encoded propositionally or in first-order logic as:

$$(Bird \wedge \neg Ab_F) \supset Fly \qquad \text{or} \qquad \forall x.(Bird(x) \wedge \neg Ab_F(x)) \supset Fly(x). \quad (1)$$

Thus, in the propositional case, if $Bird$ is true, and Ab_F is not, then one can derive Fly. The intended meaning of Ab_F is that the individual in question is not abnormal with respect to flight. In *circumscribing* the atom Ab_F, essentially if Ab_F *can* be taken to be false then it *is* taken to be false. For the predicate $Ab_F(\cdot)$, one analogously minimizes the extension of the predicate. Semantically this is carried out by defining an ordering over models of a knowledge base, preferring those models where the Ab atoms are false (or the extension of the Ab predicates is smallest, respectively), and then just considering the minimal models of a knowledge base.

It can be noted that (1) leads to the very strong conclusion that unless a bird can be shown to not fly, one concludes that it flies. Hence for example, if one knew nothing about the birds of Madagascar, one would conclude that they all fly.

The points made concerning Default Logic also apply to circumscription: Circumscription is not a logic of defaults, but rather provides a means by which defaults can be encoded. Similarly, to make default inferences have the "right" properties one needs to enhance the approach. Thus to deal with specificity information such as that implicit between penguins and birds, priorities are introduced into circumscription. Last, of course, the same comments apply to the meaning of atoms like $Bird$ and Fly in the propositional formula in (1).

Inference Relation: The area of nonmonotonic inference relations can be regarded foremost as providing a general framework whereby general principles of nonmonotonicity may be studied, and via which other formalisms may be compared. However, one might also examine a specific nonmonotonic inference relation with regard to its suitability as an approach to dealing with defaults. The base approach which has been used for representing defaults is called *preferential reasoning*. Our canonical example would be expressed via a nonmonotonic inference relation as $Bird \mathrel{\vert\!\sim} Fly$. In this case, one can specify relations between nonmonotonic inferences, for example:

$$\text{From} \quad Bird \mathrel{\vert\!\sim} Fly, \quad Bird \mathrel{\vert\!\sim} Nest \quad \text{infer} \quad Bird \wedge Nest \mathrel{\vert\!\sim} Fly. \quad (2)$$

Hence from "birds fly" and "birds build nests", one can infer that "birds that build nests fly". The resulting systems are inferentially weak, at least with regards to obtaining desirable nonmonotonic consequences; for example one cannot infer $Bird \wedge Green \mathrel{\vert\!\sim} Fly$ from $Bird \mathrel{\vert\!\sim} Fly$. This is addressed by extending the set of inferences via a (nonmonotonic) *closure* operator.

Thus, it appears that a nonmonotonic inference relation is *about* defaults, in the sense that one might read (2) as saying (despite the phrasing as an inference relation) that from defaults "birds fly" and "birds build nests" one can infer the default "birds that build nests fly". The closure operator then, so it might seem, extends reasoning to that of deduction involving individuals. We discuss this point further in the next section.

Modal Operator: Last, one might consider a default to be a "real" assertion, carrying a truth value. In this case, our example could be encoded using a new connective, as $Bird \Rightarrow Fly$. In this instance $\Rightarrow$ is a binary modal operator, with intuitive meaning "in the most normal of worlds in which $Bird$ is true, Fly is also true." The example (2) can then be expressed as a formula:

$$(Bird \Rightarrow Fly \wedge Bird \Rightarrow Nest) \supset (Bird \wedge Nest) \Rightarrow Fly. \quad (3)$$

It proves to be the case that there are very close connections between conditional logics of defaults and nonmonotonic inference relations. In fact, the central approach in each case has been shown to be translatable to the other, fully preserving inferences. As well, the central closure operator in each case has also been shown to be symmetrically translatable. It might seem that these two approaches are merely syntactic variants of each other. However, we later suggest that, despite these formal intertranslations, there are significant differences between the approaches with respect to their suitability for representing and reasoning with defaults.

None of the above schemes appears to be immediately suitable for fully dealing with defaults. In the case of Default Logic and circumscription, for example, one has powerful inference mechanisms, and the challenge is to modify the inference mechanism, or how it is applied, in order to get the "right" properties. This led to the development of variants of the basic approach, see for example the papers by Łukaszewicz [1988], Brewka [1991a], Mikitiuk and Truszczynski [1993], Delgrande, Schaub, and Jackson [1995] , and Delgrande and Schaub [1997] that concern Default Logic. This also led to a general methodology for determining suitable default inference during the 1980's and 1990's which can be called "test-and-refine": Typically a nonmonotonic inference mechanism would be proposed or modified; it would be shown to work on a set of troublesome examples; later other troublesome examples would arise; the approach would be modified, and the process continued. Thus in dealing with specificity, in Default Logic semi-normal defaults were employed, while in circumscription a notion of prioritization was introduced. The so-called Yale Shooting Problem [Hanks and McDermott, 1986] is a good example of a problem for which the obvious encodings didn't work, but that also spurring significant research and results in the area. On the other side, nonmonotonic inference relations and modal approaches provide a semantically-justified account of the notion of a default, but in this case the difficulty lies in getting a nonmonotonic counterpart that has the "right" properties.

It can also be observed that the above paradigmatic schemes for expressing defaults are very different with respect to their form. Moreover, one would expect to obtain different conclusions depending on which approach is used to express a default. This raises several key questions: *Why should one prefer one approach over another?* And: *which approach is most suitable for dealing with defaults?* And moreover: *if two approaches lead to different conclusions, how does one judge which is "correct"?* From a formal point of view,[6] the answer to these questions is clear: One needs a theory of defaults in order to be able to determine what the properties of defaults should be. Consequently, in the next subsection, we examine the question of what a default informally means.

2.2 Interpretations of Defaults

Let's reconsider what it is we're trying to deal with and, to be specific, consider various possible informal interpretations of "birds fly". Among other alternatives, the following are possible readings of "birds fly":
1. Most birds fly
2. A bird that can be consistently assumed to fly does fly.
3. Birds normally fly

---

[6]Which is to say that there are also informal considerations. In particular, any theory of defaults will need to produce plausible or commonsense conclusions. Arguably it is the job of formalization to precisely capture such informal notions.

4. The prototypical bird flies
5. Birds generally/usually fly

A fair question to ask at this point is: *Do any of these interpretations align with the encodings that we've seen in the previous subsection, and if so, which and in what fashion?* While it isn't immediately clear which, if any, of the previous approaches fit with these interpretations, we *can* consider these interpretations with respect to how they fit with an informal notion of default.

Consider the first interpretation of "birds fly" as "most birds fly". This is clearly a statistical assertion: one has some population of birds in mind, and over half of them fly. One can develop an approach where, given that some large proportion of birds fly and individual x is a bird, one accepts the belief that x flies – indeed the late Henry Kyburg has addressed this interpretation. (See for example the paper by Kyburg [1994].) Without going into detail, we will note that such probabilistic approaches may be seen as being orthogonal to nonmonotonic reasoning. Kyburg expressed the difference as follows:

Schema for probabilistic inference: $\dfrac{\mathcal{B}K, E}{C, \text{ hedged}}$

Schema for nonmonotonic inference: $\dfrac{\mathcal{B}K, E}{C}$ hedged inference

That is, in the probabilistic case, one makes a (monotonic) inference that is nonetheless "hedged". In the nonmonotonic case, a consequence is accepted while the inference itself is defeasible.

In any case, the representation schemes that we have reviewed resist a probabilistic reading. In Default Logic, for example, a rule $\dfrac{Bird(x):Fly(x)}{Fly(x)}$ applies to individuals and (roughly) rests on a notion of consistency, not probability.[7] Similar considerations apply in circumscription. As well, the nonmonotonic inference relations or conditional logics that have been proposed to represent defaults cannot be given a probabilistic reading. That is, in preferential reasoning, $Bird \mathrel{|\!\sim} Fly$ cannot be coherently interpreted as "most birds fly".[8] We return to this issue in Section 4.

The second reading, "a bird that can be consistently assumed to fly does fly" is clearly epistemic in nature. Autoepistemic logic addresses this interpretation; given the results of Denecker et al. [2003] linking autoepistemic logic and Default Logic, Default Logic can also be interpreted in this light. However, such an interpretation clearly doesn't express the meaning of "birds fly"; instead it presents a way that a reasoner may conclude a default property of an individual. Note however that in this case it is not clear how one may draw *appropriate* conclusions; for example one would have to "program" a notion of specificity between defaults.

The next interpretation ("birds normally fly" or perhaps "the normal bird flies") is arguably closer to what is *meant* by "birds fly", since it seems to be simply true

---

[7]However Reiter [1980] suggests that the default rule $\dfrac{Bird(x):Fly(x)}{Fly(x)}$ is intended to represent "most birds fly". What seems to be more appropriate is to say that if one accepts that "most birds fly" or "the large majority of birds fly" is true, then the default rule will let one jump to the conclusion that a specific bird flies.

[8]Some approaches propose the reading of "birds fly" as meaning "all but an infinitesimal number of birds fly". However, such a reading isn't just inaccurate; it seems to be simply wrong: Clearly there are significant numbers of birds that do not fly, while "birds fly" is true.

that birds normally fly. Conditional logics of defaults take this interpretation, and preferential or rational nonmonotonic inference relations can also be seen in this light. As indicated earlier, the issue here, assuming that one is happy with a given logic of defaults, is how to reason about the properties of specific individuals.

It can be noted however that this notion of normality has the following problem when it comes to reasoning about default properties: Consider yet again birds and their (default) properties. Presumably one would agree that a penguin should be concluded to have feathers. However, a penguin is clearly not a normal bird (it doesn't fly, for one thing) and so one could not use the statement "the normal bird has feathers" to reason about penguins. On the other hand, there seems to be no problem in asserting that penguins should be concluded to have feathers, since birds have feathers. This problem is pointed out by Carlson and Pelletier [1995],[9] where it is noted that presumably and hopefully every human being is exceptional in some fashion. But then no human is "normal" per se, and so one cannot directly appeal to a global notion of normality in concluding default properties. We expand on this also in Section 4.

The fourth interpretation refers to a prototype. In saying that "the prototypical bird flies", roughly one has an idea of the notion of a prototypical bird, or best or typical instance of the class of birds [Rosch, 1978]. Notions of prototypicality then are descriptive or contingent; the prototypical bird is essentially the "best" representative of the set of birds. Arguably, in reasoning about default properties, we want to go beyond notions of similarity to a prototype, which is to say, default reasoning is more than similarity to a given prototype.

The final interpretation, "birds generally or usually fly" is perhaps ambiguous. On the one hand, it can be read as a statement in qualitative probability, analogous to "most birds fly", in which case the earlier comments apply. On the other hand, it might be read as "the usual bird flies" which would seems to be roughly synonymous with "the normal bird flies".

This discussion of possible interpretations of defaults is not intended to be exhaustive; and it is quite possible that there are interpretations that have been missed. The discussion does emphasize the (obvious) point that "birds fly" is ambiguous. Moreover, given that we are interested in the meaning of a default such as "birds fly", we can rule out some of these informal interpretations. So, while we might agree that "most birds fly" is true, it doesn't capture the meaning of "birds fly" – for example if we were to arrange that all existing birds be held down, "birds fly" would still be true. Similarly, autoepistemic interpretations are inadequate to represent the meaning of "birds fly" (since certainly "birds fly" would be true even if there were no believers to hold beliefs about birds). Consequently we focus on the "normality" interpretation, and suggest that "birds normally fly" is what is *meant* by "birds fly".

3   Defaults: Two Issues

We next consider two issues regarding defaults. The first issue is representational, and concerns the dual aspects of defaults, as bearers of truth values and as things that are used for drawing inferences. As a specific consequence of this distinction, we also dis-

---

[9]See also the paper by Poole [1991].

cuss the way in which formalisms for dealing with defaults have handled individuals. The second issue concerns reasoning, specifically whether a default is best informally regarded as a "weak rule of inference" or a "weak material conditional".

3.1 Defaults: Representation vs. Reasoning

Let's reconsider the notion of a default. We can note several facts about a default such as "'birds fly". First, it asserts something about the external world: "birds fly" is clearly true while equally "cows fly" is false. As well, if one accepts that "birds fly" is true, then one rationally would accept other defaults, such as "birds fly or swim". Second, the truth or falsity of a default is independent of there being any believers. If human beings (and their knowledge bases) were to disappear, birds would still fly, and "birds fly" would still be true. Third, a default expresses a property of individuals belonging to a particular class, while not mentioning any specific individuals. Finally, a default says nothing about how one may obtain a default conclusion about an individual.

This suggests that a default asserts a general property about members of a class, and it is the task of nonmonotonic inference to draw conclusions about specific individuals based on a collection of defaults. So we can distinguish a default *assertion* or *proposition*, from an *inference* involving a default. To spell things out:

- A default is either *true* or *false*. One may derive defaults from a set of defaults; however such derivations say nothing about particular individuals nor properties of specific individuals.

- A default inference is either *sound* or *unsound* (or better perhaps, since we're dealing with nonmonotonicity, *rational* or *not rational*). A default inference ascribes a property to an individual.

Consequently we distinguish two types of reasoning:

- With defaults (as assertions).

  This is the realm of conditional logics of normality. Nonmonotonic inference relations can also be seen in this light (given the correspondence results between conditional logics and nonmonotonic inference relations).

- Applying defaults, to give conclusions about individuals.

  This is the realm of Default Logic, and circumscription, along with the rational closure for rational nonmonotonic inference relations.

Clearly, "traditional" approaches to nonmonotonic inference (as exemplified by Default Logic and circumscription) have nothing to say about the meaning of a default, and arguably *this* is what has led to issues with obtaining the "right" inference. Conditional logics, and by extension nonmonotonic inference relations, deal with the meaning of a default as an objective entity talking about classes; it is not surprising then, as we later discuss, that such approaches have difficulties when it comes to expressing default properties about individuals.

An interesting distinction that can be made concerning default assertions compared to default inference, is that the former is essentially *semantic* while the latter is *syntactic*, in the following sense: Defaults are things that are either true or false, with respect to some larger theory. Two formulas that are true under precisely the same conditions express the same thing, and the fact that they may be written differently can be seen as an irrelevant syntactic commitment. Hence, in a logic of defaults, the formulas $(Bird \Rightarrow Fly) \wedge (Bird \Rightarrow Nest)$ and $Bird \Rightarrow (Fly \wedge Nest)$ are true in exactly the same models, and so express the same *proposition*. Arguably this is as things should be.

This is not the case for default inference. In Default Logic and circumscription, one would expect quite different outcomes given the set of defaults

$$\left\{ \frac{Bird(x) \ : \ Fly(x)}{Fly(x)}, \ \frac{Bird(x) \ : \ Nest(x)}{Nest(x)} \right\}$$

and the set

$$\left\{ \frac{Bird(x) \ : \ Fly(x) \wedge Nest(x)}{Fly(x) \wedge Nest(x)} \right\},$$

or the two circumscriptive theories:

$$\{\forall x.(Bird(x) \wedge \neg Ab_F(x)) \supset Fly(x), \quad \forall x.(Bird(x) \wedge \neg Ab_N(x)) \supset Nest(x)\}$$

and

$$\{\forall x.(Bird(x) \wedge \neg Ab_{FN}(x)) \supset (Fly(x) \wedge Nest(x))\}.$$

In particular, if one knows of a bird that it doesn't fly, the former theories would allow one to conclude that it nonetheless builds nests. The overall observation then is that a default, as an instrument for inference, is a syntactic notion; "logically equivalent" sets of defaults may give different default conclusions. Again, this is as things should be.

We next examine this distinction with respect to how approaches to default reasoning address first-order issues.

3.1.1   Defaults and First-Order Concerns

We can observe that virtually all approaches to default reasoning have problems (or at least a certain awkwardness) in the first order case. Indeed, first-order issues are often ignored (with the possible exception of circumscription) in that defaults are usually expressed in a propositional language, as

$$\frac{Bird : Fly}{Fly} \qquad \text{or} \quad (Bird \wedge \neg Ab_F) \supset Fly \quad \text{or}$$

$$Bird \mathrel{|\!\sim} Fly \quad \text{or} \quad Bird \Rightarrow Fly$$

It is unclear what is meant in these cases, unless a default is understood as applying to a specific individual. That is, the rule $\frac{Bird : Fly}{Fly}$ only makes sense if $Bird$ is regarded as standing for x-is-a-$Bird$ for understood individual x (and similarly for Fly). Indeed, in the literature, this is just how such propositional glosses are taken, with the understanding that first-order issues are orthogonal to whatever a particular paper at hand is about.

In Default Logic and circumscription, there is no problem expressing a default in first-order terms. Thus in Default Logic, one can write $\frac{Bird(x):Fly(x)}{Fly(x)}$. This rule applies to instances only, and so can be regarded as a rule schema, standing for the set of its ground instances. This has been a point of criticism of Default Logic in the past. However, given the distinction between default assertions and default inference, such a criticism, at least with regards to reasoning about default properties, seems misplaced: Default Logic has nothing to say about a default as an assertion (i.e. default rules are not things that can be true or false) but rather solely concerns inference; as we suggest below, default inference is most appropriately regarded as involving individuals. Similarly, in circumscription we can write $\forall x.(Bird(x) \wedge \neg Ab_F(x)) \supset Fly(x)$. Hence every bird, except for known exceptions, flies. As an assertion about the world, this is clearly false,[10] unless one is talking about a constrained domain such as the birds at some zoo. Nonmonotonic inference relations on the other hand have representational problems in the first-order case, since if the symbol $\vdash$ stands for an inference relation, it is not clear what an expression $Bird(x) \vdash Fly(x)$ would mean. In particular, there is no formal relation between the occurrences of free variable x on either side of the $\vdash$ symbol, although informally there is.

On the other hand, in a conditional logic, since a default is part of the object language, there is no problem in adding quantification. However, it is not immediately clear how semantically this should be carried out. Something like $\forall x.Bird(x) \Rightarrow Fly(x)$ is problematic [Delgrande, 1998]; as well, intuitively this formula doesn't seem to capture the idea that birds fly since, among other things, for any bird x it isn't the case that x normally flies (a penguin doesn't for example). To address issues concerning quantification and modalities, Delgrande [1998] suggests that the conditional connective $\Rightarrow$ be a variable-binding operator, and so our canonical example would be expressed $Bird(x) \Rightarrow_x Fly(x)$.

Yet another alternative is to embrace *concepts* as objects in a domain of discourse, and declare that $B \Rightarrow F$ is a formula in some logic of concepts. Since the area of description logics can be seen as addressing (monotonic) logics of concepts, a possible course of action is to define a description logic for defaults. A goal then would be to define a suitable notion of "default subsumption", writing something like $Bird \sqsubseteq_d Fly$. However, in these cases we are back to regarding defaults as assertions, and so inference regarding individuals would be a separate issue.

So to conclude this subsection, we suggested previously that defaults are expressed at a level independent of individuals, and that inferences *about* defaults are similarly independent of individuals. Moreover, defaults concern open domains, that is they encompass all past, present, future, and possible individuals. On the other hand, default inference concerns *specific* individuals, and, putting it more strongly, default inference involves reasoning about a specific individual or individuals. Thus, the argument: "*Adults are normally employed; therefore adults are normally employed or happy*" is independent of any particular individual. A default conclusion about (adult) Chris is a different matter and is on a different level. Otherwise, if these levels are conflated, this

---

[10]In this regard then, the circumscription of such a formula has the same epistemic flavor as autoepistemic logic, in that it appears to talk about individuals not *known* to be exceptional.

can lead to undesirable conclusions such as the example from circumscription "*every bird except for the known exceptions flies*".

It can be observed that this is exactly the same situation that one has in databases. Thus for example a relation schema is defined prior to there being any database instances, and integrity constraints provide general constraints, and are also expressed independently of a database instance. Querying a database involves reasoning about individuals, in that for a simple query, a set of instances satisfying the query is returned.

3.2   Defaults: Rules vs. Conditionals

If we consider default inference, there are two distinct interpretations of a default with respect to its applicability. On the one hand, applying a default can be regarded as employing something like a defeasible rule of inference. Default logic falls into this category and, indeed, defaults in Default Logic have been referred to as "domain specific rules of inference". Hence if one knows that an individual is a bird then, lacking information to the contrary, one concludes that it flies. If one knows of an individual that it does not fly, then nothing can be concluded about birdhood.

On the other hand, applying a default can be regarded as reasoning with a weak or defeasible material conditional. Circumscription is in this category; for a formula such as $(Bird \wedge \neg Ab_F) \supset Fly$, if circumscribing yields that Ab_F is false, then one ends up effectively with a material conditional $Bird \supset Fly$ which, if $Bird$ is true, allows one to deduce Fly. And if one knows that $\neg Fly$ is true, one can conclude $\neg Bird$. Without going into details, the standard way of closing a (rational) nonmonotonic inference relation, given by the rational closure, also exhibits material-conditional-like behaviour in the absence of exceptional conditions as does, for example, conditional entailment.

Both interpretations have received criticisms or can be shown to lead to unfortunate properties. For example, in the case of Default Logic, one cannot reason by cases: given that birds normally fly, as do bats, and given that an individual is either a bird or bat, one cannot conclude that it flies. However, approaches that behave like weak material conditionals also have difficulties. Consider the defaults that if someone gets a salary increase then they're normally happy, and if they break their leg then they're not happy; also assume that $Chris$ gets a salary increase. In a circumscriptive abnormality theory this can be expressed as follows:

$$\forall x.(Raise(x) \wedge \neg Ab_R(x)) \supset Happy(x),$$
$$\forall x.(BreakLeg(x) \wedge \neg Ab_B(x)) \supset \neg Happy(x),$$
$$Raise(chris)$$

Given nothing else, we conclude $\neg BreakLeg(chris)$. This is clearly an undesirable consequence. We get similar problems with the rational closure, conditional entailment, and other such approaches.[11] This last example also appears to be fatal. In

---

[11] A possible rejoinder is that such approaches are not intended to be used for reasoning about normality properties of individuals. Such a rejoinder is well taken. However, different approaches may nevertheless be examined with respect to their overall applicability in different situations.

circumscription, for example, it is not at all clear how this behaviour can be blocked, or even if it *can* be blocked. Consequently, we take this example as being decisive and so accept that:

> *With respect to defaults, inference is rule-like and not (material) conditional-like.*

4 Defaults as Naïve Scientific Theories

The conclusion of the previous section is problematic: Most approaches to default reasoning fall into the weak-material-conditional category and those that don't, namely Default Logic and related systems, provide only a basic mechanism for inference that does not reflect how one would wish to reason with defaults. This then suggests that it would be instructive to first study the *phenomena* modelled by approaches to defaults – that is, consider what it is in the world that's being represented, and then use such a study to drive a study of what constitutes *desirable* default inferences. To this end, in this section I argue that defaults are best regarded as statements in a naïve scientific theory. (For an excellent discussion, see the paper by Putnam [1975].) From this I argue that *relevance* is the key notion needed to formalize default inference.

4.1 What *is* a Default?

Assume that we live in a Newtonian universe, and consider the following assertion.

Example 1 Planets move in ellipses. □

In our Newtonian universe, this statement would be accepted as true. However, on the other hand, no planet would *ever* be observed to move in an ellipse. Rather, if one plotted the path of a planet, it would be observed to more or less follow an ellipse. If asked about this discrepancy, an astronomer would excuse the error by saying that the measuring instruments weren't exact, or that there was atmospheric interference, or that there were other bodies whose gravitational influence needed to be taken into account, or some such conditions interfered with the observations. If pressed, the astronomer might assert that if the universe consisted only of a star and its orbiting planet, *only then* would the planet would move in a perfect ellipse. Nevertheless (back in the real world), the statement "planets move in ellipses" is nonetheless useful: it can be accepted as true, in that any deviation from an ellipse can be explained in principle by other real or hypothesized bodies, instrument errors, etc. Moreover the statement has predictive value: the orbit of the moon can still be calculated very accurately, and in fact deviations in Uranus' orbit led to the discovery of Neptune.

"Planets move in ellipses" is clearly a scientific assertion, and can be considered as part of a naïve scientific theory, in that it is a qualitative outcome of a more precise expression (using the inverse square law of gravitation) of an underlying theory. As well, it clearly has the flavor of a default.

Consider the next example:

Example 2

1. Brass doorknobs disinfect themselves of bacteria within eight hours.
2. Copper conducts electricity.
3. Copper has atomic number 29.

$\square$

The first statement certainly sounds like a default, as does the second. Both in fact are true[12] though both allow exceptions. Copper wire immersed in water does not conduct electricity, for example. However, the third statement is quite different in character. In particular, it is *definitional*, and specifies an *essential* property of copper. (Thus a mass of atoms each with atomic number 28 isn't an exceptional chunk of copper; rather it is nickel.) In fact, the properties of copper can be determined or justified via its atomic structure. That is, atomic structure allows a precise definition of an element, and this can be used to give an *account* of the previous default statements.

So arguably, a default such as "birds fly" is an assertion in a naïve scientific theory, the same way that "copper conducts electricity" is. Similarly, "birds fly" asserts that in some sense *flight* is part of the meaning of *bird*. Or, phrased differently, if we had a complete theory of birds, we could exactly account for flight as a property of birds. Note that, by this account, it is possible for "birds fly" to be true, while no existing bird in fact flies. So while this provides a means of determining the meaning of "birds fly", it has nothing to say about default inference.

4.2   Representing Defaults

The previous section raises the question: how does one reason about sentences in a "naïve scientific theory"? Arguably there is already a logic for defaults, given by the so-called "conservative core" [Pearl, 1989] and (re)discovered by various researchers or appearing under different guises, see the works by Adams [1975], Pearl [1988], Kraus, Lehmann, and Magidor [1990], Lamarre [1991], Boutilier [1992], and Dubois et al. [1994], among others. This also is the system of preferential reasoning Kraus et al. [1990] referred to earlier with respect to nonmonotonic inference relations. In the version described here, a (weak) conditional operator $\Rightarrow$ is introduced into propositional logic, as we've already seen. The operator $\Rightarrow$ is a binary modal operator, and its semantics is given in terms of a standard Kripke structure, where the accessibility relation is given by a preorder over possible worlds. This preorder reflects a notion of relative normality between possible worlds. Then, informally, $\alpha \Rightarrow \beta$ is true at a world, just if β is true at all least α worlds. The intuition then is that $\alpha \Rightarrow \beta$ is true at a world, just if, looking at the "most normal" α worlds, β is true at all these worlds. Hence "birds fly" is true just if in the least exceptional worlds (and so ignoring things like being a penguin, having a broken wing, etc.) in which there are birds, birds fly. Formally there is little to add:[13]

Sentences are interpreted in terms of a *model* $M = \langle W, \leq, P \rangle$ where:
1. W is a set (of possible worlds),

---

[12]Brass is a copper compound and copper has germicidal properties.

[13]The point in providing a sketch of a formal development isn't necessarily to establish a definitive logic for defaults; while it (or a slightly stronger logic) is the accepted account, it is possible that someone will come along with a superior account of defaults. If this were the case then the discussion here would remain unchanged.

2. the accessibility relation $\leq\ \subseteq\ W \times W$ is transitive and reflexive, and
3. $P : \mathbf{P} \mapsto 2^W$.

Truth conditions for the standard connectives are as in propositional logic, while for the weak conditional we have:

$$\models^M_w \alpha \Rightarrow \beta \text{ iff: for every } w_1 \in min(\alpha, \leq) \text{ we have } w_1 \models \beta$$

where $min(\alpha, \leq)$ is the set of least worlds according to $\leq$ in which α is true.

So based on this and the previous discussion, we can regard a default as a *counterfactual normative* statement. That is "birds fly" can be interpreted as, "for any individual bird x, in the most normal of possible affairs, x would fly" or "if x were a normal bird, then x would fly". One can then make a nonmonotonic inference by assuming that, given a set of defaults, states of affairs are ranked as normal as consistently possible with those defaults, and that given contingent information, the actual world is among those ranked least in which the contingent information is true.[14]

The difficulty is that this doesn't quite work for inference. Thus, given that one agrees that a normal bird flies, has feathers, builds a nest, etc. then if one knows only that an individual is a bird, then indeed it will be concluded that the individual flies, has feathers, builds a nest, etc. A problem arises however if one knows that an individual bird does not fly. Then this individual can't be a normal bird, and so one can't use the default that normal birds build nests. This suggests that nonmonotonic inference based on a notion of strict minimality of worlds, based in turn on aggregated normality information, is not entirely appropriate for inference involving defaults. The next subsection proposes an alternative.

4.3 Reasoning with Defaults of Normality

Consider again naïve scientific theories, and consider a length x of copper wire. We would conclude that x conducts electricity if

1. we had no further information;
2. we knew only that it was mined at Copper Mountain;[15] or
3. we knew only that it had bends in it.

We would not conclude that x conducts electricity if

1. we tested it and it didn't conduct electricity;
2. we knew it had significant impurities; or
3. it was immersed in water.

Thus, for limiting cases, if all we knew was that the antecedent of a default were true then we would apply the default; if we knew that the consequence were false, we would not apply it. Otherwise, one might note that the location where the wire was mined and the fact it has bends are *irrelevant* with respect to conducting electricity, while the presence of impurities and water are clearly *relevant*. So essentially we want to say of x that it conducts electricity if there is nothing known that is *relevant to it not*

---

[14] In fact this is a sketch of the intuitions underlying the work by Pearl [1990]. The rational closure [Lehmann and Magidor, 1992] is founded on differing intuitions, though in a strong sense the same inferences are obtained. Lehmann and Magidor [1992, p. 28] also suggest that "any reasonable system should endorse any assertion contained in the rational closure".

[15] in southern British Columbia, Canada where indeed copper is mined.

conducting electricity. This indicates that a theory of *relevance* (or perhaps *reasons* [Horty, 2010]) is the appropriate notion needed for default inference.

Relevance   An incorporation of relevance represents a shift in how defaults would be handled. Previously, for default $\alpha \Rightarrow \beta$ (or nonmonotonic inference relation $\alpha \mathrel{\vdash}\beta$), default inference was effected by assuming that the present state of affairs was as normal as consistently possible. Thus if one knew of an individual only that it was a bird, then one would conclude by default that it flies. If one knew also that it didn't build nests, then the normality assumption would no longer hold (since a non-nest-building bird isn't normal) and so either one would lose the inference about flying, or else it would have to be restored by other means, such as the lexicographic closure [Benferhat, Cayrol, Dubois, Lang, and Prade, 1993, Lehmann, 1995].

Relevance on the other hand appears most naturally, at least in this context [Delgrande and Pelletier, 1998], to be a ternary relation: one might say for example that having a broken wing is relevant to a bird flying. Intuitively, a default $\alpha \Rightarrow \beta$ provides a means of concluding β from α. Roughly, one would want to say that β can be concluded on the basis of α if there is nothing blocking the inference, or if there is no *reason* that the inference should be blocked, or if there is nothing *relevant* that would lead one to not draw the conclusion given in the consequent.

It might seem that this line of intuitions ultimately leads back to circumscriptive abnormality theories perhaps, or consistency conditions as found in Default Logic. Thus, so the argument might run, in an abnormality theory one says that a bird flies unless it is in some fashion abnormal with respect to flight and surely (so the argument might run) this is just another way of saying that there is no known reason for it to not fly, or there is nothing relevant known concerning flight. While it is true that things could be phrased in terms of abnormality (or, for that matter, consistency), there is a big difference: we are now working within a logic of conditionals, and not an augmentation of classical logic.

Consider how we might formalize a notion of relevance toward default inference. In outline, one wants to say something like:

Informal Definition:

> Given: a set of defaults $\mathcal{T}$ and facts $\mathcal{F}$.
>
> Conclude β by default if:
>
> 1. There is $\alpha \Rightarrow \beta$ where $\mathcal{T} \models \alpha \Rightarrow \beta$ and $\mathcal{F} \models \alpha$.
> 2. If $\mathcal{T} \models \gamma \Rightarrow \neg\beta$ where $\mathcal{F} \models \gamma$ then $\mathcal{T} \cup \mathcal{F} \models \alpha \supset \gamma$.

The informal definition says that β can be concluded if

1. there is a reason to do so, and

2. there is nothing *relevant* blocking the inference.

Consider where we have the defaults that birds fly, animals do not fly, and birds with broken wings don't fly, along with the fact that birds are animals:

$$Bird \Rightarrow Fly, \ Animal \Rightarrow \neg Fly,$$

$$Bird \land BrokenWing \Rightarrow \neg Fly, \ \Box(Bird \supset Animal)$$

If we are given that an individual is a bird (α in the informal definition), then we will conclude that it flies (β). Although the individual is also an animal (γ) and animals don't fly, the notion of being a bird is more specific than that of being an animal (i.e. $\mathcal{F} \models \alpha \supset \gamma$). On the other hand, if the bird has an injured wing, then clearly there is relevant information (viz. $\gamma = Bird \land BrokenWing$) as to why it does not fly by default.

Relevance: Other Conditionals In the previous section, we described the "standard" logic of defaults, given as a specific conditional logic. This logic is but one of a large family of conditional logics, where conditional logics have been used also to formalize notions including counterfactuals, deontics, hypotheticals, causality, etc. [Lewis, 1973, Chellas, 1980, Nute, 1984]. Although such notions haven't received the attention of normality defaults, it seems clear that one can reason by default in any such logic. Thus, for example, if one should not speed when driving a car, but that it is permissible to speed if it allows one to avoid an accident, then if in fact one is driving then it is a reasonable conclusion, all other things being equal, that one should not speed.

As well, a notion of *relevance* seems equally pertinent in reasoning with these other conditionals. Thus although in general one should not speed, but one may speed if speeding allows one to avoid an accident, then avoiding an accident is a relevant factor in determining how fast one may drive. On the other hand, the color of one's car is not relevant, and the fact that one is late for an appointment should not be relevant. This suggest that *relevance* is *the* appropriate mechanism for weak conditionals and default inference in general. As well, it seems that a general account of relevance may lead to a satisfactory account for default reasoning, as well as defeasible reasoning with counterfactuals, deontics, causality, etc. Last, it can be noted that the informal definition given above is expressed independently of any specific logic, and so an overarching account of relevance (as a ternary relation with respect to weak conditionals) may provide a unifying framework for an extended notion of defeasible reasoning.

5   Conclusion

This paper has examined the notion of (normality) defaults in nonmonotonic reasoning. We noted that early work, as exemplified by Default Logic and circumscription, focussed on developing inference mechanisms and then on using such mechanisms to try to suitably encode reasoning with defaults. Similar remarks apply to subsequent work, represented by applications of nonmonotonic inference relations and conditional logics to defaults, even though such work began with a semantic account of defaults. Since no extant approach satisfactorily captures default reasoning, we suggested that a suitable strategy is to step back and consider first the phenomenon that is being modelled, that is, determining what a default such as "birds fly" means. This requires distinguishing the *representation* (or assertion) of a default from an inference involving the application of the default. The former is either true of false, while the latter is (in the case of defaults) rational or not rational.

Along the way, I suggested or noted that:

1. A default is essentially a semantic notion, in that a default or set of defaults can be replaced by logically equivalent defaults without altering the meaning of the theory. Default inference on the other hand is syntactic, in that replacing a set of defaults with an equivalent set of defaults may result in different default conclusions.

2. A general first-order default, as a proposition, is independent of specific individuals and applies to open domains. Default reasoning on the other hand concerns inference of properties of individuals. Thus, "birds fly" is expressed independently of any individual, but default inference concerns specific individuals. Reasoning about a population as a whole is best approached at the level of reasoning *about* defaults. Mixing these two levels yields conclusions such as "every bird except known exceptions flies".

   This split is analogous to that in database systems where the database schema and integrity constraints corresponds to defaults (along with other general information concerning a domain), while querying a database instance corresponds to default reasoning about individuals over contingent information.

3. Default reasoning is analogous to reasoning with a rule, not a version of a weak (material) conditional. This suggests that we have a ways to go with respect to getting the inferences right, since most approaches to default inference are closer to the weak-(material-)conditional interpretation. This latter group includes approaches using circumscription, as well as conditional entailment, and the rational closure and related approaches. Default Logic obviously involves rules, but in this case one has a general inference mechanism only, but where there is little connection between the inference mechanism and how one might want to reason with defaults.

I argued that an appropriate theory of defaults involves adopting (or specifying) an appropriate *conditional logic* for representing normality defaults. This logic might well be provided by the so-called "conservative core", or a slightly stronger variant given by a conditional logic based on a notion of normality reflected by a total preorder over possible worlds.[16]

Given such a logic, one can then ask *What are the principles that justify a default inference?* For normality defaults, default inference hinges on a formalization of *relevance* or *reasons*. As well, this notion of founding default inference on relevant properties also appears applicable to the full range of conditional logics, and so applicable to approaches for reasoning with counterfactuals, norms, deontics, causality, etc.

Acknowledgements

I would like to thank Marc Denecker, whose discussions on this topic got me thinking about the subject of this paper. As well, I thank Jeff Horty and Jeff Pelletier along with

---

[16] And so corresponding to the *rational* systems of Lehmann and Magidor [1992].

the other Nonmon@30 participants (and in particular Didier Dubois, Alex Bochman and Daniel Lehmann) for helpful and often enlightening discussions. I thank a reviewer of this paper for helpful comments. The fact that "snow is white" is a generic statement is noted by Carlson and Pelletier [1995], from where I got it. Finally, thanks to Gerd, Victor, and Mirek for the terrific conference that led to this book.

References

Special issue on nonmonotonic reasoning. *Artificial Intelligence*, 13(1-2), 1980.

E. Adams. *The Logic of Conditionals.* D. Reidel Publishing Co., Dordrecht, Holland, 1975.

G. Antoniou. *Nonmonotonic Reasoning.* MIT Press, 1997.

C. Baral. *Knowledge Representation, Reasoning and Declarative Problem Solving.* Cambridge University Press, 2003.

S. Benferhat, C. Cayrol, D. Dubois, J. Lang, and H. Prade. Inconsistency management and prioritized syntax-based entailment. In *Proceedings of the International Joint Conference on Artificial Intelligence*, pages 640–645, Chambéry, Fr., 1993.

C. Boutilier. *Conditional Logics for Default Reasoning and Belief Revision.* PhD thesis, Department of Computer Science, University of Toronto, 1992.

C. Boutilier. Conditional logics of normality: A modal approach. *Artificial Intelligence*, 68(1): 87–154, 1994.

G. Brewka. Cumulative default logic: In defense of nonmonotonic inference rules. *Artificial Intelligence*, 50(2):183–205, 1991a.

G. Brewka. *Nonmonotonic Reasoning: Logical Foundations of Commonsense.* Cambridge University Press, Cambridge, 1991b.

G. Brewka, I. Niemelä, and M. Truszczynski. Nonmonotonic reasoning. In F. van Harmelen, V. Lifschitz, and B. Porter, editors, *Handbook of Knowledge Representation*, pages 239–284. Elsevier Science, San Diego, USA, 2007.

G. N. Carlson and F. J. Pelletier, editors. *The Generic Book.* University of Chicago Press, Chicago, 1995.

B.F. Chellas. *Modal Logic.* Cambridge University Press, 1980.

J.P. Delgrande. An approach to default reasoning based on a first-order conditional logic: Revised report. *Artificial Intelligence*, 36(1):63–90, 1988.

J.P. Delgrande. On first-order conditional logics. *Artificial Intelligence*, 105(1-2):105–137, 1998.

J.P. Delgrande and J. Pelletier. A formal analysis of relevance. *Erkenntnis*, 49(2):137–173, 1998.

J.P. Delgrande and T. Schaub. Compiling specificity into approaches to nonmonotonic reasoning. *Artificial Intelligence*, 90(1-2):301–348, 1997.

J.P. Delgrande, T. Schaub, and K. Jackson. Alternative approaches to default logic. *Artificial Intelligence*, 70(1-2):167–237, October 1995.

M. Denecker, V. W. Marek, and M. Truszczynski. Uniform semantic treatment of default and autoepistemic logics. *Artificial Intelligence*, 143(1):79–122, 2003.

D. Dubois, J. Lang, and H. Prade. Possibilistic logic. In D. M. Gabbay, C. J. Hogger, and J. A. Robinson, editors, *Nonmonotonic Reasoning and Uncertain Reasoning*, volume 3 of *Handbook of Logic in Artificial Intelligence and Logic Programming*, pages 439–513. Oxford, 1994.

H. Geffner and J. Pearl. Conditional entailment: Bridging two approaches to default reasoning. *Artificial Intelligence*, 53(2-3):209–244, 1992.

M. Gelfond and V. Lifschitz. The stable model semantics for logic programming. In R. Kowalski and K. Bowen, editors, *Proceedings of the Fifth International Conference and Symposium of Logic Programming (ICLP'88)*, pages 1070–1080. The MIT Press, 1988.

M. Gelfond and V. Lifschitz. Classical negation in logic programs and deductive databases. *New Generation Computing*, 9:365–385, 1991.

S. Hanks and D. McDermott. Default reasoning, nonmonotonic logics, and the frame problem. In *Proceedings of the AAAI National Conference on Artificial Intelligence*, pages 328–333, 1986.

J.F. Horty. Some direct theories of nonmonotonic inheritance. In D.M. Gabbay, C.J. Hogger, and J.A. Robinson, editors, *Nonmonotonic Reasoning and Uncertain Reasoning*, volume 3 of *Handbook of Logic in Artificial Intelligence and Logic Programming*, pages 111–187. Oxford: Clarendon Press, 1994.

J. F. Horty. Reasons as defaults. Draft, 2010.

S. Kraus, D. Lehmann, and M. Magidor. Nonmonotonic reasoning, preferential models and cumulative logics. *Artificial Intelligence*, 44(1-2):167–207, 1990.

H.E. Kyburg, Jr. Believing on the basis of evidence. *Computational Intelligence*, 10(1):3–20, 1994.

P. Lamarre. S4 as the conditional logic of nonmonotonicity. In *Proceedings of the Second International Conference on the Principles of Knowledge Representation and Reasoning*, pages 357–367, Cambridge, MA, April 1991.

D. Lehmann. Another perspective on default reasoning. *Annals of Mathematics and Artificial Intelligence*, 15(1):61–82, 1995.

D. Lehmann and M. Magidor. What does a conditional knowledge base entail? *Artificial Intelligence*, 55(1):1–60, 1992.

D. Lewis. *Counterfactuals*. Harvard University Press, 1973.

V. Lifschitz. Computing circumscription. In *Proceedings of the International Joint Conference on Artificial Intelligence*, pages 121–127, Los Angeles, 1985.

W. Łukaszewicz. Considerations on default logic: An alternative approach. *Computational Intelligence*, 4(1):1–16, Jan. 1988.

J. McCarthy. Circumscription – a form of non-monotonic reasoning. *Artificial Intelligence*, 13: 27–39, 1980.

J. McCarthy. Applications of circumscription to formalizing common-sense knowledge. *Artificial Intelligence*, 28:89–116, 1986.

A. Mikitiuk and M. Truszczynski. Rational default logic and disjunctive logic programming. In A. Nerode and L. Pereira, editors, *Proceedings of the Second International Workshop on Logic Programming and Non-monotonic Reasoning.*, pages 283–299. The MIT Press, 1993.

R.C. Moore. Semantical considerations on nonmonotonic logic. *Artificial Intelligence*, 25: 75–94, 1985.

D. Nute. Conditional logic. In D. Gabbay and F. Guenthner, editors, *Handbook of Philosophical Logic*, volume 2, pages 387–439. D. Reidel Pub. Co., 1984.

J. Pearl. *Probabilistic Reasoning in Intelligent Systems: Networks of Plausible Inference.* Morgan Kaufman, San Mateo, CA, 1988.

J. Pearl. Probabilistic semantics for nonmonotonic reasoning: A survey. In *Proceedings of the First International Conference on the Principles of Knowledge Representation and Reasoning*, pages 505–516, Toronto, May 1989. Morgan Kaufman.

J. Pearl. System Z: A natural ordering of defaults with tractable applications to nonmonotonic reasoning. In R. Parikh, editor, *Proc. of the Third Conference on Theoretical Aspects of Reasoning About Knowledge*, pages 121–135, Pacific Grove, Ca., 1990. Morgan Kaufmann Publishers.

D.L. Poole. A logical framework for default reasoning. *Artificial Intelligence*, 36(1):27–48, 1988.

D.L. Poole. The effect of knowledge on belief: Conditioning, specificity, and the lottery paradox in default reasoning. *Artificial Intelligence*, 49(1-3):281–307, 1991.

H. Putnam. The meaning of 'meaning'. In *Mind, Language and Reality: Philosophical Papers Volume II*, pages 215–271. Cambridge University Press, 1975.

R. Reiter. A logic for default reasoning. *Artificial Intelligence*, 13(1-2):81–132, 1980.

E. Rosch. Principles of categorization. In E. Rosch and B.B. Lloyds, editors, *Cognition and Categorization*. Lawrence Erlbaum Associates, 1978.

Reiter's Default Logic Is a Logic of Autoepistemic Reasoning And a Good One, Too

Marc Denecker
Department of Computer Science
K.U. Leuven
Celestijnenlaan 200A
B-3001 Heverlee, Belgium

Victor W. Marek
Mirosław Truszczyński
Department of Computer Science
University of Kentucky
Lexington, KY 40506-0633, USA

Abstract: A fact apparently not observed earlier in the literature of nonmonotonic reasoning is that Reiter, in his default logic paper, did not directly formalize *informal* defaults. Instead, he translated a default into a certain natural language proposition and provided a formalization of the latter. A few years later, Moore noted that propositions like the one used by Reiter are fundamentally different than defaults and exhibit a certain *autoepistemic* nature. Thus, Reiter had developed his default logic as a formalization of autoepistemic propositions rather than of defaults.

The first goal of this paper is to show that some problems of Reiter's default logic as a formal way to reason about informal defaults are directly attributable to the autoepistemic nature of default logic and to the mismatch between informal defaults and the Reiter's formal defaults, the latter being a formal expression of the autoepistemic propositions Reiter used as a representation of informal defaults.

The second goal of our paper is to compare the work of Reiter and Moore. While each of them attempted to formalize autoepistemic propositions, the modes of reasoning in their respective logics were different. We revisit Moore's and Reiter's intuitions and present them from the perspective of *autotheoremhood*, where theories can include propositions referring to the theory's own theorems. We then discuss the formalization of this perspective in the logics of Moore and Reiter, respectively, using the unifying semantic framework for default and autoepistemic logics that we developed earlier. We argue that Reiter's default logic is a better formalization of Moore's intuitions about autoepistemic propositions than Moore's own autoepistemic logic.

1 Introduction

In this volume we celebrate the publication in 1980 of the special issue of the Artificial Intelligence Journal on Nonmonotonic Reasoning that included three seminal

papers: *Logic for Default Reasoning* by Reiter [1980], *Nonmonotonic Logic I* by Mc-
Dermott and Doyle [1980], and *Circumscription — a form of nonmonotonic reasoning*
by McCarthy [1980]. While the roots of the subject go earlier in time, these papers
are universally viewed as the main catalysts for the emergence of nonmonotonic reasoning as a distinct field of study. Soon after the papers were published, nonmonotonic reasoning attracted widespread attention of researchers in the area of artificial
intelligence, and established itself firmly as an integral sub-area of knowledge representation. Over the years, the appeal of nonmonotonic reasoning went far beyond
artificial intelligence, as many of its research challenges raised fundamental questions
to philosophers and mathematical logicians, and stirred substantial interest in those
communities.

The groundbreaking paper by McCarthy and Hayes [1969] about ten years before
had captured the growing concern with the logical representation of *common sense
knowledge*. Attention focused on the representation of *defaults*, propositions that are
true for most objects, that commonly assume the form *"most A's are B's."*[1] Defaults
arise in all applications involving common sense reasoning and require specially tailored forms of reasoning. For instance, a default *"most A's are B's"* under suitable
circumstances should enable one to infer from the premise *"x is an A"* that *"x is a
B."* This inference is *defeasible*. Its consequent *"x is a B"* may be false even if its
premise *"x is an A"* is true. It may have to be withdrawn when new information is
obtained. Providing a general, formal, domain independent and elaboration tolerant
representation of defaults and an account of what inferences can be *rationally* drawn
from them was the artificial intelligence challenge of the time.

The logics proposed by McCarthy, Reiter, and McDermott and Doyle were developed in an attempt to formalize reasoning where defaults are present. They went about
it in different ways, however. McCarthy's *circumscription* extended a set of first-order
sentences with a second-order axiom asserting *minimality* of certain predicates, typically of *abnormality predicates* that capture the exceptions to defaults. This reflected
the assumption that the world deviates as little as possible from the "normal" state.
Circumscription has played a prominent role in nonmonotonic reasoning. In particular, it has been a precursor to preference logics [Shoham, 1987] that provided further
important insights into reasoning about defaults.

Reiter [1980] and McDermott and Doyle [1980], on the other hand, focused on the
inference pattern *"most A's are B's."* In Reiter's words [Reiter, 1980, p. 82]:

'We take it [that is, the default "Most birds can fly" — DMT to mean
something like "If an x is a bird, then in the absence of any information
to the contrary, infer that x can fly."'

Thus, Reiter (and also McDermott and Doyle) quite literally equated a default *"most
A's are B's"* with an inference rule that involves, besides the premise *"x is an A"*,
an additional premise "there is no information to the contrary" or, more specifically,
"there is no information indicating that "x is not a B." The role of this latter premise,
a *consistency* condition, is to ensure the rationality of applying the default. In logic,

---

[1]In this paper, we interpret the term "default" as an informal statement *"most A's are B's"* [Reiter,
1980]. The term is sometimes interpreted more broadly to capture *communication conventions, frame
axioms* in temporal reasoning, or statements such as *"normally* or *typically, A's are B's"*.

inference rules are meta-logical objects that are not expressed in a logical language. Reiter, McDermott and Doyle sought to develop a logic in which such meta-logical inference rules could be stated *in* the logic itself. They equipped their logics with a suitable *modal operator* (in the case of Reiter, embedded within "his" default expression) to be able to express the *consistency* condition and, in place of a default *"most A's are B's"*, they used the statement *"if x is an A and if it is consistent (with the available information) to assume that x is a B, then x is a B."* We will call this latter statement the *Reiter-McDermott-Doyle* (RMD, for short) proposition associated with the default.

Moore [1985] was one of the first, if not the first, who realized that defaults and their RMD propositions are of a different nature. This is how Moore [1985, p. 76] formulated RMD propositions in terms of theoremhood and non-theoremhood:

> '[In the approaches of McDermott and Doyle, and of Reiter — DMT] the inference that birds can fly is handled by having, in effect, a rule that says that, for any X, "X can fly" is a theorem if "X is a bird" is a theorem and "X cannot fly" is *not* a theorem.'

Moore then contended that RMD propositions are *autoepistemic* statements, that is, introspective statements referring to the reasoner's own belief or the theory's own theorems. He pointed out fundamental differences between the nature of default propositions and autoepistemic ones and argued that the logics developed by McDermott and Doyle [1980] and, in the follow-up paper, by McDermott [1982], are attempts at a logical formalization of of autoepistemic statements and not of defaults. Not finding the McDermott and Doyle formalisms quite adequate as autoepistemic logics, Moore [1984, 1985] proposed an alternative, the *autoepistemic logic*.

Unfortunately, Moore did not refer to the paper by Reiter [1980] but only to those by McDermott and Doyle [1980] and McDermott [1982], and his comments on this topic were not extrapolated to Reiter's logic. Neither did Moore explain what could go wrong if a default is replaced by its RMD proposition. Yet, if Moore is right then given the close correspondence between Reiter's and McDermott and Doyle's views on defaults, also Reiter's logic is an attempt at a formalization of autoepistemic rather than of default propositions. Moreover, if defaults are really fundamentally different from autoepistemic propositions, as Moore claimed, it should be possible to find demonstrable defects of Reiter's default logic for reasoning about defaults that could be attributed to the different nature of a default and of its Reiter's autoepistemic translation.

Our main objective in Section 2 is to argue that Moore was right. We show there two forms of such defects that (1) the RMD proposition is not always *sound* in the sense that inferences made from it are not always rational with respect to the original defaults, and (2) the RMD proposition is not always *complete*, that is, there are sometimes rational inferences from the original defaults that are not covered by this particular inference rule. In fact, both types of problems can be illustrated with examples long known in the literature.

In the remaining sections, we explain Reiter's default logic as a formalization of autoepistemic propositions and show that in fact, Reiter's default logic is a better formalization of Moore's intuitions than Moore's own autoepistemic logic. On a formal level, our investigations exploit the results on the unifying semantic framework for default logic and autoepistemic logic that we proposed earlier [Denecker, Marek, and Truszczyński, 2003]. That work was based on a algebraic fixpoint theory for non-monotone operators [Denecker, Marek, and Truszczyński, 2000]. We show that the different dialects of autoepistemic reasoning stemming from our informal analysis can be given a principled formalization using these algebraic techniques. In our overview, we will stress the view on autoepistemic logic as a logic of *autotheoremhood*, in which theories can include propositions referring to the theory's own theorems.

Some history. We mentioned that Moore's comments concerning the RMD proposition and the formalisms by McDermott and Doyle [1980] and McDermott [1982] have never been applied to Reiter's logic. For example, Konolige [1988], who was the first to investigate the formal link between autoepistemic reasoning and default logic, wrote that "*the motivation and formal character of these two systems [Reiter's default and Moore's autoepistemic logics – DMT] are different*". This bypasses the fact that Reiter, as we have seen, starts his enterprise of building default logic after translating a default into a proposition which Moore later identified as an autoepistemic proposition.

There may be several reasons why Moore's comments have never been extrapolated to Reiter's logic. As mentioned before, one is that Moore did not refer to the paper by Reiter [1980] but only to the papers by McDermott and Doyle [1980] and McDermott [1982]. In addition, the logics of Reiter and, respectively, McDermott and Doyle were quite different; the formal connection was not known at that time (mid 1980s) and was established only about five years later [Truszczyński, 1991]. Also autoepistemic and default logics seemed to be quite different [Marek and Truszczyński, 1989], and eventually turned out to be different in a certain precise sense [Gottlob, 1995]. Moreover, the intuitions underlying the nonmonotonic logics of the time had not been so clearly articulated, not even in Moore's work as we will see later in the paper, and were not easy to formalize. This was clearly demonstrated about ten years later by Halpern [1997], who reexamined the intuitions presented in the original papers of default logic, autoepistemic logic and Levesque's [1990] related logic of only knowing and showed gaps and ambiguities in these intuitions, and various non-equivalent ways in which they could be formalized.

As a result, the nature of autoepistemic propositions, its relationship to defaults and what may go wrong when the latter are encoded by the first, was never well understood. The relevance of Moore's claims for Reiter's default logic has never become generally acknowledged. Reiter's logic has never been thought of and has never been truly analyzed as a formalization of autoepistemic reasoning. The influence of Reiter's paper has been so large, that even today, the default "*most A's are B's*" and the statement "*if x is an A and if it is consistent to assume that x is a B, then x is a B*"[2] are still considered synonymous in some parts of the nonmonotonic reasoning community. Yet, in fact, they are quite different and, more importantly, a logical representation of the second is unsatisfactory for reasoning about the first.

---

[2] Or its propositional version "*if A and if it is consistent to assume B, then B*".

2 Reiter's Defaults Are Not Defaults But Autoepistemic Statements

Our goal below is to justify the claim in the title of the section. To avoid confusion, we emphasize that by a *default* we mean an informal expression of the type *most A's are B's*. In Reiter's approach (similarly in that of McDermott and Doyle), the default is first translated into an *RMD proposition if x is an A and if it is consistent with the available information to assume that x is a B, then x is a B*, which is then expressed by a *Reiter's default expression* in default logic:

$$\frac{A(x) : M\ B(x)}{B(x)}.$$

To explain the section title, let us assume a setting in which a human expert has knowledge about a domain that consists of propositions and defaults. In the approach of Reiter (the same applies to McDermott and Doyle), the expert builds a knowledge base T by including in T formal representations of the propositions (given as formulas in the language of classical logic) and of RMD propositions of the defaults (given by the corresponding Reiter's default expressions). The presence of Reiter's default expressions in T means that T contains propositions referring to its own information content, i.e., to what is consistent with T, or dually to what T entails or does not entail. Moore [1985] called such reflexive propositions *autoepistemic* and argued that they statements could be phrased in terms of theorems and non-theorems of T.

Reiter developed a default expression as a formal expression of the RMD proposition rather than of the default itself (the same holds for McDermott and Doyle). This is why this logic expression does not capture the full informal content of the default. When considered more closely, it indeed becomes apparent that a default and its RMD proposition are not equivalent or even related in a strict logical sense. A straightforward possible-world analysis reveals this. The default might be true in the actual world (say 95% of the A's are B's) but if there is just one x that is an A and not a B, and for which T has no evidence that it is not a B, the RMD proposition is false in this world and x is a witness of this. Thus, it is obvious that in many applications where a default holds, its RMD proposition does not. Conversely, the default might not hold in the actual world (few A's are in fact B's) yet the expert knows all x's that are not B's, in which case the RMD proposition is true.

A fundamental difference pointed out by Moore between defaults and autoepistemic propositions, is that the latter are naturally *nonmonotonic* but inference rules used for reasoning with them are not *defeasible*. For example, extending the knowledge base T containing an RMD proposition with new information, e.g., that some x is not a B, may indeed have a nonmonotonic effect and delete some previous inferences, e.g., that x is a B. The initial inference of x *is a B*, resulted in a fact that was false. However, that inference was not defeasible. The essential property of a defeasible inference is that it may derive a false conclusion from premises that are true in the actual world. For instance, the inference from *most A's are B's* and x is an A that x is a B is defeasible as its consequent may be false while the premises are true. In the context of our example above the theory, say T, entailed the false fact that x *is a B* from the premises (i) the RMD proposition, (ii) x is an A and (iii) T contained no evidence that

x is not a B. It was not defeasible since one of its premises was false. Indeed, the RMD proposition was false and x was a witness. The inference rules applied are not defeasible (they are, essentially, the introduction of conjunction and modus ponens). To sum up, an inference from a knowledge base involving an RMD proposition may be false but only if the RMD proposition itself is false.

To emphasize further consequences of equating defaults and RMD propositions we will look at well-known examples from the literature. First, we turn our attention to the question whether there are cases when applying the RMD proposition leads to inferences that do not seem rational (lack of "soundness" with respect to understood informally "rationality"). The Nixon Diamond example by Reiter and Criscuolo [1981] and reasoning problems with related inheritance networks illustrate the problems that arise.

Example 1 Richard M. Nixon, the 37th president of the United States, was a Republican and a Quaker. Most Republicans are hawks while most Quakers are doves (pacifists). Nobody is a hawk and a dove. Some people are neither hawks nor doves. Encoding the Reiter-McDermott-Doyle proposition of these defaults in default logic, we obtain the following theory:

$$Republican(Nixon) \wedge Quaker(Nixon)$$
$$\forall x(\neg Dove(x) \vee \neg Quaker(x))$$

$$\frac{Republican(x) : M\ Hawk(x)}{Hawk(x)} \qquad \frac{Quaker(x) : M\ Dove(x)}{Dove(x)}.$$

In default logic, this theory gives rise to two extensions. In one of them Nixon is believed to be a hawk and not a dove, in the other one, a dove and not a hawk. But is this rational? As we mentioned above, the use of an RMD-proposition is rational when it is expected to hold for most x, and hence, in absence of information, it is likely to hold for some specific x. But in the case of Nixon, we know in advance that at least one of the two "Nixon" instances of the RMD propositions has to be wrong. As to which one is wrong, without further information one could as well throw a coin. Moreover, it is not unlikely that they are both wrong and that in fact, Nixon is neither dove nor hawk. And in fact, it seems more rational not to apply any of the defaults, leading to a situation where it is not known whether Nixon is a dove, a hawk or neither. The rationale of using the RMD proposition as a substitute for the default does not hold for Nixon or any other republican quaker for that matter.   □

Example 2 Let us assume now that all quakers are republicans. In this case, the default that most quakers (say 95%) are doves is more specific than and overrules the default that most republicans (say 95%) are hawks. It is rational here to give priority to the quaker default, leading to the defeasible conclusion that Nixon is a dove. However, this conclusion cannot be derived from the RMD propositions because their consistency premise *"it is consistent to assume that x is a dove (respectively a hawk)"* is too general to take such information into account.   □

Such scenarios were studied in the context of inheritance hierarchies [Touretzky, 1986]. To reason correctly on this sort of applications using Reiter's logic, the consistency condition of the RMD propositions has to be tweaked to take the hierarchy into account and give priority to the quaker default. For example, we can reformulate the RMD proposition of the default *"most republicans are hawks"* as *"if x is known to be a republican and it is consistent to assume that he is a hawk and it is consistent to assume that he is not a quaker, then x is a hawk"*, which takes additional information into account. Such modified rules can of course be represented in default logic. After all, the logic was developed for representing (defeasible) inference rules. But, as in the examples above, they cannot be *inferred* from the RMD propositions. And the inferences that can be drawn from the RMD propositions are not always the rational ones.

The next problem that arises is of a complementary nature and concerns (lack of) completeness with respect to "rational" inferences. Are there cases where rational albeit defeasible inferences can be drawn from defaults that cannot be inferred from RMD propositions? As suggested above by our general discussion, the answer is indeed positive. After all, the RMD proposition expresses only a single and quite specific type of inference that might be associated with a default.

Example 3 As an illustration, let us consider the defaults *most Swedes are blond* and *most Japanese have black hair*. Nobody is both Swede and Japanese, or has both blond and black hair. If we learn know that Boris is a Swede or a Japanese then, given that he cannot be both Swede and Japanese, it seems rational to conclude defeasibly that Boris's hair is blond or black. In other words, defaults can (sometimes) be combined and together give rise to defeasible inference rules like:

$$\frac{\text{Boris is Swede or Japanese: } M \text{ Boris's hair is blond or black}}{\text{Boris's hair is blond or black}}.$$

If all we know is that Boris is Swede or Japanese, the conclusion of this rule cannot be drawn from the two original RMD propositions for the simple reason that for each, one of their premises is not satisfied: it is not known that Boris is a Swede, and neither is it known that he is Japanese. For instance, in the logic of Reiter, the two defaults would be encoded as

$$\frac{Swede(x) : M \, Blond(x)}{Blond(x)} \quad \text{and} \quad \frac{Japanese(x) : M \, Black(x)}{Black(x)}.$$

If we only know $Swede(Boris) \vee Japanese(Boris)$, then neither $Swede(Boris)$ nor $Japanese(Boris)$ can be established. Therefore, the premises of neither rule are established and no inference can be made. Even more, if we accept Reiter's logic as a logic of autoepistemic propositions, these conclusions *should not be drawn* from these expressions. □

This example shows a clear case of a desired defeasible inference that cannot be drawn from the rules expressed in the two RMD propositions. A default expression in Reiter's logic that would do the job has to encode explicitly the combined inference rule:

$$\frac{Swede(x) \vee Japanese(x) : M(Blond(x) \vee Black(x))}{Blond(x) \vee Black(x)}.$$

This expresses an inference rule which is not derivable from the original RMD propositions in the logics of Reiter, McDermott, Doyle, or Moore. Default logic does not support such reasoning unless the combined inference rule is explicitly encoded as well.

Example 4 Assume that we now find out that Boris has black hair. Given that he is Japanese or Swede, and given the defaults for both, it seems rational to assume that he is Japanese. Can we infer this from the combined inference rules expressed above and given that nobody can be blond and black, or Swede and Japanese? The answer is no and, consequently, yet another inference rule should be added to obtain this inference.

□

Problems of these kind were reported many times in the NMR literature and prompted attempts to "improve" Reiter's default logic so as to capture additional defeasible inferences of the informal default. This is, however, a difficult enterprise, as it starts from a logic whose semantical apparatus is developed for a very specific form of reasoning, namely autoepistemic reasoning. And while at the formal level the resulting logics [Brewka, 1991, Schaub, 1992, Lukaszewicz, 1988, Mikitiuk and Truszczyński, 1995] capture some aspects of defaults that Reiter's logic does not, also they formalize a small fragment only of what a default represents and, certainly, none has evolved into a method of reasoning about defaults. In the same time, theories in these logics entail formulas that cannot be justified from the point of view of default logic as an autoepistemic logic.

To summarize, an RMD proposition expresses one defeasible inference rule associated with a default. It often derives rational assumptions from the default but not always, and it may easily miss some useful and natural defeasible inferences. The RMD proposition is autoepistemic in nature; Reiter's original default logic is therefore a formalism for autoepistemic reasoning. As a logic in which inference rules can be expressed, default logic is quite useful for reasoning on defaults. The price to be paid is that the human expert is responsible for expressing the desired defeasible inference rules stemming from the defaults and for fine-tuning the consistency conditions of the inference rules in case of conflicting defaults. This may require substantial effort and leads to a methodology that is not elaboration tolerant.

While our discussion shows that in general, RMD propositions and Reiter's defaults do not align well with the informal concept of a default of the form *most A's are B's*, there are other nonmonotonic reasoning patterns that are correctly expressed through Reiter's defaults. In particular, patterns such as communication conventions, database or information storage conventions and policy rules in the typology of McCarthy [1986], can be expressed well by *true* autoepistemic propositions and, consequently, are correctly formalized in Reiter's logic. E.g., the convention that an airport customs database explicitly contains the nationality of only non-American passengers, is correctly specified by the Reiter default

$$\frac{: MNationality(x) = USA}{Nationality(x) = USA}.$$

Similarly, the policy rule that the departmental meetings are normally held on Wednesdays at noon, is correctly formalized by

$$\frac{: MTime(meeting) = "Wed, noon"}{Time(meeting) = "Wed, noon"}.$$

In spite of such examples, the fact remains that default logic is not a logic of defaults. Are there other logics that could be regarded as such? There have been several interesting attempts at formalizing defaults *most A's a re B's*. Most important of them focused on defaults as *conditional assertions* and on abstract nonmonotonic consequence relations [Makinson, 1989, Lehmann, 1989, Pearl, 1990, Kraus, Lehmann, and Magidor, 1990, Lehmann and Magidor, 1992]. This research direction resulted in elegant mathematical theories and deep insights into the nature of some forms of nonmonotonic reasoning. However, it is not directly related to our effort here. Thus, rather than to discuss it we refer to the papers we cited.

Instead, in the remainder of the paper, we focus on the second objective identified in the introduction. That is, we provide an informal basis to autoepistemic reasoning, we place Reiter's default logic firmly among dialects of autoepistemic reasoning, and show that Reiter's logic was a watershed point that pinpointed one of the most fundamental and most important forms of autoepistemic reasoning.

3 Studies of Relationships Between Default Logic and Autoepistemic Logic

Konolige [1988] was the first to investigate a formal link between default and autoepistemic logic. He proposed the following translation Kon from default logic to autoepistemic logic:

$$\frac{\alpha : M\beta_1, \ldots, \beta_n}{\gamma} \quad \mapsto \quad K\alpha \wedge \neg K \neg \beta_1 \wedge \cdots \wedge \neg K \neg \beta_n \to \gamma$$

and argued that Kon was equivalence preserving in the sense that default extensions of the default theory were exactly the autoepistemic expansions of its translation. This translation is intuitively appealing, essentially expressing formally the RMD proposition of the default in modal logic, and it indeed plays an important role in the story. Nevertheless, it turned out that this translation was only partially correct [Konolige, 1989]. Later, Gottlob [1995] presented a correct translation from default logic to autoepistemic logic but also proved that no *modular* translation exists. The latter result showed that these two logics are essentially different in some important aspect. As a result, the autoepistemic nature of default logic, which Moore had implicitly pointed at, and his implicit criticism on default logic as a logic of defaults were never widely acknowledged.

But Reiter's logic is just that — a logic of autoepistemic reasoning. Moreover, in many respects it is a better logic of autoepistemic reasoning than the one by Moore. Our goal now is to reconsider the intuitions of autoepistemic reasoning, to distinguish between different dialects of it and to develop principled formalizations for these dialects. In particular, we relate Reiter's and Moore's logics, and explain in what sense Reiter's logic is better than Moore's. Our discussion uses the formal results we developed in an earlier paper [Denecker et al., 2003]. There we used the algebraic fixpoint theory for arbitrary lattice operators [Denecker et al., 2000] to define four different semantics of default logic and of autoepistemic logic. This theory can be summarized as follows.

$$A : L^2 \to L^2 \qquad\qquad\qquad\qquad\qquad\qquad\qquad \text{Kripke-Kleene least fixpoint}$$
$$O_A : L \to L \qquad O_A(x) = A_1(x, x) \qquad\qquad \text{Supported fixpoints}$$
$$S_A : L \to L \qquad S_A(x) = \mathit{lfp}(A_1(\cdot, x)) \qquad \text{Stable fixpoints}$$
$$\mathcal{S}_A : L^2 \to L^2 \quad \mathcal{S}_A(x, y) = (S_A(y), S_A(x)) \qquad \text{Well-founded least fixpoint}$$

Table 8: Lattice operators and the corresponding semantics

A complete lattice $\langle L, \leq \rangle$ induces a complete bilattice $\langle L^2, \leq_p \rangle$, where $\leq_p$ is the precision order on L^2 defined as follows: $(x, y) \leq_p (u, v)$ if $x \leq u$ and $v \leq y$. Tuples (x, x) are called exact. For any $\leq_p$-monotone operator $A : L^2 \to L^2$ that is *symmetric*, that is, $A(x, y) = (u, v)$ if and only if $A(y, x) = (v, u)$, we can define three derived operators. These four operators identify four different types of fixpoints or least fixpoints (when the derived operator is monotone). They are summarized in Table 8 (where the operator $A_1(\cdot, \cdot)$ used to define O_A is the projection of A on the first coordinate).

By assumption, A is a $\leq_p$-monotone operator on L^2 and its $\leq_p$-least fixpoint is called the *Kripke-Kleene* fixpoint of A. Fixpoints of the operator O_A correspond to exact fixpoints of A (x is a fixpoint of O_A if and only if (x, x) is a fixpoint of A) and are called *supported* fixpoints of A. The operator S_A is an anti-monotone operator on L. Its fixpoints yield exact fixpoints of A (if x is a fixpoint of S_A then (x, x) is a fixpoint of A). They are called *stable* fixpoints of the operator A. It is clear that stable fixpoints are supported. The operator $\mathcal{S}_A$ is a $\leq_p$-monotone operator on L^2 and its $\leq_p$-least fixpoint is called the well-founded fixpoint of A (fixpoints of $\mathcal{S}_A$ are also fixpoints of A). The names of these fixpoints reflect the well-known semantics of logic programming, where they were first studied by means of operators on lattices. Taking Fitting's four-valued immediate consequence operator [Fitting, 1985] for A, we proved [Denecker et al., 2000] that the four different types of fixpoint correspond to four well-known semantics of logic programming: Kripke-Kleene semantics [Fitting, 1985], supported model semantics [Clark, 1978], stable semantics [Gelfond and Lifschitz, 1988] and well-founded semantics [Van Gelder et al., 1991].

This elegant picture extends to default logic and autoepistemic logic [Denecker et al., 2003]. In that paper, we identified the semantic operator $\mathcal{E}_\Delta$ for a default theory Δ, and the semantic operator $\mathcal{D}_T$ for an autoepistemic theory T. Both operators where defined on the bilattice of possible-world sets, which we introduce formally in the following section. Just as for logic programming, each operator determines three derived operators and so, for each logic we obtain four types of fixpoints, each inducing a semantics. Some of these semantics turned out to correspond to semantics proposed earlier; other semantics were new. Importantly, it turned out that the operators $\mathcal{E}_\Delta$ and $\mathcal{D}_{Kon(\Delta)}$ are identical. Hence, Konolige's mapping turned out to be equivalence preserving for *each* of the four types of semantics! Table 9 summarizes the results. The first two lines align the theories and the corresponding operators. The last four lines describe the matching semantics (the new semantics for autoepistemic and default logics obtained from this operator-based approach [Denecker et al., 2000] are in bold font).

From this purely mathematical point of view Konolige's intuition seems basically

default theory Δ	$\overset{Kon}{\longrightarrow}$	autoepistemic theory T
semantic operator $\mathcal{E}_\Delta$	$\overset{Kon}{\longrightarrow}$	semantic operator $\mathcal{D}_T$
KK-extension	$\overset{Kon}{\longrightarrow}$	KK-extension
		[Denecker et al., 1998]
Weak extensions	$\overset{Kon}{\longrightarrow}$	Moore expansions
[Marek and Truszczyński, 1989]		[Moore, 1984]
Reiter extensions	$\overset{Kon}{\longrightarrow}$	**Stable extensions**
[Reiter, 1980]		
Well-founded extension	$\overset{Kon}{\longrightarrow}$	**Well-founded extension**
[Baral and Subrahmanian, 1991]		

Table 9: The alignment of default and autoepistemic logics

right. His mapping failed to establish a correspondence between Reiter extensions and Moore expansions *only* because they are on different levels in the hierarchy of the semantics. Once we correctly align the dialects, his transformation works perfectly. Conversely, we also proved that the standard method to eliminate nested modalities in the modal logic S5 can be used to translate any autoepistemic logic theory T into a default theory that is equivalent to T under each of the four semantics.

While the non-modularity result by Gottlob [1995] had shown that default logic and autoepistemic logic are essentially different logics, our results summarized above unmistakenly point out that default and autoepistemic logics are tightly connected logical systems. They suggest that the four semantics formalize different *dialects* of autoepistemic reasoning and that Reiter and Moore formalized different dialects. Therefore, in the rest of the paper, we will view Reiter's logic simply as a fragment of modal logic, as identified by Konolige's mapping.

4 Formalizing Autoepistemic Reasoning — an Informal Perspective

In our paper [Denecker et al., 2003] we developed a purely algebraic, abstract study of semantics. The study identified the (nonmonotone) operators of autoepistemic and default logic theories, and applied the different notions of fixpoints to them. What that paper was missing was an account of what these fixpoint constructions mean at the informal level and how the different dialects in the framework differ. Being as clear as possible about the informal semantics of autoepistemic theories is essential, as it is there where problems with formal accounts start.

This is the gap that we close in the rest of this paper. To this end we first return to the original concern of Reiter, and of McDermott and Doyle. Let us suppose that we have incomplete knowledge about the actual world, represented in, say, a first order theory T, and that we know that most A's are B's. Following the Reiter, McDermott and Doyle approach, we would like to assert the following proposition:

> If for some x, $T \models A(x)$ and $B(x)$ is consistent with T (that is, $T \not\models \neg B(x)$), then $B(x)$.

In fact, we would like to express this statement *in* the logic and, moreover, to add this proposition, with its references to what T entails or does not entail, to T itself. What we obtain is a theory T that refers to its own theorems. In this view then, modal literals $K\varphi$ in an autoepistemic theory $T = \{\dots F[K\varphi]\dots\}$ are to be interpreted *informally* as statements $T \models \varphi$, and the theory T itself as having the form $T = \{\dots F[T \models \varphi]\dots\}$, emphasizing the intuition of the self-referential nature of autoepistemic theories.

This view reflects what seems to us the most precise intuition that Moore proposed: to view autoepistemic propositions as inference rules. Specializing the discussion above to the autoepistemic formula

$$K\alpha_1 \wedge \cdots \wedge K\alpha_n \wedge \neg K\beta_1 \wedge \cdots \wedge \neg K\beta_m \to \gamma \quad (1)$$

we can write it (informally) as:

$$T \models \alpha_1 \wedge \cdots \wedge T \models \alpha_n \wedge T \not\models \beta_1 \wedge \cdots \wedge T \not\models \beta_m \to \gamma,$$

and understand it (informally) as an inference rule:

> if $\alpha_1, \dots, \alpha_n$ are theorems and $\beta_1, \dots, \beta_m$ are *not* theorems (2)
> then γ holds.

which is consistent with Moore's [1985, p. 76] position we cited earlier. Alternatively, $K\varphi$ can be read as "*φ can be derived, or proven*" (again, from the theory itself), which amounts at the informal level just to a different wording. We will refer to this notion of theorem and derivation as *autotheorem* and *autoderivation*, respectively. Accordingly, we will call the basic Moore's perspective as that of *autotheoremhood*.

The autotheoremhood view can be seen as a special case of a more generic view, also proposed by Moore, based on *autoepistemic agents*. In this view which, incidentally, is the reason behind the name *autoepistemic logic*, an autoepistemic theory is seen as a set of introspective propositions, believed by the agent, about the actual world and his own beliefs about it. The crucial assumption is the one which Levesque [1990] dubbed later the *All I Know* Assumption: the assumption that all that is known by the agent is *grounded* in his theory, in the sense that it belongs to it or can be derived from it. In the case of the autotheoremhood view, the agent is nothing else than a personification of the theory itself, and what it knows is what it entails. We discuss alternative instances of this agent-based view in the next section.

But let us now focus on developing the autotheoremhood perspective. We regard it as a more precise intuition that is more amenable to formalization despite the fact that self-reference, which is evidently present in the notion of autotheoremhood, is a notoriously complex phenomenon. It plagued, albeit in a different form, the *theory of truth* in philosophical logic with millennia-old paradoxes [Tarski, 1983, Kripke, 1975, Barwise and Etchemendy, 1987]. The best known example is the famous *liar* paradox:

"This sentence is false."

An autoepistemic theory that is clearly reminiscent of this paradox is:

$$T_{liar} = \{\neg KP \to P\}.$$

In the autotheoremhood view, this theory states that if *it* does not entail P then P holds. However, if P is not entailed, then we *have* an argument for P, and if P is entailed, the unique proposition of the theory is trivially satisfied; no argument for P can be constructed. This is *mutatis mutandis* the argument for the inconsistency of the liar sentence. In view of the difficulties that self-reference has posed to the development of the theory of truth, it would be naive to hope that a crisp, unequivocal formalization of autoepistemic logic existed.

Moore [1985, p. 82] explained the difficulty of defining the semantics for autoepistemic inference rules (2) as follows. When the inference rules are monotonic, that is, when $m = 0$,

> 'once a formula has been generated at a given stage, it remains in the generated set of formulas at every subsequent stage. [...] The problem with attempting to follow this pattern with nonmonotonic inference rules [that is, when $m > 0$ (note of the authors)] is that we cannot draw nonmonotonic inferences reliably at any particular stage, since something inferred at a later stage may invalidate them.'

To put it differently, the problem is that when a rule (2) is applied to derive γ at some stage when all α_i's have been inferred to be theorems and none of the β_j's has been derived, later inferences may derive some β_j and hence, invalidate the derivation of γ. In such case, Moore argues, all we can do is to characterize the desired result as the solution of a *fixpoint equation* instead of computing it by a *fixpoint construction*:

> 'Lacking such an iterative structure, nonmonotonic systems often use nonconstructive "fixed point" definitions, which do not directly yield algorithms for enumerating the "derivable" formulas, but do define sets of formulas that respect the intent of the nonmonotonic inference rules.'

This was an extremely clear and compelling representation of intuitions behind not only the Moore's own autoepistemic logic, but also behind the formalisms of McDermott and Doyle, and of Reiter, too, for that matter.

It is useful now to look at these ideas from a more formal point of view. Let us consider a modal theory T over some vocabulary Σ. Let T consist of "inference rules" of the form (2), where for simplicity we assume that all formulas $\alpha_i, \beta_j, \gamma$ are objective (that is, contain no modal operator).[3] The inference processes that Moore had in mind are syntactic in nature and are derivations of formulas. Yet, it is straightforward to cast these inference processes in semantical terms.

Let $\mathcal{W}$ be the set of all Σ-interpretations. A state of belief is represented as a set $B \subseteq \mathcal{W}$ of possible worlds.[4] Intuitively, each element $w \in B$ represents a *possible world*, a state of affairs that satisfies the agent's beliefs. A world $w \notin B$ represents an *impossible world*, a state of affairs that violates at least one proposition of the

---

[3] Our approach works equally well for arbitrary modal theories.
[4] A *possible-world set* is a special Kripke structure in which the accessibility relation is total.

agent. Given a set B representing the worlds held possible by an agent, the following, standard, definition formalizes which (modal) formulas the agent believes (or knows — we do not distinguish between these two modalities in our discussion).

Definition 1 *We define the satisfiability relation $B, w \models \varphi$ as in the modal logic S5 by the standard recursive rules of propositional satisfaction augmented with one additional rule:*

$$B, w \models K\varphi \text{ if for every } v \in B, B, v \models \varphi.$$

We then define $B \models K\varphi$ (φ is believed or known in state B) if for every $w \in B$, $B, w \models \varphi$.

This definition extends the standard definition of truth in the sense that if φ is an objective formula then $B, w \models \varphi$ if and only if $w \models \varphi$. We define $Th(B) = \{\varphi \mid B \models K\varphi\}$ and $Th_{obj}(B)$ the restriction of $Th(B)$ to objective formulas. These sets represent all modal formulas and all objective formulas, respectively, known in the state of belief B.

It is natural to order belief states according to "how much" they believe or know. For two belief states B_1 and B_2, we define $B_1 \leq_k B_2$ if $Th_{obj}(B_1) \subseteq Th_{obj}(B_2)$ or, equivalently, if $B_2 \subseteq B_1$. The ordering $\leq_k$ is often called the *knowledge* ordering. We observe that $B_1 \leq_k B_2$ does not entail $Th(B_1) \subseteq Th(B_2)$, due to the nonmonotonicity of modal literals $\neg K\varphi$ expressing ignorance, some of which may be true in B_1 and false in B_2.

We can see Moore's inference processes as sequences $(B_i)_{i=0}^{\lambda}$ of possible-world sets such that $B_0 = \mathcal{W}$, the possible-world set of maximum ignorance in which only tautologies are known. In each derivation step $B_i \to B_{i+1}$, some worlds $w \in B_i$ might be found to be impossible and eliminated in B_{i+1}; other worlds $w \notin B_i$ might be established to be possible and added to B_{i+1}. This process is described through Moore's semantic operator D_T, which maps a possible-world set B to the possible-world set $\{w \mid B, w \models T\}$. For theories consisting of formulas (1), $D_T(B_i)$ is exactly the set of all possible worlds that satisfy the conclusions γ of all inference rules that are "active" in B_i, that is, for which $B_i \models K\alpha_j$, $1 \leq j \leq n$, and $B_i \not\models K\beta_j$, $1 \leq j \leq m$.

Let us come back to Moore's claims. The nonmonotonicity of the inference rules (2), or more precisely, formulas (1) is due to the negative conditions $\neg K\beta_j$ (β_i not known, not proved, not a theorem). So let us assume that $m = 0$ for all inference rules in T.[5] One can show that under this assumption D_T is a monotone operator with respect to $\subseteq$: if $B_1 \subseteq B_2$, then $D_T(B_1) \subseteq D_T(B_2)$. This can be rephrased in terms of knowledge ordering: if $B_1 \leq_k B_2$, then $D_T(B_1) \leq_k D_T(B_2)$. In other words, the operator D_T is also monotone in terms of the knowledge ordering $\leq_k$. Moore's inference process $(B_i)_{i=0}^{\lambda}$ is now an *increasing* sequence in the knowledge order $\leq_k$. It yields a least fixpoint B_T in the knowledge order (equivalently, the greatest fixpoint of D_T in the subset order $\subseteq$). Every other fixpoint of D_T contains more knowledge than B_T. The fixpoint B_T is the intended belief state associated with the theory T of monotonic inference rules.

---

[5] For arbitrary theories T, the corresponding assumption is that there are no modal literals $K\varphi$ occurring positively in T.

In the general case of nonmonotonic inference rules ($m > 0$, for some rules), the operator D_T may not be monotone. The inference process constructed with D_T may oscillate and never reach a fixpoint, or may reach an unintended fixpoint due to the fact that it may derive that a world is impossible on the basis of an assumption $\neg K\beta_i$ which is later withdrawn. In such case, stated Moore, all we can do is to focus on possible-world sets that "respect the intent of the nonmonotonic inference rules" as expressed by a *fixpoint equation* associated to T, rather than being the result of a *fixpoint construction*. In this way Moore arrived at his semantics of autoepistemic logic, summarized in the following definition.

Definition 2 *An autoepistemic expansion of a modal theory T over Σ is a possible-world set $B \subseteq W$ such that $B = D_T(B)$.*

We agree with Moore that the condition of being a fixpoint of D_T is a necessary condition for a belief state to be a possible-world model of T. However, it is obviously not a sufficient one, at least not in the autotheoremhood view on T. This is obvious, as this semantics does not coincide with Moore's own ideas on the semantics of monotonic inference rules. A counterexample is the following theory:

$$T = \{KP \to P\}.$$

This theory consists of a unique monotonic inference rule, albeit a rather useless one as it says "*if P is a theorem then P holds*". According to Moore's account of monotonic inference rules, the intended possible-world model of this theory is $W = \{\emptyset, \{P\}\}$ (we assume that $\Sigma = \{P\}$). Yet, T has two autoepistemic expansions, the second being the self-supported possible-world set $\{\{P\}\}$.

It is worth noting that this theory is related to yet another famous problematic statement in the theory of truth, namely the *truth sayer*:

"This sentence is true."

The truth value of this statement can be consistently assumed to be true, or equally well, to be false. Therefore, in Kripke's [1975] three-valued truth theory, the truth value of the truth sayer is *undetermined* **u**. In case of the related autoepistemic theory $\{KP \to P\}$, also Moore's semantics does not determine whether P is known or not. But in the autotheoremhood view, it is clear that P should not be known and this transpires from Moore's own explanations on monotonic inference rules.[6] We come back to the issue of self-supported expansions in Section 5, where we explore alternative perspectives on autoepistemic propositions, in which such self-supported expansions might be acceptable.

The main question then is: Can we improve Moore's method to build inference processes in the presence of nonmonotonic inference rules in T? In this respect, the situation has changed since 1984. The algebraic fixpoint theory for nonmonotone lattice operators [Denecker et al., 2000], which we developed and then used to build

---

[6]There does not seem to be an analogous strong argument why the truth sayer sentence should be false. Yet, Fitting [1997] proposed a refinement of Kripke's theory of truth in which truth is minimized and the truth sayer statement is *false*. For this, he used the same well-founded fixpoint construction that we will use below to obtain a semantics that minimizes knowledge for autotheoremhood theories.

the unifying semantic framework for default and autoepistemic logics [Denecker et al., 2003], gives us new tools for defining fixpoint constructions and fixpoint equations which can be applied to Moore's problem.

We illustrate now these tools in an informal way and refer to these intuitions later when we introduce major concepts for a formal treatment. Let us consider the theory:

$$T = \{P, \neg KP \to Q, KQ \to Q\}.$$

Informally, the theory expresses that P holds, that if P is not a theorem then Q holds, and that if Q is a theorem, then Q holds. Intuitively, it is clear what the model of this theory should be: P is a theorem, hence the second formula cannot be used to derive Q and neither can the truth sayer proposition $KQ \to Q$. Therefore, the intended possible-world set is $B_T = \{\{P\}, \{P, Q\}\}$, that is, P is entailed, Q is unknown.

It is easily verified that B_T is a fixpoint of D_T. Yet D_T has a second, unintended fixpoint $\{\{P, Q\}\}$ which contains more knowledge than B_T. This is a problem as it is this unintended fixpoint that is obtained by iterating D_T starting with $\mathcal{W}$. The reason for this mistake is that the second, nonmonotonic inference rule applies in the initial stage $B_0 = \mathcal{W}$ when $\neg KP$ holds. Later, when P is derived, the conclusion that Q is a theorem continues to reproduce itself through the third truth sayer rule.

The problem above is that at each step and for each world w an *assumption* is made of whether w is possible or impossible. Each such an assumption might be right or wrong. These assumptions are revised by iterated application of D_T. In the context of monotonic inference rules, the only wrong assumptions that might be made during the monotonic fixpoint construction starting in $\mathcal{W}$ are that some world is possible, while in fact it turns out to be impossible. But these wrong assumptions can never lead to an erroneous application of an inference rule: if a condition $K\varphi$ of an inference rule holds when w is assumed to be possible, then it will still hold when w turns out to be impossible. But in the context of nonmonotonic inference rules, an inference rule may fire due to an erroneous assumption and its conclusion might be maintained through a circular argument in all later iterations. In our scenario, it is the initial assumption that worlds in which P is false are possible that lead to the assumption that worlds in which Q is false are impossible, and this assumption is later reproduced by a circular reasoning using the third truth sayer proposition for Q.

The solution to this problem is very simple: *never make any unjustified assumption about the status of a world.* Start without any assumption about the status of any worlds and only assign a specific status when certain. We will elaborate this idea in two steps. In the first step, we illustrate this idea for a simplification T' of T, in which the third axiom $KQ \to Q$ has been deleted.

1. Initially, no world is known to be possible or impossible. At this stage, the truth value of the unique modal literal KP in T' cannot be established. Yet, some things are clear. First, all worlds in which P is false, that is, $\emptyset$ and $\{Q\}$, are certainly impossible since they violate the first formula, P, of T'. Second, the world $\{P, Q\}$ is definitely possible since no matter whether P is a theorem or not, this world satisfies the two formulas of T'. All this can be established without making a single unsafe assumption. Thus, the only world about which we are uncertain at this stage is the world $\{P\}$ in which Q is false. Due to the second axiom, this world is possible if P is known and impossible otherwise.

2. In the next pass, we first use the knowledge that we gained in the previous step to re-evaluate the modal literal KP. In particular, it can be seen that P is true in all possible worlds and in the last remaining world of unknown status, $\{P\}$. This suffices to establish that P is a theorem, that is, that KP is true.

With this newly gained information, we can establish the status of the last world and see that $\{P\}$ satisfies the two axioms of T. Hence this world is possible.

The construction stops here. The next pass will not change anything, and we obtain the possible-world set $B_{T'} = \{\{P\}, \{P,Q\}\}$. Now, let us add the third axiom $KQ \rightarrow Q$ back and consider the full theory T.

1. The first step of the construction is identical to the one above and determines the status for all worlds except $\{P\}$: $\{P,Q\}$ is possible, and $\emptyset$ and $\{Q\}$ are impossible.

2. As before, in the second pass, KP can be established to be true. The second modal literal KQ in T cannot be established yet since its truth depends on the status of the world $\{P\}$. The literal would be false if $\{P\}$ is possible, and true otherwise. Thus the truth of the third axiom in $\{P\}$ is still undetermined. We are blocked here.

3. But there is a way out of the deadlock. So far, the methods to determine whether a world is possible or impossible were perfectly symmetrical. The solution lies in breaking this symmetry. In T, we have a truth-sayers axiom: it is consistent to assume that Q is a theorem, and also to assume that Q is not a theorem. In semantical terms, both assumptions on world $\{P\}$ are consistent: if this world is chosen possible, then KQ is false and all axioms are satisfied in $\{P\}$; if it is chosen impossible, then KQ is true. Since we want to interpret the modal operator as a theoremhood modality, it is clear what assumption to make: that Q is not a theorem. We should make the assumption of *ignorance* and take it that the world is possible (and Q is not a theorem). Thus, we obtain again the possible world model $B_T = \{\{P\}, \{P,Q\}\}$.

From these two examples, we can extract the concepts necessary to formalize the above informal reasoning processes. At each step, we have *partial* information about the status of worlds that was gained so far. This naturally formalizes as a 3-valued set of worlds. We call such a set a *partial possible-world set*. Formally, a partial possible-world set B is a function

$$B : \mathcal{W} \rightarrow \{\mathbf{t}, \mathbf{f}, \mathbf{u}\},$$

where $\mathcal{W}$ is the collection of all interpretations. Standard, *total* possible-world sets can be viewed as special cases, where the only two values in the range of the function are $\mathbf{t}$ and $\mathbf{f}$. In the context of a partial possible world B, we call a world w *certainly possible* if $B(w) = \mathbf{t}$ and *potentially possible* if $B(w) = \mathbf{t}$ or $\mathbf{u}$. Likewise, we call a world w *certainly impossible* if $B(w) = \mathbf{f}$ and *potentially impossible* if $B(w) = \mathbf{f}$ or $\mathbf{u}$. If $B(w) = \mathbf{u}$, w is potentially possible and potentially impossible. We define $CP(B)$ as the set of certainly possible worlds of B, $PP(B)$ as the set of potentially possible worlds, and likewise, $CI(B)$ and $PI(B)$ as the sets of certainly impossible, respectively potentially impossible worlds of B.

At each inference step $B_i \rightarrow B_{i+1}$, we evaluated the propositions of T in one or more unknown worlds w, given the partial information available in B_i. When all propositions of T turned out to be true in w, w was derived to be possible; if some evaluated to false, w was inferred to be impossible. To capture this formally, we need a three-valued truth function to evaluate theories in the context of a world w, the one we are examining, and a partial possible-world set $\mathcal{B}$. The value of this truth function on a theory T, denoted as $|T|^{\mathcal{B},w}$, is selected from $\{\mathbf{t}, \mathbf{f}, \mathbf{u}\}$. There are some obvious properties that this function should satisfy.

1. The three-valued truth function should coincide with the standard (implicit) truth function for modal logic in total possible-world sets. In particular, when $\mathcal{B}$ is a total possible-world set, that is, $\mathcal{B}$ has no unknown worlds, then $|T|^{\mathcal{B},w}$ should be true precisely when $\mathcal{B}, w \models T$ (and false, otherwise).

2. The three-valued truth function should be monotone with respect to the *precision* of the partial possible-world sets. A more precise partial possible-world set is one with fewer (with respect to inclusion) unknown worlds.

The intuition presented in (2) can be formalized as follows. We define $\mathcal{B} \leq_p \mathcal{B}'$ if $\mathcal{B}(w) \leq_p \mathcal{B}'(w)$, where the latter (partial) order $\leq_p$ on truth values is the one generated by $\mathbf{u} \leq_p \mathbf{t}$ and $\mathbf{u} \leq_p \mathbf{f}$.

A three-valued truth function $|T|^{\mathcal{B},w}$ is monotone in $\mathcal{B}$ if $\mathcal{B}' \leq_p \mathcal{B}''$ implies that $|T|^{\mathcal{B}',w} \leq_p |T|^{\mathcal{B}'',w}$. In particular, if $|T|^{\mathcal{B},w}$ is monotone in $\mathcal{B}$ and B is a total possible-world set such that $\mathcal{B}' \leq_p B$, then $|T|^{\mathcal{B}',w} = \mathbf{t}$ implies that $B, w \models T$, and $|T|^{\mathcal{B}',w} = \mathbf{f}$ implies that $B, w \not\models T$.

Designing such a three-valued truth function is routine, the problem is that there is more than one sensible solution. One approach, originally proposed by Denecker et al. [1998], extends Kleene's [1952] three-valued truth evaluation to modal logic.

Definition 3 *For a formula φ, world $w \in \mathcal{W}$ and partial possible-world set $\mathcal{B}$, we define $|\varphi|^{\mathcal{B},w}$ using the standard Kleene truth evaluation rules of three-valued logic augmented with one additional rule:*

$$|K\varphi|^{\mathcal{B},w} = \begin{cases} \mathbf{f} & \text{if } |\varphi|^{\mathcal{B},w'} = \mathbf{f}, \text{ for some } w' \text{ such that } \mathcal{B}(w') = \mathbf{t} \\ \mathbf{t} & \text{if } |\varphi|^{\mathcal{B},w'} = \mathbf{t}, \text{ for all } w' \text{ such that } \mathcal{B}(w') = \mathbf{t} \text{ or } \mathbf{u} \\ \mathbf{u} & \text{otherwise.} \end{cases}$$

For a theory T, we define $|T|^{\mathcal{B},w}$ in the standard way of three-valued logic:

$$|T|^{\mathcal{B},w} = \begin{cases} \mathbf{f} & \text{if } |\varphi|^{\mathcal{B},w} = \mathbf{f}, \text{ for some } \varphi \in T \\ \mathbf{t} & \text{if } |\varphi|^{\mathcal{B},w} = \mathbf{t}, \text{ for all } \varphi \in T \\ \mathbf{u} & \text{otherwise.} \end{cases}$$

To illustrate the use of this truth function, let us evaluate the formula $K\varphi$, where φ is objective, in the context of a partial possible-world set $\mathcal{B}$ and an arbitrary world w. We have $|K\varphi|^{\mathcal{B},w} = \mathbf{t}$ if $PP(\mathcal{B}) \models K\varphi$, that is, if all potentially possible worlds satisfy φ. Likewise, we have $|K\varphi|^{\mathcal{B},w} = \mathbf{f}$ if $CP(\mathcal{B}) \not\models K\varphi$, that is, at least one certainly possible world violates φ. Let B be a more precise total possible world set; that is, $\mathcal{B} \leq_p B$ or equivalently, $PP(\mathcal{B}) \supseteq B \supseteq CP(\mathcal{B})$. Then, obviously, if $K\varphi$ holds true in $\mathcal{B}$, the formula is true in B, and if $K\varphi$ is false in $\mathcal{B}$ then it is false in B as well. In general this truth function is *conservative* (that is, $\leq_p$-monotone) in the sense that if a formula evaluates to true or false in some partial possible-world set, then it has the same truth value in every more precise possible-world set thus, in particular, in every total possible-world set B such that $\mathcal{B} \leq_p B$.

It is easy to see (and it was proven formally by Denecker et al. [2003]) that this truth function satisfies the two desiderata listed above. We also note that this is not the only reasonable way in which the three-valued truth function can be defined. We will come back on this topic in Section 4.5.

We now review the framework of semantics of autoepistemic reasoning we introduced in our study of the relationship between the default logic of Reiter and the autoepistemic logic of Moore [Denecker et al., 2003]. We listed these semantics in the previous section. All semantics in the framework require that a (partial) possible-world model $\mathcal{B}$ of an autoepistemic theory be justified by some type of an inference process:

$$\mathcal{B}_0 \to \mathcal{B}_1 \to \ldots \to \mathcal{B}_n = \mathcal{B}.$$

At each step i, modal literals $K\varphi$ appearing in T are evaluated in $\mathcal{B}_i$. When such literals are derived to be true or false, this might lead to further inferences in $\mathcal{B}_{i+1}$. Taking the semantic point of view, we understand an inference here as a step in which some worlds of undetermined status are derived to be possible and some others are derived impossible.

Dialects of autoepistemic logic, and so of default logic, too, differ from each other in the nature of the derivation step $\mathcal{B}_i \to \mathcal{B}_{i+1}$, and in initial assumptions $\mathcal{B}_0$ they make. Some dialects make no initial assumptions at all; in some others making certain initial "guesses" is allowed. In this way, we obtain autoepistemic logics of different degrees of *groundedness*. In the following sections, we describe inference processes underlying each of the four semantics in the framework described in Section 3.

Finally, we link the above concepts with the algebraic lattice theoretic concepts sketched in the previous section and used in the semantic framework of Denecker et al. [2003]. There, the different semantics of an autoepistemic theory T emerged as different types of fixpoints of a $\leq_p$-monotone operator $\mathcal{D}_T$ on the bilattice consisting of arbitrary pairs (B, B') of possible-world sets. The partial possible-world sets $\mathcal{B}$ correspond to the *consistent* pairs $(PP(\mathcal{B}), CP(\mathcal{B}))$ in this bilattice; a pair (B, B') is *consistent* if $B \supseteq B'$, that is, certainly possible worlds are potentially possible. Inconsistent pairs give rise to possible-world sets that in addition to truth values $\mathbf{t}$, $\mathbf{f}$ and $\mathbf{u}$ require the fourth one $\mathbf{i}$ for "inconsistency". The Kleene truth function defined above can be extended easily to a four-valued truth function on the full bilattice.

The operator $\mathcal{D}_T$ on that bilattice was then defined as follows:

$$\mathcal{D}_T(\mathcal{B}) = \mathcal{B}', \text{ if for every } w \in \mathcal{W}, \mathcal{B}'(w) = |T|^{\mathcal{B},w}.$$

We observe that this operator maps partial possible-world sets into partial possible-world sets and that it coincides with Moore's derivation operator D_T when applied on total possible-world sets.

In the sequel, we will often represent a partial possible-world set $\mathcal{B}$ in its bilattice representation, as the pair $(PP(\mathcal{B}), CP(\mathcal{B}))$ of respectively potentially possible and certainly possible worlds. For example, the least precise partial possible-world set $\perp_p$ for $\Sigma = \{P, Q\}$ will be written as $(\{\emptyset, \{P\}, \{Q\}, \{P, Q\}\}, \emptyset)$: all worlds are potentially possible; no world is certainly possible.

We will now discuss the four semantics discussed above that define different dialects of autoepistemic reasoning.

4.1 The Kripke-Kleene semantics

This semantics is a direct formalization of the discussion above. We are given a finite modal theory T (we adopt the assumption of finiteness to simplify presentation, but it can be omitted). A *Kripke-Kleene inference process* is a sequence

$$\mathcal{B}_0 \to \ldots \to \mathcal{B}_n$$

of partial possible-world sets such that:

1. $\mathcal{B}_0$ is the totally unknown partial possible-world set. That is, for every $w \in \mathcal{W}$, $\mathcal{B}_0(w) = \mathbf{u}$. We denote this partial possible-world set by $\perp_p$. This choice of the starting point indicates that Kripke-Kleene inference process does not make any initial assumptions.

2. For each $i = 0, \ldots, n - 1$, there is a set of worlds U such that for every $w \in U$, $\mathcal{B}_i(w) = \mathbf{u}$, $|T|^{\mathcal{B}_i, w} \neq \mathbf{u}$ and $\mathcal{B}_{i+1}(w) = |T|^{\mathcal{B}_i, w}$, and for every $w \notin U$, $\mathcal{B}_i(w) = \mathcal{B}_{i+1}(w)$. Thus, in each step of the derivation the status of the worlds that are certainly possible and certainly impossible does not change. All that can change is the status of some worlds of unknown status (worlds, that are potentially possible and potentially impossible). This set is denoted by U above. It is not necessary that U contains all worlds that are unknown in $\mathcal{B}_i$. In the derivation, worlds in U become certainly possible or certainly impossible, depending on how the theory T evaluates in them. If for such a potentially possible world $w \in U$, $|T|^{\mathcal{B}_i, w} = \mathbf{t}$, w becomes certainly possible. If $|T|^{\mathcal{B}_i, w} = \mathbf{f}$, w becomes certainly impossible. Otherwise, the status of w does not change. As such a derivation starts from the least precise, hence assumption-free, partial possible-world set $\perp_p$, all these derivations are assumption-free.

3. The halting condition: no more inferences can be made once we reach the state $\mathcal{B}_n$. Here this means that for each unknown $w \in \mathcal{W}$, $|T|^{\mathcal{B}_n, w} = \mathbf{u}$. The process terminates.

This precise definition formalizes and generalizes the informal construction we presented in the previous section. When applied to the theory we considered there,

$$T' = \{P, \neg KP \to Q\},$$

one Kripke-Kleene inference process that might be produced is (we represent here worlds, or interpretations, as sets of atoms they satisfy, and partial possible-world sets $\mathcal{B}$ as pairs $(PP(\mathcal{B}), CP(\mathcal{B}))$):

$$
\begin{aligned}
\perp_p &\to \mathcal{B}_1 = (\{\emptyset, \{P\}, \{P, Q\}\}, \emptyset) & &\{Q\} \text{ certainly impossible} \\
&\to \mathcal{B}_2 = (\{\{P\}, \{P, Q\}\}, \emptyset) & &\emptyset \text{ certainly impossible} \\
&\to \mathcal{B}_3 = (\{\{P\}, \{P, Q\}\}, \{\{P, Q\}\}) & &\{P, Q\} \text{ certainly possible} \\
&\to \mathcal{B}_4 = (\{\{P\}, \{P, Q\}\}, \{\{P\}, \{P, Q\}\}) & &\{P\} \text{ certainly possible.}
\end{aligned}
$$

The first derivation can be made since $|P \wedge (\neg KP \to P)|^{\perp_p, w} = \mathbf{f}$, for $w = \{Q\}$ (in fact, for every w, in which P is false). The second derivation is justified similarly as the first one. The third derivation follows as $|P \wedge (\neg KP \to Q)|^{\mathcal{B}_2, w} = \mathbf{t}$, for $w = \{P, Q\}$, and the forth one as $|P \wedge (\neg KP \to Q)|^{\mathcal{B}_3, w} = \mathbf{t}$, for $w = \{P\}$. Let us explain one more detail of the last of these claims. Here, $|P|^{\mathcal{B}_3, w} = \mathbf{t}$ holds because P holds in $w = \{P\}$. Moreover, $|KP|^{\mathcal{B}_3, w} = \mathbf{t}$ as P holds in every world that is potentially possible in $\mathcal{B}_3$. Thus, $|\neg KP|^{\mathcal{B}_3, w} = \mathbf{f}$ and so indeed, $|\neg KP \to Q|^{\mathcal{B}_3, w} = \mathbf{t}$.

The shortest derivation sequence that corresponds exactly to the informal construction of the previous section is:

$$\perp_p \to (\{\{P\}, \{P, Q\}\}, \{\{P, Q\}\}) \to (\{\{P\}, \{P, Q\}\}, \{\{P\}, \{P, Q\}\}).$$

The fact that there may be multiple Kripke-Kleene inferences processes is not a problem as all of them end in the same partial possible-world.

Proposition 1 *For every modal theory T, all Kripke-Kleene inference processes converge to the same partial possible-world set, which is the $\leq_p$-least fixpoint of the operator $\mathcal{D}_T$.*

We call this special partial possible-world set the *Kripke-Kleene extension* of the modal theory T.

While the Kripke-Kleene construction is an intuitively sound construction, it has an obvious disadvantage: in general, its terminating partial belief state may not match the intended belief state even if T consists of "monotonic" inference rules (no negated modal atoms in the antecedents of formulas of the form (1)). An example where this happens is the truth sayer theory:

$$T = \{KP \to P\}.$$

It consists of a single monotonic inference rule, and its intended total possible-world set is $\{\{\emptyset, \{P\}\}$, which in the current (PP, CP) notation corresponds to

$$(\{\{\emptyset, \{P\}\}, \{\{\emptyset, \{P\}\}).$$

However, the one and only Kripke-Kleene construction is

$$\perp_p \to (\{\emptyset, \{P\}\}, \{\{P\}\}).$$

Then the construction halts. No more Kripke-Kleene inferences on the status of worlds can be made and the intended possible-world set is not reached.

We conclude with a historical note. The name Kripke-Kleene semantics was used for the first time in the context of the semantics of logic programs by Fitting [1985]. Fitting built on ideas in an earlier work by Kleene [1952], and on Kripke's [1975] theory of truth, where Kripke discussed how to handle the liar paradox.

4.2 Moore's autoepistemic logic

Moore's autoepistemic logic has a simple formalization in our framework. A possible-world set B is an autoepistemic expansion of T if there is a *one-step* derivation for it:

$$\mathcal{B}_0 \to \mathcal{B}_1,$$

where $\mathcal{B}_0 = \mathcal{B}_1 = B$. Clearly, here we allow the inference process to make initial assumptions. Moreover, in the derivation step $\mathcal{B}_0 \to \mathcal{B}_1$ we simply verify that we made no incorrect assumptions and that no additional inferences can be drawn. The inference (more accurately here, the verification) process works as follows:

1. A world w is derived to be possible if $\mathcal{B}_0, w \models T$.
2. A world w is derived to be impossible if $\mathcal{B}_0, w \not\models T$.

Thus, formally, $\mathcal{B}_1 = \{w \mid \mathcal{B}_0, w \models T\} = D_T(\mathcal{B}_0)$. Consequently, the limits of this derivation process are indeed precisely the fixpoints of the Moore's operator D_T (we stress that we talk here only about total possible-world sets).

Since $\mathcal{D}_T$ coincides with D_T on total possible-world sets, all autoepistemic expansions are fixpoints of $\mathcal{D}_T$. Thus, we have the following result.

Proposition 2 *The Kripke-Kleene extension is less precise than any other autoepistemic expansion of T. If the Kripke-Kleene extension is total, then it is the unique autoepistemic expansion of T.*

The weakness of Moore's logic from the point of view of modeling the autotheoremhood view has been argued above. In Section 5, we will discuss another interpretation of autoepistemic logic in which his semantics may be more adequate.

4.3   The well-founded knowledge derivation

The problem with the Kripke-Kleene derivation is that it treats ignorance and knowledge in the same way. Ignorance is reflected by the presence of possible worlds. Knowledge is reflected by the presence of impossible worlds. In the Kripke-Kleene derivation, both possible and impossible worlds are derived in a symmetric way, by evaluating the theory T in the context of a world w, given the partial knowledge $\mathcal{B}$.

What we would like to do is to impose ignorance as a default. That a world is possible should not have to be derived. A world should be possible *unless* we can show that it is impossible. In other words, we need to impose a principle of *maximizing ignorance*, or equivalently, *minimizing knowledge*. Under such a principle, it is obvious that the possible-world set $\{\{P\}\}$ cannot be a model of the truth sayer theory $T = \{KP \to P\}$. It does not minimize knowledge while the other candidate for a model, the possible-world set $\{\emptyset, \{P\}\}$, does.

To refine the Kripke-Kleene construction of knowledge, we need an additional derivation step that allows us to introduce the assumption of ignorance. Intuitively, in such a derivation step, we consider a set U of unknown worlds, which are turned into certainly possible worlds to maximize ignorance.

Formally, a *well-founded inference process* is a derivation process $\mathcal{B}_0 \to \ldots \to \mathcal{B}_n$ that satisfies the same conditions as a Kripke-Kleene inference process except that some derivation steps $\mathcal{B}_i \to \mathcal{B}_{i+1}$ may also be justified as follows (by the *maximize-ignorance* principle):

MI: There is a set U of worlds such that $\mathcal{B}_{i+1}(w) = \mathcal{B}_i(w)$ for all $w \notin U$ and for all $w \in U$, $\mathcal{B}_i(w) = \mathbf{u}$, $\mathcal{B}_{i+1}(w) = \mathbf{t}$ and $|T|^{\mathcal{B}_{i+1}, w} = \mathbf{t}$.

In other words, in such a step we pick a set U of unknown words, assume that they are certainly possible, and verify that this assumption was justified, that is, under the increased level of ignorance, all of them turn out to be certainly possible. To put it yet differently, we select a set U of unknown worlds, for which it is consistent to assume that they are certainly possible, and we turn them into certainly possible worlds (increasing our ignorance). By analogy with the notion of an unfounded set of atoms [Van Gelder et al., 1991], we call the set of worlds U, with respect to which the maximize-ignorance principle applies at the partial belief state $\mathcal{B}_i$, an *unfounded* set for $\mathcal{B}_i$.

We also note that the halting condition of a well-founded inference process is stronger than that for a Kripke-Kleene process. This means that for each unknown world w of $\mathcal{B}_n$, $|T|^{\mathcal{B}_n,w} = \mathbf{u}$ and in addition, $\mathcal{B}_n$ does not allow a MI inference step, that is, it has no non-empty unfounded set.

There are two properties of well-founded inference processes that are worth noting.

Proposition 3 *All well-founded inference processes converge to the same (partial) possible-world set.*

This property gives rise to the *well-founded extension* of the modal theory T defined as the limit of *any* well-founded inference process. This limit can be shown to coincide with the well-founded fixpoint of $\mathcal{D}_T$, that is, the $\leq_p$-least fixpoint of the operator $\mathcal{S}_{\mathcal{D}_T}$ defined in the previous section.

Another important property concerns theories with no positive occurrences of the modal operator (for instance, theories consisting of formulas (1) with no modal literals $\neg K\beta_j$ in the antecedent).

Proposition 4 *If T contains only negative occurrences of the modal operator, then the well-founded extension is the $\leq_k$-least fixpoint of D_T.*

This property shows that the well-founded extension semantics has all key properties of the desired semantics of sets of "monotonic inference rules." Let us revisit the truth sayer theory:

$$T = \{KP \rightarrow P\}.$$

The Kripke-Kleene construction is

$$\perp_p \rightarrow (\{\emptyset, \{P\}\}, \{\{P\}\}).$$

The inference that $\{P\}$ is possible is also sanctioned under the rules of the well-founded inference process. However, while there is no Kripke-Kleene derivation that applies now, the maximize-ignorance principle does apply and the well-founded inference process can continue. Namely, in the belief state given by $(\{\emptyset, \{P\}\}, \{\{P\}\})$, there is one world of unknown status (neither certainly impossible, nor certainly possible): $\emptyset$. Taking $U = \{\emptyset\}$ and applying the maximize-ignorance principle to U, we see that the well-founded inference process extends and yields $(\{\emptyset, \{P\}\}, \{\emptyset, \{P\}\})$. This possible-world set is total and so, necessarily, the limit of the process. Thus, this (total) possible-world set $\{\emptyset, \{P\}\}$ is the well-founded extension of the theory $\{KP \rightarrow P\}$.

The well-founded extension is total not only for monotonic theories. For instance, let us consider the theory:

$$T = \{KP \leftrightarrow Q\} \quad \text{or equivalently,} \quad \{KP \to Q, \neg KP \to \neg Q\}.$$

Intuitively, there is nothing known about P, hence Q should be false. The unique Kripke-Kleene inference process ends where it starts, that is, with $\perp_p$. Indeed, when KP is unknown, no certainly possible or certainly impossible worlds can be derived. However, the possible-world set $U = \{\emptyset, \{P\}\}$ is unfounded with respect to $\perp_p$. Indeed, if both worlds are assumed possible, KP evaluates to false, and both worlds satisfy T. Thus, in the well-founded derivation we can establish that and then, in the next two steps, we can derive the impossibility of the two remaining unknown worlds, first of $\{Q\}$ and then of $\{P, Q\}$. This yields the following well-founded inference process:

$$\begin{aligned}
\perp_p \quad &\to \mathcal{B}_1 = (\{\emptyset, \{P\}, \{Q\}, \{P, Q\}\}, \{\emptyset, \{P\}\}) \\
&\to \mathcal{B}_2 = (\{\emptyset, \{P\}, \{P, Q\}\}, \{\emptyset, \{P\}\}) \\
&\to \mathcal{B}_3 = (\{\emptyset, \{P\}\}, \{\emptyset, \{P\}\}).
\end{aligned}$$

In other cases, the well-founded extension is a partial possible-world set. An example is the theory:

$$\{\neg KP \to Q, \neg KQ \to P\}\}.$$

In this case, there is only one well-founded inference process, which derives that $\{P, Q\}$ is a certainly possible world and derives no certainly impossible worlds. That is, the well-founded extension is: $(\{\emptyset, \{P\}, \{Q\}, \{P, Q\}\}, \{\{P, Q\}\})$.

4.4   Stable possible-world sets

We recall that a partial possible-world set $\mathcal{B}$ corresponds to the pair of total possible-worlds sets: $(PP(\mathcal{B}), CP(\mathcal{B}))$, where $PP(\mathcal{B})$ is the set of potentially possible worlds and $CP(\mathcal{B})$ is the set of certainly possible worlds.

We now define a *stable derivation* for a possible-world set B as a sequence of partial belief states of the form:

$$(\mathcal{W}, B) \to (PP_1, B) \to \ldots \to (PP_{n-1}, B) \to (PP_n, B),$$

where:

1. $PP_n = B$

2. For every $i = 0, \ldots, n-1$, and for every $w \in PP_i \setminus PP_{i+1}$, $|T|^{(PP_i, B), w} = \mathbf{f}$. That is, some worlds w in which T is false with respect to $\mathcal{B}_i = (PP_i, B)$ become certainly impossible and are removed from PP_i to form PP_{i+1}.

3. Halting condition: for every $w \in PP_n$, $|T|^{(PP_n, B), w} = \mathbf{t}$ or $\mathbf{u}$.

If a total belief set B has a stable derivation then we call B a *stable extension*. This concept captures the idea of the Reiter's extension of a default theory.

We recall that an inference rule (1) evaluates to false in world w with respect to (PP_i, B) if $w \not\models \gamma$, $PP_i \models K\alpha_i$, for all i, $0 \leq i \leq n$, and $B \not\models K\beta_j$, for all j, $0 \leq j \leq m$. We see here an asymmetric treatment of prerequisites α_i and justifications β_j which are evaluated in two different possible world sets. The same feature shows up, not coincidentally, in Reiter's definition of extension of a default theory.

The intuition underlying a stable derivation comes from a different implementation of the idea that ignorance does not need to be justified and that only knowledge must be justified. In a partial possible-world set $\mathcal{B}$, the component sets $PP(\mathcal{B})$ and $CP(\mathcal{B})$ have different roles. Since $PP(\mathcal{B})$ determines the certainly impossible worlds, this is the possible-world set that determines what is definitely known. On the other hand the set $CP(\mathcal{B})$ of certainly possible worlds determines what is definitely not known by $\mathcal{B}$.

A stable derivation for B is a justification for each impossible world of B (each world is initially potentially possible but eventually determined not to be in B, that is, determined impossible in B). The key point is that this justification may use the assumption of the ignorance in B. By fixing $CP(\mathcal{B}_i)$ to be B, it takes the ignorance in B for granted. What is justified in a stable inference process is the impossible worlds of B, not the possible worlds.

We saw above that the theory

$$\{\neg KP \rightarrow Q, \neg KQ \rightarrow P\}$$

has a partial well-founded extension. It turns out that it has two stable extensions $\{\{P\}, \{P, Q\}\}$ and $\{\{Q\}, \{P, Q\}\}$. For instance, the following stable derivation reconstructs $B = \{\{P\}, \{P, Q\}\}$. Note that in any partial possible-world set $(\cdot, B)$ (that is, where the worlds of B are certainly possible), KQ evaluates to false. In all such cases, T evaluates to false in any world in which P is false. Hence we have the following very short stable derivation:

$$(\mathcal{W}, B) \quad \rightarrow (B, B).$$

We now have two key results. The first one links up well-founded and stable extensions.

Proposition 5 *If the well-founded extension is a total possible-world set, it is the unique stable extension.*

The second result shows that indeed, the Konolige's translation works if the semantics of default logic of Reiter and the autoepistemic logic of Moore are correctly aligned. Here we state the result for the most important case of default extensions and stable extensions, but it extends, as we noted earlier, to all semantics we considered.

Proposition 6 *For every default theory Δ, B is an extension of Δ if and only B is a stable extension of $Kon(\Delta)$.*

4.5 Discussion

We have obtained a framework with four different semantics. This framework is parameterized by the truth function. We have concentrated on the Kleene truth function but other viable choices exist. One is super-valuation [van Fraassen, 1966] which defines $|T|^{\mathcal{B},w}$ in terms of the evaluation of T in all possible world sets $B \geq_P \mathcal{B}$ approximated by $\mathcal{B}$. In particular,

$$|T|^{\mathcal{B},w} = Min_{\leq_p}\{|T|^{B,w} \mid \mathcal{B} \leq_p B\}.$$

In this way we obtain another instance of the framework, the family of *ultimate* semantics [Denecker et al., 2004]. For many theories, the corresponding semantics of the two families coincide but ultimate semantics are sometimes more precise. An example is the theory $\{KP \vee \neg KP \to P\}$. It's Kripke-Kleene and well-founded extension is the partial possible world set $(\{\emptyset, \{P\}\}, \{\{P\}\})$ and there are no stable extensions. But the premise $KP \vee \neg KP$ is a propositional tautology, making $|T|^{\perp_P,w}$ true if $w \models P$ and false otherwise. As a consequence, the ultimate Kripke-Kleene, well-founded and unique stable extension is $\{\{P\}\}$.

For a scientist interested in the formal study of the informal semantics of a certain type of (informal) propositions this diversity is troubling. Indeed, what is then the nature of autoepistemic reasoning, and which of the semantics that we defined and that can be defined by means of other truth functions is the "correct" one? It is necessary to bring some order to this diversity.

In the autotheoremhood view, the formal semantics should capture the information content of an autoepistemic theory T that contains propositions referring to T's own information content; the semantics should determine whether a world is possible or impossible, or equivalently, whether a formula is or is not entailed by T. As we saw, Moore's semantics of expansions and the Kripke-Kleene extension semantics are arguably less suited in the case of monotonic inference rules with cyclic dependencies (cf. the truth sayer theory). This leaves us with four contenders only: the well-founded and the stable extension semantics and their ultimate versions. All employ a technique to maximize ignorance and correctly handle autoepistemic theories with monotonic inference rules. Which of these semantics is to be preferred?

Let us first consider the choice of the truth function. The semantics based on the Kleene truth function and the ones induced by super-valuation make different trade-offs: the higher precision of the ultimate semantics, which is good, comes at the price of higher complexity of reasoning, which is bad [Denecker et al., 2004]. When there is a trade-off between different desired characteristics, there is per definition no *best* solution. Yet, when looking closer, the question of the choice between these two truth functions turns out to be largely *academic* and without much practical relevance. There are classes of autoepistemic theories for which the Kleene and the super-valuation truth functions coincide, and hence, so do the semantics they induce. Denecker et al. [2004, Proposition 6.14] provide an example of such a class. Even more importantly, the semantics induced by Kleene's truth function and by super-valuation differ only when case-based reasoning on modal literals is necessary to make certain inferences. Except for our own artificial examples introduced to illustrate the formal difference between both semantics [Denecker et al., 2004], we are not aware of any reasonable autoepistemic or default theory in the literature where such reasoning would be necessary. They may exist, but if they do, they will constitute an insignificant fringe. The take-home message here is that in all practical applications that we are aware of, the Kleene truth function suffices and there is no need to pay for the increased complexity of super-valuation. This limits the number of semantics still in the running to only two. Of the remaining two, the most faithful formalization of the autotheoremhood view seems to be the well-founded extension semantics. As we view a theory as a set of inference rules, the construction of the well-founded extension formalizes the process of the application of the inference rules more directly than the construction of the stable extension semantics.

Nevertheless, there are some commonsense arguments for not overemphasizing the differences between these semantics. First, we should keep in mind that theories of interest are those that are developed by human experts, and hence, are meaningful to them. What are the meaningful theories in the autotheoremhood? Not every syntactically correct modal theory makes sense in this view. "Paradoxical" theories such as the liar theory T_{liar} can simply not be ascribed an information content in a consistent manner and are not a sensible theory in the autotheoremhood view. For theories T viewed as sets of inference rules, the inference process associated with the theory should be able to determine the possibility of each world and hence, for each proposition, whether it is a theorem or not of T. In particular, this is the case when the well-founded extension is total. We view theories with theorems that are subject to ambiguity and speculation with suspicion. And so, methodologies based on the autotheoremhood view will naturally tend to produce theories with a total well-founded extension. From a practical point of view, the presence of a unique, constructible state of belief for an autoepistemic theory is a great advantage. For instance, unless the polynomial hierarchy collapses, for such theories the task to construct the well-founded extension and so, also the unique stable expansion, is easier than that of computing a stable expansion of an arbitrary theory or to determine that none exists. Further, for such theories, skeptical and credulous reasoning (with respect to stable extension) coincide and are easier, again assuming that the polynomial hierarchy does not collapse, than they are in the general case.

For all the reasons above, a human expert using autoepistemic logic in the autotheoremhood view, will be naturally inclined to build an autoepistemic knowledge base with a well-founded extension that is total. When the well-founded semantics induced by the Kleene truth function is total, the four semantics — the two stable semantics and the two well-founded semantics — coincide! It is so, in particular for the class of theories built of formulas (1) with no recursion through negated modal literals (the so-called *stratified* theories [Gelfond, 1987]). Hence, such a methodology could be enforced by imposing syntactical conditions.

All these arguments notwithstanding, the fact is that many default theories discussed in the literature or arising in practical settings do not have a unique well-founded extension and that the stable and well-founded extension semantics do not coincide[7]. We have seen it above in the Nixon Diamond example. More generally, it is the case whenever the theory includes conflicting defaults and no guidance on how to resolve conflicts. Such conflicts may arise inadvertently for the programmer, in which case a good strategy seems to be to analyze the conflicts (potentially by study-

---

[7]Some researchers believe that multiple extensions are *needed* for reasoning in the context of incomplete knowledge. Our point of view is different. The essence of incomplete knowledge is that different states of affairs are possible. Therefore, the natural — and standard — representation of a belief state with incomplete knowledge is by one possible-world set with multiple possible worlds, and not by multiple possible-world sets, which to us would reflect the state of mind of an agent that does not know what to believe.

ing the stable extensions) and to refine the theory by building in conflict-resolution in the conditions of default rules. Otherwise, when conflicts are a deliberate decision of the programmer who indeed does not want to offer rules to resolve conflicts, all we can do is to accept each of the multiple stable extensions as a possible model of the theory and also accept that none of them is in any way preferred to others.

In conclusion, rather than pronouncing a strong preference for the well-founded extension over stable extensions or vice versa, what we want to point out is the attractive features of theories for which these two semantics coincide, and advantages of methodologies that lead to such theories.

5 Autoepistemic Logics in a Broader Landscape

In this section, we use the newly gained insights on the nature of autoepistemic reasoning to clarify certain aspects of autoepistemic logic and its position in the spectrum of logics, in particular in the families of logics of nonmonotonic reasoning and classical modal logics.

A good start for this discussion is Moore's "second" view on autoepistemic logic. Later in his paper, when developing the expansion semantics, Moore rephrased his views on autoepistemic reasoning in terms of the background concept of an *autoepistemic agent*. Such an agent is assumed to be ideally rational and have the powers of perfect introspection. An autoepistemic theory T is viewed as a set of propositions that are known by this agent. Modal literals $K\varphi$ in T now mean *"I (that is, the agent) know φ"*. The most important assumption, the one on which this informal view of autoepistemic logic largely rests, is that the agent's theory T represents *all the agent knows* [Levesque, 1990] or, in Moore's terminology, what the agent knows is *grounded* in the theory. We will call this implicit assumption the *All I Know* Assumption.

Without the *All I Know* Assumption, the theory T would be just a list of believed introspective propositions. The state of belief of the agent might then correspond to any possible-world set B such that $B \models K\varphi$, for each $\varphi \in T$ (where $B \models K\varphi$ if for all $w \in B, B, w \models \varphi$). But in many such possible-world sets B, the agent would know much more than what can be derived from T. In this setting, nonmonotonic inference rules such as $KA(x) \wedge \neg K\neg B(x) \rightarrow B(x)$ would not be useful for default reasoning since conclusions drawn from them would not be derived from the information given in T. So the problem is to model the *All I Know* Assumption in the semantics. Moore implemented this condition by imposing that for any model B, if $B, w \models T$, then w is possible according to B, i.e., $w \in B$. Combining both conditions, models that satisfy the *All I Know* Assumption are fixpoints of D_T, that is Moore's expansions.

Moore's expansion semantics does not violate the assumptions underlying the autoepistemic agent view. Expansions do correspond to belief states of an ideally rational, fully introspective agent that believes all axioms in T and, in a sense, does not believe more that what he can *justify* from T. But the same can be said for the autotheoremhood view as implemented in the well-founded and stable extension semantics. We may identify the theory with what the agent knows, and the theoremhood operator with the agent's epistemic operator K, and see the well-founded extension (if it is total) or stable extensions as representing belief states of an agent that can be *justified* from T.

As we stated in the previous section, Moore's expansion semantics does not formalize the autotheoremhood view, but it formalizes a dialect of autoepistemic reasoning, based on an autoepistemic agent that accepts states of belief with a weaker notion of justification, allowing for *self-supporting* states of belief. While not appropriate for modeling default reasoning, the semantics may work well in other domains. Indeed, humans sometimes do hold self-supporting beliefs. For example, self-confidence, or lack of self-confidence often are to some extend self-supported. Believing in one's own qualities makes one perform better. And a good performance supports self-confidence (and self-esteem). Applied to a scientist, this loop might by represented by the theory consisting of the following formulas:

$$K(ICanSolveHardProblems) \rightarrow Happy$$
$$Happy \rightarrow Relaxed$$
$$Relaxed \rightarrow ICanSolveHardProblems.$$

Along similar lines, the placebo effect is a medically well-researched fact often attributed to self-supporting beliefs. The self-supporting aspect underlying the placebo-effect can be described by the theory consisting of the rules:

$$K(IGetBetter) \rightarrow Optimistic$$
$$Optimistic \rightarrow IGetBetter.$$

Taking a placebo just flips the patients into the belief that they are getting better. In this form of autoepistemic reasoning of an agent, self-supporting beliefs are justified and Moore's expansion semantics, difficult to reconcile with the notion of derivation and theorem, may be suitable.

There are yet other instances of the *All I Know* Assumption in the autoepistemic agent view. For example, let us consider the theory $T = \{KP\}$. In the autotheorem-hood view, this theory is clearly inconsistent, for there is no way this theory can prove P. The situation is not so clear-cut in the agent view. We see no obvious argument why the agent could not be in a state of belief in which he believes P and its consequences and nothing more than that. In fact, the logic of *minimal knowledge* [Halpern and Moses, 1984] introduced as a variant of autoepistemic logic accepts this state of belief for T.

What our discussion shows is that the *All I Know* Assumption in Moore's autoepistemic agent view is a rather vague intuition, which can be worked out in more than one way, yielding different formalizations and different dialects. It may explain why Moore built a semantics that did not satisfy his own first intuitions (inference rules) and why Halpern [1997] could build several formalizations for the intuitions expressed by Reiter and Moore. In contrast, the autotheoremhood view eliminates the agent from the picture and hence, the difficult tasks to specify carefully the key concepts such as ideal rationality, perfect introspection and, most of all, the *All I Know* Assumption. Instead, it builds on more solid concepts of inference rules, theoremhood and entailment which yields a more precise intuition.

6 Conclusions

We presented here an analysis of informal foundations of autoepistemic reasoning. We showed that there is principled way to arrive at all major semantics of logics of autoepistemic reasoning taking as the point of departure the autotheoremhood view of a theory. We see the main contributions of our work as follows.

First, extending Moore's arguments we clarified the different nature of defaults and autoepistemic propositions. Looking back at Reiter's intuitions, we now see that, just as Moore had claimed about McDermott and Doyle, also Reiter built an autoepistemic logic and not a logic of defaults. We showed that some long-standing problems with default logic can be traced back to pitfalls of using the autoepistemic propositions to encode defaults. On the other hand, we also showed that once we focus theories understood as consisting of autoepistemic propositions and adopt the autotheoremhood perspective, we are led naturally to the Kripke-Kleene semantics, the semantics of expansions by Moore, the well-founded semantics and the semantics of extensions by Reiter.

Second, we analyzed what can be seen as the center of autoepistemic logic: the *All I Know* Assumption. We showed that this rather fuzzy notion leads to multiple perspectives on autoepistemic reasoning and to multiple dialects of the autoepistemic language, induced by different notions of what can be derived from (or is *grounded in*) a theory. One particularly useful informal perspective on autoepistemic logic goes back to Moore's truly insightful view of autoepistemic rules as inference rules. This view, which we called the autotheoremhood view, was the main focus of our discussion. In this view, theories "contain" their own entailment operator and "I" in the *All I Know* Assumption is understood as the theory itself. The most faithful formalization of this view is the well-founded extension semantics but the stable-extension semantics, which extends Reiter's semantics to autoepistemic logic, coincides with the well-founded extension semantics wherever the autotheoremhood view seems to make sense. Thus, it was Reiter's default logic that for the first time incorporated into the reasoning process the principle of knowledge minimization, resulting in a better formalization of Moore's intuitions than Moore's own logic.

Fifteen years ago Halpern [1997] analyzed the intuitions of Reiter, McDermott and Doyle, and Moore, and showed that there are alternative ways, in which they could be formalized. Halpern's work suggested that the logics proposed by Reiter, McDermott and Doyle, and Moore are not necessarily "determined" by these intuitions. We argue here that by looking more carefully at the informal semantics of those logics, they do indeed seem "predestined" and can be derived in a systematic and principled way from a few basic informal intuitions.

Acknowledgments

The work of the first author was partially supported by FWO-Vlaanderen under project G.0489.10N. The work of the third author was partially supported by the NSF grant IIS-0913459.

References

C. Baral and V. S. Subrahmanian. Duality between alternative semantics of logic programs and nonmonotonic formalisms. In A. Nerode, W. Marek, and V. Subrahmanian, editors, *Proceedings of the 1st Internaational Workshop on Logic Programming and Nonmonotonic Reasoning*, pages 69–86. MIT Press, 1991.

J. Barwise and J. Etchemendy. *The Liar: An Essay on Truth and Circularity.* Oxford University Press, 1987.

G. Brewka. Cumulative default logic: in defense of nonmonotonic inference rules. *Artificial Intelligence*, 50(2):183–205, 1991.

K. L. Clark. Negation as failure. In H. Gallaire and J. Minker, editors, *Logic and Data Bases*, pages 293–322. Plenum Press, 1978.

M. Denecker, V. W. Marek, and M.ław Truszczyński. Approximating operators, stable operators, well-founded fixpoints and applications in non-monotonic reasoning. In J. Minker, editor, *Logic-based Artificial Intelligence*, The Kluwer International Series in Engineering and Computer Science, pages 127–144. Kluwer Academic Publishers, Boston, 2000.

M. Denecker, V.W. Marek, and M. Truszczyński. Uniform semantic treatment of default and autoepistemic logics. *Artificial Intelligence*, 143(1):79–122, 2003.

M. Denecker, V. Marek, and M. Truszczyński. Ultimate approximation and its application in nonmonotonic knowledge representation systems. *Information and Computation*, 192:84–121, 2004.

M. Denecker, V. Marek, and M. Truszczyński. Fixpoint 3-valued semantics for autoepistemic logic. In *Proceedings of AAAI 1998*, pages 840–845. MIT Press, 1998.

M. Fitting. A Kripke-Kleene Semantics for Logic Programs. *Journal of Logic Programming*, 2 (4):295–312, 1985.

M. Fitting. A theory of truth that prefers falsehood. *Journal of Philosophical Logic*, 26:477–500, 1997.

M. Gelfond. On stratified autoepistemic theories. In *Proceedings of AAAI 1987*, pages 207–211. Morgan Kaufmann, 1987.

M. Gelfond and V. Lifschitz. The stable model semantics for logic programming. In R. A. Kowalski and K. A. Bowen, editors, *Proceedings of the International Joint Conference and Symposium on Logic Programming*, pages 1070–1080. MIT Press, 1988.

G. Gottlob. Translating default logic into standard autoepistemic logic. *Journal of the ACM*, 42 (4), 1995.

J. Y. Halpern. A critical reexamination of default logic, autoepistemic logic, and only knowing. *Computational Intelligence*, 13(1):144–163, 1997.

J. Y. Halpern and Y. Moses. Towards a theory of knowledge and ignorance: Preliminary report. In R. Reiter, editor, *Proceedings of the Workshop on Non-Monotonic Reasoning*, pages 125–143, 1984.

S. C. Kleene. *Introduction to Metamathematics.* Van Nostrand, 1952.

K. Konolige. On the relation between default and autoepistemic logic. *Artificial Intelligence*, 35:343–382, 1988.

K. Konolige. Errata: On the relation between default and autoepistemic logic. *Artificial Intelligence*, 41(1):115, 1989.

S. Kraus, D. Lehmann, and M. Magidor. Nonmonotonic reasoning, preferential models and cumulative logics. *Artificial Intelligence*, 44:167–207, 1990.

S. Kripke. Outline of a theory of truth. *Journal of Philosophy*, 72:690–712, 1975.

D. Lehmann and M. Magidor. What does a conditional knowledge base entail? *Artificial Intelligence*, 55:1–60, 1992.

D. J. Lehmann. What does a conditional knowledge base entail? In R. J. Brachman, H. J. Levesque, and R. Reiter, editors, *Proceedings of the 1st International Conference on Principles of Knowledge Representation and Reasoning, KR 1989*, pages 212–222. Morgan Kaufmann, 1989.

H. J. Levesque. All I know: a study in autoepistemic logic. *Artificial Intelligence*, 42(2-3): 263–309, 1990.

W. Lukaszewicz. Considerations on default logic: an alternative approach. *Computational Intelligence*, 4:1–16, 1988.

D. Makinson. General theory of cumulative inference. In M. Reinfrank, J. de Kleer, M.L. Ginsberg, and E. Sandewall, editors, *Nonmonotonic reasoning*, volume 346 of *Lecture Notes in Computer Science*, pages 1–18, Berlin-New York, 1989. Springer.

W. Marek and M. Truszczyński. Relating autoepistemic and default logics. In R. J. Brachman, H. J. Levesque, and R. Reiter, editors, *Proceedings of the 1st International Conference on Principles of Knowledge Representation and Reasoning, KR 1989*, pages 276–288. Morgan Kaufmann, 1989.

J. McCarthy. Circumscription - a form of nonmonotonic reasoning. *Artifical Intelligence*, 13: 27–39, 1980.

J. McCarthy. Applications of circumscription to formalizing common-sense knowledge. *Artificial Intelligence*, 28(1):89–116, 1986.

J. McCarthy and P.J. Hayes. Some philosophical problems from the standpoint of artificial intelligence. In B. Meltzer and D. Michie, editors, *Machine Intelligence 4*, pages 463–502. Edinburgh University Press, 1969.

D. McDermott. Nonmonotonic Logic II: Nonmonotonic Modal Theories. *Journal of the ACM*, 29(1):33–57, 1982.

D. McDermott and J. Doyle. Nonmonotonic Logic I. *Artificial Intelligence*, 13(1-2):41–72, 1980.

A. Mikitiuk and M. Truszczyński. Constrained and rational default logics. In C. S. Mellish, editor, *Proceedings of IJCAI 1995*, pages 1509–1515. Morgan Kaufmann, 1995.

R. C. Moore. Possible-world semantics for autoepistemic logic. In R. Reiter, editor, *Proceedings of the Workshop on Non-Monotonic Reasoning*, pages 344–354, 1984. Reprinted in: M. Ginsberg, editor, *Readings on Nonmonotonic Reasoning*, pages 137–142, Morgan Kaufmann, 1990.

R. C. Moore. Semantical considerations on nonmonotonic logic. *Artificial Intelligence*, 25(1): 75–94, 1985.

J. Pearl. System Z: A natural ordering of defaults with tractable applications to nonmonotonic reasoning. In R. Parikh, editor, *Proceedings of the 3rd Conference on Theoretical Aspects of Reasoning about Knowledge, TARK 1990*, pages 121–135. Morgan Kaufmann, 1990.

R. Reiter. A logic for default reasoning. *Artificial Intelligence*, 13(1-2):81–132, 1980.

R. Reiter and G. Criscuolo. On interacting defaults. In P. J. Hayes, editor, *Proceedings of IJCAI 1981*, pages 270–276. William Kaufman, 1981.

T. Schaub. On constrained default theories. In B. Neumann, editor, *Proceedings of the 11th European Conference on Artificial Intelligence*, pages 304–308. Wiley and Sons, 1992.

Y. Shoham. Nonmonotonic logics: meaning and utility. In J. P. McDermott, editor, *Proceedings of IJCAI 1987*, pages 388–393. Morgan Kaufmann, 1987.

A. Tarski. The concept of truth in formalized languages. In J. Corcoran, editor, *Logic, Semantics, Metamathematics: Papers from 1923-38*, pages pp. 152–278. Indianapolis: Hackett, 1983. Translated from Tarski 1935 by J. H. Woodger.

D. Touretzky. *The mathematics of inheritance systems*. Pitman, London, 1986.

M. Truszczyński. Modal interpretations of default logic. In R. Reiter and J. Mylopoulos, editors, *Proceedings of IJCAI 1991*, pages 393–398. Morgan Kaufmann, 1991.

B. van Fraassen. Singular terms, truth-value gaps and free logic. *Journal of Philosophy*, 63(17): 481–495, 1966.

A. Van Gelder, K. A. Ross, and J. S. Schlipf. The well-founded semantics for general logic programs. *Journal of the ACM*, 38(3):620–650, 1991.

Non-Monotonic Reasoning and Uncertainty Theories

Didier Dubois
Henri Prade
IRIT
CNRS and Université de Toulouse
118 Route de Narbonne
31062 Toulouse Cedex 9, France

Abstract: The connections between uncertainty and nonmonotonic reasoning are not so widely known although there are a number of works bridging the two frameworks. This paper provides a survey of these connections, showing that a certain view of nonmonotonic reasoning, mainly devoted to the handling of exceptions in rule-based systems, is more akin to probabilistic reasoning than to nonmonotonic logic programming. Interestingly, this approach has roots in early works on probabilistic reasoning by George Boole, Bruno De Finetti and Ernest Adams, as well as conditional logics of David Lewis. At the syntactic level, suitable inference rules for deriving conditionals from sets of conditionals were proposed by Gabbay, followed up by Lehmann and colleagues. We show that central to these works is the notion of three-valued conditional event whose probability is a conditional probability, and that provides a simple semantics for the preferential logic of conditionals. We show that inference from conditional probability statements is in agreement with preferential non-monotononic inference, and that resorting to possibility theory leads to a less conservative approach that unifies rational closure, system Z of Pearl and other related approaches.

1   Introduction

From its inception, there seems to have been no consensus on what nonmonotonic reasoning deals with, except for the fact that it deals with logical systems where the monotonicity property: *from A implies B, infer $A \wedge C$ implies B* does not follow. Historically, the area is heterogeneous in its formalisms and motivations:

- The use of negation by default in logic programming, which comes down to assuming more knowledge than what is actually available via additional assumptions, such as the closed-world assumption, in order to derive useful conclusions.

- The issue of handling inconsistencies in knowledge bases and databases, which leads to consider several possible extensions that are not consistent with one another.

- Assumption-based reasoning systems with application to diagnosis, where one tries to minimize the number of faults, given observations that contradict assumptions of good behavior.

- The frame problem for the logical description of evolving systems, whereby it is not easy to express that while some aspects of the world change, other don't, unless inertia assumptions are explicitly made.

- Plausible reasoning in rule-based systems, whereby generic rules have exceptions but can still be applied when information is incomplete, the price paid being the defeasibility of such conclusions when new information comes up.

Moreover, the development of uncertainty theories and the emergence of nonmonotonic logic seem to have been rather unrelated for the most part.

- Uncertainty theories (an extensive review is provided by Dubois and Prade [2009]) have privileged a numerical approach to uncertainty. The origin of uncertainty in Artificial Intelligence can be traced to the use, in the early seventies, of confidence factors in expert systems like MYCIN [Shortliffe and Buchanan, 1975], and PROSPECTOR [Duda et al., 1976]. Those systems were basically ad hoc, even if they tried to address a real issue. This trend has been made more rigorous by the development of Bayesian networks, counterparts of which exist in possibility theory, belief function theory and imprecise probability theory.

- Nonmonotonic reasoning was developed within the logic-based artificial intelligence community under a symbolic approach, referring to epistemic logic, logic programming and inconsistency handling in logical databases.

This cultural gap between uncertainty and non-monotonic reasoning was mitigated by the existence of bridges between the two traditions soon noticed by scholars like Judea Pearl [1988] who, in the very first book dedicated to Bayesian networks points out a connection between probability and non-monotonicity. However, reading the book, it became quite clear that this connection only dealt with one specific problem where nonmonotonicity is present: exception-tolerant reasoning. Pearl was clearly motivated by the pioneering works of Ernest Adams [1975], whose logic of conditionals is clearly non-monotonic and in full agreement with probability theory. Adams idea is to interpret actual conditionals as infinitesimal conditional probabilities, and to derive the basic properties of inferences made from such conditionals. Non-monotonicity in probability is patent once it is noticed that $P(B|A \wedge C)$ can be arbitrarily close to 0, while at the same time $P(B|A)$ is arbitrarily close to 1.

In fact, the tradition of considering conditional probabilities as modeling rule-style pieces of knowledge can be traced to Bruno De Finetti's approach to probability [De Finetti, 1936]. While the statistical tradition of probability, eventually formalized by Kolmogorov, considers the probability distribution as the primitive object from which other notions, like conditional probability, are derived, De Finetti [1974] considered conditional probability knowledge as primitive and checked the existence of probability measures consistent with this knowledge (this is the coherence condition). In particular, in De Finetti's approach, zero probability conditional events are allowed. The idea that conditionals (modelled by conditional probabilities) are the actual data to be processed, rather than random observations, is also present in the philosophy of Bayesian networks understood as modeling causal knowledge coming from experts, based on attaching conditional probability tables to a directed acyclic graph representing dependencies. The main difference is that with Bayesian networks only a prescribed set of conditional probabilities yielding a single distribution is allowed, while this condition is not requested by De Finetti and his followers [Coletti and Scozzafava, 2002, Biazzo and Gilio, 2000].

Interestingly, Adams logic bears striking similarity with David Lewis's modal approach to counterfactual conditional statements [Lewis, 1973]. The latter is based on a qualitative notion of similarity, and a comparative possibility order between events. This possibility relation is actually closely related to set functions called possibility measures by Zadeh [1978], whose basic property precisely is what becomes of the additivity of probabilities when they are infinitesimal. This framework is also the one of integer-valued kappa functions of Spohn [1988], who explicitly interprets such integers as exponents of infinitesimal probabilities.

Independently of this uncertainty framework for conditionals, there is a syntactic trend for conditional knowledge bases initiated by Dov Gabbay [1985], and culminating with the works of Daniel Lehmann and his colleagues [Kraus, Lehmann, and Magidor, 1990, Lehmann and Magidor, 1992], and David Makinson [1994]. This approach manipulates so-called conditional assertions $A \mathrel{\vert\!\sim} B$ relating two propositional formulas A and B, and expressing the idea that in general if A collects what is known to be true, then B usually follows. Such entities are viewed as atoms in a higher order logic, and Gabbay proposed inference rules allowing to derive conditional assertions from conditional assertions, that were taken over and completed by other scholars. Interestingly these properties already appear in the conditional logics of Adams and Lewis. It is not surprising, then, that this syntactic framework turns out to possess various semantics, in terms of preference relations among propositional interpretations, infinitesimal probabilities, and possibility measures.

This conditional logic framework to non-monotonic reasoning did not appear in the famous special issue of the Artificial Intelligence journal (Vol. 13(1-2), 1980) that launched research in nonmonotonic logic, not even in later reviews [Sombé, 1990], nor does it seem to have triggered much research in the last ten years or so, contrary to the other nonmonotonic tradition related to logic programming [Lifschitz, 1996], which has blossomed from a theoretical point of view (unifying Reiter's default logic [Reiter, 1980] and Moore's autoepistemic logic [Moore, 1985] among others) and a practical one (with the impressive development of answer-set programming [Niemelä, 1999, Gelfond, 2008]). There are important differences between this trend and the conditional assertion framework:

- Non-monotonic logic programs consist of a set of rules, where facts are encoded as special types of rules as well, and the aim is to produce facts (especially literals) that form the solution to a problem. On the contrary, the framework of Lehmann and colleagues aims at producing conditionals from conditionals, and on this basis derive plausible conclusions in a given context.

- There is a dissymmetry between positive and negative literals in non-monotonic logic programs, while, in exception-handling plausible reasoning, they are treated on a par in the setting of propositional logic.

- The output of a non-monotonic logic program is a set of possible extensions (the answer sets) that contradict each other, while the output of conditional logic à la Lehmann is a single set of plausible conclusions, likely to be revised if more factual information comes up.

- Non-monotonic logic programs seem to be tailored for problem-solving tasks, when default assumptions have to be added in order to generate solutions. Such solutions are valid if they do not contradict the assumptions made. The conditional framework is more adapted to exception-tolerant reasoning, a set of rules modeling what is normal and what is less normal in the world.

This paper surveys the conditional assertion approach to nonmonotonic reasoning and its connections with uncertainty theories, especially probability theory. The aim is to facilitate communication with the nonmonotonic logic programming school, explicating the above mentioned differences, with a view to trigger a cross-fertilization. In Section 2 we provide a more detailed view on the problem of belief construction from generic knowledge and evidential information, highlighting the similarity of paradigms between belief networks in the Bayesian tradition, and sets of conditional assertions, in the symbolic tradition. Section 3 presents the syntactic framework for the logic of conditional assertions after Lehmann and colleagues. Section 4 lays bare a three-valued semantics of conditional assertions, that provides a definition of the conditional appearing in a conditional probability. The next section then presents probabilistic semantics to the logic of conditional assertions. It discusses the lottery paradox, originally introduced by Kyburg [1974], and that is often presented as a major objection against symbolic approaches to plausible reasoning. Finally, Section 6 surveys ordinal preference relations semantics of the logic of conditional assertions, based on confidence relations between events. This paper can be seen as updating a previous similar survey [Dubois and Prade, 1996].

2   Belief construction and the exception-handling problem

In the following, capital letters A and B denote subsets of possible states of the world and represent propositions in propositional logics, or events, and form a set $\mathcal{E}$. We do not use different notations for syntax and semantics of those propositions. We use a propositional language $\mathcal{L}$ with sets of interpretations Ω and $A, B \subseteq \Omega$. In other words $A, B, \ldots \in 2^{\Omega} = \mathcal{E}$. We use $\wedge, \vee, \neg$ to denote conjunction, disjunction and negation of propositions respectively, unambiguously interpreted as intersection, union and complement of the corresponding sets. A central question for plausible reasoning is how to model conditional statements of the form *generally if A then B* and how to check their validity? There are at least two interpretations of this kind of statements.

- **Most As are Bs**: This kind of formulation is precisely represented by a relative frequency (of B's among the A's) or more generally a conditional probability

$P(B|A)$; checking its validity comes down to having a probability distribution on possible worlds and checking that $P(B|A)$ is high enough.

- **Typical As are Bs**: This kind of formulation only requires some states of the world where A is true be preferred to others, which can be modelled by means of a preference relation among possible worlds expressing plausibility, normality or usuality. The validity of the conditional statement can be asserted by checking that the most plausible elements in A do satisfy B, a proposal originally due to Shoham [1988].

In fact, there is a formal agreement between both understandings, as will be clear later on. While non-monotonicity is patent in the first formulation, since a large $P(B|A)$ is compatible with a small $P(B|A \wedge C)$, it is also obvious in the second formulation, where the fact that maximal elements in A satisfy B does not forbid the maximal elements in $A \wedge C$ to satisfy its negation $\neg B$. Research in this area then aimed at characterizing the notion of *typical inference*, relating it to probabilistic reasoning and to ordinal frameworks for representing normality, such as possibilistic logic [Dubois, Lang, and Prade, 1994b].

Reasoning with this kind of conditionals presupposes a distinction to be formally made between generic knowledge and evidential information. Basically, we assume that an agent usually possesses three kinds of information on the world

- *Generic information* (or equivalently background knowledge). Its characteristic feature is that it pertains to a class of situations that is not necessarily referred to explicitly. It may consist of statistics on a well-defined population or be based on commonsense knowledge[1]. In the latter case, the underlying population is often ill-defined (birds generally fly, but which birds are we speaking about?).

- *Evidential information* on a particular situation (or equivalently, evidence). It often consists of observed facts about the current world or a case at hand, for instance, results of medical tests, sensor measurement, testimonies.

- *Beliefs* about the current situation. Such beliefs are built by applying generic knowledge to observed facts and making plausible conclusions.

So plausible reasoning techniques provide an explanation on how beliefs are constructed. For instance, the opinion of a medical doctor on the disease of a patient is built by restricting medical knowledge to rules whose conditions correspond to symptoms similar to the ones of the patient (or in a statistical approach, the population of diagnosed patients having similar symptoms and for whom the doctor had some test results). The test results form a proposition A, and the question is whether the patient suffers from disease B. The medical doctor starts believing B if (s)he knows of a generic rule of the form *if (all that is known is) A, conclude B.*

To automatize such inferences, we need a language that treats generic and evidential information separately. For instance, propositional logic is not enough as both would be encoded likewise. Of course, the distinction between general statements and

---

[1] In this paper, we interpret knowledge as generic background information, not as true belief, in contrast with the epistemic logic tradition.

evidential ones is already at work in first order logic, distinguishing between quantified statements and ground formulas, but it is clear that this logic is not exception-tolerant.

A first example of this scheme is probabilistic reasoning in the traditional setting. In this setting it is assumed that a single probability on the state space is available, that represents statistical information, which stands for the generic knowledge. Typically it can be encoded as a Bayesian network. The available evidence takes the form of a Boolean proposition A: it gathers all the agent knows about the current case or situation. A first inference step is then to compute the conditional probability $P(B|A)$ for the reference class A, that is the frequency of B in the subpart of the population that satisfy A. A second inference step (often implicit) is to estimate the resulting degree of belief of B in the epistemic context described by A: we let $Bel_A(B) = P(B|A)$, that is we equate the degree of belief with the frequency derived from the population (this is the so-called frequency principle of Hacking [1965]). Subjective probabilists may directly define such conditional belief degrees in terms of betting behavior, or define a probability distribution on possible worlds.

3   The syntactic approach to exception-prone rules

Interestingly the same methodology is at work with syntactic exception-tolerant plausible reasoning formalisms. Basically the background knowledge is described by a set of commonsense exception-prone rules[2] generally valid for a certain population (it is not necessary to indicate which one precisely). Each rule is a syntactic entity of the form $A \rightsquigarrow B$, where the arrow is a specific symbol not standing for material implication. Let Δ be a set of such rules, of the form $A_i \rightsquigarrow B_i, i = 1, \dots, n$ relating two Boolean propositions. By convention here, we consider two rules $A_i \rightsquigarrow B_i$ and $A_i' \rightsquigarrow B_i'$ to be the same if A_i and A_i' on the one hand, B_i and B_i' on the other hand, are logically equivalent (they refer to the same set of possible worlds). As will be clear later on in Section 6, a set of rules can be understood at the semantic level as a preference relation between propositions, whereby some propositions are in general more likely the case than others. In general, you can define such a preference relation from a set of rules and conversely.

A set of rules usually goes along with an inference machinery according to which you can derive rules from rules. This machinery can be monotonic, and then yields a deductive closure $Cl(\Delta)$ that is unique. The available evidence again takes the form of a Boolean proposition A, with the same understanding as above. The first inference step is to check if the rule $A \rightsquigarrow B$ lies in $Cl(\Delta)$, which comes down to asking whether B is generally the case in context where all that is known is A. The second inference step is to add B to the set of current accepted beliefs of the agent. Note that this set of beliefs is exactly $\mathcal{A}(A) = \{B : A \rightsquigarrow B \in Cl(\Delta)\}$.

If a new piece of evidence C becomes available (we assume it is consistent with A), then A becomes $A \wedge C$, and one must check that $(A \wedge C) \rightsquigarrow B$ lies in $Cl(\Delta)$ in order to still keep B as an accepted belief, now that the epistemic situation has changed. The new set of beliefs is $\mathcal{A}(A \wedge C) = \{B : (A \wedge C) \rightsquigarrow B \in Cl(\Delta)\}$. Note

---

[2]We avoid saying default rules, not mix up with Reiter style defaults, that turned out to be closely related to non-monotonic logic programs.

that even if the inference machinery on Δ is monotonic, nothing forbids the situation where $A \rightsquigarrow B \in Cl(\Delta)$ but $(A \wedge C) \rightsquigarrow \neg B \in Cl(\Delta)$, that is, beliefs change with new evidence. For instance, if someone has a bird at home you think it flies, but if you hear that, to your surprise, it is a penguin[3], you finally believe it does not fly.

This approach is the path followed by Lehmann and colleagues [Kraus et al., 1990, Lehmann and Magidor, 1992], in opposition with early approaches to nonmonotonic reasoning. Their idea is to interpret rules as consequence relations on a propositional language. Namely, the rule $A \rightsquigarrow B$ constrains the pair (A, B) to belong to a consequence relationship denoted by $\vdash\!\!\sim$, and they use syntactic entities of the form $A \vdash\!\!\sim B$, they call conditional assertions, to represent knowledge. The problem is then to define what kind of inference machinery is needed to properly define the deductive closure of Δ. It comes down to prescribing what the properties of this consequence relationship $\vdash\!\!\sim$ should be. They propose the following postulates:

- *Left Logical Equivalence*: from $A \equiv B$ then $A \vdash\!\!\sim C$ if and only if $B \vdash\!\!\sim C$ (LLE)[4]

- *Right Weakening*: from $B \subseteq C$ and $A \vdash\!\!\sim B$ deduce $A \vdash\!\!\sim C$ (RW)

- *Reflexivity*: $A \vdash\!\!\sim A$

- *Left OR*: from $A \vdash\!\!\sim C$ and $B \vdash\!\!\sim C$ deduce $A \vee B \vdash\!\!\sim C$ (LOR)

- *Cautious Monotony*: from $A \vdash\!\!\sim B$ and $A \vdash\!\!\sim C$ deduce $A \wedge B \vdash\!\!\sim C$ (CM)

- *Weak Transitivity*: from $A \vdash\!\!\sim B$ and $A \wedge B \vdash\!\!\sim C$ deduce $A \vdash\!\!\sim C$ (Cut)

The three last rules already appear in a paper of Gabbay [1985], and even earlier in the work of Adams [1975]. Kraus et al. [1990] call *preferential* a non-monotonic consequence relation $\vdash\!\!\sim$ satisfying the above postulates and name **P** the logic whose atoms are of the form $A \vdash\!\!\sim B$, whose unique connective is conjunction, unique axiom is reflexivity, and inference rules are the above postulates. They prove that the following rules of inference can be derived from the above set of postulates:

- from $A \vdash\!\!\sim B$ and $A \vdash\!\!\sim C$ deduce $A \vdash\!\!\sim B \wedge C$ (Right AND)

- from $A \wedge B \vdash\!\!\sim C$ deduce $A \vdash\!\!\sim \neg B \vee C$ (S)

It is interesting to be convinced of the relevance of the above inference rules for plausible reasoning. Reflexivity is obvious when A is not the contradiction (for the contradiction it is a matter of convention). CM restricts the use of monotonicity to when B is plausibly inferred from A already, i.e., property B is not exceptional for models of A, so that if C is believed in context A, it should be so too when B is heard of. The Cut rule restricts transitivity from A to C to when models of A are normal models of B with respect to property C. LLE is only a consistency condition with respect to classical logic in which propositions are written. The rules RAND and RW ensure that the set $\mathcal{A}(A)$ of plausible consequences of A is deductively closed.

---

[3] As in the Russian novel *Death and the Penguin* by Andrei Kurkov.
[4] This axiom is taken for granted by our convention not to tell propositions apart from sets of models.

Property (S) looks like one half of the classical deduction theorem $A \wedge C \vdash B \iff A \vdash \neg B \vee C$. The direction that works here is when the condition part is relaxed via inference.

Kraus et al. [1990] define a syntactic deduction operation, denoted by $\vdash_P$ in the following, acting from a set Δ of conditional assertions of the form $A_i \hspace{0.5mm}\mid\!\sim\hspace{0.5mm} B_i$. Namely, $\Delta \vdash_P A \hspace{0.5mm}\mid\!\sim\hspace{0.5mm} B$ if and only if $A \hspace{0.5mm}\mid\!\sim\hspace{0.5mm} B$ can be derived from using $A \hspace{0.5mm}\mid\!\sim\hspace{0.5mm} A$ as an axiom schema and the inference rules LLE, RW, LOR, CM and Cut of logic **P**. $Cl_P(\Delta)$ is the set of conditional assertions deduced from Δ in **P**.

So plausible inference in system **P** follows the plausible reasoning scheme outlined above: in the presence of evidence A, to check that B is believed comes down to trying to prove $\Delta \vdash_P A \hspace{0.5mm}\mid\!\sim\hspace{0.5mm} B$, just like, in the Bayesian approach, one has to compute $P(B|A)$ to estimate the degree of belief in B. Note that it is a two-tiered logic: at the bottom level 1, there is propositional logic at work (we need it for LLE and Right weakening, and more generally because the propositions appearing in the postulates are understood up to logical equivalence); then, at level 2, we have the logic of conditional assertions, and the metalanguage is at level 3. Clearly, this inference mode is radically different from the expert system / logic programming tradition (here, you do not directly produce a fact B from a fact A and a rule $A \hspace{0.5mm}\mid\!\sim\hspace{0.5mm} B \in \Delta$).

As it turns out, the logic **P** is very weak. It does not solve the problem of irrelevant properties. For instance, if $\Delta = \{B \hspace{0.5mm}\mid\!\sim\hspace{0.5mm} F, P \hspace{0.5mm}\mid\!\sim\hspace{0.5mm} \neg F, P \hspace{0.5mm}\mid\!\sim\hspace{0.5mm} B\}$ it does not follow that $B \wedge R \hspace{0.5mm}\mid\!\sim\hspace{0.5mm} F \in Cl_P(\Delta)$ e.g. *red birds fly*. The cautious monotony property is too weak because in the example, a rule $B \hspace{0.5mm}\mid\!\sim\hspace{0.5mm} R$ does not exist as explicit knowledge and it would be strange to declare it for birds anyway. Several proposals have been made to strengthen it [Lehmann and Magidor, 1992, Makinson, 1994], by changing cautious monotonicity into a stronger property. The main proposal is to assume that a set of conditional assertions should be closed under the previous inference rules plus the so-called Rational Monotony. The idea is to augment the preferential closure $Cl_P(\Delta)$ by other conditional assertions, and construct a so-called *Rational Extension* of Δ, denoted by $Ex_R(\Delta)$, which satisfies the following property called *Rational Monotony*:

  if $A \hspace{0.5mm}\mid\!\sim\hspace{0.5mm} B \in Ex_R(\Delta)$ and $A \hspace{0.5mm}\mid\!\sim\hspace{0.5mm} \neg C \notin Ex_R(\Delta)$ then $A \wedge C \hspace{0.5mm}\mid\!\sim\hspace{0.5mm} B \in Ex_R(\Delta)$.
(RM)

Unfortunately, extensions $Ex_R(\Delta)$ cannot be computed using (RM) as it is not an inference rule like others. Actually it has a striking similarity with a rule in an answer-set program, of the form

$$A \wedge C \hspace{0.5mm}\mid\!\sim\hspace{0.5mm} B \leftarrow (A \hspace{0.5mm}\mid\!\sim\hspace{0.5mm} B) \wedge \sim (A \hspace{0.5mm}\mid\!\sim\hspace{0.5mm} \neg C),$$

where $\sim$ stands for negation as failure, and conditional assertions stand for positive literals[5]. Unsurprisingly, it gives rise to multiple extensions. There are many supersets of $C_P(\Delta)$ that are closed under (RM), and, worse, the intersection of two rational extensions is generally not rational [Lehmann and Magidor, 1992]. More specifically, the intersection of all rational extensions of $Cl_P(\Delta)$ is $Cl_P(\Delta)$ itself. Actually one of

---

[5]with the proviso that in the answer-set programming conventions, $A \hspace{0.5mm}\mid\!\sim\hspace{0.5mm} \neg C \notin Ex_R(\Delta)$ would then mean that *it is assumed* that $A \hspace{0.5mm}\mid\!\sim\hspace{0.5mm} \neg C$ does not hold.

the rational extensions can be viewed as less debatable because less committed than other ones: the so-called Rational Closure. This point will be discussed in Section 6.

The above paradigm of commonsense exception tolerant reasoning enables the bridge with belief revision [Gärdenfors, 1988] to be laid bare: the belief kinematics is captured by the transformation from $\mathcal{A}(A)$ to $\mathcal{A}(A \wedge C)$ upon arrival of the new piece of evidence C. However this process makes it clear that [Dubois, 2008]:

- The construction of the revised belief set $\mathcal{A}(A \wedge C)$ does not depend on the content of the previous one $\mathcal{A}(A)$: The former only depends on the current total evidence $A \wedge C$, and the generic knowledge $Cl(\Delta)$. You do not use the fact that you thought the bird was flying in order to conclude it does not, now that you know it is a penguin.

- What changes in this process is the agent's epistemic state about the current situation, not the background knowledge $Cl(\Delta)$.

Kraus et al. [1990] have proposed an ordinal semantics for the logic **P** that is based on a two-level structure involving a set X of states, a mapping f from X to the set of interpretations of the propositional language and a preference relation among states which is asymmetric and transitive. A state x satisfies A if and only if $f(x) \in A$. Then $A \mathrel{|\!\sim} B$ is true when all the preferred states satisfying A satisfy B. The logic **P** is sound and complete with respect to this semantics, which has clear connection to the idea of "typical" inference based on a preference relation. However this semantics is somewhat contrived (what do states in X stand for in practice ?) A much simpler semantics easily related to conditional probability has been laid bare, and is described in the next section. Besides it can be proved [Lehmann and Magidor, 1992, Gärdenfors and Makinson, 1994, Benferhat, Dubois, and Prade, 1992] that any rational extension corresponds to a unique complete preordering of interpretations. It then comes close to possibility theory [Dubois and Prade, 1988], as discussed in Section 6.

Finally, it is interesting to point out the connection between postulates of plausible inference and *choice functions* in the social choice literature, that is, functions which, in each set of possible worlds, select a subset of preferred ones. Under suitable axioms on such functions, the selection process is driven by a preference relation on possible worlds, for instance a weak order, as for rational extensions. The connection was pointed out by Lehmann [2001] and a bibliography can be found in the paper by Schlechta [2007].

4 The logic of conditional events

At this point it is useful to come back to the question of providing a mathematical model of exception-prone rules in agreement with conditional probability. Indeed, the two views of exception-prone rules, i.e. with a numerical quantifier or in terms of typicality must be reconciled. Viewing an exception-prone rule as an inference rule, the logic **P** clearly shows it is not a classical one, so that if defined by means of a connective, it cannot be the material implication. So the question is: what is this object a conditional probability is the probability of?

4.1   Rules are three-valued entities

The issue of knowing what is the proper connective standing for a conditional in a conditional probability has been a philosophical debate for years [Lewis, 1976]. The way out of this puzzle is to admit that what De Finetti [1936] calls a conditional event of the form "$B|A$" is not a Boolean, but a three-valued entity. Indeed, a rule *if A then B* shares the set of possible worlds in 3 parts:

- *Examples of the rule*: interpretations where $A \wedge B$ is true;

- *Counterexamples of the rule*: interpretations where $A \wedge \neg B$ is true;

- *Irrelevant cases*: interpretations where A is false.

The satisfiability of a conditional event $B|A$ can thus be defined by means of a three-valued valuation $t : \mathcal{E} \times (\mathcal{E} \setminus \{\emptyset\}) \rightarrow \{T, F, I\}$. If $\omega \in \Omega$, then:

- ω verifies $B|A$ if and only if $\omega \in B \wedge A$ (ω is an *example* of $B|A$) and we write $t(B|A) = T$;

- ω falsifies $B|A$ if and only if $\omega \in \neg B \wedge A$ (ω is a *counterexample* of $B|A$) and we write $t(B|A) = F$;

- otherwise $B|A$ *does not apply* to ω because $\omega \notin A$ and we write $t(B|A) = I$.

   The last case corresponds to a third truth-value. Besides ω satisfies $B|A$ if and only if it does not falsify it, that is, if $\omega \in \neg A \vee B$, which then corresponds to two possible truth-values. This approach, originally proposed by De Finetti [1936], has been rediscovered several times in the literature (see bibliographies in the papers by Goodman, Nguyen, and Walker [1991] and Dubois and Prade [1994]), and especially by Adams [1975] and Calabrese [1987].

   A conditional event $B|A$ can be viewed as a *set* of propositions rather than a single one, namely $\{C : A \wedge B \subseteq C \subseteq \neg A \vee B\}$. The conjunction $A \wedge B$ is obtained when interpreting the third truth-value I as *false* (F) and the material implication $\neg A \vee B$ is obtained when interpreting I as *true* (T). In other words the conditional event is interpreted as the set of Boolean solutions X to the equation $B \wedge A = X \wedge A$, i.e., the following Boolean form of Bayes rule holds

$$B \wedge A = (B|A) \wedge A.$$

This set of solutions $B|A = \{C : A \wedge B \subseteq C \subseteq \neg A \vee B\}$ forms an interval in the Boolean algebra of propositions, lower bounded by the conjunction $B \wedge A$ and upper bounded by material implication $\neg A \vee B$.

4.2   Black ravens and white swans

In contrast with conditional events, the set of models of the material implication $\neg A \vee B$ excludes exceptions to the rule "*if A then B*" but does not single out its examples. In fact, the three-valued representation of a rule may also "solve" some well-known paradoxes of confirmation discussed by scholars, casting new light on a question posed a long time ago by Hempel [1945]. Consider the statement "all ravens are black" whose validity is to be experimentally tested. Observing black ravens indeed confirms this statement while the observation of a single non-black one would refute it. However since this sentence is modeled in first order logic as $\forall x, \neg \text{Raven}(x) \vee \text{Black}(x)$, it is logically equivalent to its contrapositive form "*if not Black then not Raven*". So observing a white swan should also confirm the claim that all ravens are black, since it does not violate the sentence. This is hardly defendable.

Using a three-valued representation of the rule, it is clear that while black ravens are examples of the rule "all ravens are black", white swans are irrelevant items for this rule. Conversely, white swans are examples of the rule "all non-black items are non-ravens", and black ravens are in turn irrelevant. Both rules have the same set of counterexamples (all ravens that are not black, if any) but their examples differ, and only examples should meaningfully confirm rules. Hence the three-valued representation enables a rule to be distinguished from its contrapositive form. Note that, while counterexamples cannot tell a rule from its contrapositive form, examples cannot tell a rule "*if A then B*" from the rule "*if B then A*", hence cannot capture any form of causality.

4.3 A simple semantics for preferential logic of conditional assertions

Since the main issue of reasoning with exception-prone rules is to infer rules from sets thereof, we must now explain what semantic inference of a conditional event from a conditional event means; and what a conjunction of conditional events means. The rest of this section is borrowed from Dubois and Prade [1994].

A rule $A \rightsquigarrow B$ semantically implies another rule $C \rightsquigarrow D$, if the latter has more examples and less exceptions than the former. Formally, we interpret rules as conditional events and this intuition is expressed as:

$$A \rightsquigarrow B \models C \rightsquigarrow D \text{ if and only if } A \wedge B \subseteq C \wedge D \text{ and } C \wedge \neg D \subseteq A \wedge \neg B.$$

Noticing that $C \wedge \neg D \subseteq A \wedge \neg B$ is equivalent to $\neg A \vee B \subseteq \neg C \vee D$, this is the canonical extension of the classical semantic inference relation between propositions to intervals $[A \wedge B, \neg A \vee B]$ and $[C \wedge D, \neg C \vee D]$ in the Boolean algebra[6]. If $A = C = \Omega$ are tautologies, then this entailment comes down to $B \subseteq D$. Equipping the truth-set $\{T, F, I\}$, interpreted as $2^{\{0,1\}} \setminus \{\emptyset\}$, with the logical ordering $T = \{1\} > I = \{0, 1\} > F = \{0\}$, the entailment between rules also reads:

$$A \rightsquigarrow B \models C \rightsquigarrow D \text{ if and only if } t(B|A) \leq t(C|D).$$

Now, consider a set Δ made of two rules $A_i \rightsquigarrow B_i, i = 1, 2$.

- Δ is said to be verified by a propositional interpretation ω if ω verifies at least one rule and does not falsify the other; this is naturally an example of Δ.

---

[6]The analogy with the comparisons of standard interval on the real line is as follows: $[a, b] \leq [c, d]$ if and only if $a \leq c$ and $b \leq d$.

&	F	I	T
F	F	F	F
I	F	I	T
T	F	T	T

Table 10: Conjunctive uninorm

- The rule base Δ is falsified by an interpretation ω if ω falsifies at least one rule in it; this is naturally a counterexample of Δ.

- The rule base Δ does not apply to an interpretation ω if none of the two rules apply to ω.

These requirements completely determine the conjunction & of two rules $(A_1 \rightsquigarrow B_1)$ and $(A_2 \rightsquigarrow B_2)$ as standing for another rule of the form

$$(A_1 \rightsquigarrow B_1)\&(A_2 \rightsquigarrow B_2) \equiv (A_1 \vee A_2) \rightsquigarrow (\neg A_1 \vee B_1) \wedge (\neg A_2 \vee B_2).$$

This is called quasi-conjunction by Adams [1975]. This conjunction is associative. The three-valued truth-table of quasi-conjunction is given in Table 10. Mathematically, operation & is known as a conjunctive idempotent *uninorm* [De Baets et al., 2009]. It is a monotonically increasing semigroup operation on the ordered set $\{T > I > F\}$ with identity I, that coincides with conjunction on $\{T, F\}$. It is the most elementary example of such an operation. This quasi-conjunction is actually the one of Sobociński's three-valued logic [Sobociński, 1952], that pioneered relevance logic [Anderson and Belnap, 1975].

The validity of Δ is the one of the quasi-conjunction of its rules. If Δ has no example, the rule-base is considered to be inconsistent. This is when the sets of classical propositions $\mathcal{B}_i = \{A_i \wedge B_i\} \cup \{\neg A_j \vee B_j : A_j \rightsquigarrow B_j \in \Delta \setminus \{A_i \rightsquigarrow B_i\}\}$ are all inconsistent for all rules $A_i \rightsquigarrow B_i \in \Delta$. This coincides with the so-called *toleration test* of Pearl [1990]. When $\mathcal{B}_i$ is consistent, Δ is said to tolerate $A_i \rightsquigarrow B_i$ in the sense that there is an example of $A_i \rightsquigarrow B_i$ that falsifies no other rule in Δ. For instance, it is obvious that the two rules $A \rightsquigarrow B$ and $A \rightsquigarrow \neg B$ form an inconsistent rule set.

However there are other cases when a rule base Δ can be considered inconsistent. Indeed, the peculiarity of this three-valued logic makes it possible to have inconsistencies between rules hidden by the presence of other rules. To check consistency of Δ, Pearl [1990] proposed the following procedure:

- Extract from Δ the subset Δ_1 of rules tolerated by the other rules in Δ;

- Extract from $\Delta \setminus \Delta_1$ the subset Δ_2 of rules tolerated by the other rules in $\Delta \setminus \Delta_1$;

- And so on, till all remaining rules tolerate each other.

Δ will be called inconsistent if and only if $\Delta_k = \emptyset$ for some $k \geq 1$. More formally, a rule base Δ is consistent if and only if $\forall \emptyset \neq \Sigma \subseteq \Delta$, the quasi-conjunction $QC(\Sigma)$ of all rules in Σ has one example, that is $t(QC(\Sigma))$ is not confined to $\{F, I\}$.

At this point we can define the semantic inference of rules from rules based on this three-valued semantics: A consistent set of rules Δ is said to semantically imply another rule $A \rightsquigarrow B$, denoted by $\Delta \models_3 A \rightsquigarrow B$ if and only if

$$\exists \Sigma \subseteq \Delta, QC(\Sigma) \models A \rightsquigarrow B.$$

Note that this is not equivalent to $QC(\Delta) \models A \rightsquigarrow B$ due to interference between rules.

It is possible to show that this three-valued semantic inference validates all postulates of system **P** enumerated in the previous section. And the following inference rule related to quasi-conjunction is valid in system **P**:

if $A \mid\!\sim B$ and $C \mid\!\sim D$, then $A \vee C \mid\!\sim (\neg A \vee B) \wedge (\neg C \vee D)$.

It is easy to see that some usual deduction patterns fail in the three-valued logic:

- *Monotony*: $A \rightsquigarrow B \not\models_3 (A \wedge C) \rightsquigarrow B$: indeed the latter rule has less examples than the first one.

- *Transitivity*: $\{A \rightsquigarrow B, B \rightsquigarrow C\} \not\models_3 A \rightsquigarrow C$; indeed, an example to $B \rightsquigarrow C$ that falsifies A verifies $\{A \rightsquigarrow B, B \rightsquigarrow C\}$ but it is not an example of the conclusion.

- (The other) *half deduction theorem*: $A \rightsquigarrow (\neg B \vee C) \not\models_3 (A \wedge B) \rightsquigarrow C$; indeed models of $A \wedge \neg B$ verify the premise, not the conclusion.

In fact it was proved by Dubois and Prade [1994] that system **P** is a syntactic counterpart of the three-valued logic of conditional events:

$$\Delta \vdash_P A \mid\!\sim B \text{ if and only if } \Delta \models_3 A \rightsquigarrow B$$

So we can reason in system P by means of a 3-valued logic with truth set $\{T, F, I\}$ equipped with ordering $T > I > F$, and the conjunctive uninorm of Sobociński, using truth-tables. As shown in the next section, this logic actually seems to provide symbolic underpinnings for reasoning with conditional probabilities as well.

5 Probabilistic semantics of reasoning with exception-prone rules

The conditional probability $P(B|A)$ is the probability of a conditional event describing a rule $A \rightsquigarrow B$; indeed the former can be expressed in terms of $P(A \wedge B)$ and $P(\neg A \vee B)$ only as follows

$$P(B|A) = \frac{P(A \wedge B)}{P(A \wedge B) + 1 - P(\neg A \vee B)} \tag{1}$$

In other words, a conditional probability $P(B|A)$ can be computed from a data table only if the proportion of examples and counterexamples is known. It explains why in data-mining you need two evaluation indices to extract association rules. These evaluation indices rely on some combination of $P(A \wedge B)$ and $P(\neg A \vee B)$, usually [Dubois, Hüllermeier, and Prade, 2006].

It is clear that the function $\frac{x}{x+1-y}$ is increasing in both places as soon as $x > 0, y < 1$ (i.e., if the rule has examples and counterexamples). In such conditions, the monotonicity of probability measures with respect to inclusion is extended to conditional events as

$$\text{If } A \rightsquigarrow B \models_3 C \rightsquigarrow D \text{ then } P(B|A) \leq P(C|D). \tag{2}$$

It becomes clear that the extension of a rule base Δ to the probabilistic setting should take the form of a family of constraints on conditional probabilities, namely, $\Delta_P = \{A_i \rightsquigarrow_{a_i} B_i, i = 1, \ldots, n\}$ where the weighted rule $A_i \rightsquigarrow_{a_i} B_i$ corresponds to the constraint $P(B_i|A_i) \geq a_i > 0.5$ using lower probability bounds. In this section, we provide an overview of such a kind of probabilistic reasoning and its relation to system P. Note the similarity between a Bayes net and a conditional probabilistic database Δ_P, since both settings presuppose information in the form of conditional probabilities. However there are significant differences:

- In a Bayesian network, supplied probabilities are precise. Here lower bounds are enough (actually any kind of bounds will do).

- In a Bayesian network, the conditionals form a directed acyclic graph (possibly expressing commonsense causality). There is no such constraint in Δ_P: any set of conditionals will do.

- In a Bayesian network, the numerical probabilities supplied are such that a unique probability distribution is characterized (a Bayes network is neither incomplete, nor inconsistent). On the contrary, Δ_P defines a family of probability measures consistent with the constraints. This family may be empty if the constraints in Δ_P are in conflict.

- A Bayesian network heavily relies on conditional independence assumptions so as to cut down the computational complexity of computing degrees of belief. No independence assumptions are made along with the specification of Δ_P. Such independence assumptions can be added but will make the computation of inferences more difficult as it will introduce non-linear constraints.

A compromise between a probabilistic rule base and a Bayes network is the credal network [Cozman, 2000], that adopts the acyclicity assumption of Bayes nets, while relaxing the requirement of precise probabilistic data. It is developed in the setting of coherent previsions of Walley [1991].

In this section we review various kinds of probabilistic conditional knowledge bases and their relation to the preferential logic of Kraus et al. [1990].

5.1   Reasoning with conditional probabilities

A probabilistic knowledge base can take the more general form of a set of interval constraints of the form $P(B_i|A_i) \in [a_i, b_i]$. The inference problem consists of deciding if another conditional probability of the form $P(B|A)$ lies in a prescribed interval

$[a, b] \subseteq [0, 1]$ [Paass, 1988]. Let Ω be the (finite) set of interpretations of the propositional language in which events are described, and suppose the cardinality of Ω is N. The above problem can be expressed in terms of finding whether a linear system of constraints is satisfied. Namely, let x_i be the (ill-known) probability of $\omega_i \in \Omega$. The constraint $P(B_i | A_i) \in [a_i, b_i]$ clearly takes the form of linear constraints (two per conditional statement in Δ_P, plus the normalization constraint) on a probability distribution $(x_1, x_2, \ldots, x_N)$ of the form:

$$\sum_{\omega_k \in A_i \wedge B_i} x_k \geq a_i \left(\sum_{\omega_j \in A_i} x_j \right);$$

$$\sum_{\omega_k \in A_i \wedge B_i} x_k \leq b_i \left(\sum_{\omega_j \in A_i} x_j \right);$$

$$\sum_{j=1}^{N} x_j = 1.$$

If events are described by means of Boolean variables, the number of unknowns is clearly exponential in the number of atoms of the underlying propositional language. A more demanding kind of inference consists in finding the tightest interval $[a^*, b_*]$ containing the conditional probability $P(B|A)$. It comes down to two optimization problems of the linear fractional type (respectively maximizing the lower bound, and minimizing the upper bound of $P(B|A)$ under the above constraints). This problem has been studied and proved NP-hard even when rules relate two atomic propositions (Lukasiewicz [1999b] provides a bibliography and results). In any case, this is the natural semantics of a probabilistic conditional knowledge base to which one must refer. In order to cope with the complexity, special cases have been studied. For instance one may assume that not only the A_i's and B_i's are atomic propositions, but the set of conditionals also form a non-directed tree whose vertices are atoms A, B and edges represent the two rules $A \leadsto B$ and $B \leadsto A$.

Another approach derives local inference rules [Thöne, Güntzer, and Kießling, 1992, Amarger, Dubois, and Prade, 1991, Frisch and Haddawy, 1994, Lukasiewicz, 1999a] that enable bracketings of the probability of interest to be derived. In the following, we shall denote by $A \leadsto_a B$ rules that express the constraint $P(B|A) \geq a$. A good example of this kind of inference rule is a probabilistic extension of the syllogism given by Dubois, Godo, de Mántaras, and Prade [1993]:

Probabilistic Syllogism: $\{A \leadsto_a B, B \leadsto_b A, B \leadsto_c C\} \models A \leadsto_d C,$

where the tightest lower bound d is of the form $P(C|A) \geq \max(0, a\frac{b+c-1}{b})$. Note that deleting the rule $B \leadsto_b A$ (letting $b = 0$) prevents any informative conclusion from being derived (lack of transitivity). This bound reduces to $\max(0, 2a - 1)$ when $a = b = c$. The main advantage of using local inference rules is that one can explain the resulting conclusions using a trace of the reasoning steps; moreover the complexity is reduced. Unfortunately, even if locally complete, such rules will generally not deliver the tightest probability bounds for the conclusion obtained from a set of conditionals, contrary to the above optimization approach. The most advanced stage of such a kind of local reasoning is the work of Lukasiewicz [1999a].

The probabilistic syllogism is a good example of degenerated inference pattern from system P. Indeed, $\{A \hspace{1pt}\vert\hspace-3pt\sim B, B \hspace{1pt}\vert\hspace-3pt\sim A, B \hspace{1pt}\vert\hspace-3pt\sim C\} \vdash_P A \hspace{1pt}\vert\hspace-3pt\sim C$ is valid (using CM to deduce $B \wedge A \hspace{1pt}\vert\hspace-3pt\sim C$, and then Cut). More generally all inference rules of system P still hold for sets of lower conditional probability bounds, albeit in a degenerated form. This has been explored systematically by Gilio [2002] (partial results had been previously obtained by Dubois and Prade [1991a]). In the following we provide the counterparts to properties of preferential inference for weighted probabilistic rules of the form $A \leadsto_a B$ (Gilio provides more general results for interval-weighted rules):

- *Reflexivity*: $A \leadsto_1 A$ always holds if $A \neq \emptyset$.

- *Right weakening*: $A \leadsto_a B \models A \leadsto_a B \vee C$

- *And*: $\{A \leadsto_b B, A \leadsto_c C\} \models A \leadsto_{\max(0, b+c-1)} B \wedge C$

- *Cautious monotony*: $\{A \leadsto_b B, A \leadsto_c C\} \models A \wedge B \leadsto_{\max(0, \frac{b+c-1}{b})} C$

- *Cut*: $\{A \leadsto_a B, A \wedge B \leadsto_b C\} \models A \leadsto_{ab} C$

- *Or*: $\{A \leadsto_a C, B \leadsto_b C\} \models A \vee B \leadsto_{\frac{ab}{a+b-ab}} C$

Following the steps of the proof of the quasi-conjunction in System P [Dubois and Prade, 1994], from $A \leadsto_a B$ and $C \leadsto_c D$ one can deduce $A \vee C \leadsto_{\max(0,a+c-1)} (\neg A \vee B) \wedge (\neg C \vee D)$. The lower bound for the probabilistic syllogism can be retrieved by application of the Cautious Monotony rule followed by the Cut rule.

5.2   Exact probabilistic semantics for System P

In the above local inference patterns, it is easy to see that in many cases the lower bound of the conclusion is less than the lower bound of the weakest premise. However, when the latter gets close to 1, so does the probability of the conclusion. It explains why there is an infinitesimal probability semantics for System P [Lehmann and Magidor, 1992, Pearl and Goldszmidt, 1996]: suppose the rule $A_i \leadsto B_i$ is interpreted as $P(B_i|A_i) \geq 1 - \epsilon$ where ϵ is a number that is arbitrary close to 0, and denote by $\models_\epsilon$ the inference of the rule $A \leadsto B$ from the rule base Δ:

$$\Delta \models_\epsilon A \leadsto B \text{ if and only if } P(B_i|A_i) \geq 1 - \epsilon, \forall i = 1, \ldots, n, \text{ then}$$
$$P(B|A) \geq 1 - O(\epsilon)$$

where $\lim_{\epsilon \to 0} O(\epsilon) = 0$. Then it can be proved that $\Delta \vdash_P A \hspace{1pt}\vert\hspace-3pt\sim B$ if and only if $\Delta \models_\epsilon A \leadsto B$. This logic was first studied by Adams [1975] who seems to also be the first one to come up with patterns of inference of System P.

However it is possible to equip system P with standard probabilistic semantics. There are two ways to proceed: using so-called big-stepped probabilities, and using De Finetti's coherence approach to interpret extreme conditional probability statements (probabilistic rules of the form $A \leadsto_1 B$).

5.2.1 Big-stepped probabilities

A big-stepped (also called atomic-bound [Snow, 1999]) probability distribution on a finite set Ω is a $N = |\Omega|$-tuple of positive values $p_1 > p_2 > \ldots p_N$ such that $\sum_{i=1}^{N} p_i = 1$, with moreover the additional constraints $p_i > \sum_{j=i+1}^{N} p_j, \forall i = 1, \ldots,$ $N - 1$. Mathematically the probabilities $p_i = P(\{\omega_i\})$ form a super-decreasing sequence: a state ω_i is always more probable than the event formed by the disjunction of other less probable states. It captures an idea of usuality, namely ω_1, being more probable than all the other states as a whole appears as a normal state of facts. This pattern is preserved by conditioning: within any subset of facts there is a single normal state. For instance, the sequence $0.6, 0.3, 0.06, 0.03, 0.01$ is big-stepped. A big-stepped probability assignment is the discrete counterpart of a continuous exponential law.

Now, consider an exception-prone rule $A \rightsquigarrow_{0.5} B$, i.e., $P(B|A) > 0.5$. This weak semantics of a probabilistic rule can be counterbalanced by another constraint: we restrict probability measures to the family $\mathcal{P}_{bs}$ of big-stepped probabilities. Namely a set of conditionals semantically entails another conditional in the big-stepped probability semantics, denoted by $\Delta \models_{bs} A \rightsquigarrow B$ if and only if $P(B|A) > 0.5$ follows from the set of constraints $P(B_i|A_i) > 0.5, i = 1, \ldots, n, \forall P \in \mathcal{P}_{bs}$. Then Benferhat, Dubois, and Prade [1999] have proved that $\Delta \vdash_P A \rightsquigarrow B$ if and only if $\Delta \models_{bs} A \rightsquigarrow B$.

The big-stepped probability semantics sheds some light on the so-called lottery paradox offered by Kyburg [1974] on the notion of probabilistic acceptance. It has been used to question the significance of the axioms of preferential entailment for reasoning about beliefs, and of nonmonotonic logic at large [Poole, 1991]. Namely suppose there is a threshold θ such that a fact A is accepted as true if $P(A) > \theta$. Then because of the weakening of the threshold, when combining events, accepting A and accepting B does not imply we should accept their conjunction since $P(A \wedge B)$ may well be less that θ. This fact is at work with the probabilistic rendering of the And property of System P, which only holds in a degenerate fashion [Gilio, 2002], explained in Section 5.1. For instance, if $1,000,000$ people buy a lottery ticket for which there will be a single winner, each individual player is basically sure to lose, but not all will. The reason why this property of adjunction fails is because it is used in cases when there is no normal state of facts: especially in the lottery game, all tickets are equally likely, and the underlying probability is supposedly uniform. In contrast, if the world is ruled by a big-stepped probability, one state of the world is expected and the other ones are exceptions. This is precisely when plausible reasoning makes sense. This paradox just shows that we cannot apply the patterns of non-monotonic reasoning universally. Yet, the fact that people comply with these inference patterns in situations were some states of fact are much more likely than other ones has been experimentally tested and validated [Da Silva Neves, Bonnefon, and Raufaste, 2002].

5.2.2 De Finetti coherence and system P

Finally, another probabilistic semantics for system P has been proposed by Biazzo, Gilio, Lukasiewicz, and Sanfilippo [2002], relying on the notion of coherence after De Finetti [Biazzo and Gilio, 2000]. In this case, an exception-prone rule $A \rightsquigarrow B$ is understood as $P(B|A) = 1$. The system $\mathcal{S}_\Delta$ of equations induced by conditional probability constraints Δ reads

$$\sum_{\omega_k \in A_i \wedge B_i} x_k = \sum_{\omega_j \in A_i} x_j;$$

$$\sum_{j=1}^{N} x_j = 1.$$

Such a system can be used in two ways: looking for standard probability models or for lexicographic ones, depending on whether $P(B|A)$ makes sense when $P(A) = 0$ or not.

The classical model-theoretic approach to probabilistic reasoning [Hansen, Jaumard, Douanya Nguetsé, and Poggi de Aragão, 1995] considers that $A \rightsquigarrow B$ follows from Δ, which we denote by $\Delta \models_{CP} A \rightsquigarrow B$, if $P(B|A) = 1, \forall P$ satisfying equations in $\mathcal{S}_\Delta$ and moreover $P(A_i) > 0, \forall i = 1, \ldots, n$. Let $\mathcal{P}_\Delta$ be this set of probability functions.

The conditions $P(A_i) > 0$ make sense because the usual definition $P(B_i|A_i) = \frac{P(A_i \wedge B_i)}{P(A_i)}$ needs this condition. In particular, $P(B_i|A_i) = 1$ is then assimilated to $P(B_i \vee \neg A_i) = 1$, i.e., $\mathcal{P}_\Delta = \{P, P(B_i \vee \neg A_i) = 1, P(A_i) > 0, \forall i = 1, \ldots, n\}$.

In the penguin triangle example, consider the three atoms PE(nguin), BI(rd) and $FL(y)$, and the set of rules $\Delta = \{PE \rightsquigarrow \neg FL, BI \rightsquigarrow FL, PE \rightsquigarrow BI\}$, the three constraints $P(FL \wedge BI) = P(BI), P(BI \wedge PE) = P(PE) = P(\neg FL \wedge PE)$ imply $P(PE) = 0$, so $\mathcal{P}_\Delta = \emptyset$ and the same difficulties as in classical logic appear using this view. By default, one can deduce the trivial conclusion $P(B|A) \in [0, 1]$ if $P(A) = 0$ (here, $P(\neg FL|PE) \in [0, 1]$).

Clearly the set of equations $\mathcal{S}_\Delta$ usually will not always yield a probability vector $(x_1, x_2, \ldots, x_N)$ with all components positive: it may force some conditions A_i to have probability 0. While this may be interpreted as the presence of impossible worlds, De Finetti [1974] considered that zero probability does not mean impossible. It just means sufficiently abnormal for being neglected in the first stance. So, once the set of conditionals in Δ with $P(A_i) = 0$ is detected, the equations corresponding to such rules are extracted from $\mathcal{S}_\Delta$, and form, along with the normalization constraint, a smaller system of equations $\mathcal{S}_\Delta^1$ restricted to the interpretations of $\bigvee \{A_i : \mathcal{S}_\Delta \Rightarrow P(A_i) = 0\}$. Let Ω_1 be the complement to this set of interpretations (they can be of positive probability in the first equation system). Of course the same feature ($P(A_i) = 0$ for some A_i) may be observed again at the next round, and the procedure is then repeated until the remaining system has solutions with positive probabilities on all remaining interpretations, thus leading us to consider several smaller and smaller subsystems of equations $\mathcal{S}_\Delta^j, j = 0, \ldots k$, where $\mathcal{S}_\Delta^0 = \mathcal{S}_\Delta$.

We obtain a partitioning of the set of interpretations, say $\Omega_1, \Omega_2, \ldots, \Omega_{k+1}$, and $k + 1$ corresponding sets of probabilities $\mathcal{P}_j$, where $\mathcal{P}_1$ is the set of solutions to the largest system $\mathcal{S}_\Delta$, and $\mathcal{P}_j$ is the set of probability distributions P_j whose support is Ω_j, that are solutions to the subsystem of equations $\mathcal{S}_\Delta^j$. The latter corresponds to a set of rules $\Delta_j \subset \Delta$, where $\Delta_k \cup \Delta_{k-1} \cup \ldots \cup \Delta_1 = \Delta$. The rule-base is thus partitioned into as many subsets of rules as subsets of equations. The set of probability functions satisfying Δ can be extended by collecting solutions to all systems encountered in the above procedure. A probabilistic solution to the system is then made of a sequence of probability distributions P_j whose support is Ω_j. Note that the partitioning of the rule-base is precisely the one of the toleration method described in the previous section.

Each probability P_j is a classical probabilistic model of the subset of equations corresponding rule base Δ_j. Indeed, the probability function P_j corresponding to the subset of rules Δ_j, satisfies any equation $P(B_i|A_i) = 1$, with $A_i \rightsquigarrow B_i \in \Delta_j$ of the form

$$\sum_{\omega_\ell \in A_i \wedge B_i} x_\ell = \sum_{\omega_\ell \in A_i} x_\ell > 0.$$

Other equations (inside S_Δ^{j+1}) are verified in the form $0 = 0$ since the conditions A_i of the rules $A_i \rightsquigarrow B_i$ correspond to interpretations out of Ω_j.

The inference of $P(B|A) = 1$ from Δ in the sense of De Finetti (called g-coherent entailment) comes down to inferring it in the model-theoretic sense from the largest subset Σ of Δ such that $P(A) > 0, \forall P$ for which $P(\bigvee_{i=1}^n A_i) > 0$. This maximal subset of rules is shown by Biazzo et al. [2002] to be unique and can be obtained using the toleration algorithm. They show that the inference of $P(B|A) = 1$ from the set of constraints of the same kind induced by Δ, following this principle is equivalent to reasoning in system P.

6   The ordinal approach to defeasible acceptance

In the previous section, we have bridged the gap between nonmonotonic reasoning and probabilistic reasoning. It suggests addressing the exception-tolerant approach from a more general point of view, starting from a representation of generic information by means of a confidence relation among events, namely by pieces of knowledge expressing that some propositions are normally more likely than others. This approach, initiated by Friedman and Halpern [1996], enables properties of system P to be retrieved by extracting propositions accepted in the sense of the confidence relation, and requesting that such accepted propositions form a deductively closed set. In his book, Halpern [2003] provides details on this approach to non-monotonic reasoning, based on so-called *plausibility measures*[7]. It also leads to relate nonmonotonic reasoning with other uncertainty theories such as possibility theory [Dubois and Prade, 1988, 1998]. It also brings the issue of belief revision into play [Gärdenfors, 1988]. A given set of accepted propositions can be revised upon the arrival of new pieces of evidence by conditioning the confidence relation and extracting new accepted beliefs. The idea relating belief revision to conditionals is that if a proposition C summarizes the available evidence and leads an agent to believe a proposition A by conditioning the confidence relation, then the generic rule $C \rightsquigarrow A$ should be present in the knowledge base of the agent.

6.1   Acceptance relations and Preferential Entailment

Consider a set of propositions $\mathcal{E}$ expressible in a propositional language $\mathcal{L}$ (typically equated to the power set of the set of interpretations Ω). A *confidence relation* denoted

---

[7]Not to be confused with Shafer's older plausibility measures [Shafer, 1976]. To avoid this confusion, we shall not follow Halpern's terminology here.

by $\geq_L$ can be defined on $\mathcal{E}$, whereby $A \geq_L B$ means A is at least as likely as B, in the sense that the agent as at least as much confidence in the truth of A as in the truth of B[8]. The strict part of this relation is denoted by $>_L$ and $A >_L B$ stands for $A \geq_L B$ but not $B \geq_L A$. The indifference relation is defined as $A \sim_L B$ if and only if both $A \geq_L B$ and $B \geq_L A$ hold.

Natural properties of the confidence relation are as follows

- *Non-triviality*: $\Omega >_L \emptyset$.

- *Reflexivity*: $A \geq_L A$.

- *Monotony 1*: if $A \subseteq C$ then $C \geq_L A$.

- *Quasi-transitivity*: if $A >_L B$ and $B >_L C$ then $A >_L C$.

- *Monotony 2*: if $A \subseteq C$ and $D \subseteq B$ then $A >_L B$ implies $C >_L D$

A consequence of Monotony 1 (of relation $\geq_L$) is that limit conditions $S \geq_L A \geq_L \emptyset$ hold. Monotony 2 (of the strict part of the confidence relation), was introduced by Halpern [1997] under the name "orderly property". The transitivity of the strict part of $\geq_L$ is a minimal requirement, and does not imply the transitivity of the confidence relation. Then we have the following definition

Definition 1 *A confidence relation is a relation on $\mathcal{E}$, that is non-trivial, reflexive, quasi-transitive, and monotonic 1 and 2.*

Given a confidence relation on events $\geq_L$, the set of beliefs induced by this relation is defined as

$$\mathcal{A}_L = \{A : A >_L \neg A\}.$$

Note that the condition $A >_L \neg A$ looks like the weakest approach to the notion of accepted belief extracted from the confidence relation, in the sense that A cannot be believed if $\neg A \geq_L A$. We assume that the confidence relation represents generic knowledge. A set of observations representing evidence on the current state of facts is modelled by an event C. It is assumed $C >_L \emptyset$ (observed facts should not be considered impossible). Then, the set of beliefs induced by the confidence relation $\geq_L$ in the context C is naturally defined as:

$$\mathcal{A}_L(C) = \{A : A \wedge C >_L \neg A \wedge C\}.$$

This is a natural way of conditioning the confidence relation $\geq_L$ on the context C by restricting all events to this context. It again highlights examples and counterexamples of rules of the form $C \rightsquigarrow A$. Note that when $C = \emptyset$, $\mathcal{A}_L(\emptyset) = \emptyset$, because, like in probability theory, it makes no sense to condition on a contradictory proposition. If $\mathcal{A}_L(C)$ is to be viewed as consequences of C in some appropriate logic, this feature contrasts with classical logic where everything follows from contradiction: here supraclassicality cannot be assumed.

The same problem will occur if $C \neq \emptyset$, but $C \sim_L \emptyset$. To avoid it, the usual assumption is that no non-contradictory proposition is impossible. This can be expressed by the following axiom:

---

[8]We do not restrict the use of the word *likely* to a strict probabilistic understanding.

$$Non\text{-}dogmatism: \forall C \neq \emptyset, C >_L \emptyset$$

If you assume that people reason in classical logic with accepted beliefs, then the set $\mathcal{A}_L(C)$ should be closed under deduction. It means it should satisfy the following properties [Dubois, Fargier, and Prade, 2004]:

- *Success* (SUC): $\forall C \neq \emptyset, C \in \mathcal{A}_L(C)$

- *Conditional Entailment Closure* (CEC): If $A \subseteq B$ and $C \wedge A >_L C \wedge \neg A$ then $C \wedge B >_L C \wedge \neg B$;

- *Conditional Conjunction Closure* (CCC): If $C \wedge A >_L C \wedge \neg A$ and $C \wedge B >_L C \wedge \neg B$ then $C \wedge A \wedge B >_L C \wedge (\neg A \vee \neg B)$.

The latter condition (also called adjunction property) is of course very strong, but one example of it is a comparative possibility ordering described below. We are now in a position to define acceptance relations, that are confidence relations which enable closed sets of accepted beliefs to be derived from them.

Definition 2 : *A strict acceptance relation (SAR) is an irreflexive, transitive and non-dogmatic relation $>_a$ on $\mathcal{E}$ that satisfies Monotony 2 and CCC.*

Note that property CEC follows from Monotony 2 and property SUC is a consequence of the non-dogmatism assumption. Noticeably, the properties that make the set of accepted beliefs (according to a confidence relation) a consistent non-empty deductively closed set only involve the strict part of the confidence relation, moreover restricted to disjoint subsets (we call this part of the relation its *disjoint graph*).

A confidence relation $\geq_L$ also gives rise to a set of rules Δ, based on the intuition that if in context C the proposition A is believed by the agent, this is because the latter considers that in general, *if C then A* holds:

$$\Delta_L = \{C \rightsquigarrow A : C \wedge A >_L C \wedge \neg A\}$$

Note that Δ_L will not contain $C \rightsquigarrow C$ if $C = \emptyset$, nor $C \rightsquigarrow \emptyset$ if $C \neq \emptyset$. In line with this remark, we shall restrict the reflexivity axiom of system P to non contradictory propositions and exclude rules with satisfiable conditions and conflicting conclusions. These two conditions, called Restricted Reflexivity and Consistency Preservation taken together weaken and innocuously replace the reflexivity axiom.

The connection between the properties of acceptance relations and those of preferential entailment from rule base is expressed by the following result Dubois et al. [2004]:

Theorem 1 *If Δ is a conditional knowledge base closed under axioms AND, OR, RW, CM, Consistency Preservation ($C \rightsquigarrow \emptyset \notin \Delta$), and Restricted Reflexivity ($A \rightsquigarrow A \in \Delta$, for $A \neq \emptyset$), then the confidence relation defined by*

$$A \wedge B >_L A \wedge \neg B \iff A \rightsquigarrow B \in \Delta$$

is (the disjoint graph of) a Strict Acceptance Relation. Conversely, any SAR $>_a$ induces a rule base Δ_a as

$$A \vee B \rightsquigarrow \neg B \in \Delta \iff A >_a B$$

for disjoint sets $A, B \in \mathcal{E}$ that is closed under RW, CM, CUT, OR, Consistency Preservation and Restricted Reflexivity.

In other words, the family of conditional assertions corresponding to Δ_a satisfies system P but for axiom $\emptyset \hspace{0.2em}\sim\joinrel\hspace{-0.5em} \emptyset$. The latter is purely linked to conventions in classical logic regarding the status of the contradiction. Assuming $\emptyset \hspace{0.2em}\sim\joinrel\hspace{-0.5em} \emptyset$ is in conflict with conventions in probability theory, where conditioning on the empty set is not allowed. It can be claimed that Restricted Reflexivity and Consistency Preservation are more natural in non-monotonic plausible reasoning (in general, from a contradiction, we preferentially infer nothing).

6.2 Transitive acceptance relations: comparative probability and possibility

In this section we supplement properties of a confidence relation with additional ones that make them amenable to a representation by a numerical set-function representing uncertainty. Doing so leads to a non-monotonic reasoning system strictly less conservative than system P, where Rational Monotony is central. The two additional properties are

- *Completeness*: $A \geq_L B$ or $B \geq_L A$

- *Transitivity*: if $A \geq_L B$ and $B \geq_L C$ then $A \geq_L C$

These properties enable $\geq_L$ to be represented (in a non-unique way) by a numerical set-function g in the sense that: $g(A) \geq g(B) \iff A \geq_L B$. There are a number of such transitive confidence relations well-known in the literature.

- **Comparative probability** [De Finetti, 1937][9]: they were proposed as a natural way of defining subjective probabilities from first principles. A comparative probability relation is a confidence relation $\geq_P$ that is complete, transitive, non-trivial and monotone; moreover, it also satisfies the *preadditivity* axiom:

  Preadditivity: if $A \wedge (B \vee C) = \emptyset$ then $B \geq_P C \iff A \vee B \geq_P A \vee C$.

  Interestingly, in the finite setting, this property is insufficient to ensure a representation by numerical probability functions, while probability functions do satisfy these axioms [Kraft, Pratt, and Seidenberg, 1958]: Given a probability measure P, the relation $P(A) \geq P(B)$ is preadditive, the converse is not true. A comparative probability relation is more complex than a probability function because the latter is completely determined by a set of weights summing to 1 attached to each possible world, while relation $\geq_P$ is not characterized by its restriction to possible worlds. Nevertheless, comparative possibility orderings are self-conjugate, in the sense that $A \geq_P B$ if and only if $\neg B \geq_P \neg A$.

- **Possibility orderings** [Lewis, 1973, Dubois, 1986] They were motivated in the setting of modal logics of counterfactuals. A possibility ordering is a confidence relation $\geq_\Pi$ that is complete, transitive and that satisfies a property similar to preadditivity:

---

[9]translated by Kyburg and Smokler [1980].

Disjunctive Stability: $\forall A, B \geq_\Pi C \Rightarrow A \vee B \geq_\Pi A \vee C$.

$B \geq_\Pi C$ reads B is at least as possible as C. Possibility orderings are totally characterized by their restriction to possible worlds: let $\omega_1 \geq_\pi \omega_2$ stand for $\{\omega_1\} \geq_\Pi \{\omega_2\}$. The relation $\geq_\pi$ is a complete preordering of possible worlds expressing their relative plausibility or normality. Then:

$$B \geq_\Pi C \iff \exists \omega_1 \in B, \forall \omega_2 \in C, \omega_1 \geq_\pi \omega_2.$$

The one and only numerical counterparts to possibility orderings are so-called possibility measures Π such that $\Pi(A \vee B) = \max(\Pi(A), \Pi(B))$ [Dubois, 1986]. Possibility measures were introduced by Zadeh [1978] as set-functions induced by possibility distributions, mathematically fuzzy sets on Ω with membership functions $\pi : \Omega \to [0, 1]$. Namely, $\Pi(A) = \max_{\omega \in A} \pi(\omega)$, which expresses a degree of consistency between the proposition A and the background knowledge described by π. Possibility measures actually date back to the economist G.L.S. Shackle [1961] for whom worlds are all the more impossible as they are potentially surprising. Possibility orderings are also used by Grove [1988] in the belief revision setting. Halpern [1997] studies the case where the plausibility relation $\geq_\pi$ is a partial order, which yields several ways of defining the corresponding relation $\geq_\Pi$.

• **Necessity orderings**: as possibility measures and orderings are not self-conjugate, the conjugate of a possibility ordering is a necessity ordering [Dubois, 1986], defined by $A \geq_N B$ if and only if $\neg B \geq_\Pi \neg A$. A necessity ordering $\geq_N$ is a complete and transitive confidence relation that satisfies

Conjunctive Stability: $B \geq_N C \Rightarrow A \wedge B \geq_N A \wedge C$.

By duality with possibility orderings, and in agreement with modal logics, $B \geq_N C$ reads "B is at least as certain as C." Necessity orderings also subsume so-called epistemic entrenchment relations [Gärdenfors, 1988], instrumental in the theory of belief revision. Necessity orderings are totally defined by the plausibility ordering π. Their one and only numerical counterparts are necessity measures N such that $N(A \wedge B) = \min(N(A), N(B))$ [Dubois and Prade, 1988]. Necessity measures and possibility measures are conjugate set-functions in the sense that necessity measures are of the form $N(A) = 1 - \Pi(\neg A)$ where Π is a possibility measure.

• **Confidence orderings representable by Shafer belief and plausibility functions**. A weaker form of conjunctive stability for a transitive and complete confidence relation we shall denote by $\geq_b$ (referring to Shafer's belief functions [Shafer, 1976]) is as follows:

If A, B, C are disjoint sets then $B \vee C >_b C \Rightarrow A \vee B \vee C >_b A \vee C$.

This axiom is actually a slight reinforcement of the Monotony 1 axiom, since under the latter, the property $B \vee C >_L C \Rightarrow A \vee B \vee C \geq_L A \vee C$ holds.

Wong, Yao, Bollman, and Burger [1993] have proved that such orderings were always representable by belief functions [Shafer, 1976], namely set functions $Bel : \mathcal{E} \to [0,1]$ of the form $Bel(A) = \sum_{E \subseteq A} m(E)$, where m is a probability distribution over $\mathcal{E}$ such that $m(\emptyset) = 0$. Both necessity and comparative probability orderings satisfy this weak conjunctive stability property. The dual orderings defined $A \geq_{pl} B$ if and only if $\neg B \geq_b \neg A$ satisfy the converse property:

If A, B, C are disjoint sets then $A \vee B \vee C >_{pl} A \vee C \Rightarrow B \vee C >_{pl} C$

These confidence relations, we can call "plausibility orderings"[10] are always representable by Shafer plausibility functions of the form $Pl(A) = 1 - Bel(\neg A)$. Pl is the conjugate set-function of Bel. Both possibility and comparative probability orderings satisfy this weak stability property. Note that these two properties can be joined into a symmetric axiom

$$A \vee B \vee C >_{bpl} A \vee C \iff B \vee C \geq_{bpl} C.$$

Transitive and complete relations that obey this property are not self-conjugate but they are nevertheless representable by both a belief function and a plausibility function (that are not numerically conjugate). This property is indeed more general than the pre-additivity of comparative probabilities. Comparative probabilities are thus always representable by belief or plausibility functions (but not always by probability functions).

In the remainder of this section, we check whether such transitive relations corresponding to well-known uncertainty theories qualify as acceptance relations.

It is easy to check that the set $\mathcal{A}_\Pi(C)$ of accepted beliefs in the non-contradictory context C, induced by a possibility ordering such that $C >_\Pi \emptyset$ is deductively closed; this is because the CCC property holds for possibility orderings. Using the numerical representation by a possibility measure, it reads: if $\Pi(A \wedge C) > \Pi(\neg A \wedge C)$ and $\Pi(B \wedge C) > \Pi(\neg B \wedge C)$, then $\Pi(A \wedge B \wedge C) > \Pi((\neg A \vee \neg B) \wedge C)$.

Proof $\Pi(A \wedge C) > \Pi(\neg A \wedge C)$ and $\Pi(B \wedge C) > \Pi(\neg B \wedge C)$, imply
$\max(\Pi(A \wedge C), \Pi(B \wedge C)) > \max(\Pi(\neg A \wedge C), \Pi(\neg B \wedge C)) = \Pi((\neg A \vee \neg B) \wedge C)$
But $\max(\Pi(A \wedge C), \Pi(B \wedge C)) = \Pi((A \vee B) \wedge C)$
$= \max(\Pi(A \wedge B \wedge C), \Pi(\neg A \wedge B \wedge C), \Pi(A \wedge \neg B \wedge C)) = \Pi(A \wedge B \wedge C)$
since maximal elements with respect to $\geq_\pi$ on C are within A and within B by assumption.
QED

Non-dogmatic possibility orderings rely on a plausibility ordering of possible worlds such that none is viewed as impossible ($\forall \omega_i \in \Omega, \{\omega_i\} >_\Pi \emptyset$.) Note that this property does not properly apply to necessity orderings. Namely, when a possibility ordering is non-dogmatic, its conjugate necessity ordering may be such that $A \sim_N \emptyset$, for some $A \neq \emptyset$. In particular one may have that $\forall \omega_i \in \Omega, \{\omega_i\} \sim_N \emptyset$ while

---

[10]Not to be confused with Halpern's notion of plausibility functions, which are more general confidence orderings.

the corresponding possibility ordering is non-dogmatic. It intuitively means that all possible worlds are somewhat possible while none is whatsoever certain. Clearly, $A \sim_N \emptyset$ does not mean that A is an impossible proposition. A definition of an impossible proposition A valid for any confidence ordering is: a proposition that does not affect the plausibility of other propositions when forming their disjunction with them, in the sense that

$$\forall B \in \mathcal{E} \setminus \{\emptyset\}, B \wedge A = \emptyset \Rightarrow A \vee B \sim_L B.$$

It is clear that for a non-dogmatic possibility ordering $\forall A \neq \emptyset \in \mathcal{E}, A >_\Pi \emptyset$, no event is impossible in the above sense for its conjugate relation $\geq_N$. Since $\geq_\Pi$ and $\geq_N$ contain the same information, also expressed by the plausibility ordering $\geq_\pi$, the set of accepted beliefs induced by the plausibility ordering $\geq_\pi$ should then be defined in terms of $\geq_\Pi$, not in terms of $\geq_N$, since even with no impossible event, one may have that $\mathcal{A}_N(C) = \emptyset$, while $\mathcal{A}_\Pi(C) \neq \emptyset, \forall C \neq \emptyset$.

This is confirmed by the following result showing that possibility orderings are unavoidable when defining deductively closed sets of accepted beliefs induced by a complete and transitive confidence relation $\geq_L$ [Dubois et al., 2004]:

Theorem 2 : *For any non-dogmatic, transitive and complete confidence relation, the following propositions are equivalent:*

- $\mathcal{A}_L(C)$ *is closed in any non-contradictory context C;*

- $\geq_L$ *is a possibility ordering*

- *The conditional knowledge base Δ_L it generates is closed under restricted reflexivity, consistency preservation, RW, Right AND and Left OR and Rational Monotony: if $A \rightsquigarrow B \in \Delta_L$ and $A \rightsquigarrow \neg C \notin \Delta_L$ then $A \wedge C \rightsquigarrow B \in \Delta_L$.*

As a consequence, possibility theory is *the* theory of defeasible acceptance in a classical logic setting. This should not come as a surprise. Strict acceptance orderings satisfy the negligibility property proposed by Halpern [1997]:

if $A \wedge B = \emptyset, A \wedge C = \emptyset, A >_a B$ and $A >_a C$, then $A >_a B \vee C$.

It means that an acceptance ordering is not compensatory: the disjunction of several implausible events cannot be more plausible than an event more likely than each of them. Possibility orderings satisfy a slightly stronger property: $A \geq_\Pi B$ and $A \geq_\Pi C$, then $A \geq_\Pi B \vee C$, and this property is characteristic of max-decomposable set-functions.

Besides, acceptance relations are hardly compatible with probabilistic orderings: if an acceptance relation is a comparative probability relation on Ω, and the latter has more than two elements ($N > 2$), it can be proved [Dubois et al., 2004] that there is a permutation of Ω such that

$$\omega_1 >_\pi \omega_1 >_\pi \dots >_\pi \omega_{N-1} \geq_\pi \omega_N$$

for some plausibility ordering $\geq_\pi$ of possible worlds. Very few comparative probability relations are concerned: the only recovered family is the one of big-stepped probabilities. In fact, an acceptance relation induced by a big-stepped probability has the same disjoint graph as a (linear) possibility ordering. Alternatively, we can interpret the qualitative setting of acceptance functions in terms of infinitesimal probabilities [Pearl and Goldszmidt, 1996, Goldszmidt and Pearl, 1996].

More generally, acceptance orderings are partial orderings that can be refined by possibility orderings (restoring completeness and transitivity of the partial relation $\geq_a$). In fact it can be proved that, for any acceptance relation, there exists a family of possibility relations such that the derived set of accepted beliefs $\mathcal{A}_a(C)$ in any context C is the intersection of the sets of accepted beliefs $\mathcal{A}_\Pi(C)$ derived from the possibility relations Π that refine it [Dubois et al., 2004]. This result comes close to the result by Lehmann and Magidor [1992] that any preferentially closed set of conditional assertions is the intersection of sets closed under Rational Closure (that is, for which Rational Monotony holds).

6.3   Possibilistic logic, kappa functions and system Z

It is more natural to start with a rule base than with a confidence ordering when representing knowledge in practice. Given that the qualitative notion of acceptance leads to preferential inference and Rational Closure, a major question is how to compute a plausibility ordering from a rule base. A similar question is at the root of De Finetti's approach to probability. He viewed knowledge as a bunch of conditional statements to which probabilities are attached, and from which a probability measure (or a sequence thereof) must be constructed. Under preferential entailment, it is clear that a rule base constrains a family of plausibility orderings, and these questions can then be spelled out as follows:

- How to select a plausibility ordering from a rule base ?

- How to efficiently compute the set of accepted beliefs in a given context ?

One way of selecting a plausibility ordering from a rule base is to prioritize the rules in terms of toleration. If a rule prescribes the behavior of exceptions to another rule, the latter cannot be applied if the former is. Hence the former should have priority over the latter. This idea leads to an algorithm first suggested by Pearl [1990]. The toleration algorithm, already outlined in section 4.3, is now given in more details. Suppose a rule base Δ is given.

1. For each rule $A_i \leadsto B_i \in \Delta$, check the consistency of the propositional knowledge base $\mathcal{B}_i = \{A_i \wedge B_i, \neg A_j \vee B_j, j \neq i\}$.

2. If $\mathcal{B}_i$ is consistent, it means that there are examples of $A_i \leadsto B_i$ that are not counterexamples to any other rule: $A_i \leadsto B_i$ is tolerated by $\Delta \setminus \{A_i \leadsto B_i\}$.

3. If $\mathcal{B}_i$ is inconsistent it means that all examples of $A_i \leadsto B_i$ are counterexamples to some other rule: $A_i \leadsto B_i$ is not tolerated by $\Delta \setminus \{A_i \leadsto B_i\}$.

4. Delete from Δ a set Δ_1 containing tolerated rules and redo the above procedure until all rules in the current Δ are tolerated by each other.

The result is a ranking of the rule base $\Delta = \bigcup_{j=1}^{k} \Delta_i$, where Δ_j contains rules that have less priority than Δ_{j+1}. Pearl [1990] showed that it corresponds to the most compact ranking whereby rules in Δ_i are tolerated by those in $\bigcup_{j=i+1}^{k} \Delta_j$.

It may happen that the procedure fails at some point, whereby no rule is tolerated by other ones, this is a case when the rule base is inconsistent (as explained in section 4.3). Such a kind of inconsistency also prevents inferences in the sense of System P from being made. The obtained ranking of rules, when it exists, is the same as the one obtained when solving systems of equations induced by conditional probability assessments in the De Finetti approach as mentioned in Section 5.2.2.

Another view of the above algorithm consists in interpreting each rule $A_i \rightsquigarrow B_i \in \Delta$ as a constraint of the form $c_i : A_i \wedge B_i >_\Pi A_i \wedge \neg B_i$ expressing that examples of the rule are more plausible than counterexamples in the sense of a possibility ordering. Then a unique ranking of possible worlds can be obtained by computing the least specific possibility distribution $\pi : \Omega \rightarrow L$, where L is a totally ordered possibility scale of the form $\{\lambda_1 = 1 > \lambda_2 >, \ldots, \lambda_i, \ldots, 0\} \subseteq [0,1]^{11}$. A possibility distribution π_1 is less specific than another π_2 if and only if $\forall \omega \in \Omega, \pi_1(\omega) \geq \pi_2(\omega)$: in other words π_1 is less informative than π_2 [Dubois and Prade, 1988]. Minimizing specificity means maximizing possibility degrees, and yields the most compact ranking of possible worlds.

Given the set of constraints $\mathcal{C} = \{c_i, i = 1, \ldots, N\}$ the possibilistic algorithm, proposed by Benferhat et al. [1992] goes as follows:

1. Consider the proposition $\bigvee_{i=1}^{N} A_i \wedge \neg B_i$: its models are all exceptions to all rules. In order to respect the constraints, the condition $\Pi(A_i \wedge \neg B_i) \leq \lambda_2, \forall i$ must be respected.

2. Compute Ω_1 as the set of models of $\neg \bigvee_{i=1}^{N} A_i \wedge \neg B_i$: these possible worlds are exceptions to no rules. They can have maximal possibility degree $\pi(\omega) = \lambda_1$.

3. For each rule $A_i \rightsquigarrow B_i$ check if $A_i \wedge B_i$ is satisfiable by some $\omega \in \Omega_1$. If so, then constraint c_i is satisfied and can be deleted. Delete the satisfied constraints from $\mathcal{C}$ (they form set $\mathcal{C}_1$) and redo the above procedure until all constraints in the current $\mathcal{C}$ are satisfied.

The procedure yields a well-ordered partition $\Omega_1, \Omega_2, \ldots, \Omega_k$ of the set of worlds, where Ω_j contains worlds that are more normal than those in Ω_{j+1}, in the sense that $\forall \omega \in \Omega_j, \pi(\omega) = \lambda_j$. This is the same partition as when checking g-coherence of probabilistic conditionals in the sense of De Finetti (Section 5.2.2). If Δ is inconsistent, the set of constraints $\mathcal{C}$ is itself inconsistent. It can be checked that the set of constraints $\mathcal{C}_j$ corresponds to the set of rules Δ_j.

In order to reflect priorities expressed by the partitioning of the rule base, we can attach to each rule $A_i \rightsquigarrow B_i \in \Delta_j$ the propositional clause $\neg A_i \vee B_i$ and a degree of necessity $N(\neg A_i \vee B_i) = 1 - \lambda_{j+1} = 1 - \Pi(A_i \wedge \neg B_i)$ (since some exceptions to rule $\neg A_i \vee B_i$ lie in Ω_{j+1}). Interestingly, the priority ordering of the set of interpretations obtained above is the same as the so called zero-layers obtained by De Finetti's coherence algorithm applied to a set of conditional probabilities [Coletti and Scozzafava, 2002] (see also Section 5.2.2).

---

[11] The absolute values of the $\lambda_i, i > 1$ are immaterial.

The semantic consequence from Δ according to the priority ranking thus constructed can then be defined as follows:

$$\Delta \models_\pi A \leadsto B \text{ if and only if } \Pi(A_i \wedge B_i) > \Pi(A_i \wedge \neg B_i),$$

where Π is the least specific qualitative possibility distribution satisfying $\mathcal{C}$, generated by the above procedure. The closure $Cl_\pi(\Delta) = \{A \leadsto B : \Delta \models_\pi A \leadsto B\}$ is precisely the same as the Rational Closure by Lehmann and Magidor [1992]: the ranking of possible worlds provided by π is the same ranking as the one they obtain on possible worlds. It is also the same inference as the one at work in system Z of Pearl [1988]. It satisfies Rational Monotony.

The set of weighted clauses $\mathcal{K}_\Delta = \bigvee_{j=1}^n \{(\neg A_i \vee B_i, 1 - \lambda_{j+1}), A_i \leadsto B_i \in \Delta_j\}$ is a knowledge base in possibilistic logic [Dubois et al., 1994b]. Possibilistic logic is a convenient framework for implementing the inference of rules from a rule base Δ in the sense of Rational Closure. Let $\mathcal{K} = \{(C_i, \nu_i), i = 1, \ldots, N\}$ be a possibilistic knowledge base. Syntactic inference of a weighted sentence (D, ν) from $\mathcal{K}$ is defined as follows. Let $\mathcal{K}_\nu = \{C_i : (C_i, \nu_i) \in \mathcal{K}, \nu_i \geq \nu\}$ be a classical propositional base named the ν-cut of $\mathcal{K}$. Then $\mathcal{K} \vdash (D, \nu)$ if and only if $\mathcal{K}_\nu \vdash D$ (in the usual sense). The degree of inconsistency of $\mathcal{K}$ is $Inc(\mathcal{K}) = \max\{\nu : \mathcal{K}_\nu \text{ is inconsistent}\}$. A proposition D is a non-trivial consequence of $\mathcal{K}$ if and only if $\mathcal{K} \vdash (D, \nu)$ with $\nu > Inc(\mathcal{K})$ (it is a consequence of the largest consistent sub-base of $\mathcal{K}$ made of the most prioritary propositions). It is clear that the complexity of this type of inference is at worst n times the complexity of a SAT problem.

A possibilistic knowledge base induces a complete preordering of possible worlds encoded by the possibility distribution $\pi(\omega) = \inf\{1 - \nu_i : \omega \not\models C_i\}$, and it holds that D is a non-trivial consequence of $\mathcal{K}$ if and only if $\Pi(D) > \Pi(\neg D)$. Applying these results to the possibilistic knowledge base $\mathcal{K}_\Delta$ issued from the rule base, Δ, it holds that $\Delta \models_\pi A \leadsto B$ if and only if B is a non-trivial consequence of $\mathcal{K}_\Delta \cup \{(A, 1)\}$ (the condition A is given the highest priority). In fact it turns out that any preferential inference (in the sense of system P) that obeys Rational Monotony can be simulated by non-trivial inference in possibilistic logic [Benferhat, Dubois, and Prade, 1997].

Any well-ordered partition of the set of interpretations can be modelled by a set of rules closed under Rational Closure using the above algorithms. Hence, in some sense, the approach exhausts the possible qualitative representations of epistemic states, and should be sufficient for this kind of plausible reasoning. Yet, other kinds of closures have been proposed that try to solve apparent paradoxes of Rational Closure (for instance, entropy-based rankings [Goldszmidt, Morris, and Pearl, 1993, Kern-Isberner, 2001], integer priorities like system Z^+ [Goldszmidt and Pearl, 1991], lexicographic rankings [Benferhat et al., 1993, Lehmann, 1995], the conditional entailment method [Geffner, 1992] rankings based on belief functions [Benferhat, Saffiotti, and Smets, 2000], etc.). Besides, default rules allow for implicit exceptions, but do not include any assessment of the certainty with which their conclusion holds when the rule applies. This can be expressed by so-called *uncertain default rules*, where the distinction between defeasibility and uncertainty semantics is highlighted by means of a two steps processing (both handled in possibility theory) [Dupin de Saint Cyr-Bannay and Prade, 2008].

To figure out why other types of closure may be searched for, consider again the penguin triangle example $\Delta = \{PE \rightsquigarrow \neg FL, BI \rightsquigarrow FL, PE \rightsquigarrow BI\}$: if we add that birds lay eggs ($BI \rightsquigarrow LE$) as a new rule, the above possibilistic/Rational Closure approach applied to the set of four rules will not be able to infer that penguin lay eggs, i.e, $PE \rightsquigarrow LE$ (because penguins are exceptional birds for flying, we can no longer conclude about eggs) and it remains agnostic about it, while some other approaches may be able to infer it. Other techniques that select a different best ranking as listed above may cope with this problem.

But one may admit that, in the example, being careful about inheritance concerning eggs is reasonable as penguins may be exceptional birds for properties other than flying as well. Moreover the formal schema of the penguin triangle applies to many other examples where the "natural" conclusion will differ, because the domain knowledge is different. One cannot rely on the inference machinery to embed all common sense knowledge in the world. Clearly, one piece of information is missing in the penguin story, namely that laying eggs and flying are independent features for birds (Rational Closure cannot guess this). Independence information can be expressed by adding new conditionals [Benferhat, Dubois, and Prade, 2002] and the resulting knowledge base will deliver the correct results. In fact, it is enough to add the information that penguin lay eggs, and this rule is tolerated by the other ones. More generally, non-intuitive conclusions coming from Rational Closure can provably be repaired by adding the proper default information [Benferhat, Dubois, and Prade, 1998].

The same approach can be developed and the same results can be obtained replacing non-dogmatic plausibility rankings and possibility distributions by so-called kappa-rankings introduced by Spohn [1988] and exploited by Pearl and colleagues [Goldszmidt and Pearl, 1996]. They are mappings denoted by κ from Ω to the set of integers, containing levels of impossibility (or disbelief): $\kappa(\omega) = 0$ if ω is fully possible, and ω is all the more impossible as $\kappa(\omega)$ increases. By a simple rescaling function, it is easy to change kappa-rankings into possibility distributions [Dubois and Prade, 1991b]. In particular they satisfy the characteristic property $\kappa(A \wedge B) = \min(\kappa(A), \kappa(B))$, and $\kappa(A) = \min_{\omega \in A} \kappa(\omega)$ exactly evaluates the degree of potential surprise of A coined by Shackle [1961]. The appealing feature of this encoding mode lies in the simplicity of using integers (rather than abstract qualitative values or real numbers), and the fact that they can be related to infinitesimal probability in a rather direct way: $\kappa(A) = n$ corresponds to the infinitesimal probability $P(A) = \epsilon^n$, another semantics of Rational Closure [Lehmann and Magidor, 1992, the appendix]. This connection explains why some scholars have tried to maximize entropy when selecting the proper ranking of possible worlds induced by a set of rules. Moreover, the fact that these values are integers strongly suggests using addition and subtraction and getting approaches more expressive than the one based on simple qualitative rankings (like Goldszmidt and Pearl's system Z^+ [Goldszmidt and Pearl, 1991]; Goldszmidt and Pearl [1996] provide an overview). For instance, conditioning a kappa-function comes down to computing $\kappa(A|C) = \kappa(A \wedge C) - \kappa(C)$, as a counterpart to probabilistic conditioning. Hence a rule $A \rightsquigarrow B$ can be attached a strength $n \in \mathbb{N}$ modelled as a (numerical) constraint of the form $n \cdot \kappa(A|B) < \kappa(A|\neg B)$, for instance, meaning that in context A, proposition B is all the more plausible than its negation as n is greater.

7 Conclusion

There is one form of symbolic non-monotonic reasoning that seems to be the back-bone of probabilistic reasoning with imprecise conditional probabilities. This kind of non-monotonic reasoning, syntactically captured by system P due to Lehmann and colleagues is accounted for at the semantic level by a three-valued logic conditional events whose probabilities are conditional probabilities. It answers the old question of probabilities of conditionals vs. conditional probabilities [Lewis, 1976]. This kind of non-monotonic reasoning is dedicated to exception-tolerant plausible reasoning, and as such has little to do with the non-monotonic logic programming tradition.

Possibility theory seems to be the unifying framework to the extension of system P by addition of the Rational Monotony axiom. The computation of Rational Closure can be encoded in possibilistic logic. It also accounts for reasoning with very small or very high probabilities, so that in this setting, qualitative possibility theory appears as a kind of coarse form of probabilistic reasoning where some events are much more probable than other ones. Note that there is another interpretation of quantitative possibility measures in terms of upper probability bounds, where conditional reasoning can be studied as well [Dubois and Prade, 1997].

In fact, there seem to be several equivalently expressive languages that have been discussed in the literature, namely

- Exception-tolerant reasoning based on conditional knowledge bases (the Kraus Lehmann Magidor theory) (using conditional assertions $A \mathrel{|\!\sim} B$).

- Reasoning with probabilistic conditionals under De Finetti coherence, where a rule can be expressed as $P(B|A) = 1$, while supporting non-monotonicity.

- Qualitative possibility theory (using conditional possibility and representing exception-prone rules as constraints $B \wedge A >_\Pi B \wedge \neg A$ on a possibility relation).

- Belief revision and contraction (the AGM theory [Gärdenfors, 1988]), where the existence of a rule $C \rightsquigarrow B$ means that when all that is known is C, the proposition B is accepted in the revised belief set under input information A.

- Qualitative independence which can be expressed by means of conditional assertions being verified simultaneously [Benferhat et al., 1998, 2002].

This paper has focused on the four first items. Regarding belief revision, it has been stressed that it was the other side of the non-monotonic coin [Makinson and Gärdenfors, 1989, Gärdenfors and Makinson, 1994]. However, what this means has been sometimes misinterpreted. A belief set in the sense of the AGM theory seems to underlie a set of conditionals Δ that in some sense guides the revision operation. This set of conditionals representing generic knowledge is not much highlighted in the book dedicated to this theory. What is revised in this approach is the set of accepted beliefs in a certain context. Namely the initial belief set can be understood as $K = \{A, \Delta \models \Omega \rightsquigarrow A\}$ where the inference is to be taken in the sense of System P or Rational Closure. The belief set K contains accepted beliefs when no evidential information is present. The revised belief set under input C is then

$K * C = \{A, \Delta \models C \rightsquigarrow A\}$. Then the correspondence between the AGM revision axioms and the properties of the corresponding conditional logic is perfect. In particular, the 8 AGM axioms correspond to the axioms of Rational Closure, and the epistemic entrenchment relation between propositions as induced by the AGM axioms is precisely a necessity ordering [Dubois and Prade, 1991b]. Dubois [2008] offers an extensive discussion of this kind of revision (that corresponds to querying the rule base Δ, based on evidential information) as opposed to merging operations, and the revision of a rule-base [Boutilier, 1996].

In the case of independence, the parallel between probabilistic and possibilistic independence proves fruitful, as possibilistic qualitative counterparts of equivalent expressions of probabilistic independence turn out not to be equivalent any longer [Dubois, Fariñas Del Cerro, Herzig, and Prade, 1994a, de Campos and Huete, 1999a,b]. In particular, qualitative independence need not be symmetric nor invariant with respect to negation. Moreover statements of qualitative independence can be expressed by means of the addition of rules to a conditional knowledge base. For instance the conditional independence of A with respect to B in context C (which reads $P(A|B \wedge C) = P(A|C)$ in the probabilistic case) can be expressed by adding both rules $C \rightsquigarrow A$ and $B \wedge C \rightsquigarrow A$ and corresponds to possibility orderings satisfying the simultaneous constraints $A \wedge B \wedge C >_\Pi \neg A \wedge B \wedge C$ and $A \wedge C >_\Pi \neg A \wedge C$. This notion can be useful to solve some counterexamples to Rational Closure [Benferhat et al., 1998, 2002]. It has been recently applied to the study of causality in the framework of System P [Bonnefon, Da Silva Neves, Dubois, and Prade, 2008].

References

E. W. Adams. *The Logic of Conditionals*. D. Reidel, Dordrecht, The Netherlands, 1975.

S. Amarger, D. Dubois, and H. Prade. Constraint propagation with imprecise conditional probabilities. In *Proc. 7th Conf. on Uncertainty in AI, Los Angeles*, pages 26–34, San Francisco, 1991. Morgan Kaufmann Publishers.

A. Anderson and N. Belnap. *Entailment - The Logic of Necessity and Relevance*, volume 1. Princeton University Press, Princeton, N. J., 1975.

S. Benferhat, D. Dubois, and H. Prade. Representing default rules in possibilistic logic. In *Proc. of the 3rd Inter. Conf. on Principles of Knowledge Representation and Reasoning (KR'92), Cambridge, MA*, pages 673–684, San Francisco, 1992. Morgan Kaufmann.

S. Benferhat, C. Cayrol, D. Dubois, J. Lang, and H. Prade. Inconsistency management and prioritized syntax-based entailment. In *Proc. of the 13th Inter. Joint Conf. on Artificial Intelligence (IJCAI'93), Chambéry*, pages 640–645. AAAI Press, 1993.

S. Benferhat, D. Dubois, and H. Prade. Nonmonotonic reasoning, conditional objects and possibility theory. *Artificial Intelligence*, 92(1–2):259–276, 1997.

S. Benferhat, D. Dubois, and H. Prade. Practical handling of exception-tainted rules and independence information in possibilistic logic. *Applied Intelligence*, 9:101–127, 1998.

S. Benferhat, D. Dubois, and H. Prade. Possibilistic and standard probabilistic semantics of conditional knowledge bases. *Journal of Logic and Computation*, 9(6):873–895, 1999.

S. Benferhat, A. Saffiotti, and P. Smets. Belief functions and default reasoning. *Artificial Intelligence*, 122:1–69, 2000.

S. Benferhat, D. Dubois, and H. Prade. The possibilistic handling of irrelevance in exception-tolerant reasoning. *Annals of Mathematics and Artificial Intelligence*, 35:29–61, 2002.

V. Biazzo and A. Gilio. A generalization of the fundamental theorem of De Finetti for imprecise conditional probability assessments. *Int. J. Approximate Reasoning*, 24(2-3):251–272, 2000.

V. Biazzo, A. Gilio, T. Lukasiewicz, and G. Sanfilippo. Probabilistic logic under coherence, model-theoretic probabilistic logic, and default reasoning in System P. *Journal of Applied Non-Classical Logics*, 12(2):189–213, 2002.

J.-F. Bonnefon, R. Da Silva Neves, D. Dubois, and H. Prade. Predicting causality ascriptions from background knowledge: model and experimental validation. *International Journal of Approximate Reasoning*, 48:752–765, 2008.

C. Boutilier. Iterated revision and minimal change of conditional beliefs. *Journal of Philosophical Logic*, 25:263–305, 1996.

P. Calabrese. An algebraic synthesis of the foundations of logic and probability. *Information Sciences*, 42(3):187–237, 1987.

G. Coletti and R. Scozzafava. *Probabilistic Logic in a Coherent Setting*. Kluwer Academic Pub, 2002.

F. Cozman. Credal networks. *Artificial Intelligence*, 120(2):199–233, 2000.

R. Da Silva Neves, J.-F. Bonnefon, and E. Raufaste. An empirical test of patterns for non-monotonic inference. *Annals of Mathematics and Artificial Intelligence*, 34(1-3):107–130, 2002.

B. De Baets, J. C. Fodor, D. Ruiz-Aguilera, and J. Torrens. Idempotent uninorms on finite ordinal scales. *International Journal of Uncertainty, Fuzziness and Knowledge-Based Systems*, 17(1):1–14, 2009.

L. M de Campos and J. F. Huete. Independence concepts in possibility theory, part I. *Fuzzy Sets and Systems*, 103:127–152, 1999a.

L. M de Campos and J. F. Huete. Independence concepts in possibility theory, part II. *Fuzzy Sets and Systems*, 103:487–505, 1999b.

B. De Finetti. La logique de la probabilité. In *Actes du Congrès Inter. de Philosophie Scientifique (1935)*, pages IV1–IV9, Paris, France, 1936. Hermann et Cie Editions.

B. De Finetti. La prévision : ses lois logiques, ses sources subjectives. *Ann. Inst. Poincaré*, 7: 1–68, 1937.

B. De Finetti. *Theory of probability*. Wiley, New York, 1974.

D. Dubois. Belief structures, possibility theory and decomposable confidence measures on finite sets. *Computers and Artificial Intelligence (Bratislava)*, 5:403–416, 1986.

D. Dubois. Three scenarios for the revision of epistemic states. *J. Log. Comput.*, 18(5):721–738, 2008.

D. Dubois and H. Prade. *Possibility Theory: An Approach to Computerized Processing of Uncertainty.* Plenum Press, New York, 1988.

D. Dubois and H. Prade. Semantic considerations on order of magnitude reasoning. In M. G. Singh and L. Travé-Massuys, editors, *Decision Support Systems and Qualitative Reasoning*, pages 223–228. North-Holland, 1991a.

D. Dubois and H. Prade. Epistemic entrenchment and possibilistic logic. *Artificial Intelligence*, 50(2):223–239, 1991b.

D. Dubois and H. Prade. Conditional objects as nonmonotonic consequence relationships. *Special issue on Conditional Event Algebra, IEEE Trans. on Systems, Man and Cybernetics*, 24 (12):1724–1740, 1994.

D. Dubois and H. Prade. Non-standard theories of uncertainty in knowledge representation and reasoning. In G. Brewka, editor, *Handbook of Logic in Artificial Intelligence and Logic Programming*, pages 1–32. CLSI Publications and Folli, Stanford, CA, 1996.

D. Dubois and H. Prade. Bayesian conditioning in possibility theory. *Fuzzy Sets and Systems*, 92:223–240, 1997.

D. Dubois and H. Prade. Possibility theory: qualitative and quantitative aspects. In P. Smets, editor, *Handbook on Defeasible Reasoning and Uncertainty Management Systems - Volume 1: Quantified Representation of Uncertainty and Imprecision*, pages 169–226. Kluwer Academic Publ., Dordrecht, The Netherlands, 1998.

D. Dubois and H. Prade. Formal representations of uncertainty. In D. Bouyssou, D. Dubois, M. Pirlot, and H. Prade, editors, *Decision-making - Concepts and Methods*, chapter 3, pages 85–156. Wiley, New York, 2009.

D. Dubois, L. Godo, R. López de Mántaras, and H. Prade. Qualitative reasoning with imprecise probabilities. *J. Intell. Inf. Syst.*, 2(4):319–363, 1993.

D. Dubois, L. Fariñas Del Cerro, A. Herzig, and H. Prade. An ordinal view of independence, plausible reasoning and belief revision. In R. Lopez de Mantaras and D. Poole, editors, *Proc. of the 10th Conference on Uncertainty in Artificial Intelligence (UAI-94), Seattle, USA*, pages 195–203. Morgan Kaufmann, 1994a.

D. Dubois, J. Lang, and H. Prade. Possibilistic logic. In D. M. Gabbay, C. J. Hogger, J. A. Robinson, and D. Nute, editors, *Handbook of Logic in Artificial Intelligence and Logic Programming*, volume 3, pages 439–513. Oxford University Press, 1994b.

D. Dubois, H. Fargier, and H. Prade. Ordinal and probabilistic representations of acceptance. *J. Artificial Intelligence Res. (JAIR)*, 22:23–56, 2004.

D. Dubois, E. Hüllermeier, and H. Prade. A systematic approach to the assessment of fuzzy association rules. *Data Mining and Knowledge Discovery*, 13(2):167–192, 2006.

R. O. Duda, P. E. Hart, N. J. Nilsson. Subjective Bayesian methods for rule-based systems. In *Proc. US Nat. Computer Conf.*, volume 45, pages 1075–1082. AFIPS Conf. Proceedings, 1976.

F. Dupin de Saint Cyr-Bannay and H. Prade. Handling uncertainty and defeasibility in a possibilistic logic setting. *Inter. J. of Approximate Reasoning*, 49:67–82, 2008.

N. Friedman and J. Halpern. Plausibility measures and default reasoning. In *Proc. of the 13th National Conf. on Artificial Intelligence (AAAI-96)*, pages 1297–1304, Portland, OR, 1996.

A. Frisch and P. Haddawy. Anytime deduction for probabilistic logic. *Artificial Intelligence*, 69 (1-2):93–122, 1994.

D. Gabbay. Theoretical foundations for non-monotonic reasoning in expert systems. In *Logics and models of Concurrent Systems (K.R. Apt, ed.)*, pages 439–457. Springer Verlag, Berlin, 1985.

P. Gärdenfors. *Knowledge in Flux: Modeling the Dynamics of Epistemic States*. Bradford Books. MIT Press, Cambridge, 1988.

P. Gärdenfors and D. Makinson. Nonmonotonic inference based on expectations. *Artificial Intelligence*, 65(2):197–245, 1994.

H. Geffner. *Default Reasoning:Causal and Conditional Theories*. MIT Press, Cambridge, MA, 1992.

M. Gelfond. Answer sets. In V. Lifschitz F. van Harmelen and B. Porter, editors, *Handbook of Knowledge Representation*, chapter 7, pages 285–316. Elsevier, Amsterdam, 2008.

A. Gilio. Probabilistic reasoning under coherence in System P. *Annals of Mathematics and Artificial Intelligence*, 34(1-3):5–34, 2002.

M. Goldszmidt and J. Pearl. System Z+: A formalism for reasoning with variable-strength defaults. In *Proc of the National Conf. on Artificial Intelligence (AAAI-91)*, pages 399–404, 1991.

M. Goldszmidt and J. Pearl. Qualitative probabilities for default reasoning, belief revision, and causal modeling. *Artificial Intelligence*, 84:57–112, 1996.

M. Goldszmidt, P. Morris, and J. Pearl. A maximum entropy approach to nonmonotonic reasoning. *IEEE Trans. Pattern Analysis and Machine Intelligence*, 15(3):220–232, 1993.

I. R. Goodman, H. T. Nguyen, and E. A. Walker. *Conditional Inference and Logic for Intelligent Systems*. North-Holland, Amsterdam, 1991.

A. Grove. Two modelings for theory change. *Journal of Philosophical Logic*, 17(157-180), 1988.

I. Hacking. *The Logic of Statistical Inference*. Cambridge University Press, Cambridge UK, 1965.

J. Y. Halpern. Defining relative likelihood in partially-ordered preferential structures. *Journal of A.I. Research*, 7:1–24, 1997.

J. Y. Halpern. *Reasoning about Uncertainty*. The MIT Press, Cambridge, Massachusetts, London, England, 2003.

P. Hansen, B. Jaumard, G.-B. Douanya Nguetsé, and M. Poggi de Aragão. Models and algorithms for probabilistic and Bayesian logic. In *Proc. of the 14th Inter. Joint Conf. on Artificial Intelligence, Montreal*, pages 1862–1868, 1995.

C. G. Hempel. Studies in the logic of confirmation. *Mind*, 54:97–121, 1945.

G. Kern-Isberner. *Conditionals in Nonmonotonic Reasoning and Belief Revision - Considering Conditionals as Agents*, volume 2087 of *Lecture Notes in Computer Science*. Springer, 2001. ISBN 3-540-42367-2.

C. H. Kraft, J. W. Pratt, and A. Seidenberg. Intuitive probability on finite sets. *Annals of Mathematical Statistics*, 30:408–419, 1958.

S. Kraus, D. Lehmann, and M. Magidor. Nonmonotonic reasoning, preferential models and cumulative logics. *Artificial Intelligence*, 44:167–207, 1990.

H. Kyburg. *The Logical Foundations of Statistical Inference*. Reidel, Dordrecht, the Netherlands, 1974.

H. Kyburg and H. Smokler. *Studies in Subjective Probability*. Krieger Pub. Co., Huntington, N.Y., 1980.

L. Sombé. *Reasoning Under Incomplete Information in Artificial Intelligence*. Wiley, New York, 1990.

D. Lehmann. Another perspective on default reasoning. *Annals of Mathematics and Artificial Intelligence*, 15(1):61–82, 1995.

D. Lehmann. Nonmonotonic logics and semantics. *J. Log. Comput.*, 11(2):229–256, 2001.

D. Lehmann and M. Magidor. What does a conditional knowledge base entail? *Artificial Intelligence*, 55:1–60, 1992.

D. Lewis. *Counterfactuals*. Basil Blackwell, 1973.

D. Lewis. Probabilities of conditionals and conditional probabilities. *The Philosophical Review*, 85:297–315, 1976.

V. Lifschitz. Foundations of logic programming. In G. Brewka, editor, *Handbook of Logic in Artificial Intelligence and Logic Programming*, pages 69–127. CLSI Publications and Folli, Stanford, Ca., 1996.

T. Lukasiewicz. Local probabilistic deduction from taxonomic and probabilistic knowledge-bases over conjunctive events. *Int. J. Approximate Reasoning*, 21(1):23–61, 1999a.

T. Lukasiewicz. Probabilistic deduction with conditional constraints over basic events. *J. Artificial Intelligence Res. (JAIR)*, 10:199–241, 1999b.

D. Makinson. General patterns in nonmonotonic inference. In *Handbook of Logic in Artificial Intelligence and Logic Programming, Vol. 3 (D.M. Gabbay, C.J. Hogger, J.A. Robinson, D. Nute, eds.)*, pages 35–110. Oxford University Press, 1994.

D. Makinson and P. Gärdenfors. Relations between the logic of theory change and nonmonotonic logic. In *The Logic of Theory Change, Workshop, Konstanz, FRG*, volume 465 of *LNAI*, pages 185–205, 1989.

R. C. Moore. Semantical considerations on nonmonotonic logic. *Artificial Intelligence*, 25(1): 75–94, 1985.

I. Niemelä. Logic programs with stable model semantics as a constraint programming paradigm. *Annals of Mathematics and Artificial Intelligence*, 25(3-4):241–273, 1999.

G. Paass. Probabilistic logic (with discussions). In P. Smets, E.H. Mamdani, D. Dubois, and H. Prade, editors, *Non-Standard Logics for Approximate Reasoning*, pages 213–252. Academic Press, New York, 1988.

J. Pearl. *Probabilistic Reasoning in Intelligent Systems: Networks of Plausible Inference*. Morgan Kaufmann, 1988.

J. Pearl. System Z: a natural ordering of defaults with tractable applications to nonmonotonic reasoning. In *Proceedings of the Third Conference on Theoretical Aspects of Reasoning About Knowledge (TARK'90)*, 1990.

J. Pearl and M. Goldszmidt. Probabilistic foundations of reasoning with conditionals. In G. Brewka, editor, *Handbook of Logic in Artificial Intelligence and Logic Programming*, pages 33–68. CLSI Publications and Folli, Stanford, Ca., 1996.

D. Poole. The effect of knowledge on belief: Conditioning, specificity and the lottery paradox in default reasoning. *Artificial Intelligence*, 49(1-3):281–307, 1991.

R. Reiter. A logic for default reasoning. *Artificial Intelligence*, 13:81–132, 1980.

K. Schlechta. Nonmonotonic reasoning: a preferential approach. In D. Gabbay and J. Woods, editors, *Handbook of the History of Logic*, volume 8, pages 451–516. Elsevier, 2007.

G. L. S. Shackle. *Decision, Order and Time in Human Affairs*. (2nd edition), Cambridge University Press, UK, 1961.

G. Shafer. *A Mathematical Theory of Evidence*. Princeton University Press, 1976.

Y. Shoham. Reasoning about change. In *MIT Press, Cambridge, MA*, 1988.

E. Shortliffe and B. Buchanan. A model of inexact reasoning in medicine. *Mathematical Biosciences*, 23:351–379, 1975.

P. Snow. Diverse confidence levels in a probabilistic semantics for conditional logics. *Artificial Intelligence*, 113(1–2):269–279, 1999.

B. Sobociński. Axiomatization of a partial system of three-value calculus of propositions. *Journal of Computing Systems*, 1:23–55, 1952.

W. Spohn. Ordinal conditional functions: a dynamic theory of epistemic states. In W. L. Harper and B. Skyrms, editors, *Causation in Decision, Belief Change, and Statistics*, volume 2, pages 105–134. D. Reidel, 1988.

H. Thöne, U. Güntzer, and W. Kießling. Towards precision of probabilistic bounds propagation. In D. Dubois and M. P. Wellman, editors, *Proc. of the 8th Conference on Uncertainty in Artificial Intelligence*, pages 315–322. Morgan Kaufmann, 1992.

P. Walley. *Statistical Reasoning with Imprecise Probabilities*. Chapman and Hall, 1991.

S. M. K. Wong, Y. Y. Yao, P. Bollman, and H. C. Burger. Axiomatization of qualitative belief structure. *IEEE Trans. Systems Man and Cybernetics*, 21:726–734, 1993.

L. A. Zadeh. Fuzzy sets as a basis for a theory of possibility. *Fuzzy Sets and Systems*, 1:3–28, 1978.

Personal Perspective on the Development of Logic Programming Based KR Languages

Michael Gelfond
Computer Science Department
Texas Tech University
Lubbock, TX 79409, USA

Abstract: The paper is a personal perspective on research and development of languages for knowledge representation based on logic programming. The emphasis is on the methodology of this research, which is illustrated by the author's work on the stable models semantics of logic programs, Answer Set Prolog, and CR-Prolog.

1   Introduction

This paper is a short comment on some of my work on the development of knowledge representation (KR) languages. The story it presents can be viewed as a mixture of real history and rational reconstruction of this work. Since my goal is simply to articulate the methodological problems I encountered in this research and my attempts at their solutions, the paper almost entirely ignores contributions made by other researchers. The hope, of course, is that sharing even an incomplete story can shed some light on the current state of the field and may even be helpful to other people confronted with similar problems. The paper comments on the development of stable models semantics of logic programs and on extensions of the original language of logic programming such as Answer Set Prolog and CR-Prolog. A number of methodological points are illustrated by discussing the use of these languages for representing defaults and causal effects of actions. We also comment on the role of applications in this work.

I first became interested in the problem of knowledge representation in the early eighties when Vladimir Lifschitz came back to El Paso from Stanford Summer School with fascinating stories about defaults, non-monotonic logics and circumscription he learned in John McCarthy's course on AI. I attended the same school the following year and took the same course together with a course on Logic Programming taught by Kenneth Bowen. In that class I wrote my first Prolog program and got introduced to Prolog's treatment of negation. Even though in the beginning I felt that all the problems of non-monotonic reasoning could be solved in the framework of classical first-order logic, I soon realized that we may be dealing with a new phenomenon requiring different mathematical models and techniques. Since that time I have been looking for ways of formalizing various aspects of non-mathematical (especially common-sense) knowledge and reasoning.

My first encounter with methodological problems in this research occurred when I read the conference version of the paper by Hanks and McDermott [1987]. In this paper the authors introduced a simple temporal projection problem, which later became known as the Yale Shooting Problem. They considered several possible formalizations of this problem in known non-monotonic formalisms and showed that these formalizations did not produce the expected results. Based on this the authors concluded that temporal projection could not be naturally represented in these formalisms. I was surprised by the conclusion since it was not clear to me if the *problems were caused by the formalisms or by the particular methods of representing knowledge chosen by the authors*. The second possibility seemed to be supported by the simple and natural representations of the Yale Shooting problem in Reiter's Default Logic [Reiter, 1980], in Moore's Autoepistemic Logic [Moore, 1983], and in Logic Programming, which were found almost immediately after the publication of the paper (in particular, by Morris [1988], Gelfond [1989] and Evans [1989]). This alerted me early to the danger of premature conclusions. In particular it became clear that the *syntax and semantics of non-monotonic formalisms should be developed in conjunction with the development of the corresponding knowledge representation methodology*.

The need to better understand the phenomenon of non-monotonicity has not been limited to the area of knowledge representation. In the eighties logic programing language Prolog, developed by Colmerauer, Roussel, Kowalski, Warren and others, became a serious and powerful programming tool used for multiple practical applications. The language was especially attractive since it had a strong declarative component. A program in the original, "pure" Prolog could be seen as a collection of definite clauses of first-order logic with the minimal Herbrand model of the program providing its semantics. This minimal model could be viewed as the *intended model* of the program. From the beginning, however, Prolog had important non-logical features, including new non-monotonic logical connective "*not*" often referred to as *negation as failure* or *default negation*. Its original semantics was defined procedurally — *not p* was understood as finite failure to prove p by a particular proof algorithm called *SLDNF* resolution. To maintain the declarative character of Prolog, it was important to find declarative semantics of this connective. By the mid eighties there was a substantial amount of work in this direction. One approach, suggested by Clark [1978], viewed a logic program Π as a shorthand for a first-order theory called Clark's completion of Π with models of the completion defining the semantics of Π. Another approach aimed at expanding the definition of intended model from pure Prolog to programs with default negation. The influential paper by Apt, Blair, and Walker [1988] succeeded in defining such an intended model for so called stratified logic programs.

Perhaps somewhat surprisingly work on non-monotonic logic was developing more or less independently from that on the semantics of default negation in logic programming. However, there was a growing conviction among a small group of researchers in both areas that there were deep connections between them.

In 1987 I made some progress in discovering one such connection. A simple mapping of logic programs into Autoepistemic Logic [Gelfond, 1987] allowed the use of logic programming to answer queries to a non-trivial class of autoepistemic theories and to expand the intended model semantics of stratified programs. The resulting semantics became known as Stable Model semantics of logic programs [Gelfond and Lifschitz, 1988].

At about the same time, Well-Founded Semantics of logic programs was introduced by Van Gelder, Ross, and Schlipf [1991]. Both semantics coincided for stratified programs but, in more general cases, there were substantial differences between them. *Immediately after this introduction, discussion began about the relative merits of both semantics from the KR stand point. Differences of opinion were caused not only by personal intuitions and tastes, but also by different views on the goals and proper methodology for KR research and on the proper role of KR languages.* These differences, however, were often hidden in the background and very rarely articulated (at least outside of the anonymous reviewing process). This paper is an attempt to partially fill this gap. We do not necessarily need to reach consensus on the subject — in fact I believe that such a consensus may be harmful. But I also believe that asking and answering these questions is essential for every researcher, and that seeing how they were answered by others can be helpful.

2   Stable versus Well-Founded Semantics

In this section I will recall the discussions on comparative merits of stable and well-founded semantics from the standpoint of knowledge representation which started around 1988. (Of course a large part of what was happening in research on semantics of logics programs at the time, including all other interesting semantics, will be omitted. It will be great to see other people's recollections of this.) I will start with (a possibly incomplete) *list of different criteria which were used to evaluate KR languages*:

1. Clarity: logical connectives of a language should have a reasonably clear intuitive meaning.

2. Elegance: the corresponding mathematics should be simple and elegant.

3. Expressiveness: a KR language should suggest systematic and elaboration-tolerant representations of a broad class of phenomena of natural language. This includes basic categories such as belief, knowledge, defaults, causality, etc.

4. Relevance: a large number of interesting computational problems should be reducible to reasoning about theories formulated in this language.

5. Efficiency: Reasoning in the language should be efficient.

6. Regularity of entailment: entailment relation $\models$ of the language should satisfy some natural properties, e.g. cautious monotonicity: if $T \models F$ and $T \models G$ then $T \cup \{F\} \models G$.

7. Consistency: every program written in the language should be, in some sense, consistent.

8. Supra-classicality: the new language should extend first-order classical logic.

To decide which of the above criteria were important and which were less so, I needed to have a clear understanding of why I wanted to represent knowledge. Even though the details of my views on the subject evolved substantially, the basic answer seems to remain unchanged. I had two closely interrelated but distinct goals. *As a logician I wanted to better understand the basic commonsense notions we use to think about the world: beliefs, knowledge, defaults, causality, intentions, probability, etc., and to learn how one ought to reason about them. As a computer scientist I wanted to understand how to build software components of agents — entities which observe and act upon an environment and direct its activity towards achieving goals.* It is worth noticing that at the time I viewed even the simplest programs as agents. A program usually gets information from the outside world, performs an appointed reasoning task, and acts on the outside world by printing the output, making the next chess move, or starting a car. If the reasoning tasks it performs are complex and lead to nontrivial behavior, we call the program intelligent. If the program nicely adapts its behavior to changes in its environment, it is called adaptive, etc. It is possible that I developed this view because in my first real programming work I dealt with a large control system for paper production, and my first programming assignment was to figure out why some bulb did not light when it should, and to make it function properly. Clearly my program was supposed to observe, think, and act on the environment to achieve certain goal. To my surprise I later learned that for many people this view of programs seemed unnatural.

Applied to the area of logic programming and deductive databases, this view lead to the notion of *agent* which maintains its knowledge base (KB), represented as a logic program, and is capable of expanding it by new information and of answering questions about its knowledge. The Closed World Assumption [Reiter, 1978] built into the semantics of logic programs made the agent's answers defeasible. An agent with a simple KB = $\{p(a)\}$ will answer *no* to a query $?p(b)$[1]. If in communicating with the outside world the agent will learn that $p(b)$ is true, it will be able to nicely incorporate this information into its knowledge base. The new KB will consist of $\{p(a), p(b)\}$. Now the query $?p(b)$ will be answered with *yes*. This is a *typical example of non-monotonicity which prompted me to interpret the agent's conclusions as statements about its beliefs (as opposed to statements about actual truth or falsity of propositions).* Other people had different views. The fact that none of these views were normally clearly articulated in print caused a substantial amount of misunderstanding. *My general assumption about the nature of agent and its knowledge base influenced the definition of stable model.* Let us recall that, informally, stable models are collections of ground atoms which:

• Satisfy the rules of Π.

• Adhere to the *rationality principle* which says: *Believe nothing you are not forced to believe.*

The mathematical definition of stable models captures this intuition.

---

[1] I assume here that b belongs to the signature of the program. Similar assumptions will be made throughout the paper.

Now it was *time to evaluate semantics of logic programs with respect to the above criteria*. I was satisfied with the intuitive *clarity* and mathematical *elegance* of the stable model semantics. In this respect I had more difficulties with the well-founded semantics. For my taste the original definition of well-founded semantics looked too complex and not as declarative as I wanted. Later it was demonstrated that well-founded semantics can be viewed as a three-valued version of stable model semantics, but I still had difficulties with the declarative meaning of the third value. I think that reasonable people can disagree with this evaluation, but it was clear that in both semantics we only started to address the criterion of *expressiveness* of our languages. We better understood rational beliefs and reasoning with defaults, but even this substantial advance was hampered by *inability of the language to adequately represent incomplete information*. Existence of multiple stable models allowed some representation of incompleteness, but it seemed ad hoc. Moreover, the Closed World Assumption was a built-in feature of the stable model semantics. From my standpoint it was the major drawback of the language. Despite its third value the well-founded semantics did not seem to fare any better. Not much was available to us at the time to satisfy the fourth criterion —*relevance* of our languages for computing. We were able to use datalog and logic programming algorithms to build agents capable of answering sophisticated queries but that was it. The situation changed dramatically with the development of powerful answer set solvers such as Smodels [Niemela and Simons, 1997] and DLV [Leone, Pfeifer, Faber, Eiter, Gottlob, Perri, and Scarcello, 2006] and with mathematical results reducing planning [Subrahmanian and Zaniolo, 1995], diagnostics [Balduccini and Gelfond, 2003a], and many other non-trivial problems to computing (parts of) answer sets of logic programs. In 1989 all I could do was to give the stable model semantics an "incomplete" on the fourth criterion. The fifth requirement — that of the *efficiency* of reasoning — was interpreted differently by different people. One group strongly believed that stable model semantics did not satisfy this criterion since even for programs with finite Herbrand models the problem of finding a stable model of a program is NP-complete. (Computing the well-founded model is quadratic.) Another view understood complexity in a less stringent way. Even though Pure Prolog is undecidable in the presence of function symbols, many practical programs were successfully written and run with the standard Prolog interpreter (which, under reasonable conditions, is sound with respect to both semantics). For me the second view was obviously correct. (I have to confess that I still do not really understand the arguments for the first one.) Requirement six – the *regularity of entailment* – has been much more appealing. Unfortunately, as was shown in Van Gelder, Ross, and Schlipf [1991] stable model semantics does not satisfy even the simple property of cautious monotonicity. This is unfortunate but not very surprising. When I first studied calculus, my intuition revolted against continuous functions that are nowhere differentiable, but I was glad that existence of such counter-intuitive examples did not force the mathematicians to change the notion of continuous function. Still the fact that the entailment relation under the well-founded semantics is cautiously monotonic was rather appealing. Well-founded semantics also satisfies criterion of *consistency*. Every logic program has the well-founded model.[2] This is not true for the stable model semantics. Program $p \leftarrow not\ p$ has no stable models. This fact, however, did

---

[2]This property, of course, disappears if the language allows rules with the empty head.

not seem unnatural to me. I thought that the rule which says something like "if there is no reason to include p in your set of beliefs then include it in this set" can be viewed by a rational reasoner as non-sensical. The last criterion, *supra-classicality*, was not satisfied by the language of logic programs under any type of semantics. Overall *it was clear that making a definite choice between the two semantics was premature. We needed to expand the original language of logic programs to allow better expressiveness.* This was done by Gelfond and Lifschitz [1991] who came up with the language of extended logic programs which is now known as A-Prolog, ANS-Prolog, or Answer Set Prolog.

3   Answer Set Prolog (ASP)

In order to better understand the new language, let us recall that the stable model semantics was defined for programs consisting of rules of the form

$$a_0 \leftarrow a_1, \ldots, a_m, not\ a_{m+1}, \ldots, not\ a_n. \tag{1}$$

where a's are atoms of the program's signature. Rules with variables are viewed as shorthands for the set of their ground instantiations, so we assume that our atoms and rules are ground.

Definition 1 (Stable Models)
An atom a is true in a set of ground atoms S if $a \in S$. Otherwise, a is false in S. An expression not *a is true in S if $a \notin S$. Otherwise,* not *a is false in S. A set S satisfies a rule (1) if a_0 is true in S whenever all the statements in the body of the rule are true in S. For programs without default negation* not*, the stable model of the program is defined as the minimal set of atoms satisfying the program's rules. A set S of atoms is a stable model of an arbitrary program Π if it is a stable model of the reduct Π^S of Π with respect to S, which is obtained from Π by*

1. *removing all the rules whose body contain* not *a such that $a \in S$;*

2. *removing the remaining occurrences of expressions of the form* not *a.*

A program Π entails p if p is true in every stable model of Π; Π entails $\neg p$ if p is false in every stable model of Π. In the former case, Π answers *yes* to query $?p$. In the latter, the answer to this query is *no*. Otherwise, the answer is *unknown*.

Rules of Answer Set Prolog are substantially more general. They have the form:

$$l_0\ or\ \ldots\ or\ l_i \leftarrow l_{i+1}, \ldots, l_m, not\ l_{m+1}, \ldots, not\ l_n. \tag{2}$$

where l's are literals of the program's signature. (Note that the head of the rule can be empty.) A literal is defined as an atom a or its classical negation $\neg a$. (A literal possibly preceded by default negation is often referred to as an extended literal.) Statement $\neg a$ says that a is false, while *not* a only states that there is no rational support for believing in a. Symbol *or* denotes a new logical connective, called *epistemic disjunction*. Statement $a\ or\ b$ indicates that the reasoner must believe a or must believe b. (Hence "$a\ or\ \neg a$" is not a tautology.) The new language generalizes classical Prolog programs with default negation as well as disjunctive logic programs in the style of Minker [1982] which consist of rules of the form

$$a_0 \text{ or } \ldots \text{ or } a_i \leftarrow a_{i+1}, \ldots, a_m. \tag{3}$$

As before the rules of the program are viewed as constraints on the sets of beliefs which can be formed by a reasoner who adheres to the rationality principle. But this time beliefs are represented by consistent sets of ground literals called *answer sets* of the program. The *definition of answer set requires a minimal change in the definition of stable model.* All we need is a natural definition of truth and the consistency requirement.

Definition 2 (Answer Set)
A literal l is true in a set of literals S if $l \in S$; not l is true in S if $l \notin S$; l_1 or l_2 is true in S if either l_1 or l_2 is true in S. An answer set of a program without default negation not *is a minimal consistent set of literals satisfying the rules of Π. For an arbitrary program it is an answer set of the reduct Π^S defined as in Definition 1.*

The definition of entailment and that of answer to a query remain unchanged.

Despite this similarity in the definitions, introduction of new logical connectives and the shift from sets of atoms to sets of literals had a substantial impact on the language. Consider for instance a simplest possible program

$$\Pi_1 = \{p(a)\}.$$

The answer set of this program is $S = \{p(a)\}$. The same set S is its stable model. In the latter case $p(b)$ is false in S and, hence, the agent's answer to query $?p(b)$ will be *no*. In the former case, however, $p(b)$ is neither true nor false in S and, hence, the answer to $p(b)$ under the answer set semantics will be *unknown*. In other words the *Closed World Assumption is no longer part of the semantics of the new language.* It is important to see that in Answer Set Prolog this assumption for a relation p can be expressed by a statement:

$$\neg p(X) \leftarrow not\ p(X).$$

The statement says that $p(x)$ which is not believed to be true should be false. Program Π_2 obtained by adding this rule to Π_1 will answer *no* to a query $?p(b)$. Now the agent associated with the program believes that $p(b)$ is false. Of course, if $p(b)$ is learned to be true, the agent will simply add it to its knowledge base. The previous conclusion will be withdrawn. This simple example shows that the agent using a knowledge base written in ASP can gracefully accommodate new information and change its beliefs.

After completion of the language design part of our work, we needed to clearly understand what should be done to evaluate our language. It became clear rather early that ASP preserves the degree of clarity of intuition and mathematical elegance of the original language. But much time and effort was required to show that:

1. ASP increased our ability to express basic commonsense notions we use to think about the world.

2. A large number of interesting computational problems can be reduced to reasoning in ASP.

3. Efficient ASP reasoning algorithms can be found and built into the systems capable of solving these problems.

To test the expressibility of the new language, we first concentrated on representing defaults — statements of the form "*Normally elements of class c satisfy property p.*" The importance of defaults for knowledge representation and AI has been known for a long time. A large part of our education seems to consist of learning various defaults, their exceptions, and the skill of reasoning with them. Since defaults did not occur in the language of mathematics, they were not studied in classical mathematical logic. However, they play a very important role in everyday commonsense reasoning, and present a considerable challenge to AI researchers. Even though in the eighties remarkable progress had been made addressing this challenge, I thought that more work was needed to find a fully satisfactory solution. The first question was how to represent a default. Chitta Baral and I decided on the following general representation:

$$p(X) \leftarrow c(X),$$
$$not\ ab(d(X)),$$
$$not\ \neg p(X).$$

where d is the default's name [Baral and Gelfond, 1994]. The representation uses the abnormality predicate used in work by McCarthy [1990] for representing exceptions to defaults, and the default representation used in Reiter's Default Logic [Reiter, 1980]. (Since ASP programs without disjunction can be viewed as theories of Reiter's Default Logic, such a representation was rather natural.) We also considered two types of exceptions to defaults. A subclass c_1 of c is called a *strong exception* to default $d(X)$ if elements of c_1 do not satisfy property p; c_1 is called a *weak exception* if the default shall not be applied to its elements. Strong exceptions can be represented by the rules:

$$\neg p(X) \leftarrow c_1(X).$$

$$ab(d(X)) \leftarrow not\ \neg c_1(X).$$

In our representation $ab(d(X))$ holds if the default d shall not be applied to X. The second rule says that if x may belong to c_1, then application of default d to X should be stopped. Of course if the information about membership in class c_1 is complete, the second rule is unnecessary. The weak exceptions to d are represented by the rule:

$$ab(d(X)) \leftarrow not\ \neg c_1(X).$$

If information about membership in class c_1 is complete, then $not\ \neg c_1(X)$ in the body of the rule can be replaced by $c_1(X)$. From the mapping of ASP programs without disjunction into Reiter's default theories, we know that all of the above rules can be expressed in Reiter's default logic. The situation changes when we decide to consider theories containing disjunctions. Encoding of weak exceptions in ASP can also be done using the rule:

$$p(X)\ \text{or}\ \neg p(X) \leftarrow not\ \neg c_1(X).$$

which, in some situations, may be preferable to the one using ab; however, it becomes less natural in Reiter's logic. In addition, representing alternatives using classical disjunction in default logic does not allow for default reasoning by cases. Consider, for instance, a program

$$
\begin{aligned}
p_1(X) &\leftarrow c_1(X), \\
&\quad not \ \neg p_1(X). \\
p_2(X) &\leftarrow c_2(X), \\
&\quad not \ \neg p_2(X). \\
q(X) &\leftarrow p_1(X). \\
q(X) &\leftarrow p_2(X).
\end{aligned}
$$

used together with a disjunction

$$c_1(a) \text{ or } c_2(a).$$

Clearly the program entails $q(a)$ which corresponds to our intuition about proper reasoning with defaults. If, however, the above disjunction was understood as a classical statement in the first-order logic part of the corresponding default theory, the conclusion would not be reachable.

The above method of encoding exceptions to defaults could also be used to specify preferences between conflicting defaults. Assuming that preferences and conflicts are fully specified this can be done, say, by the rule:

$$
\begin{aligned}
ab(d_1(X)) &\leftarrow prefer(d_1(X), d_2(X)), \\
&\quad in\_conflict(d_1(X), d_2(X)).
\end{aligned}
$$

Gelfond and Son [1997] tested this representation of defaults and their exceptions using SLG [Chen and Warren, 1996] — one of the first reasoning systems capable of dealing with logic programs with multiple stable models. The system was still a prototype and not very easy to use, but the ability to implement such reasoning was in itself exciting. Of course more-powerful systems, like Smodels and DLV, followed quickly and writing programs capable of sophisticated default reasoning became almost a routine task.

Unfortunately, our representation of defaults in ASP had its difficulties. *While we were able to represent weak and strong exceptions to defaults, we still were not able to deal with situations when contradictions were found not with the conclusion of the default, but with the consequences of this conclusion.* We refer to such exceptions as *indirect*. To see the problem let us consider program I_1 consisting of rules

$$
\begin{aligned}
p(X) &\leftarrow c(X), \\
&\quad not \ ab(d(X)), \\
&\quad not \ \neg p(X). \\
q(X) &\leftarrow p(X). \\
c(a).
\end{aligned}
$$

Clearly, I_1 entails $q(a)$. But this program can not accept an update of the form $\neg q(a)$. The attempt to add this fact to the program leads to inconsistency. This problem was one of the reasons for the development of an extension of ASP called CR-Prolog [Balduccini and Gelfond, 2003b, Balduccini, 2004a], which we will discuss in the next section.

Gelfond and Lifschitz [1993] also investigated the possibility of *using ASP for representing causal effects of actions*. This started important work on action languages and lead to the establishment of a close relationship between reasoning about actions and ASP, which showed that ASP is capable of elegantly expressing direct and indirect effects of actions, and addressing the frame and ramification problems. For instance, the sentence

$$a \text{ causes } f$$

of action language $\mathcal{B}$ [McCain and Turner, 1995] says that if a were executed in some state, then fluent f would be true in the system's successor states. This causal law can be written as the ASP rule

$$holds(f, I + 1) \leftarrow occurs(a, I).$$

where I ranges over steps of the system's trajectory. Of course if a were to make f false, we would simply replace the head of the rule by $\neg holds(f, I + 1)$. Another typical causal law of $\mathcal{B}$ expresses the relationship between fluents. It has a form

$$l_0 \text{ if } l_1, \ldots, l_n$$

and says that any state of the system in which fluent literals $l_1, \ldots, l_n$ are true should also satisfy fluent literal l_0. This law can be naturally represented by the ASP rule

$$h(l_0, I) \leftarrow h(l_1, I), \ldots, h(l_n, I).$$

where for every fluent f, $h(f, I)$ denotes $holds(f, I)$ and, similarly, $h(\neg f, I)$ denotes $\neg holds(f, I)$. The classical frame problem can be solved in ASP by the Inertia Axiom:

$$holds(F, I + 1) \leftarrow holds(F, I), not \ \neg holds(F, I + 1).$$

$$\neg holds(F, I + 1) \leftarrow \neg holds(F, I), not \ holds(F, I + 1).$$

which formalizes the default "*Things tend to stay as they are.*" The formalization of dynamic domains outlined above successfully solved the frame and ramification problems which caused substantial difficulties in many previous attempts to formalize reasoning about actions and change. The success was mainly due to the ability of ASP to represent defaults and the use of non-contrapositive[3] rules which seem to better capture causal relations than classical implication. *This work clarified the nature of reasoning about dynamic domains and allowed to reduce classical AI problems such as planning and reasoning to computing answer sets of logic programs.*

Of course the two examples above do not cover all the basic concepts we would like to be able to represent in ASP. Later, substantial progress was made in using ASP and its extensions to reason about knowledge, intentions, probabilities, etc. But at the time *it*

---

[3]To see that ASP rules are non-contrapositive it is enough to note that programs $\Pi_1 = \{p \leftarrow q. \ \neg p\}$ and $\Pi_2 = \{\neg q \leftarrow \neg p. \ \neg p\}$ are not equivalent, i.e. have different answer sets.

was important to go beyond the first step of our methodology and actually check if the language could be made useful for the design of intelligent agents. Our initial work concentrated on tasks which could be reduced to answering queries to a knowledge base. A typical example of such work included a paper by Traylor and Gelfond [1994] in which ASP was used to *expand deductive databases with the ability to reason about various forms of null values.* I believe that, conceptually, we succeeded in showing that this can be done in a principled way, but the absence of access to programs with efficient interfaces between deductive and relational databases precluded us from trying to use these ideas to build practical prototypes. (Part of the difficulty was probably related to the fact that some work on deductive databases was done by companies. As a result we often heard rumors about such systems but I am still not sure if they really existed at the time. Recently I was happy to learn that the DLV group implemented and used an efficient interface with standard relational databases [Terracina, Leone, Lio, and Panetta, 2008]).

Our second attempt to investigate applicability of ASP to the design of intelligent systems was more successful. It started in the mid nineties in cooperation with United Space Alliance (USA) — the company responsible at the time for day-to-day operations of the Space Shuttle. The USA people came to us with the following problem. The Shuttle has a reactive control system (RCS) that has primary responsibility for maneuvering the aircraft while it is in space. It consists of fuel and oxidizer tanks, valves and other plumbing needed to provide propellant to the maneuvering jets of the Shuttle. It also includes electronic circuitry: both to control the valves in the fuel lines and to prepare the jets to receive ring commands. Overall, the system is rather complex, in that it includes 12 tanks, 44 jets, 66 valves, 33 switches, and around 160 computer commands (computer-generated signals). When an orbital maneuver is required, the astronauts must configure the RCS accordingly. This involves changing the position of several switches, which are used to open or close valves or to energize the proper circuitry. Normally, the sequences of actions required to configure the RCS are pre-determined before the beginning of the mission, and the astronauts simply have to search for the sequence in a manual. However, faults (e.g. the inability to move a switch) may make these pre-scripted sequences of actions inapplicable. The number of possible sets of failures is too large to plan in advance for all of them. In this case, the astronauts communicate the problem to the ground flight controllers, who come up with a new sequence of actions to perform the desired task. The USA wanted to *develop software to automatically check if the plans found by controllers are correct and do not cause dangerous side-effects.* Developing this software required methodology for modeling of and reasoning about complex dynamic systems. To do the checking the system had to have knowledge of the initial situation, the causal laws that rule the evolution of the domain and the operational knowledge of the controllers. We were delighted to discover that Answer Set Prolog proved to be suitable for representing this knowledge. Moreover, the representation was written in the form of an acyclic logic program which allowed us to check correctness of plans by running a standard Prolog interpreter [Watson, 1998]. Mathematical results establishing the relationship between descriptions of dynamic domains in high-level action languages and their logic programming counterparts, together with the work done in the Prolog community on termination and soundness of the interpreter, allowed us to significantly reduce the risk of programming errors. Overall our first experiment was successful — the resulting system was comparatively small, efficient, elaboration tolerant, and understandable to the USA people. However, our attempts to expand the system's functionality by teaching it to automatically search for the plans using a Prolog interpreter failed. In the original system plans were represented by terms and hence, in principle, could be found by the resolution mechanism. We were, however, *not able to overcome technical difficulties (including those related to floundering) and, hence, could not use our experiment to conclude that Answer Set Prolog can indeed be used as a practical tool for building multi-purpose intelligent agents.*

Fortunately the situation changed very substantially with the new breakthroughs in the area. The late nineties witnessed the appearance of new powerful algorithms and systems for computing stable models/answer sets of logic programs with finite Herbrand universes. This was accompanied by the change in emphasis from Prolog-style methodology of query-answering to the new paradigm [Marek and Truszczynski, 1999, Niemela, 1998] of reducing computational problems to finding answer sets of a program and computing these sets using answer set solvers. For me this development was a very pleasant surprise. I expected the development of really efficient systems of this sort to take much more time. However, after talking to Victor Marek and Mirek Truszczynski at the Kentucky workshop in 1998, I changed my mind. Together with my students, we decided to try the new technique on our Space Shuttle project. Reduction of planning to computing answer sets was already discussed in earlier papers and proving correctness of this approach in our context was not too difficult. The method worked. Reasonable plans were normally found in a matter of seconds. (From the USA standpoint, ten minutes was a good time.) We substantially expanded the system by adding more knowledge about RCS as well as the ability to find diagnosis for unexpected observations. From then on we used the system, called the USA advisor, to test our new reduction and reasoning algorithms and new KR languages.

From my standpoint the *work on USA advisor ended the first stage of our research program of evaluating the quality of ASP as a knowledge representation language.* The work went through several stages which can be formulated as follows:

1. Development of the syntax and semantics of the language accompanied by the investigation of methodology of the language used for knowledge representation.

   Here the emphasis is on the ability to model the basic conceptual notions, faithfulness to the intuition, and mathematical accuracy and elegance.

2. Evaluation of the language and its KR methodology by its use in designing small experimental systems capable of performing intelligent tasks.

   The emphasis here is on the ability of the language to guide our design, generalizability of solutions, and discovery of new phenomena (or a new perspective on an old one).

3. Evaluation of the language by its use in design and implementation of mid-size practical intelligent software systems.

   Emphasis here is on efficiency, correctness, and degree of elaboration tolerance.

Of course none of these steps was possible without the development of a mathematical theory of the language. The last two steps were also impossible without development of reasoning systems for the language. We were fortunate that such systems were made available to us thanks to the first class work done by other researchers.

4 Expanding ASP by abduction: CR-Prolog

Now it was time to concentrate on the KR problems which were not solved by ASP. One such problem, already discussed above, was the inability of ASP to represent indirect exceptions to defaults. This observation led to the development of a simple but powerful extension of ASP called CR-Prolog (or ASP with consistency-restoring rules). To illustrate the basic idea, let us go back to program I_1 from Section 2. We have already seen that this program, extended by an observation $\neg q(a)$, is inconsistent. There, however, seems to exists a commonsense argument which may allow a reasoner to avoid inconsistency, and to conclude that a is an indirect exception to the default. The argument is based on the **Contingency Axiom** for default $d(X)$ which says that "*Any element of class c can be an exception to the default $d(X)$, which says that elements of class c normally have property p, but such a possibility is very rare and, whenever possible, should be ignored.*" One may informally argue that since the application of the default to a leads to a contradiction, the possibility of a being an exception to $d(a)$ cannot be ignored and, hence, a must satisfy this rare property.

To allow formalization of this type of reasoning, we expand the syntax of ASP by rules of the form

$$l_0 \overset{+}{\leftarrow} l_1, \ldots, l_k, not\ l_{k+1}, \ldots, not\ l_n. \tag{4}$$

where l's are literals. Intuitively, rule (4), referred to as *consistency restoring* rule (cr-rule), says that if the reasoner associated with the program believes the body of the rule, then it "may possibly" believe its head[4]. However, this possibility may be used only if there is no way to obtain a consistent set of beliefs by using only regular rules of the program. The semantics of CR-Prolog is given with respect to a partial order, $\leq$, defined on sets of cr-rules. This partial order is often referred to as a **preference relation**. We will need the following notation.

The set of regular rules of a CR-Prolog program Π is denoted by Π^r; by $\alpha(r)$ we denote a regular rule obtained from a consistency restoring rule r by replacing $\overset{+}{\leftarrow}$ by $\leftarrow$; α is expanded in a standard way to a set R of cr-rules, i.e. $\alpha(R) = \{\alpha(r)\ :\ r \in R\}$. As in the case of ASP, the semantics of CR-Prolog is defined for ground programs. A rule with variables is viewed as shorthand for a schema of ground rules.

Definition 3 (Answer Sets of CR-Prolog)
A minimal (with respect to the preference relation of the program) collection R of cr-rules of Π such that $\Pi^r \cup \alpha(R)$ is consistent (i.e. has an answer set) is called an abductive support *of Π.*

---

[4]It may be worth noting that intuitively, the rule l_0 or $l_1 \overset{+}{\leftarrow} body$ can be replaced by two rules $l_0 \overset{+}{\leftarrow} body$ and $l_1 \overset{+}{\leftarrow} body$. Hence we do not allow disjunction in the heads of cr-rules.

A set A is called an answer set of Π if it is an answer set of a regular program $\Pi^r \cup \alpha(R)$ for some abductive support R of Π.

Consider, for instance, the following CR-Prolog program:

$p(a) \leftarrow not\ q(a).$
$\neg p(a).$
$q(a) \stackrel{+}{\leftarrow}.$

It is easy to see that the regular part of this program (consisting of the program's first two rules) is inconsistent. The third rule, however, provides an abductive support which allows to resolve inconsistency. Hence the program has one answer set $\{q(a), \neg p(a)\}$.

Now let us show how CR-Prolog can be used to represent defaults and their indirect exceptions. The CR-Prolog representation of default $d(X)$ may look as follows

$$
\begin{aligned}
p(X) \quad &\leftarrow \quad c(X), \\
&\quad\quad not\ ab(d(X)), \\
&\quad\quad not\ \neg p(X). \\
\neg p(X) \quad &\stackrel{+}{\leftarrow} \quad c(X).
\end{aligned}
$$

The first rule is the standard ASP representation of the default, while the second rule expresses the Contingency Axiom for default $d(X)$[5]. Consider now a program obtained by combining these two rules with an atom

$$c(a).$$

Assuming that a is the only constant in the signature of this program, the program's answer set will be $\{c(a), p(a)\}$. Of course this is also the answer set of the regular part of our program. (Since the regular part is consistent, the Contingency Axiom is ignored.) Let us now expand this program by the rules

$$
\begin{aligned}
q(X) \quad &\leftarrow \quad p(X). \\
\neg q(a).
\end{aligned}
$$

The regular part of the new program is inconsistent. To avoid the problem we need to use the Contingency Axiom for $d(a)$ to form the abductive support of the program. As a result the new program has the answer set $\{\neg q(a), c(a), \neg p(a)\}$. The new information does not produce inconsistency as in the analogous case of ASP representation. Instead the program withdraws its previous conclusion and recognizes a as a (strong) exception to default $d(a)$.

The above examples had only one possible resolution of the conflict and, hence, its abductive support did not depend on the preference relation of the program. When this

---

[5]In this form of Contingency Axiom, we treat X as a strong exception to the default. Sometimes it may be useful to also allow weak indirect exceptions; this can be achieved by adding the rule: $ab(d(X)) \stackrel{+}{\leftarrow} c(X).$

is not the case, preferred abductive supports are used to form the program's answer sets.

The ability to encode rare events which could serve as possible exceptions to defaults proved to be very useful for various knowledge representation tasks. For instance, in reasoning about actions we assume that normally the agent is aware of all the relevant actions occurring in the domain. This assumption can be expressed as

$$\neg occurs(A, I) \leftarrow not\ occurs(A, I).$$

Here A is a variable for actions and I ranges over the steps of the agent's trajectory. The consistency restoring rule

$$occurs(A, I) \stackrel{+}{\leftarrow} agent\_action(A), I > n.$$

where n is the last step of the current trajectory says that any agent action may occur in the future. Used together with the usual planning constraint which states that failing to achieve the goal is not an option, this rule can be used to find optimal plans. Similar (but more sophisticated) techniques were used by Balduccini [2004b] to improve quality of plans found by the USA advisor. A similar rule

$$occurs(A, I) \stackrel{+}{\leftarrow} exogenous\_action(A), I < n.$$

where by exogenous action we mean actions performed by nature or other agents in the domain, can be used to do diagnostics. In this case the discrepancy between the predicted value of the fluent and its observed value will be explained by some missed occurrences of unobserved actions which restore the program's consistency. Despite the fact that CR-Prolog proved to be a sufficiently simple and useful tool for knowledge representation which nicely combine traditional ASP reasoning with some form of abduction, the CR-Prolog-related research program is far from complete. Even though there is a meta-level implementation of CR-Prolog which was sufficiently efficient to be used for some practical applications, it is still too slow for others. The proper implementation should be tightly coupled with efficient ASP solvers — something we currently cannot do at TTU. A more important and substantially more difficult problem is related to finding proper ways of defining preference relation $\leq$. Should preference be defined between pairs of cr-rules and expanded to the sets of such rules? Should the preference be dynamic (i.e. allowed to be defined by the program's rules) or static? Should we have the preference built into the semantics of CR-Prolog? In the original papers we considered a particular (rather cautious) dynamic preferences relation built into the semantics of the language. In this approach we tried to avoid allowing ambiguity in the specification of preferences, preferring inconsistency of a program to obtaining results not necessarily meant to by the program designer. It seems that for some applications this was a good idea, while for others it was clearly inadequate. This is, of course, a general problem of representing preferences which, in my judgment, is not yet sufficiently understood (despite very substantial progress in this area). There are other languages which are somewhat similar to CR-Prolog. The most similar probably are papers by Inoue and Sakama [1996] and Buccafurri, Leone, and Rullo [1997]. The first introduced a preference relation on rules. The second formalized weak constraints implemented in DLV. Unfortunately, none of these languages has a "definitive solution" to the problem of preferences. This is an important topic for further research.

5   Recent work on extensions of ASP

Finally, I'd like to briefly mention some of my more recent work on expanding ASP. First, in 2004 my colleagues and I introduced a new KR language P-log [Baral, Gelfond, and Rushton, 2004, 2009]. Our goal was to create a language which would

- allow elegant formalizations of non-trivial combinations of logical and probabilistic reasoning,

- help the language designers (and hopefully others) to better understand the meaning of probability and probabilistic reasoning,

- help to design and implement knowledge-based software systems.

The logic part of P-Log is based on ASP (or its variants such as CR-Prolog). On the probabilistic side, we adopted the view which understands probabilistic reasoning as *commonsense reasoning about degrees of belief of a rational agent*. This matches well with the ASP-based logic side of the language. The ASP part of a P-log program can be used for describing possible beliefs, while the probabilistic part would allow knowledge engineers to quantify the degrees of these beliefs. Another important influence on the design of P-log is the separation between *doing* and *observing* and the notion of *Causal Bayesian Net* [Pearl, 2000]. The language has a number of other attractive features which do not normally occur in probability theory. P-log probabilities are defined with respect to an explicitly stated knowledge base. In addition to having logical non-monotonicity, P-log is "probabilistically non-monotonic" — new information can add new possible worlds and change the original probabilistic model. Possible updates include defaults, rules introducing new terms, observations, and deliberate actions in the sense of Pearl. Even though much more research is needed to better understand mathematical properties of P-log and develop the methodology of its use, I am confident that we have already succeeded in our first two goals. To satisfy the third goal we need to develop algorithms that truly combine recent advances in probabilistic reasoning with that of ASP — a non-trivial and fascinating task.

The second extension of ASP addresses a computational bottleneck of "classical" ASP solvers. These solvers are based on grounding — the process which replaces rules with variables by the sets of their ground instantiations. If a program contains variables ranging over large (usually numerical) domains, its ground instantiation can be huge (despite the best efforts of intelligent grounding algorithms employed by such solvers). This causes both memory and time problems and makes ASP solvers practically useless for a given task. Baselice, Bonatti, and Gelfond [2005] suggest an extension of the standard ASP syntax. The new syntactic construct facilitates the creation of a new type of ASP solver which partially avoids grounding and combines standard ASP reasoning techniques with that used by constraint logic programming algorithms. The work was extended in a number of different directions reported, for instance, by

Mellarkod, Gelfond, and Zhang [2008], Balduccini [2009] and Gebser, Ostrowski, and Schaub [2009], which substantially broadened the scope of applicability of the ASP paradigm.

6 Conclusion

This paper contains several examples of applying a particular methodology of research in knowledge representation and reasoning to the design of KR languages. In all these examples the research started with a theoretical question aimed at understanding some form of reasoning. Attempts to solve such a question normally lead to the design of a KR language which was then evaluated on its ability to model basic conceptual notions, faithfulness to intuition, and mathematical accuracy and elegance. At this stage the language could be of substantial interest to a logician who tries to understand correct modes of thinking. This first step is usually followed by the design and implementation of a naive reasoning engine which is used for the development of small experimental systems capable of performing carefully selected, intelligent tasks. This allows the researcher to find and tune proper knowledge representation methodology and to test the ability of the language to guide the design. At this stage the work may become even more interesting to the logician, but may also have a practical use as a specification language for a software designer. If the researchers want to use the language as an executable specification for real applications, they may need to spend a very substantial amount of time on discovering and implementing non-trivial reasoning algorithms, learning the application domain, and getting involved in a serious knowledge representation project. This type of experimental engineering not only shows practicality of the approach, but also helps a researcher to collect valuable feedback which can be used in the continuous process of improvement and extension. Even though at different points in time I was involved in work on most of these stages, I believe that, as a rule, they are quite different in nature, require different talents, and are sometimes impossible without a well assembled group of people. The clear realization of this fact by the community may help researchers to successfully publish their work even if it does not contain experimental results comparing efficiency of different implementations or non-trivial mathematical theorems establishing practically important properties of known languages. Finally, it may be interesting to add that the development of the above methodology was guided by the desire to satisfy the first four criteria for the knowledge representation language from section 2. *I found another four criteria substantially less important and sometimes even harmful.*

Overall I believe that the ASP research program of which I was a small part has very impressive accomplishments. We better understand the intuition behind such basic notions as belief, defaults, probability, causality, intentions, etc. This was done by the development of new mathematical theory and the methodology of its applications, and by experimental engineering. This foundational work helped to put our science on solid ground.

We are learning how to use our theories to build transparent, elaboration tolerant, provably correct, and efficient software systems. Twenty years ago I did not believe that such systems would be possible in my lifetime. I am obviously happy to be proved wrong. For me, this work has been, and (I hope) will continue to be, deeply satisfying. I also hope that it will be equally satisfying for new generations of researchers.

Acknowledgments

I'd like to thank my colleagues and collaborators who are, to a large extent, responsible for the work described in this paper (but not, of course, for the author's opinions). Special thanks to Vladimir Lifschitz for many useful discussions and for his comments on the paper.

References

K. Apt, A. Blair, and A. Walker. *Towards a theory of declarative knowledge*, pages 89–148. Foundations of deductive databases and logic programming. Morgan Kaufmann, 1988.

M. Balduccini. CR-MODELS: An inference engine for CR-prolog. In *International Conference on Logic Programming and Nonmonotonic Reasoning, LPNMR-7*, pages 18–30, 2004a.

M. Balduccini. USA-Smart: Improving the Quality of Plans in Answer Set Planning. In *PADL'04*, Lecture Notes in Artificial Intelligence (LNCS), pages 135-147, Springer 2004b.

M. Balduccini. CR-Prolog as a Specification Language for Constraint Satisfaction Problems. In E. Erdem, F. Lin, and T. Schaub, editors. *Logic Programming and Nonmonotonic Reasoning, 10th International Conference, LPNMR 2009, Potsdam, Germany, September 14-18, 2009. Proceedings*, volume 5753 of *Lecture Notes in Computer Science*, pages 402-408, Springer 2009.

M. Balduccini and M. Gelfond. Diagnostic reasoning with A-Prolog. *Journal of Theory and Practice of Logic Programming (TPLP)*, 3(4–5):425–461, 2003a.

M. Balduccini and M. Gelfond. Logic Programs with Consistency-Restoring Rules. In Patrick Doherty, John McCarthy, and Mary-Anne Williams, editors, *International Symposium on Logical Formalization of Commonsense Reasoning*, AAAI 2003 Spring Symposium Series, pages 9–18, 2003b.

C. Baral and M. Gelfond. Logic Programming and Knowledge Representation. *Journal of Logic Programming*, 19(20):73–148, 1994.

C. Baral, M. Gelfond, and N. Rushton. Probabilistic Reasoning with Answer Sets. In *Proceedings of LPNMR-7*, pages 21–33. Springer-Verlag, 2004.

C. Baral, M. Gelfond, and N. Rushton. Probabilistic reasoning with answer sets. *Journal of Theory and Practice of Logic Programming (TPLP)*, 9(1):57–144, 2009.

S. Baselice, P.A. Bonatti, and M. Gelfond. Towards an integration of answer set and constraint solving. In *Proceedings of ICLP-05*, pages 52–66, 2005.

F. Buccafurri, N. Leone, and P. Rullo. Strong and weak constraints in Disjunctive Datalog. In *Proceedings of LPNMR-4*, pages 2–17, 1997.

W. Chen and D. S. Warren. Computation of stable models and its integration with logical query processing. *IEEE Transactions on Knowledge and Data Engineering*, 8(5):742–757, 1996.

K. Clark. Negation as failure. In H. Gallaire and J. Minker, editors, *Logic and Data Bases*, pages 293–322. Plenum Press, 1978.

C. Evans. Negation-as-failure as an approach to the Hanks and McDermott problem. In *Proceedings Second International Symposium on Artificial Intelligence*, 1989.

M. Gebser, M. Ostrowski, and T. Schaub. Constraint answer set solving. In *ICLP*, pages 235–249, 2009.

M. Gelfond. On stratified autoepistemic theories. In *Proceedings of AAAI87*, pages 207–211, 1987.

M. Gelfond. Autoepistemic logic and formalization of commonsense reasoning. In *Nonmonotonic Reasoning*, volume 346 of *Lecture Notes in Artificial Intelligence*, pages 176–187, 1989.

M. Gelfond and V. Lifschitz. The stable model semantics for logic programming. In *Proceedings of ICLP-88*, pages 1070–1080, 1988.

M. Gelfond and V. Lifschitz. Classical negation in logic programs and disjunctive databases. *New Generation Computing*, pages 365–385, 1991.

M. Gelfond and V. Lifschitz. Representing Action and Change by Logic Programs. *Journal of Logic Programming*, 17(2–4):301–321, 1993.

M. Gelfond and T. C. Son. Reasoning with Prioritized Defaults. In *Third International Workshop, LPKR'97*, volume 1471 of *Lecture Notes in Artificial Intelligence (LNCS)*, pages 164–224, 1997.

S. Hanks and D. McDermott. Nonmonotonic logic and temporal projection. *Artificial Intelligence Journal*, 33(3):379–412, 1987.

K. Inoue and C. Sakama. Representing Priorities in Logic Programs. In *Proceedings of the Joint International Conference and Symposium on Logic Programming (JICSLP'96)*, pages 82–96. MIT Press, 1996.

N. Leone, G. Pfeifer, W. Faber, T. Eiter, G. Gottlob, S. Perri, and F. Scarcello. The dlv system for knowledge representation and reasoning. *ACM Transactions on Computational Logic*, 7: 499–562, 2006.

V. W. Marek and M. Truszczynski. *Stable models and an alternative logic programming paradigm*, pages 375–398. The Logic Programming Paradigm: a 25-Year Perspective. Springer Verlag, Berlin, 1999.

N. McCain and H. Turner. A causal theory of ramifications and qualifications. *Artificial Intelligence*, 32:57–95, 1995.

J. McCarthy. *Formalization of common sense, papers by John McCarthy edited by V. Lifschitz*. Ablex, 1990.

V. S. Mellarkod, M. Gelfond, and Y. Zhang. Integrating answer set programming and constraint logic programming. *Ann. Math. Artif. Intell.*, 53(1-4):251–287, 2008.

J. Minker. On indefinite data bases and the closed world assumption. In *Proceedings of CADE-82*, pages 292–308, 1982.

R. C. Moore. Semantical considerations on nonmonotonic logic. In *Proceedings of the 8th International Joint Conference on Artificial Intelligence*, pages 272–279. Morgan Kaufmann, 1983.

P. Morris. The anomalous extension problem in default reasoning. *Artificial Intelligence Journal*, 35(3):383–399, 1988.

I. Niemela. Logic Programs with Stable Model Semantics as a Constraint Programming Paradigm. In *Proceedings of the Workshop on Computational Aspects of Nonmonotonic Reasoning*, pages 72–79, 1998.

I. Niemela and P. Simons. Smodels - an implementation of the stable model and well-founded semantics for normal logic programs. In *Proceedings of the 4th International Conference on Logic Programming and Non-Monotonic Reasoning (LPNMR '97)*, volume 1265 of *Lecture Notes in Artificial Intelligence (LNCS)*, pages 420–429, 1997.

J. Pearl. *Causality*. Cambridge University Press, 2000.

R. Reiter. On Closed World Data Bases. In H. Gallaire and J. Minker, editors, *Logic and Data Bases*, pages 119–140. Plenum Press, 1978.

R. Reiter. A Logic for Default Reasoning. *Artificial Intelligence*, 13(1–2):81–132, 1980.

V. Subrahmanian and C. Zaniolo. Relating stable models and AI planning domains. In Leon Sterling, editor, *Proceedings of ICLP-95*, pages 233–247. MIT Press, 1995.

G. Terracina, N. Leone, V. Lio, and C. Panetta. Experimenting with recursive queries in database and logic programming systems. *Journal of Theory and Practice of Logic Programming (TPLP)*, 8(2):129–165, 2008.

B. Traylor and M. Gelfond. Representing null values in logic programming. In Anil Nerode and Yuri Matiyasevich, editors, *Proceedings of LFCS '94*, volume 813 of *LNCS*, pages 341–352. Springer, 1994.

A. Van Gelder, K. Ross, and J. Schlipf. The well-founded semantics for general logic programs. *Journal of ACM*, 38(3):620–650, 1991.

R. Watson. An application of action theory to the space shuttle. In *Proceedings of the First International Workshop on Practical Aspects of Declarative Languages (Lecture Notes in Computer Science 1551*, pages 290–304. Springer-Verlag, 1998.

A Unified Approach to First-Order Non-Monotonic Logics

Michael Kaminski
Department of Computer Science
Technion – Israel Institute of Technology
Haifa 32000, Israel

Abstract: This paper deals with the first-order non-monotonic logics that are defined semantically as the first-order counterpart of the corresponding propositional non-monotonic logics. An important feature of the logics' definitions is a *unified* approach to first-order propositional default, (ground) non-monotonic modal, and autoepistemic logics. Thus, in particular, first-order (ground) non-monotonic modal and autoepistemic logics well comply with first-order default logic, providing a substantial evidence for their adequacy. Also, all the above mentioned first-order nonmonotonic logics well comply with circumscription and possess many of the basic properties of their propositional counterparts, which supplies an additional support to their appropriateness.

1 Introduction

Non-monotonic logic is intended to simulate the process of human reasoning by providing a formalism for deriving consistent conclusions from an incomplete description of the world. Roughly speaking, the main idea lying behind non-monotonic logic is an inference from what is possible to what should actually hold. In particular, non-monotonic modal logic use the inference rule of a form

$$\text{"if } \neg\varphi \text{ is not derivable, conclude that } \varphi \text{ is possible"} \tag{1}$$

that corresponds to *negative introspection* of a rational agent, whereas *positive introspection* is provided by classical modal *necessitation*.

For a long time the main effort has been invested into investigation of propositional non-monotonic logics. A few attempts to extend to the first-order case did not succeed. For example, the definition of first-order non-monotonic modal logic in the paper by McDermott [1982] is unsatisfactory, because it does not allow formula φ in (1) to contain free variables. However, interesting cases of non-monotonic reasoning usually deal with "open versions" of (1), because the intended use of the rule is to determine whether an object possesses a given property rather than accepting or rejecting a "fixed statement." Adequate definitions of first-order default logic, non-monotonic modal logic, ground non-monotonic modal logic, and autoepistemic logic appeared in the papers by Kaminski [1995], Kaminski and Rey [2000], Kaminski and Grimberg [2008], Kaminski and Rey [2002], respectively. Each of the above logics is defined *semantically* by extending an appropriate semantical description of the corresponding propositional non-monotonic logic to the first-order case.

This paper deals with the first-order version of propositional non-monotonic logics and is based on the works of the author with his former M.Sc. students Benjamin Grimberg, Guy Rey, and Yael Zbar. The definition of first-order non-monotonic logics is based on an equivalent semantical description of propositional default, (ground) non-monotonic modal, and autoepistemic logics. This description was first introduced in the paper of Kaminski and Rey [2002] for first-order autoepistemic logic and then was extended by Zbar [2000] to first-order (ground) non-monotonic logics. Its extension to first-order default logic is presented in Section 3. Roughly speaking, the equivalent semantical descriptions of first-order non-monotonic logics result from the corresponding syntactical definitions in the propositional case by replacing provability relation $\vdash$ with semantical entailment $\models$. These new descriptions are equivalent to those in the papers of Kaminski [1995] and Kaminski and Rey [2000].

An important feature of the above descriptions is that they do not rely on a particular semantics of any of the above mentioned first-order non-monotonic logics. That is, these descriptions provide a *unified* approach to these logics.

The paper is organized as follows. In Section 2 we recall the definitions of propositional default logic, propositional non-monotonic modal logic, propositional ground non-monotonic modal logic, and propositional autoepistemic logic and state their basic properties. Section 3 contains the original and an equivalent alternative definitions of first-order default logic. In Section 4 we recall the Kripke semantics of first-order (monotonic) modal logic and compare some of its (relevant) properties with those of the propositional one. In Section 5 we present the original and an equivalent alternative definitions of first-order non-monotonic modal logic. Section 6 deals with first-order autoepistemic logic. The definition of first-order ground non-monotonic modal logic is presented in Section 7. We end the paper with Section 8 consisting of a summary of the uniform approach. We omit the proofs which can be found in the literature, always giving the exact reference, and, for the sake of completeness, unpublished proofs are presented in the appendix.

2   Propositional non-monotonic logic

In this section we recall the definitions of propositional default logic, propositional autoepistemic logic, propositional non-monotonic modal logic, and propositional ground non-monotonic modal logic and state their basic properties and inter-translatability results.

2.1   Propositional default logic

Propositional default logic logic deals with rules of inference called *defaults* which are expressions of the form

$$\frac{\alpha : \beta_1, \ldots, \beta_m}{\gamma},$$

where α, $\beta_1, \ldots, \beta_m$, $m \geq 0$, and γ are propositional formulas. The formula α is called the *prerequisite* of the default rule, the formulas $\beta_1, \ldots, \beta_m$ are called the *justifications*, and the formula γ is called the *conclusion*. Roughly speaking, the intuitive meaning of a default is as follows. If α is believed, and the β_is are consistent with ones beliefs, then one is permitted to deduce γ and add it to the "belief set." A *default theory* is a pair (D, A), where D is a set of defaults and A is a set of propositional formulas (axioms).

Definition 1 (Reiter [1980]) *Let (D, A) be a default theory. For a set of formulas X let $\Gamma_{(D,A)}(X)$ be the smallest set of formulas B (beliefs) that includes A, is deductively closed and satisfies the following* default closure *condition.*

$$\text{If } \frac{\alpha : \beta_1, \ldots, \beta_m}{\gamma} \in D, \alpha \in B, \text{ and } \neg\beta_i \notin X, i = 1, 2, \ldots, m, \text{ then } \gamma \in B.$$

A consistent set of formulas E is an extension *for (D, A) if $\Gamma_{(D,A)}(E) = E$, i.e., if E is a fixed point of the operator $\Gamma_{(D,A)}$.*

Remark 1 By soundness and completeness of the propositional semantics, one can define $\Gamma_{(D,A)}(X)$ as the smallest set of formulas B (beliefs) that includes A, is closed under semantical entailment[1] and satisfies the *default closure* condition of Definition 1.

Below is an equivalent semantical definition of extensions for default theories.

Definition 2 (Guerreiro and Casanova [1990]) *Let (D, A) be a closed default theory. For a class of propositional interpretations W let $\Sigma_{(D,A)}(W)$ be the largest class V of propositional interpretations satisfying all formulas from A for which the following holds.*

$$\text{If } \frac{\alpha : \beta_1, \ldots, \beta_m}{\gamma} \in D, V \models \alpha, \text{ and } W \not\models \neg\beta_i, i = 1, \ldots, m, \text{ then } V \models \gamma[2].$$

Guerreiro and Casanova [1990] showed that the definition of extensions as the theories of the fixed points of Σ is equivalent to Reiter's original definition (Definition 1). That is, a set of sentences E is an extension for a closed default theory (D, A) if and only if $E = \{\varphi : W \models \varphi\}$ for some non-empty fixed point W of $\Sigma_{(D,A)}$.

2.2 Propositional modal logic

Propositional modal logic is obtained from the classical one by adding to it the modal connective L (believed). The dual connective M (consistent) is defined by $\neg L\neg$. Formulas not containing L are called *ground* or *modal-free*. The set of all propositional modal formulas will be denoted $\boldsymbol{Fm}$ and the set of all ground propositional formulas will be denoted $\boldsymbol{GFm}$.

In this paper we shall mainly deal with the weakest *normal* modal logic K that results from the classical propositional logic by adding the rule of inference

---

[1] We say that a set of formulas A *semantically entails* a formula φ, if each propositional interpretation that satisfies all formulas from A also satisfies φ.

[2] This largest class $\Sigma_{(D,A)}(W)$ always exists [Lifschitz, 1990, Proposition 1].

NEC $\varphi \vdash L\varphi$

called *necessitation* and the axiom scheme

k $L(\varphi \supset \psi) \supset (L\varphi \supset L\psi)$[3].

The "classical" modal logics are obtained by adding to K all instances of some axiom schemes, e.g.,

t $L\varphi \supset \varphi$,

d $L\varphi \supset M\varphi$,

4 $L\varphi \supset LL\varphi$,

f $(\varphi \wedge ML\psi) \supset L(M\varphi \vee \psi)$,

5 $M\varphi \supset LM\varphi$.

Adding **t** to K results in T, adding **4** to T results in S4, adding **f** to S4 results in S4F, and adding **5** to S4 results in S5, etc., as discussed by Marek and Truszczyński [1993, page 197].

For a modal logic S and a set of formulas A, called (proper) *axioms*, we define the (monotonic) *theory of A*, denoted $\boldsymbol{Th}_S(A)$, as $\boldsymbol{Th}_S(A) = \{\varphi : A \vdash_S \varphi\}$, whereas the unsubscripted symbols $\boldsymbol{Th}$ and $\vdash$ indicate deductions without *necessitation*. As usual, we write $A \vdash_S \varphi$, if there exists a sequence of formulas $\varphi_1, \varphi_2, \ldots, \varphi_n$ containing φ such that each φ_i is an axiom from S or belongs to A or is obtained from some of the formulas $\varphi_1, \varphi_2, \ldots, \varphi_{i-1}$ by *modus ponens* or *necessitation*. Note that any normal modal logic S can be embedded in K by extending the set of proper axioms with S. That is, $A \vdash_S \varphi$ if and only if $A, S \vdash_K \varphi$. This is one of the reasons for which the logic K is of a special interest.

The Kripke semantics of propositional modal logic is defined as follows.

A *Kripke interpretation* is a triple $\mathfrak{M} = \langle U, R, I \rangle$, where U is a non-empty set of *possible worlds*, R is an *accessibility* relation on U, and I is an assignment to each world in U of a set of propositional variables.

Definition 3 *Let* $\mathfrak{M} = \langle U, R, I \rangle$ *be a Kripke interpretation,* $u \in U$, *and let* $\varphi \in \boldsymbol{Fm}$. *We say that the pair* $(\mathfrak{M}, u)$ *satisfies* φ, *denoted* $(\mathfrak{M}, u) \models \varphi$, *if the following holds.*

- *If* φ *is a propositional variable p, then* $(\mathfrak{M}, u) \models \varphi$, *if* $p \in I(u)$;[4]

- $(\mathfrak{M}, u) \models \neg\varphi$, *if* $(\mathfrak{M}, u) \not\models \varphi$;

- $(\mathfrak{M}, u) \models \varphi \supset \psi$, *if* $(\mathfrak{M}, u) \not\models \varphi$ *or* $(\mathfrak{M}, u) \models \psi$;

- $(\mathfrak{M}, u) \models L\varphi$, *if for each v such that* uRv, $(\mathfrak{M}, v) \models \varphi$.

---

[3] The results of this paper, similarly to those in the paper by Kaminski and Rey [2000, Section 8], extend to the *pure logic of necessitation* N introduced by Fitting et al. [1992].

[4] That is, p is satisfied by the propositional interpretation $I(u)$.

We say that a Kripke interpretation $\mathfrak{M}$ *satisfies* a formula φ, denoted $\mathfrak{M} \models \varphi$, if for every $u \in U$, $(\mathfrak{M}, u) \models \varphi$, and we say that $\mathfrak{M}$ *satisfies* a set of formulas X or $\mathfrak{M}$ is a *model* of X, denoted $\mathfrak{M} \models X$, if $\mathfrak{M} \models \varphi$ for every $\varphi \in X$. The set of all formulas satisfied by $\mathfrak{M}$ is called the *theory* of $\mathfrak{M}$ and is denoted $Th(\mathfrak{M})$. That is, $Th(\mathfrak{M}) = \{\varphi : \mathfrak{M} \models \varphi\}$. Finally, we say that a set of formulas X *semantically entails* a formula φ, denoted $X \models \varphi$ (respectively, $X \models_S \varphi$), if each Kripke model of X (respectively, each Kripke model of X that satisfies S) satisfies φ.

The above semantics is sound and complete for K. That is, $X \vdash_K \varphi$ if and only if φ is satisfied by all Kripke interpretations which satisfy X. Kripke interpretations with a reflexive accessibility relation are sound and complete for T, Kripke interpretations with a reflexive and transitive accessibility relation are sound and complete for S4, and Kripke interpretations where the accessibility relation is an equivalence relation are sound and complete for S5 [Hughes and Cresswell, 1972, Section 2]. Kripke interpretations $\mathfrak{M} = \langle U, R, I \rangle$ with the *total* accessibility relation $R = U \times U$ are called clusters.

2.3   Propositional stable theories

We precede the definitions of propositional autoepistemic logic, propositional non-monotonic modal logic, and propositional ground non-monotonic modal logic with the definition of propositional stable theories and statements of some of their basic properties.

The notion of a stable theory naturally arises in context of non-monotonic reasoning and, in propositional logic, stable theories are tightly related to *all* of the above mentioned non-monotonic logics.

Definition 4 (Stalnaker [1980]) *A set of propositional modal formulas X is called* stable *if it satisfies the following three conditions.*

1. *X is closed under propositional consequence $\vdash$.*

2. *For every modal formula φ, if $\varphi \in X$, then $L\varphi \in X$.*

3. *For every modal formula φ, if $\varphi \notin X$, then $\neg L\varphi \in X$.*

Stable theories are supposed to represent belief sets of a rational agent possessing a full power of introspection. A set of believes of such an agent should be closed under

- propositional consequence, which is reflected by clause 1 of Definition 4;

- positive introspection – if an agent believes in something, then (s)he must believe that (s)he believes in this, which is reflected by clause 2 of Definition 4; and

- negative introspection – if an agent does not believe in something, then (s)he must not believe that (s)he does believe in this, which is reflected by clause 3 of Definition 4.

The basic properties of propositional stable theories we are interested in are given by Theorems 1 and 2 below.

Theorem 1 (Marek and Truszczyński [1993, Theorems 8.10, p. 228 and Theorem 8.12, p. 229]) *A consistent set of propositional modal formulas is stable if and only if it is a theory of a cluster.*

Theorem 2 (Marek and Truszczyński [1993, Corollary 8.19, p. 233]) *If X_1 and X_2 are stable theories such that $X_1 \cap \boldsymbol{GFm} = X_2 \cap \boldsymbol{GFm}$, then $X_1 = X_2$.*

Loosely speaking, Theorem 2 states that propositional stable theories (or, equivalently, clusters) are uniquely defined by their ground parts. The example in the appendix we show that first-order stable theories do not possess this property. That is, the first-order counterpart of Theorem 2 is not true.

2.4 Propositional non-monotonic modal logic and propositional ground non-monotonic modal logic

In this section we recall the definition of *propositional* non-monotonic modal logic and propositional ground non-monotonic modal logic based on the McDermott and Doyle fixed point equation. We start with the definition of propositional non-monotonic modal logic. This definition is a relativization of the original definition of McDermott [1982] that was restricted to the classical modal logics T, S4, or S5 to any modal logic S. A general form of McDermott's definition is as follows.

Definition 5 *Let S be a modal logic and let A be a sets of propositional modal formulas (axioms). For a set of formulas X we denote the S-theory of $A \cup \{M\varphi : X \not\vdash_{\mathrm{S}} \neg\varphi\}$ by $\mathbf{NM}_{\mathrm{S}}^{A}(X)$. That is,*

$$\mathbf{NM}_{\mathrm{S}}^{A}(X) = \boldsymbol{Th}_{\mathrm{S}}(A \cup \{M\varphi : X \not\vdash_{\mathrm{S}} \neg\varphi\}).$$

Consistent fixed points of the operator $\mathbf{NM}_{\mathrm{S}}^{A}$ are called S-expansions for A.

Remark 2 Similarly to Remark 1, by soundness and completeness of the Kripke semantics, $\mathbf{NM}_{\mathrm{S}}^{A}(X)$ can be defined as $\boldsymbol{Th}_{\mathrm{S}}(A \cup \{M\varphi : X \not\models_{\mathrm{S}} \neg\varphi\})$.

Roughly speaking, fixed points of the operator $\mathbf{NM}_{\mathrm{S}}^{A}$ correspond to the "deductive closure" of A in S extended with "rule of inference" (1) that also is referred to as *possibilitation*[5]. In presence of the modal scheme $\boldsymbol{k}$, *possibilitation* is "equivalent" to

$$\text{"if } \not\vdash_{\mathrm{S}} \varphi, \text{ then } \neg L\varphi\text{"} \tag{2}$$

that corresponds negative introspection, whereas positive introspection is provided by **NEC**. Thus, a fixed point of $\mathbf{NM}_{\mathrm{S}}^{A}$ can be considered as an acceptable set of beliefs that a rational agent may hold about an incompletely specified world. This well-complies with intuition of stable theories, see Theorem 3 below.

Theorem 3 (Marek and Truszczyński [1993, Theorem 9.4, p. 253]) *For any modal logic S, S-expansions are stable.*

---

[5]Like all non-monotonic rules of inference, (1) is self-referring and, therefore, it is ill-defined.

Next we turn to the semantics of propositional non-monotonic modal logic proposed by Schwarz [1992].

Definition 6 *Let* $\mathfrak{M}' = \langle U', R', I' \rangle$ *and* $\mathfrak{M}'' = \langle U'', R'', I'' \rangle$ *be Kripke interpretations such that* $U' \cap U'' = \emptyset$[6]. *The* concatenation *of* $\mathfrak{M}'$ *and* $\mathfrak{M}''$, *denoted* $\mathfrak{M}' \odot \mathfrak{M}''$, *is the Kripke interpretation* $\mathfrak{M} = \langle U, R, I \rangle$, *where* $U = U' \cup U''$, $R = R' \cup (U' \times U'') \cup R''$, *and* I *is defined by*

$$I(u) = \left\{ \begin{array}{ll} I'(u) & \text{if } u \in U' \\ I''(u) & \text{if } u \in U'' \end{array} \right. .$$

Definition 7 (Schwarz [1992]) *Let* $\mathfrak{M} = \langle U, R, I \rangle$ *be a Kripke interpretation and let* $\mathfrak{M}' = \langle U', U' \times U', I' \rangle$ *be a cluster. We say that* $\mathfrak{M} \odot \mathfrak{M}'$ *is* preferred *over* $\mathfrak{M}'$, *denoted* $\mathfrak{M} \odot \mathfrak{M}' \sqsubset \mathfrak{M}'$, *if there is a world* $u \in U$ *and a (modal-free) formula* $\theta \in \mathbf{GFm}$ *such that* $I(u) \not\models \theta$, *but* $\mathfrak{M}' \models \theta$.

Definition 8 (Schwarz [1992]) *Let* $\mathcal{C}$ *be a class of Kripke interpretations. A cluster* $\mathfrak{M}' \in \mathcal{C}$ *is called* $\mathcal{C}$-minimal *for a set of propositional modal formulas* A, *if* $\mathfrak{M}' \models A$ *and for every Kripke interpretation* $\mathfrak{M}$ *such that* $\mathfrak{M} \odot \mathfrak{M}' \in \mathcal{C}$, $\mathfrak{M} \odot \mathfrak{M}' \models A$ *implies* $\mathfrak{M} \odot \mathfrak{M}' \not\sqsubset \mathfrak{M}$.

If the class $\mathcal{C}$ *consists of all Kripke interpretations, then* $\mathcal{C}$-minimal models of A *are called* minimal.

Definition 9 (Schwarz [1992]) *A class* $\mathcal{C}$ *of Kripke interpretations is called* cluster closed, *if it contains all clusters and at least one of the two following conditions is satisfied.*

1. *For every Kripke interpretation* $\mathfrak{M} \in \mathcal{C}$ *and every cluster* $\mathfrak{M}'$, *the Kripke interpretation* $\mathfrak{M} \odot \mathfrak{M}'$ *belongs to* $\mathcal{C}$.

2. *For every Kripke interpretation from* $\mathcal{C}$ *of the form* $\mathfrak{M} \odot \mathfrak{M}'$, *where* $\mathfrak{M}'$ *is a cluster, and every cluster* $\mathfrak{M}''$, *the Kripke interpretation* $\mathfrak{M} \odot \mathfrak{M}''$ *belongs to* $\mathcal{C}$.

Definition 10 *Let* $\mathcal{C}$ *be a class of Kripke interpretations and let* S *be a modal logic. We say that* S *is* characterized by $\mathcal{C}$, *if the following holds. For every set of propositional modal formulas* X *and every propositional modal formula* φ, $X \vdash_{\text{S}} \varphi$ *if and only if for every Kripke interpretation* $\mathfrak{M} \in \mathcal{C}$, $\mathfrak{M} \models X$ *implies* $\mathfrak{M} \models \varphi$.

At last, we have arrived at Schwarz's description of S-expansions.

Theorem 4 (Schwarz [1992]) *Let* S *be a modal logic characterized by a cluster closed class* $\mathcal{C}$ *of Kripke interpretations and let* A *be a set of formulas. A set of formulas* E *is an* S-expansion *for* A *if and only if there exists a cluster* $\mathfrak{M}$ *such that* $\mathfrak{M}$ *is* $\mathcal{C}$-minimal *for* A *and* $E = \mathbf{Th}(\mathfrak{M})$.

Remark 3 In fact, it was shown by Kaminski and Rey [2000] that, if $\mathcal{C}$ contains all Kripke models of S, then the cluster closure requirement of $\mathcal{C}$ is redundant.

---

[6] In the sequel, when dealing with a number of Kripke interpretation, by renaming their worlds, if necessary, we may always assume that sets of worlds of Kripke interpretations under consideration are pairwise disjoint.

Next we recall the definition of *ground non-monotonic modal logic* that bounds negative introspection (2) to *ground* (i.e., modal-free) formulas as described below.

Definition 11 (Tiomkin and Kaminski [1991]) *Let* S *be a modal logic, and let* A *be a set of propositional modal formulas (axioms). For a set of formulas* X *we denote the* S*-theory of* $A \cup \{M\varphi : X \nvdash_S \neg\varphi, \ \varphi \in \mathbf{GFm}\}$ *by* $\mathbf{GNM}_S^A(X)$[7]. *That is,*

$$\mathbf{GNM}_S^A(X) = \mathbf{Th}_S(A \cup \{M\varphi : X \nvdash_S \neg\varphi, \ \varphi \in \mathbf{GFm}\}).$$

Consistent fixed points of the operator $\mathbf{GNM}_S^A$ *are called* ground S-expansions *for* A *or* S_G-expansions *for short.*

Remark 4 Similarly to Remarks 1 and 2, $\mathbf{GNM}_S^A(X)$ can be defined as $\mathbf{Th}_S(A \cup \{M\varphi : X \nvDash_S \neg\varphi, \ \varphi \in \mathbf{GFm}\})$.

It turns out that for any modal logic S and any set of axioms A, S_G-expansions for A are stable [Marek and Truszczyński, 1993, Theorem 11.36, p. 344]. That is, even though, in propositional modal logic negative introspection (2) of a rational agent is bounded to ground formulas, the belief set of the agent is still stable.

Finally, we recall a semantical description of a class of propositional ground non-monotonic modal logics presented by Donini, Nardi, and Rosati [1997]. Similarly to the definition of Schwarz [1992], the authors define a preference relation on Kripke interpretations that is based on the minimization of knowledge expressed by modal-free formulas.

Definition 12 (Donini et al. [1997]) *Let* $\mathfrak{M}' = (U', R', V')$ *and* $\mathfrak{M}'' = (U'', R'', V'')$ *be Kripke interpretations. We write* $\mathfrak{M}'' \supset_G \mathfrak{M}'$, *if* $U' = U''$, $V' = V''$, *and* $R' \subset R''$.

Definition 13 (Donini et al. [1997]) *Let* $\mathfrak{M}$ *and* $\mathfrak{M}'$ *be Kripke interpretations. We say that* $\mathfrak{M}'$ *is* groundly preferred *over* $\mathfrak{M}$, *denoted* $\mathfrak{M}' \sqsubset_G \mathfrak{M}$, *if there exists a Kripke interpretation* $\mathfrak{M}''$ *such that* $\mathfrak{M}' \supset_G \mathfrak{M}'' \odot \mathfrak{M}$ *and there exists a formula* $\theta \in \mathbf{GFm}$ *such that* $\mathfrak{M} \models \theta$, *but* $\mathfrak{M}'' \not\models \theta$.

Definition 14 (Donini et al. [1997]) *Let* A *be a set of formulas and let* $\mathcal{C}$ *be a class of Kripke interpretations. An interpretation* $\mathfrak{M} \in \mathcal{C}$ *is called* ground $\mathcal{C}$-minimal *for* A, *if* $\mathfrak{M} \models A$ *and for every interpretation* $\mathfrak{M}' \in \mathcal{C}$, $\mathfrak{M}' \models A$ *implies* $\mathfrak{M}' \not\sqsubset_G \mathfrak{M}$.

It is easy to see that every Kripke interpretation which is minimal in the class of interpretations $\mathcal{C}$ according to the ground criterion, is minimal in $\mathcal{C}$ according to Schwarz's criterion as well, while the converse, in general, does not hold.

Definition 15 (Donini et al. [1997, Clause 2 of Definition 9]) *A class* $\mathcal{C}$ *of Kripke interpretations is called* cluster decomposable, *if every interpretation in* $\mathcal{C}$ *is either a cluster or is of the form* $\mathfrak{M} \odot \mathfrak{M}'$, *where* $\mathfrak{M}'$ *is a cluster, and for every such interpretation* $\mathfrak{M} \odot \mathfrak{M}' \in \mathcal{C}$ *and every cluster* $\mathfrak{M}''$, *the interpretation* $\mathfrak{M} \odot \mathfrak{M}''$ *is also in* $\mathcal{C}$[8].

---

[7]Recall that $\mathbf{GFm}$ denotes the set of all propositional ground formulas.

[8]An axiomatization of the propositional modal logic characterized by the class of all cluster-decomposable Kripke interpretations can be found in the paper by Tiomkin and Kaminski [2007].

Theorem 5 (Donini et al. [1997, Theorem 3.19]) *Let S be a modal logic characterized by a cluster decomposable class of Kripke interpretations $\mathcal{C}$ and let A be a set of formulas. A set of formulas E is an S_G-expansion for A if and only if there exists a cluster $\mathfrak{M}$ such that $\mathfrak{M}$ is ground $\mathcal{C}$-minimal for A and $E = Th(\mathfrak{M})$[9].*

2.5 Propositional autoepistemic logic

The syntax of propositional autoepistemic logic is that of propositional modal logic, i.e., it is obtained from the syntax of propositional logic by augmenting it with the unary modal connective L (believed). A propositional *modal* interpretation is a set of propositional variables and formulas of the form $L\varphi$ called (propositional) *modal atoms* [Moore, 1985]. Note that a propositional modal formula is a propositional combination of propositional variables and propositional modal atoms. Satisfiability of a propositional modal formula by a propositional modal interpretation is defined in the standard propositional way. Namely, let w be a propositional modal interpretation and let φ be a propositional modal formula. We say that w *satisfies* φ, denoted $w \models \varphi$ if the following holds.

- If φ is a propositional variable or a propositional modal atom, then $w \models \varphi$, if $\varphi \in w$;

- $w \models \varphi \supset \psi$, if $w \not\models \varphi$ or $w \models \psi$; and

- $w \models \neg\varphi$, if $w \not\models \varphi$.

Let X be a set of propositional modal formulas. We say that w satisfies X (or w is a *model* of X), denoted $w \models X$, if for every $\varphi \in X$, $w \models \varphi$. Finally, we denote by $Mod(X)$ the set of all models of X. That is,

$$Mod(X) = \{w : w \models X\}.$$

Definition 16 (Moore [1985]) *Let w be a propositional modal interpretation and let X be a set of propositional modal formulas. We say that w respects X, if for every propositional modal formula φ, $L\varphi \in w$ if and only if $\varphi \in X$.*

Let X, Y, and φ be sets of propositional modal formulas and a propositional modal formula, respectively. We say that Y *entails* φ with respect to X, denoted $Y \models_X \varphi$, if every propositional modal interpretation w that respects X and satisfies Y also satisfies φ.

Definition 17 (Moore [1985]) *Let A be a set of propositional modal formulas. A consistent set of propositional modal formulas E is called an* autoepistemic expansion *of A, if $E = \{\varphi : A \models_E \varphi\}$.*

Theorem 6 (Shvarts [1990, Proposition 2.1]) *A set of propositional modal formulas E is an autoepistemic expansion of a set of propositional modal formulas A if and only if E is a K45-expansion for A[10].*

---

[9] It is not known whether the cluster decomposability requirement of $\mathcal{C}$ is redundant, if $\mathcal{C}$ contains all Kripke models of S, cf. Remark 3.

[10] Of course, by K45 we mean K + **4** + **5**. In fact, it can be readily seen that K45 can be replaced with a weaker logic N + **5**, where N is the *pure logic of necessitation* – the extension of classical propositional logic with NEC only.

2.6   Interpretation of non-monotonic modal logic, ground non-monotonic modal logic, and autoepistemic logic in default logic

In this section we recall Janhunen's interpretation of non-monotonic modal logic in default logic [Janhunen, 1996, 1998, 1999], and present a similar interpretation of ground non-monotonic modal logic in default logic. Since classical propositional logic does not contain the modal connective L, the default theories under consideration are over the "purely propositional fragment" of propositional modal logic. That is, even though we deal with formulas of propositional *modal* logic, we restrict ourselves to the ordinary propositional inference $\vdash$, only. In other words, like in autoepistemic logic, modal formulas of the form $L\varphi$ (where φ may contain L) are thought of as *new* propositional variables also called *modal atoms*, see Section 2.5, and the language of default logic consists of the set of propositional variables of non-monotonic modal logic augmented with the set of modal atoms, i.e., new propositional variables of the form $L\varphi$. Thus, extensions for default theories over the purely propositional part of propositional modal logic are well defined via the operator Γ.

To state Janhunen's interpretation theorem we need the following notation. We denote the set of all default rules of the form $\frac{\varphi:}{L\varphi}$ by D^+ and the set of all default rules of the form $\frac{:\varphi}{M\varphi}$ by D^-. These sets of defaults naturally simulate positive and negative introspections, respectively. Recall that the language of default logic is that of propositional modal logic, where modal atoms $L\varphi$ are treated as *new* propositional variables, or, equivalently, new atomic formulas.

Theorem 7   (Janhunen [1996, 1998, 1999]) *Let* S *be a modal logic and let* A *be a set of modal formulas. A set of modal formulas* E *is an* S-*expansion for* A *if and only if* E *is an extension for the default theory* $(D^+ \cup D^-, A \cup S)$.

Remark 5 In fact, Janhunen [1996,1998,1999] deals with the (non-normal) modal logic 5, only. However, an easy inspection of the proofs shows that they hold for any modal logic S.

The interpretation of ground non-monotonic modal logic in default logic is a straightforward modification of that given by Theorem 7. To state it we need one more bit of notation. We denote the set of all default rules of the form $\frac{:\varphi}{M\varphi}$, where $\varphi \in \boldsymbol{GFm}$ by D_G^-. This set of defaults simulates negative introspection restricted to ground formulas.

The proof of Theorem 8 below is similar to that of Theorem 7.

Theorem 8 *Let* S *be a modal logic and let* A *be a set of modal formulas. A set of modal formulas* E *is an* S_G-*expansion for* A *if and only if* E *is an extension for the default theory* $(D^+ \cup D_G^-, A \cup S)$.

2.7   Interpretation of default logic in non-monotonic modal logic and non-monotonic modal logic

In this section we recall Truszczyński's interpretation of non-monotonic modal logic and ground non-monotonic modal logic in default logic [Truszczyński, 1991a,b]. For a default $d = \frac{\alpha : \beta_1, \ldots, \beta_m}{\gamma}$ we denote the modal formula $(L\alpha \wedge \bigwedge_{i=1}^{m} LM\beta_i) \supset \gamma$ by $\tau(d)$.

Theorem 9 (Truszczyński [1991a,b]) *Let (D, A) be a propositional default theory and let a modal logic S be S4F or weaker. A set of formulas E is an extension for (D, A) if and only if there exists an S-expansion (S_G-expansion) E^L for $A \cup \{\tau(d) : d \in D\}$ such that $E = E^L \cap \mathbf{GFm}$*[11].

3   First-order default logic

In this section we recall the definition of first-order default logic. We start with the Herbrand semantics of first-order logic that lies in the basis of our semantical approach to *all* first-order non-monotonic logics addressed in this paper.

3.1   Herbrand semantics of first-order logic

The language of the underlying first-order logic will be denoted $\mathcal{L}$. Let b be a set that contains no symbols of $\mathcal{L}$. We denote by $\mathcal{L}_b$ the language obtained from $\mathcal{L}$ by augmenting its set of constant symbols with all elements of b. The set of all closed terms of $\mathcal{L}_b$, denoted $\mathbf{Tr}_b$, is called the *Herbrand universe* of $\mathcal{L}_b$, and closed formulas over $\mathcal{L}_b$, denoted $\mathbf{GSt}_b$, will be referred to as b-*sentences*[12]. Note that b-sentences are of the form $\varphi(t_1, \ldots, t_n)$, where $\varphi(x_1, \ldots, x_n)$ is a formula of $\mathcal{L}$ whose free variables are among $x_1, \ldots, x_n$, and $t_1, \ldots, t_n \in \mathbf{Tr}_b$. A *Herbrand b-interpretation* of $\mathcal{L}$ is a set of atomic b-sentences.

Definition 18 *Let w be a Herbrand b-interpretation and let φ be a b-sentence. We say that w satisfies φ, denoted $w \models \varphi$, if the following holds.*

- *If φ is an atomic formula, then $w \models \varphi$, if $\varphi \in w$;*

- *$w \models \neg\varphi$, if $w \not\models \varphi$;*

- *$w \models \varphi \supset \psi$, if $w \not\models \varphi$ or $w \models \psi$; and*

- *$w \models \forall x\varphi(x)$, if for all $t \in \mathbf{Tr}_b$, $w \models \varphi(t)$.*

For a Herbrand b-interpretation w we define the $\mathcal{L}_b$-*theory* of w, denoted $\mathbf{Th}_b(w)$, as the set of all b-sentences satisfied by w:

$$Th_b(w) = \{\varphi \in \mathbf{GSt}_b : w \models \varphi\}.$$

Let X be a set of b-sentences. We say that Herbrand b-interpretation w is a (Herbrand) b-*model* of X, denoted $w \models X$, if $X \subseteq \mathbf{Th}_b(w)$, and for a set of Herbrand b-interpretations W we write $W \models X$, if for every $w \in W$, $w \models X$. Also, for a

---

[11] Recall that $\mathbf{GFm}$ denotes the set of all classical propositional formulas, i.e., the set of all formulas not containing the modal connective L.

[12] We reserve $\mathbf{St}_b$ to denote the set of all modal sentences over $\mathcal{L}_b$. Thus, when dealing with modal logic, $\mathbf{GSt}_b$ refers to the subset of $\mathbf{St}_b$ consisting of all *ground* (i.e., modal-free) b-sentences.

set of Herbrand b-interpretations W we denote by $\boldsymbol{Th}_b(W)$ the set of all b-sentences satisfied by all elements of W. That is, $\boldsymbol{Th}_b(W) = \bigcap_{w \in W} \boldsymbol{Th}_b(w)$. Thus, $W \models X$ if and only if $X \subseteq \boldsymbol{Th}_b(W)$.

Next, for a set of b-sentences X we denote by $\boldsymbol{Mod}_b(X)$ the set of all Herbrand b-models of X.

Finally, we say that a set of b-sentences X b-*entails* a b-sentence φ, denoted $X \models_b \varphi$, if every Herbrand b-interpretation that satisfies X also satisfies φ. The set of all b-sentences b-entailed by X will be denoted $\boldsymbol{Th}_b(X)$[13].

Remark 6 It can be readily seen that for a set of b-sentences X the following holds.

- $\boldsymbol{Th}_b(\boldsymbol{Th}_b(X)) = \boldsymbol{Th}_b(X)$,

- $\boldsymbol{Th}_b(\boldsymbol{Mod}_b(X)) = \boldsymbol{Th}_b(X)$, and

- $\boldsymbol{Mod}_b(\boldsymbol{Th}_b(X)) = \boldsymbol{Mod}_b(X)$.

Remark 7 It is well-known that for an infinite set of new constant symbols b, Herbrand b-interpretations are sound and complete for first-order logic. That is, for a set of $\mathcal{L}$ sentences X and a $\mathcal{L}$ sentence φ, $X \vdash \varphi$ if and only if $X \models_b \varphi$. In particular, Herbrand b-interpretations with an infinite set of new constant symbols b naturally arise in the Henkin proof of the completeness theorem [Mendelson, 1997, Lemma 2.16, p. 70].

3.2   Why Herbrand b-interpretations?

In this section we present the intuition lying behind the definition of first-order nonmonotonic logics.

There are two types of the domain objects of a logic over $\mathcal{L}$. One type consists of the *fixed* built in objects which correspond to closed terms of $\mathcal{L}$ and must be present in any interpretation, and the other type consist of implicitly defined *unknown* objects which may vary from one interpretation to other[14]. These objects generate other unknown objects by means of the function symbols of $\mathcal{L}$. Thus, it seems natural to assume that the theory domain is a Herbrand universe of the original language augmented with a set b of new constant symbols which are supposed to represent the unknown objects (Lloyd [1993, Chapter 1, §3] provides more comments on this issue). Note that, in general, it is impossible to describe a Herbrand universe by means of a proof theory. The only exception is the case when the theory domain is "explicitly" finite. We also assume that the set of new constant symbols b is infinite. This is because if the set of new constant symbols finite, then, in some cases, non-monotonic theories of the underlying set of axioms may contain consequences which do not follow by non-monotonic (and, of course, monotonic) rules of inference. Kaminski [1995] provides examples and a general discussion[15].

---

[13]Note that the operator $\boldsymbol{Th}_b$ applies both to sets of Herbrand b-interpretations and sets of b-sentences. In the former case, when applied to a set of Herbrand b-interpretations W, the result is the set of all b-sentences satisfied by all elements of W; and in the latter case, when applied to a set of b-sentences X, the result is the set of all b-sentences b-entailed by X.

[14]For example, the latter objects may be defined by existentially quantified formulas.

[15]An additional motivation for the requirement that b is infinite is Remark 7 in the previous section.

3.3 First-order default logic: semantical approach

First-order defaults are expressions of the form

$$\frac{\alpha(\boldsymbol{x}) : \beta_1(\boldsymbol{x}), \dots, \beta_m(\boldsymbol{x})}{\gamma(\boldsymbol{x})},$$

where $\alpha(\boldsymbol{x})$, $\beta_1(\boldsymbol{x}), \dots, \beta_m(\boldsymbol{x})$, $m \geq 0$, and $\gamma(\boldsymbol{x})$ are formulas of first-order logic whose free variables are among $\boldsymbol{x} = x_1, \dots, x_n$. A default is *closed* if none of $\alpha, \beta_1, \dots, \beta_m$, and γ contains a free variable. Otherwise, a default is called *open*. Similarly to the propositional case, the intuitive meaning of a first-order default is as follows. For every n-tuple of objects $\boldsymbol{t} = t_1, \dots, t_n$, if $\alpha(\boldsymbol{t})$ is believed, and the $\beta_i(\boldsymbol{t})$s are consistent with ones beliefs, then one is permitted to deduce $\gamma(\boldsymbol{t})$ and add it to the "belief set." Thus, an open default can be thought of as a kind of a "default scheme," where free variables $\boldsymbol{x}$ can be replaced by any of the theory's objects. Various examples of deduction by defaults can be found in the paper by Reiter [1980]. We remind the reader that a *default theory* is a pair (D, A), where D is a set of defaults and A is a set of first-order sentences (axioms). A default theory is called *closed*, if all its defaults are closed. Otherwise, it is called *open*.

Definition 19 (Reiter [1980]) *Let (D, A) be a closed default theory. For a set of sentences X let $\Gamma_{(D,A)}(X)$ be the smallest set of sentences B (beliefs) that includes A, is deductively closed, and satisfies the following condition.*

For any $\frac{\alpha : \beta_1, \dots, \beta_m}{\gamma} \in D$, if $\alpha \in B$ and $\neg\beta_1, \dots, \neg\beta_m \notin S$, then $\gamma \in B$.

A consistent set of sentences E is an extension *for (D, A) if $\Gamma_{(D,A)}(E) = E$, i.e., if E is a fixed point of the operator $\Gamma_{(D,A)}$.*

Now, following Lifschitz [1990] and Kaminski [1995], we define extensions for open default theories. The definition of extensions for open default theories below is a relativization of Definition 2 to Herbrand b-interpretations.

Definition 20 (Kaminski [1995]) *Let (D, A) be a default theory and let b be an infinite set of new constant symbols. For a set of Herbrand b-interpretations (possible worlds) W let $\Delta^b_{(D,A)}(W)$ be the largest set V of b-models of A (belief worlds) that satisfies the following condition.*

For any $\frac{\alpha(\boldsymbol{x}) : \beta_1(\boldsymbol{x}), \dots, \beta_m(\boldsymbol{x})}{\gamma(\boldsymbol{x})} \in D$, and any tuple $\boldsymbol{t}$ of elements of $\boldsymbol{Tr}_b$, if $V \models \alpha(\boldsymbol{t})$ and $W \not\models \neg\beta_i(\boldsymbol{t})$, $i = 1, \dots, m$, then $V \models \gamma(\boldsymbol{t})$[16].

A consistent set of b-sentences E is called a Δ^b-extension for (D, A) if $E = \boldsymbol{Th}_b(W)$ for some non-empty fixed point W of $\Delta^b_{(D,A)}$.

Kaminski [1995, Theorem 42] proved that for closed default theories, Definition 20 is equivalent to Reiter's original definition (Definition 19). Namely, Reiter's extensions are exactly the restrictions of Δ^b-extensions to the original language $\mathcal{L}$.

---

[16]Similarly to the proof by Lifschitz [1990, Proposition 1], see also the proof of Proposition 1, it can be shown that this largest class $\Delta_{(D,A)}(W)$ always exists.

Remark 8 Let (D, A) be a default theory and let W' and W'' be sets of Herbrand b-interpretations. It immediately follows from Definition 20 that $\boldsymbol{Th}_b(W') = \boldsymbol{Th}_b(W'')$ implies $\Delta^b_{(D,A)}(W') = \Delta^b_{(D,A)}(W'')$.

Remark 9 In Definition 20, the requirement for b to be infinite is essential, because for a finite b, in some cases, extensions might contain an upper bound on the number of elements in the domain that does not follow from the axioms and defaults of the default theory, see the paper by Kaminski [1995, p. 295].

In view of Remarks 7 and 9, *for the rest of this paper we assume that the set of new constant symbols b added to $\mathcal{L}$ is infinite.*

Remark 10 In the above definitions we implicitly assume that there is no equality relation between the domain elements. Any such equality in an extension must follow from axioms and defaults. To some extent, this can be considered as a weak form of the *unique name assumption* as discussed by Kaminski [1995, p. 304]. Note that assignments to equality in Herbrand interpretations are binary relations which do not have to be identity in the domain of the interpretation, but satisfy the equality first-order axioms. That is, equality is treated as an ordinary dyadic predicate[17].

At last, we present an alternative definition of extensions for open default theories.

Definition 21 *Let (D, A) be a default theory and let b be an infinite set of new constant symbols. For a set of b-sentences X, let $\Gamma^b_{(D,A)}(X)$ be the smallest set of b-sentences B (b-beliefs) that includes A, is closed under semantical b-entailment $\models_b$,[18] and satisfies the following condition.*

For any $\dfrac{\alpha(\boldsymbol{x}) : \beta_1(\boldsymbol{x}), \ldots, \beta_m(\boldsymbol{x})}{\gamma(\boldsymbol{x})} \in D$, *and any tuple t of elements of $\boldsymbol{Tr}_b$, if* $\alpha(t) \in B$ *and* $\neg\beta_i(t) \notin X$, $i = 1, \ldots, m$, *then* $\gamma(t) \in B$.

A consistent set of b-sentences E is called a Γ^b-extension for (D, A) if E is a fixed point of the operator $\Gamma^b_{(D,A)}$.

Note the similarity between Definitions 19 and 21: the latter results from the former in replacing $\vdash$ with $\models_b$, cf. Remark 1, and replacing the set of defaults with the set of all its closed "b-instances." The precise relationship between $\Gamma^b_{(D,A)}$ and $\Delta^b_{(D,A)}$ is as follows.

Proposition 1 (See the appendix for the proof) *Let (D, A) be an open default theory and b be an infinite set of new constant symbols. Let W and X be a set of Herbrand b-interpretations and a set of b-sentences, respectively. Then*
(a) $\boldsymbol{Th}_b(\Delta^b_{(D,A)}(W)) = \Gamma^b_{(D,A)}(\boldsymbol{Th}_b(W))$, *and*
(b) $\boldsymbol{Mod}_b(\Gamma^b_{(D,A)}(X)) = \Delta^b_{(D,A)}(\boldsymbol{Mod}_b(X))$.

Corollary 1 *Let (D, A) be an open default theory, b be an infinite set of new constant symbols, and let E be a set of sentences over $\mathcal{L}$. Then E is a Δ^b-extension for (D, A) if and only if it is a Γ^b-extension for (D, A).*

---

[17]Interpretations where the assignment to the equality relation is identity in the domain of the interpretation are called *normal* [Mendelson, 1997, p. 78].

[18]That is, $\boldsymbol{Th}_b(B) = B$.

In view of the above corollary (its proof is in the appendix), both Δ^b- and Γ^b-extensions will be simply referred to as b-extensions.

4   First-order monotonic modal logic

In this section we recall the Kripke semantics of first-order modal logic and point out some differences between the notion of preference and minimality in the case of propositional and first-order modal logics. In particular, we present a different notion of minimality, called *strong minimality* that is equivalent to minimality in the case of propositional modal logic, but becomes non-equivalent to it in the first-order case. The semantical definition of first-order non-monotonic modal logic in the next section is based on strong minimality that, as we argue, is more appropriate when dealing with first-order non-monotonic modal logic.

4.1   Kripke semantics of first-order modal logic

Like propositional modal logic, first-order modal logic is obtained from the classical one by adding to it the modal connective L. For the compliance of first-order non-monotonic modal logics with first-order default theories, we need the worlds of first-order Kripke interpretations to be Herbrand b-interpretations for the same infinite set of new constant symbols b. This is imposed by the axiom scheme

BF $\forall x L \varphi \supset L \forall x \varphi,$

called the *Barcan formula* that will be contained in all (monotonic) modal logics under consideration

Similarly to the propositional case, any modal logic S containing **BF** can be embedded in K+**BF** by extending the set of proper axioms with S[19]. That is, $A \vdash_S \varphi$ if and only if $A, S \vdash_{K+BF} \varphi$. This is one of the reasons for which the logic K+**BF** is of a special interest. The set of all modal sentences over $\mathcal{L}_b$ will be referred to as (modal) b-*sentences* and will be denoted $\boldsymbol{St}_b$. Recall that the set of all ground (i.e., modal-free) b-sentences is denoted $\boldsymbol{GSt}_b$.

The Kripke semantics of first-order normal modal logic with **BF** is defined as follows.

A *Kripke b-interpretation* is a triple $\mathfrak{M} = \langle U, R, I \rangle$, where U is a non-empty set of *possible worlds*, R is an *accessibility* relation on U, and I is an assignment to each world in U of an Herbrand b-interpretation (a set of atomic b-sentences).

Let $\mathfrak{M} = \langle U, R, I \rangle$ be a Kripke b-interpretation, $u \in U$, and let φ be a b-sentence. We say that the pair $(\mathfrak{M}, u)$ *satisfies* φ, denoted $(\mathfrak{M}, u) \models \varphi$ if the following holds.

- If φ is an atomic formula $P(t_1, \ldots, t_n)$, then $(\mathfrak{M}, u) \models \varphi$, if $P(t_1, \ldots, t_n) \in I(u)$;

- $(\mathfrak{M}, u) \models \neg\varphi$, if $(\mathfrak{M}, u) \not\models \varphi$;

---

[19]We shall identify a first-order modal logic S with the set of the universal closures of its axioms, also denoted S.

- $(\mathfrak{M}, u) \models \varphi \supset \psi$, if $(\mathfrak{M}, u) \not\models \varphi$ or $(\mathfrak{M}, u) \models \psi$;

- $(\mathfrak{M}, u) \models \forall x \varphi(x)$, if for every $t \in \mathbf{Tr}_b$, $(\mathfrak{M}, u) \models \varphi(t)$; and

- $(\mathfrak{M}, u) \models L\varphi$, if for each v such that uRv, $(\mathfrak{M}, v) \models \varphi$.

We say that a Kripke b-interpretation $\mathfrak{M}$ *satisfies* a b-sentence φ, denoted $\mathfrak{M} \models \varphi$, if for all $u \in U$, $(\mathfrak{M}, u) \models \varphi$, and we say that $\mathfrak{M}$ *satisfies* a set of b-sentences X or that $\mathfrak{M}$ is a Kripke b-*model* of X, denoted $\mathfrak{M} \models X$, if $\mathfrak{M} \models \varphi$ for every $\varphi \in X$. The set of all b-sentences satisfied by $\mathfrak{M}$ will be denoted $\mathbf{Th}_b(\mathfrak{M})$. That is,

$$\mathbf{Th}_b(\mathfrak{M}) = \{\varphi \in \mathbf{St}_b : \mathfrak{M} \models \varphi\}.$$

We say that a set of modal b-sentences X b-*entails* a modal b-sentence φ, denoted $X \models_b \varphi$, if every Kripke b-model of X satisfies φ. Finally, the set of all modal b-sentences b-entailed by X will be denoted $\mathbf{Th}_b(X)$[20].

Like in the proofs in the mongraph by Hughes and Cresswell [1972, Section 9] one can show that the Kripke semantics is sound and complete for K+**BF**. That is, $X \vdash_{K+\mathbf{BF}} \varphi$ if and only if φ is satisfied by all Kripke b-models of X (for any infinite set of new constant symbols b). Similarly, Kripke b-interpretations with a reflexive accessibility relation are sound and complete for T+**BF**, Kripke b-interpretations with a reflexive and transitive accessibility relation are sound and complete for S4+**BF**, and Kripke b-interpretations where the accessibility relation is an equivalence relation are sound and complete for S5. Kripke b-interpretations $\mathfrak{M} = \langle U, R, I \rangle$ with the total accessibility relation are called b-*clusters*.

Remark 11 Let $\mathfrak{M} = \langle U, R, I \rangle$ be a Kripke b-interpretation and let $u \in U$. Let $\mathfrak{M}^u = \langle U^u, R^u, I^u \rangle$ denote the Kripke b-interpretation whose set of worlds U^u consists of the worlds of U which are reachable from u by means of R^*, where R^* denotes the reflexive and transitive closure of R, and R^u and I^u are the restrictions of R and I on U^u, respectively. Then for a b-sentence φ and $u \in U$, $(\mathfrak{M}, u) \models \varphi$ if and only if $(\mathfrak{M}^u, u) \models \varphi$ [Segerberg, 1971, Theorem 3.10, p. 37].

For the proof of Theorem 14 in the next section we shall need the notion of *canonical* Kripke models that involves the definitions below.

Definition 22 *A set of modal b-sentences X is called* K-*consistent, if for no finite subset X' of X,* $\vdash_K \neg \bigwedge_{\varphi \in X'} \varphi$.

Definition 23 *A set modal of b-sentences X is called* maximal, *if it satisfies the following conditions.*

1. *For each b-sentence φ, either $\varphi \in X$ or $\neg\varphi \in X$.*

2. *If $\exists x \varphi(x) \in X$, then for some $t \in \mathbf{Tr}_b$, $\varphi(t) \in X$.*

---

[20] Note that the same notation $\models_b$ for b-entailment was also used in the ordinary first-order case. The reason for extending it to modal logic is that ground b-sentence φ is b-entailed by a set of ground b-sentences X according to the Herbrand semantics if and only if it is b-entailed by X according to the Kripke semantics, see Corollary 2 in the next section.

For Definition 24 below we need one more bit of notation. Let X be a set of modal b-sentences. The set of modal b-sentences $\{\varphi : L\varphi \in X\}$ is denoted by X^-.

Definition 24 *The* canonical *Kripke b-model is the Kripke b-interpretation* $\mathfrak{M}_b = \langle U_b, R_b, I_b \rangle$, *where*

- U_b *is the set of all maximal consistent sets of b-sentences;*[21]

- $uR_b v$, $u, v \in U_b$, *if* $u^- \subseteq v$; *and*

- *for a world* $u \in U_b$, *the Herbrand b-interpretation* $I_b(u)$ *consists of all atomic b-sentences which belong to* u.

The proof of Theorem 10 below is similar to that given by Hughes and Cresswell [1984, Theorem 9.8, p. 176].

Theorem 10 *For the canonical Kripke b-model* $\mathfrak{M}_b$, *every* $u \in U_b$, *and every b-sentence* φ, $(\mathfrak{M}_b, u) \models \varphi$ *if and only if* $\varphi \in u$.

4.2 Preference and minimality

Exactly like in Section 2.4, we define the notions of concatenation, preference, and minimality for the case of first-order modal logic. In particular, we call a cluster b-minimal, if it is minimal with respect to the class of all Kripke b-interpretations. For example, Definition 25 below is the first-order counterpart of of Definition 7.

Definition 25 *Let* $\mathfrak{M} = \langle U, R, I \rangle$ *be a Kripke b-interpretation and let* $\mathfrak{M}' = \langle U', U' \times U', I' \rangle$ *be a b-cluster. We say that* $\mathfrak{M} \odot \mathfrak{M}'$ *is* preferred over $\mathfrak{M}'$, *denoted* $\mathfrak{M} \odot \mathfrak{M}' \sqsubset \mathfrak{M}'$, *if there is a world* $u \in U$ *and a ground b-sentence* θ *such that* $I(u) \not\models \theta$, *but* $\mathfrak{M}' \models \theta$.

The definition of *strong minimality* mentioned in the beginning of this section involves the notion of *weak preference* defined below.

Definition 26 (Kaminski and Rey [2000]) *Let* $\mathfrak{M} = \langle U, R, I \rangle$ *be a Kripke b-interpretation and let* $\mathfrak{M}' = \langle U', U' \times U', I' \rangle$ *be a b-cluster. We say that* $\mathfrak{M} \odot \mathfrak{M}'$ *is* weakly preferred over $\mathfrak{M}'$, *denoted* $\mathfrak{M} \odot \mathfrak{M}' \sqsubset_w \mathfrak{M}'$, *if there is a (not necessary ground) b-sentence* θ *such that* $\mathfrak{M} \odot \mathfrak{M}' \not\models \theta$, *but* $\mathfrak{M}' \models \theta$.

Obviously, $\mathfrak{M} \odot \mathfrak{M}' \sqsubset \mathfrak{M}'$ implies $\mathfrak{M} \odot \mathfrak{M}' \sqsubset_w \mathfrak{M}'$. Moreover, as shows Proposition 2 below, in propositional logic, the converse implication is also true.

Proposition 2 (Kaminski and Rey [2000, Proposition 4.1]) *Let* $\mathfrak{M}$ *be a propositional Kripke interpretation and let* $\mathfrak{M}'$ *be a propositional cluster such that* $\mathfrak{M} \odot \mathfrak{M}' \sqsubset_w \mathfrak{M}'$. *Then* $\mathfrak{M} \odot \mathfrak{M}' \sqsubset \mathfrak{M}'$.

In contrast to Proposition 2, it can be shown that in the first-order case $\mathfrak{M} \odot \mathfrak{M}' \sqsubset_w \mathfrak{M}'$ does not imply $\mathfrak{M} \odot \mathfrak{M}' \sqsubset \mathfrak{M}'$ [Levesque, 1990, Theorem 3.6] and [Kaminski and Rey, 2000, Example 4.1]. The reason for which Proposition 2 is not true in first-order

---

[21] It is well-known that $U_b \neq \emptyset$ [Hughes and Cresswell, 1984, Section 9].

logic is that, by the example in the appendix, the theory of a first-order cluster is not determined by the ground b-sentences it satisfies, in contrast with Corollary 8.19 on p. 233 in the book by Marek and Truszczyński [1993]. In particular, the first-order counterpart of Theorem 7.3 in the book by Marek and Truszczyński [1993] is not true either.

Definition 27 (Kaminski and Rey [2000]) *Let $\mathcal{C}$ be a class of Kripke b-interpretations. A b-cluster $\mathfrak{M}' \in \mathcal{C}$ is called strongly $\mathcal{C}$-minimal for a set of b-sentences A, if $\mathfrak{M}' \models A$ and for every Kripke b-interpretation $\mathfrak{M}$ such that $\mathfrak{M} \odot \mathfrak{M}' \in \mathcal{C}$ and $\mathfrak{M} \odot \mathfrak{M}' \models A$, $\mathfrak{M} \odot \mathfrak{M}' \not\sqsubseteq_w \mathfrak{M}'$.*

If the class $\mathcal{C}$ consists of all Kripke b-interpretations, then strongly $\mathcal{C}$-minimal clusters are called strongly minimal.

Obviously, each strongly minimal b-interpretation is also minimal, but not vice versa [Kaminski and Rey, 2000, Example 4.1]. However, by Proposition 2, in propositional logic minimal interpretations are strongly minimal.

5   First-order non-monotonic modal logic

We start this section with the (original) definition of first-order non-monotonic modal logic "via minimal model semantics" from the paper by Kaminski and Rey [2000]. Then we present a "uniform" definition of first-order non-monotonic modal logic from the M.Sc. thesis by Zbar [2000] and state the equivalence of the definitions.

Definition 28 *Let $\mathcal{C}$ be a class of Kripke b-interpretations and let S be a modal logic. We say that S is* characterized *by $\mathcal{C}$, if the following holds. For every set of sentences X and every sentence φ, all over the original language $\mathcal{L}$, $X \vdash_S \varphi$ if and only if for every Kripke b-interpretation $\mathfrak{M} \in \mathcal{C}$, $\mathfrak{M} \models X$ implies $\mathfrak{M} \models \varphi$.*

Definition 29 (Cf. Definition 9) *A class $\mathcal{C}$ of Kripke b-interpretations is called* cluster closed, *if it contains all b-clusters and at least one of the two following conditions is satisfied.*

1. *For every Kripke b-interpretation $\mathfrak{M} \in \mathcal{C}$ and every b-cluster $\mathfrak{M}'$, the Kripke b-interpretation $\mathfrak{M} \odot \mathfrak{M}'$ belongs to $\mathcal{C}$.*

2. *For every Kripke b-interpretation from $\mathcal{C}$ of the form $\mathfrak{M} \odot \mathfrak{M}'$, where $\mathfrak{M}'$ is a cluster, and for every b-cluster $\mathfrak{M}''$, the Kripke b-interpretation $\mathfrak{M} \odot \mathfrak{M}''$ belongs to $\mathcal{C}$.*

Theorem 11 below motivates our extension of the corresponding definition in the paper by Schwarz [1992] to modal logic which are not characterized by a cluster closed class of Kripke b-interpretations. This theorem states that for a cluster closed class $\mathcal{C}$ of Kripke b-interpretations that characterizes a modal logic S and a set of sentences A, the $\mathcal{C}$-minimality for A is equivalent to minimality for $A \cup S$.

Theorem 11 (Kaminski and Rey [2000, Theorem 5.1]) *Let $\mathcal{C}$ be a cluster closed class of Kripke b-interpretations, S be a modal logic characterized by $\mathcal{C}$, A be a sets* *of first-order modal sentences over* $\mathcal{L}$*, and and let* $\mathfrak{M}$ *be a cluster. Then* $\mathfrak{M}$ *is strongly* $\mathcal{C}$*-minimal for* A *if and only if it is strongly minimal for* $A \cup S$[22].

At last we have arrived at the definition of non-monotonic expansions in the first-order case.

Definition 30 (Kaminski and Rey [2000]) *Let* S *be a normal modal logic (i.e.,* $K \subseteq$ S*),* A *be a sets of first-order modal sentences (axioms) over* $\mathcal{L}$*, and let* b *be an infinite set of new constant symbols. A set of first-order modal* b*-sentences* E *is called an* S-b-*expansion for* A*, if there exists a Kripke* b*-interpretation* $\mathfrak{M}$ *strongly minimal for* $A \cup S$ *such that* $E = \boldsymbol{Th}_b(\mathfrak{M})$[23].

The motivation for replacing minimality with strong minimality stems from a relationship between first-order non-monotonic modal logic and first-order default logic. Namely, were the definition of first-order non-monotonic modal logic be based on just minimality, it would not embed into first-order default logic, cf. Theorem 12 below.

Since classical first-order logic does not contain the modal connective L, like in the propositional case, we pass to "default theories of first-order modal formulas." In other words, formulas of the form $L\varphi$ are thought of as new atomic formulas whose free variables are those of the modal formula $L\varphi$ (this approach is similar to that by Levesque [1990, Section 3]). This is the first-order counterpart of the "purely propositional fragment" of propositional modal logic from Section 2.6. To deal with default theories of first-order modal formulas we need to extend the notion of Herbrand b-interpretation to formulas containing the modal connective L.

A Herbrand *modal* b-interpretation of $\mathcal{L}$ is a set of atomic b-sentences and b-sentences of the form $L\varphi$, where φ is a modal sentence over $\mathcal{L}_b$. The definition of satisfiability (of modal b-sentences) by Herbrand modal b-interpretations results from Definition 18 by adding to it the clause

- $w \models L\varphi$, if $L\varphi \in w$.

In particular, satisfiability of $L\varphi$ *does not depend* on satisfiability of φ at all.

A kind of a relationship between Kripke b-interpretations and Herbrand modal b-interpretations is illustrated by the following proposition and corollary (the proofs are in the appendix).

Proposition 3 *Let* $\mathfrak{M} = \langle U, R, I \rangle$ *be a Kripke* b*-interpretation,* $u \in U$*, and let Herbrand modal* b*-interpretation* $u^{\mathfrak{M}}$ *consist of all atomic* b*-sentences and all* b*-sentences of the form* $L\psi$ *satisfied by the pair* $(\mathfrak{M}, u)$*. Then for all* b*-sentences* φ*,* $(\mathfrak{M}, u) \models \varphi$ *if and only if* $u^{\mathfrak{M}} \models \varphi$*.*

Corollary 2 *If a set of modal* b*-sentences* X b*-entails a modal* b*-sentences* φ *with respect to Herbrand modal* b*-interpretations, then it also* b*-entails* φ *with respect to Kripke* b*-interpretations.*

---

[22]Recall that by $A \cup S$ we mean the union of A and the universal closures of the axioms of S.

[23]It follows from Theorem 11 that, in propositional modal logic, for a cluster closed class $\mathcal{C}$ of Kripke interpretations that characterizes a modal logic S and a set of formulas A, both $\mathcal{C}$-minimality and strong $\mathcal{C}$-minimality are equivalent to $A \cup S$ minimality.

When dealing with Herbrand modal b-interpretations, extensions for default theories over formulas of first-order modal logic are well defined semantically via the operator Δ.

Theorem 12 (Kaminski and Rey [2000, Theorem 6.1]) *A set of first-order modal sentences E is an S-b-expansion for A if and only if E is a b-extension for the default theory $(D^+ \cup D^-, A \cup S)$.*

Note that the defaults in D^+ and D^- are of the form $\dfrac{\varphi(\boldsymbol{x}) :}{L\varphi(\boldsymbol{x})}$ and $\dfrac{: \varphi(\boldsymbol{x})}{M\varphi(\boldsymbol{x})}$, respectively, i.e., the defaults in $D^+ \cup D^-$ are *open*. Like in the propositional case, these defaults reflect *full* introspection of a rational agent.

To embed first-order default logic into first-order non-monotonic modal logic we shall need one more bit of notation. For a default

$$d(\boldsymbol{x}) = \frac{\alpha(\boldsymbol{x}) : \beta_1(\boldsymbol{x}), \ldots, \beta_m(\boldsymbol{x})}{\gamma(\boldsymbol{x})}$$

we denote by $\tau(d(\boldsymbol{x}))$ the modal formula

$$(L\alpha(\boldsymbol{x}) \wedge \bigwedge_{i=1}^{m} LM\beta_i(\boldsymbol{x})) \supset \gamma(\boldsymbol{x})$$

and for a set of defaults D we denote the set of modal sentences $\{\forall \boldsymbol{x} \tau(d(\boldsymbol{x})) : d(\boldsymbol{x}) \in D\}$ by $\tau(D)$.

Theorem 13 (Kaminski and Rey [2000, Theorem 7.1]) *Let (D, A) be a default theory and let a modal logic S be S4F or weaker. A set of sentences E is a b-extension for (D, A) if and only if there exists an S-b-expansion E^L for the set of first-order modal sentences $A \cup \tau(D)$ such that E is the restriction of E^L to its modal-free sentences[24].*

Remark 12 An argument similar to one used in the proof by Kaminski and Rey [2000, Theorem 7.2] shows that extensions for first-order default logic can also be embedded into theories of minimal models by the same translation τ.

Next we present an alternative description of first-order non-monotonic modal logic that is equivalent to Definition 30. The alternative definition is a modification of the corresponding definition from the M. Sc. thesis by Zbar [2000] (Definition 38 below).

Definition 31 *Let S be a modal logic, A be a sets of first-order modal sentences (axioms), and let b be an infinite set of new constant symbols. For a set of sentences X we denote by $\mathbf{NM}_S^{b,A}(X)$ the set of b-sentences b-entailed by $A \cup S \cup \{M\varphi : X \not\models_b \neg\varphi\}$. That is,*

$$\mathbf{NM}_S^{b,A}(X) = \boldsymbol{Th}_b(A \cup S \cup \{M\varphi : X \not\models_b \neg\varphi\}).$$

Note that the form of $\mathbf{NM}_S^{b,A}(X)$ resembles that of $\mathbf{NM}_S^A(X)$ (Definition 5, see also Remark 2).

---

[24]That is, E consists of all sentences of E^L which do not contain the modal connective L.

Theorem 14 (Zbar [2000], see the appendix for the proof) *Let* S *be a modal logic and let* A *be a set of first-order modal sentences (axioms) over* $\mathcal{L}$. *A set of first-order modal* b- *sentences is an* S-b-*expansion for* A *if and only if it is a consistent fixed point of* $\mathbf{NM}_{\mathrm{S}}^{b,A}$.

6   First-order autoepistemic logic

The idea lying behind our definition first-order autoepistemic logics is to replace propositional modal interpretations with "Herbrand modal b-interpretations." The motivation for our definition is similar to the motivation for the definition of open default theories in Section 3.3, see also the paper by Kaminski [1995] for a general discussion.

The syntax of first-order autoepistemic logic is the syntax of first-order modal logic, i.e., it is obtained from the syntax of first-order logic by extending it with the unary modal connective L (believed).

Definition 32 (Kaminski and Rey [2000]) *Let* w *be a Herbrand modal* b-*interpretation and let* X *be a set of modal* b-*sentences. We say that* w *respects* X, *if for every modal* b-*sentence* φ, $L\varphi \in w$ *if and only if* $\varphi \in X$.

Let X, Y, and φ be sets of modal b-sentences and a modal b-sentence, respectively. We say that Y *b-entails* φ *with respect to* X, denoted $Y \models_{X,b} \varphi$, if every Herbrand modal b-interpretation w that respects X and satisfies Y also satisfies φ.

Definition 33 (Kaminski and Rey [2000]) *Let* A *be a set of first-order modal sentences over* $\mathcal{L}$. *A consistent set of modal* b-*sentences* E *is called an* autoepistemic b-*expansion of* A, *if* $E = \{\varphi : A \models_{E,b} \varphi\}$.

Note that if $\mathcal{L}$ is a propositional language, then Definition 33 degenerates to Definition 17.

Alternatively, autoepistemic b-expansions can be described as follows[25].

Theorem 15 (Kaminski and Rey [2002, Theorem 5.3]) *Let* A *be a set of first-order modal sentences over* $\mathcal{L}$. *A set of modal* b-*sentences* E *is an autoepistemic* b-*expansion of* A *if and only if*

$$Mod_b(E) = Mod_b(A \cup \{L\varphi : \varphi \in E\} \cup \{\neg L\varphi : \varphi \notin E\}). \qquad (3)$$

Theorem 16 below is the first-order counterpart of Theorem 7 and provides a strong support for acceptance of our definition of first-order autoepistemic logics (Definition 33). However the proof of Theorem 16 is much more involved. The reason is that, since Definitions 20 and 33 have no syntactical equivalent,[26] we cannot base the proof on the syntactical argument from the paper of Janhunen [1996]. Of course, the semantical proof holds for the propositional case as well.

---

[25]This description is similar to the description of propositional autoepistemic expansions established in the paper by Moore [1985].

[26]The differences between the syntactical and the semantical definitions are discussed by Kaminski et al. [1998].

Theorem 16 (Kaminski and Rey [2002, Theorem 5.2]) *Let A be a set of first-order modal sentences over $\mathcal{L}$. A set of first-order modal sentences E is an autoepistemic expansion of A if and only if E is an extension for the default theory $(D^+ \cup D^-, A \cup \mathbf{5})$[27].*

Similarly to Theorem 12, default theory $(D^+ \cup D^-, X \cup \mathbf{5})$ is over the formulas of first-order modal logic and Herbrand b-interpretations in the semantical definition of extensions are Herbrand modal b-interpretations. Namely, we deal with the ordinary first-order logic over the formulas of first-order modal logic. In other words, like in Theorem 12, formulas of the form $L\varphi$ are thought of as new atomic formulas whose free variables are those of modal formula $L\varphi$ (this approach is similar to that by Levesque [1990, Section 3]). Thus, extensions for $(D^+ \cup D^-, X \cup \mathbf{5})$ are well defined semantically via the operator $\Delta^b_{(D^+ \cup D^-, X \cup \mathbf{5})}$.

Theorems 12 and 16 indicate a strong relationship between first-order non-monotonic modal logics and first-order autoepistemic logic. Namely, similarly to the proof of Theorem 16 one can show that autoepistemic expansions for A coincide with the extensions for $(D^+ \cup D^-, A \cup \mathrm{K5})$, $(D^+ \cup D^-, A \cup \mathrm{K45})$, or $(D^+ \cup D^-, A \cup \mathrm{KD45})$. Thus, by Theorem 12, *a set of sentences E is an autoepistemic expansion for a set of modal sentences A if and only if E is a* 5, K5, K45 *or* KD45*-expansion for A*, which extends the result by Shvarts [1990] to the first-order case.

The converse relationship is given by Propositions 4 and 5 below.

Proposition 4 (Kaminski and Rey [2002, Proposition 6.2]) *Let (D, A) be a default theory. If a set of sentences E is an extension for (D, A), then there exists an autoepistemic expansion E^L of the set of first-order modal sentences $A \cup \tau(D)$ such that E is the restriction of E^L to its modal-free sentences.*

Defaults of the form

$$\delta(\boldsymbol{x}) = \frac{: \beta_1(\boldsymbol{x}), \ldots, \beta_m(\boldsymbol{x})}{\gamma(\boldsymbol{x})}$$

are called *prerequisite-free*. A default theory (D, A) is called prerequisite-free if each default in D is prerequisite free.

Proposition 5 (Kaminski and Rey [2002, Proposition 6.3]) *Let (D, A) be a prerequisite-free default theory. If E^L is an autoepistemic expansion of the set of first-order modal sentences $A \cup \tau(D)$, then the restriction of E^L to its modal-free sentences is an extension for (D, A).*

Corollary 3 (Kaminski and Rey [2002, Corollary 6.5]) *Let (D, A) be a prerequisite-free default theory. A set of sentences E is an extension for (D, A) if and only if there exists an autoepistemic expansion E^L of the set of first-order modal sentences $A \cup \tau(D)$ such that E is the restriction of E^L to its modal-free sentences.*

---

[27]Recall that the defaults in D^+ and D^- are of the form $\dfrac{\varphi(\boldsymbol{x}) :}{L\varphi(\boldsymbol{x})}$ and $\dfrac{: \neg\varphi(\boldsymbol{x})}{\neg L\varphi(\boldsymbol{x})}$, respectively, i.e., the defaults in $D^+ \cup D^-$ are *open*. Also, the first-order version of **5** consists of the universal closures of all instances of modal scheme $\neg L \neg L\varphi \supset L\varphi$.

Of course, the best possible result would be an extension of Corollary 3 to general default theories, cf. Theorem 13. However, as it was shown by Gottlob [1995], there is no *modular* translation from default logic into first-order autoepistemic logic. Also, the translation by Gottlob [1995] is based on finiteness of the underlying autoepistemic theory and, therefore, is inapplicable in the first-order case. Summing up, an embedding of default logic into first-order autoepistemic logic, if exists, seems to be very non-trivial. Alternatively, similarly to Konolige [1994], who showed that the extensions for (D, A) correspond to *strongly grounded* autoepistemic expansions of $A \cup \tau(D)$, one might try to refine Corollary 3 by giving an independent description of autoepistemic expansions of $A \cup \tau(D)$ which correspond to the extensions for (D, A). Here, the main difficulty is to find a *semantical* equivalent of strong groundedness.

We conclude this section with an embedding of *circumscription* formalism[28] [McCarthy, 1980, Lifschitz, 1987] into first-order autoepistemic logic. Since circumscription is the first (and independent) first-order non-monotonic formalism, its embedding into first-order autoepistemic logic provides an additional support for acceptability of our approach to the latter.

Roughly speaking, the circumscription of a set of predicates symbols $\mathcal{P}$ in a first-order sentence A with the set of predicate symbols $\mathcal{C}$ *fixed* ($\mathcal{P} \cap \mathcal{C} = \emptyset$) picks out those models of A in which the interpretations of the predicates in $\mathcal{P}$ are minimal with respect to the models where the interpretations of the predicates in $\mathcal{C}$ coincide.

The connection between circumscription and first-order autoepistemic logic is given by Theorem 17 below. This result is stronger than the similar result by Konolige [1994, Proposition 6.3.2, p. 277]. It should be noted, however, that our definition of autoepistemic logic is not equivalent to the definition in the paper by Konolige cited above.

Theorem 17 (Kaminski and Rey [2002, Theorem 6.1]) *Let $\mathcal{P}$ and $\mathcal{C}$ be disjoint subsets of predicate symbols of $\mathcal{L}$ such that $= \notin \mathcal{P}$. A sentence φ is entailed by the circumscription of $\mathcal{P}$ in A with the fixed set of predicate symbols $\mathcal{C}$ and the fixed set of all function symbols of $\mathcal{L}$ if and only if φ belongs to all autoepistemic expansions of*

$$A \cup \{\forall \boldsymbol{x}(\neg LP(\boldsymbol{x}) \supset \neg P(\boldsymbol{x}))\}_{P \in \mathcal{P}} \cup$$
$$\{\forall \boldsymbol{x}(\neg LP(\boldsymbol{x}) \supset \neg P(\boldsymbol{x})), \forall \boldsymbol{x}(\neg L \neg P(\boldsymbol{x}) \supset P(\boldsymbol{x}))\}_{P \in \mathcal{C} \cup \{=\}}.$$

7 First-order ground non-monotonic modal logic

Similarly to the propositional case, one might try to define expansions for first-order ground non-monotonic modal logic as b-theories of *first-order* ground minimal Kripke interpretations. However, in the first-order case, the restriction of negative introspection (1) to ground formulas

$$\frac{\nvdash \neg \varphi}{M\varphi}, \text{where } \varphi \text{ is a ground first-order } b\text{-sentence} \tag{4}$$

is not strong enough to force first-order ground expansions to be theories of clusters. In addition, expansions defined in such way do not embed into first-order default logic,

---

[28] We assume that the reader is acquainted with the notion of circumscription.

cf. Theorem 18 below. Also, the example in the appendix shows that, in contrast to Theorem 2, b-clusters are not uniquely determined by the modal-free parts of their theories.

Definition 34 below is the "ground counterpart" of Definition 31.

Definition 34 (Zbar [2000]) *Let* S *be a modal logic,* A *be a sets of first-order modal sentences (axioms), and let* b *be an infinite set of new constant symbols. For a set of sentences* X *we denote by* $\mathbf{GNM}_S^{b,A}(X)$ *the of* b-*sentences* b-*entailed by* $A \cup S \cup \{M\varphi : X \not\models_b \neg\varphi,\ \varphi \in \mathbf{GSt}_b\}$. *That is,*

$$\mathbf{GNM}_S^{b,A}(X) = \mathbf{Th}_b(A \cup S \cup \{M\varphi : X \not\models_b \neg\varphi,\ \varphi \in \mathbf{GSt}_b\}).$$

A consistent set of first-order modal sentences E *is called a* ground S-b-expansion *for* A *(or* S_G-b-expansions *for short), if it is a fixed point of* $\mathbf{GNM}_S^{b,A}$.

Note that the form of $\mathbf{GNM}_S^{b,A}(X)$ resembles that of $\mathbf{GNM}_S^A(X)$ (Definition 11), see also Remark 4. It also reflects the ground non-monotonic rule of inference (4) that corresponds to negative introspection bounded to ground first-order formulas.

Remark 13 Actually, Zbar [2000] defined S_G-b-expansions as fixed points of the operator

$$\widehat{\mathbf{GNM}_S}^{b,A}(X) = \mathbf{Th}_b(A \cup S \cup \{M\varphi : \neg\varphi \notin X,\ \varphi \in \mathbf{GSt}_b\}),$$

cf. Definition 38 in the appendix. However. similarly to the proof of Proposition 6, it can be shown that $\mathbf{GNM}_S^{b,A}$ and $\widehat{\mathbf{GNM}_S}^{b,A}$ have the same fixed points. In this paper, we decided to choose Definition 34, because $\mathbf{GNM}_S^{b,A}$ resembles the operator $\mathbf{GNM}_S^A$ (Definition 11).

Next we present an embedding of first-order ground non-monotonic modal logic into first-order default logic and vice versa. The embeddings are similar to the embedding in the propositional case (Theorems 8 and 9, respectively) and to embeddings of first-order non-monotonic modal logic and first-order default logic into each other (Theorems 12 and 13).

Theorem 18 (Zbar [2000]) *A set of first-order (modal) sentences* E *is an* S_G-*expansion for* A *if and only if* E *is an extension for default theory* $(D^+ \cup D_G^-, A \cup S)$.

Recall that the defaults in D^+ are of the form $\dfrac{\varphi(\boldsymbol{x}) :}{L\varphi(\boldsymbol{x})}$ and the defaults in D_G^- are of the form $\dfrac{: \varphi(\boldsymbol{x})}{M\varphi(\boldsymbol{x})}$, where $\varphi(\boldsymbol{x})$ is a ground formula. In particular, the defaults in $D^+ \cup D_G^-$ are *open*. Similarly to Theorem 12, defaults in D^+ and in D_G^- reflect positive and "negative ground" introspections of a rational agent.

The proof of Theorem 18 is similar to that of Theorem 12 and is omitted.

Theorem 19 (Kaminski and Grimberg [2008, Theorem 5.1]) *Let* (D, A) *be a default theory and let a modal logic* S *be S4F or weaker. A set of sentences* E *is an extension for* (D, A) *if and only if there exists an* S_G-*expansion* E^L *for* $A \cup \tau(D)$ *such that* E *is the restriction of* E^L *to its modal-free sentences.*

We conclude this section with minimal model semantics of first-order ground non-monotonic modal logic. For this semantics we shall need the following definitions and notation.

Definition 35 (Cf. Definition 15) *A class $\mathcal{C}$ of Kripke b-interpretations is called* cluster-decomposable, *if every interpretation in $\mathcal{C}$ is either a cluster or of the form $\mathfrak{M} \odot \mathfrak{M}'$, where $\mathfrak{M}'$ is a b-cluster (called the* final *cluster of $\mathfrak{M} \odot \mathfrak{M}'$), and for every b-interpretation $\mathfrak{M} \odot \mathfrak{M}' \in \mathcal{C}$, where $\mathfrak{M}'$ is a b-cluster, and every b-cluster $\mathfrak{M}''$, the Kripke b-interpretation $\mathfrak{M} \odot \mathfrak{M}''$ is in $\mathcal{C}$*[29].

Let $\mathcal{C}$ be a class of cluster-decomposable Kripke b-interpretations. With each element $\mathfrak{M} \in \mathcal{C}$ we associate the following b-cluster, denoted $\widehat{\mathfrak{M}}$.

- If $\mathfrak{M}$ is a b-cluster, then $\widehat{\mathfrak{M}}$ is $\mathfrak{M}$ itself.

- Otherwise, $\widehat{\mathfrak{M}}$ is the final b-cluster of $\mathfrak{M}$[30].

Definition 36 (Cf. Kaminski [1991, Definition 2], Marek and Truszczyński [1993, Definition 11.33, p. 343]) *Let b be an infinite set of new constant symbols and let $\mathfrak{M}'$ and $\mathfrak{M}''$ be Kripke b-interpretations. We write $\mathfrak{M}' \lhd \mathfrak{M}''$, if $\boldsymbol{Th}_b(\mathfrak{M}') \cap \boldsymbol{GSt}_b \subset \boldsymbol{Th}_b(\mathfrak{M}'') \cap \boldsymbol{GSt}_b$.*

Definition 37 (Kaminski and Grimberg [2008]) *Let b be an infinite set of new constant symbols, $\mathcal{C}$ be a cluster-decomposable class of Kripke b-interpretations, and let A be a set of modal sentences. A Kripke b-interpretation $\mathfrak{M} \in \mathcal{C}$ is called* ground minimal *for A, if it satisfies conditions G1–G4 below.*

G1. $\mathfrak{M} \models A$.

G2. $\boldsymbol{Th}_b(\mathfrak{M}) \cap \boldsymbol{GSt}_b = \boldsymbol{Th}_b(\widehat{\mathfrak{M}}) \cap \boldsymbol{GSt}_b$.

G3. There is no Kripke b-interpretation $\mathfrak{M}' \in \mathcal{C}$ such that $\mathfrak{M}' \models A$ and $\widehat{\mathfrak{M}'} \lhd \widehat{\mathfrak{M}}$.

G4. There is no Kripke b-interpretation $\mathfrak{M}' \in \mathcal{C}$ such that $\mathfrak{M}' \models A$, $\mathfrak{M}' \lhd \mathfrak{M}$, and $\boldsymbol{Th}_b(\widehat{\mathfrak{M}'}) \cap \boldsymbol{GSt}_b = \boldsymbol{Th}_b(\widehat{\mathfrak{M}}) \cap \boldsymbol{GSt}_b$.

Remark 14 (Cf. Kaminski [1991, Definition 2]) Since every b-cluster $\mathfrak{M}$ trivially satisfies clauses G2 and G4 of Definition 37, this definition can be simplified for the modal logic S5 as follows: *A b-cluster $\mathfrak{M}$ is ground minimal for A if it satisfies A and is minimal with respect to the partial ordering $\lhd$ among all b-clusters which satisfy A.*

At last, we have arrived at a connection between minimality and ground expansions.

---

[29]The axiomatization of the propositional modal logic characterized by the class of all cluster-decomposable Kripke interpretations in the paper of Tiomkin and Kaminski [2007] naturally extends to first-order modal logic containing **BF**.

[30]That is, $\mathfrak{M} = \mathfrak{M}' \odot \widehat{\mathfrak{M}}$, for some Kripke b-interpretation $\mathfrak{M}'$.

Theorem 20 (Kaminski and Grimberg [2008, Theorem 6.1]) *Let* S *be a modal logic characterized by a cluster-decomposable class* $\mathcal{C}$ *of Kripke interpretations, and let* A *be a set of modal sentences (axioms). A set of sentences* E *is an* S_G-*bexpansion for* A *if and only if there exists a Kripke b-interpretation* $\mathfrak{M} \in \mathcal{C}$ *that is ground minimal for* A *such that*

$$E = \boldsymbol{Th}_b(A \cup S \cup \{M\varphi : \mathfrak{M} \not\models \neg\varphi, \ \varphi \in \boldsymbol{GSt}_b\}).^{[31]}$$

8 Summary: what is the unified approach?

We conclude this paper with a comparison of propositional and first-order non-monotonic logics. Whereas propositional default and non-monotonic modal logics are defined *syntactically* by fixed point equations, see Definitions 1 and 5 respectively, their first-order counterparts are defined *semantically*. Namely, first-order default logic is defined by fixed points of the "semantic" equation, see Definition 20, and first-order non-monotonic modal logic is defined via its minimal model semantics, see Definition 30.

By Corollary 1 and Theorem 14, respectively, the last two definitions are equivalent to fixed point definitions that result from the propositional case in replacing the provability relation $\vdash_S$ with semantical entailment $\models_b$.

Based on the above equivalence, first-order ground non-monotonic logics are defined by replacement of $\vdash_S$ with $\models_b$ in Definition 11 (see Definition 34)[32].

Summing up, the uniform approach to all first-order non-monotonic logics above is the (uniform) replacing of the provability relation $\vdash_S$ in their respective "propositional" definitions with semantical entailment $\models_b$ (and augmenting the set of axioms A with the axioms of the underlying modal logic S, of course.)

Appendix: Proofs

Proof of Proposition 1

It can be readily seen that $\Delta^b_{(D,A)}(W)$ is the union of all sets V of b-models of A which satisfy the condition of Definition 20 and $\Gamma^b_{(D,A)}(X)$ is the intersection of all $\models_b$-closed sets B of b-sentences which include A and satisfy the condition of Definition 21. Moreover, if a set V of b-models of A satisfies the condition of Definition 20, then $\boldsymbol{Th}_b(V)$ is $\models_b$-closed, includes A, and satisfy the condition of Definition 21; and if a $\models_b$-closed set B of b-sentences that includes A satisfies the condition of Definition 21, then $\boldsymbol{Mod}_b(B)$ is a set of b-models of A that satisfies the condition of Definition 20. Thus, the result follows from the identities $\boldsymbol{Th}_b(\bigcup_{i \in I} V_i) = \bigcap_{i \in I} \boldsymbol{Th}_b(V_i)$ and $\boldsymbol{Mod}_b(\bigcap_{i \in I} B_i) = \bigcup_{i \in I} \boldsymbol{Mod}_b(B_i)$.

---

[31] Note that this is no longer a fixed point equation, cf. Definition 34.

[32] This is because the minimal model semantics of propositional grand non-monotonic modal logic does not extend to the first order case.

Proof of Corollary 1

Let W be a fixed point of $\Delta^b_{(D,A)}$ such that $E = \boldsymbol{Th}_b(W)$. Then, by part a) of Proposition 1,

$$\Gamma^b_{(D,A)}(E) = \Gamma^b_{(D,A)}(\boldsymbol{Th}_b(W)) = \boldsymbol{Th}_b(\Delta^b_{(D,A)}(W)) = \boldsymbol{Th}_b(W) = E.$$

Conversely, let E be a fixed point of $\Gamma^b_{(D,A)}$ and let $W = \boldsymbol{Mod}_b(E)$. Since $\boldsymbol{Th}_b(E) = E$, by Remark 6, $\boldsymbol{Th}_b(W) = E$. Therefore, by part b) of Proposition 1,

$$\Delta^b_{(D,A)}(W) = \Delta^b_{(D,A)}(\boldsymbol{Mod}_b(E)) = \boldsymbol{Mod}_b(\Gamma^b_{(D,A)}(E)) = \boldsymbol{Mod}_b(E) = W.$$

Proof of Proposition 3

The proof is by induction on the complexity of φ. The basis, i.e., the case where φ is atomic or of the form $L\psi$, immediately follows from the definition of $u^{\mathfrak{M}}$. The induction step is equally easy, because it does not depend on the accessibility relation R.

Proof of Corollary 2

Assume to the contrary that X does not b-entail φ with respect to Kripke b-interpretations. Then, there exist a Kripke b-model $\mathfrak{M} = \langle U, R, I \rangle$ of X and a world $u \in U$ such that $(\mathfrak{M}, u) \not\models \varphi$, implying $(\mathfrak{M}, u) \models \neg\varphi$. Therefore, by Proposition 3, $u^{\mathfrak{M}} \models X \cup \{\neg\varphi\}$, which contradicts $X \models \varphi$ with respect to Herbrand modal b-interpretations.

Proof of Theorem 14

The proof of Theorem 14 is based on a "kind of equivalence" between the operators $\mathbf{NM}^{b,A}_S$ and $\Gamma^b_{(D+\cup D^-, A\cup S)}$. However, an additional preliminary step is required, because the definitions of the sets of b-sentences $\mathbf{NM}^{b,A}_S(X)$ and $\Gamma^b_{(D+\cup D^-, A\cup S)}(X)$ are asymmetric: whereas the former introduces modal b-sentences of the form $M\varphi$ according to $X \not\models_b \neg\varphi$, the latter introduces those sentences according to $\neg\varphi \notin X$. Therefore, we shall replace the operator $\mathbf{NM}^{b,A}_S$ with $\widehat{\mathbf{NM}}^{b,A}_S$ that is defined as follows.

Definition 38 (Zbar [2000]) *Let* S *be a modal logic,* A *be a sets of first-order modal sentences (axioms), and let* b *be an infinite set of new constant symbols. For a set of formulas* X *we denote the set of* b-*sentences* b-*entailed by* $A \cup S \cup \{M\varphi : \neg\varphi \notin X\}$ *by* $\widehat{\mathbf{NM}}^{b,A}_S(X)$. *That is,*

$$\widehat{\mathbf{NM}}^{b,A}_S(X) = \boldsymbol{Th}_b(A \cup S \cup \{M\varphi : \neg\varphi \notin X\}). \tag{5}$$

Proposition 6 below states that fixed points of $\widehat{\mathbf{NM}}^{b,A}_S$ coincide with those of $\mathbf{NM}^{b,A}_S$. Thus, in Theorem 14 we can replace the operator $\mathbf{NM}^{b,A}_S$ with $\widehat{\mathbf{NM}}^{b,A}_S$.

Proposition 6 *Let E be a set of modal b-sentences. Then $E = \mathbf{NM}_S^{b,A}(E)$ if and only if $E = \widehat{\mathbf{NM}}_S^{b,A}(E)$.*

Proof If $E = \mathbf{NM}_S^{b,A}(E)$ or $E = \widehat{\mathbf{NM}}_S^{b,A}(E)$, by Remark 6, $E = \boldsymbol{Th}_b(E)$. Therefore, $E \not\models_b \neg\varphi$ if and only if $\neg\varphi \notin E$, and the result follows from the definition of $\mathbf{NM}_S^{b,A}$ and $\widehat{\mathbf{NM}}_S^{b,A}$ (Definitions 31 and 38, respectively). ∎

Proposition 7 *Let A be a set of modal sentences over $\mathcal{L}$ and b be an infinite set of new constant symbols. Then for a set of b-sentences X, $\widehat{\mathbf{NM}}_S^{b,A}(X) = \Gamma^b_{(D^+ \cup D^-, A \cup S)}(X)$.*

Proof For the inclusion $\Gamma^b_{(D^+ \cup D^-, A \cup S)}(X) \subseteq \widehat{\mathbf{NM}}_S^{b,A}(X)$ it suffices to show that (the belief set $B =$) $\widehat{\mathbf{NM}}_S^{b,A}(X)$ is closed under $\models_b$ (with respect to Herbrand modal b-interpretations), includes $A \cup S$, and satisfies the condition of Definition 21 for $D = D^+ \cup D^-$ stating that

$$\{L\varphi : \varphi \in \widehat{\mathbf{NM}}_S^{b,A}(X)\} \cup \{M\varphi : \neg\varphi \notin X\} \subseteq \widehat{\mathbf{NM}}_S^{b,A}(X).$$

The inclusions $A \cup S \subseteq \widehat{\mathbf{NM}}_S^{b,A}(X)$ and $\{M\varphi : \neg\varphi \notin X\} \subseteq \widehat{\mathbf{NM}}_S^{b,A}(X)$ follow from the definition of $\widehat{\mathbf{NM}}_S^{b,A}$ (Definition 38), the inclusion $\{L\varphi : \varphi \in \widehat{\mathbf{NM}}_S^{b,A}(X)\} \subseteq \widehat{\mathbf{NM}}_S^{b,A}(X)$ follows from soundness of Kripke semantics, and the closure under $\models_b$ (with respect to Herbrand modal b-interpretations) is by Corollary 2.

For the proof of the converse inclusion $\widehat{\mathbf{NM}}_S^{b,A}(X) \subseteq \Gamma^b_{(D^+ \cup D^-, A \cup S)}(X)$ assume to the contrary that there exists a modal b-sentence $\varphi \in \widehat{\mathbf{NM}}_S^{b,A}(X)$ such that $\varphi \notin \Gamma^b_{(D^+ \cup D^-, A \cup S)}(X)$. Since $\Gamma^b_{(D^+ \cup D^-, A \cup S)}(X)$ is closed under $\models_b$ (with respect to Herbrand modal b-interpretations), there exists a Herbrand modal b-interpretation w such that $w \models \Gamma^b_{(D^+ \cup D^-, A \cup S)}(X)$, but $w \not\models \varphi$. By the definition of $\models$, $\boldsymbol{Th}_b(\{w\})$ is maximal (see Definition 23). We contend that it is also consistent. Since $\boldsymbol{Th}_b(\{w\})$ is closed under conjunction, it suffices to show that for no modal b-sentence ψ satisfied by w, $\vdash_K \neg\psi$.

So, assume to the contrary that for some modal b-sentence ψ satisfied by w, $\vdash_K \neg\psi$. Then

$$\neg\psi \in \Gamma^b_{(D^+ \cup D^-, A \cup S)}(X), \tag{6}$$

because

- $K \subseteq S \subseteq \Gamma^b_{(D^+ \cup D^-, A \cup S)}(X)$;

- $\Gamma^b_{(D^+ \cup D^-, A \cup S)}(X)$ is closed under first-order provability relation $\vdash$, by soundness of the Herbrand semantics; and

- $\Gamma^b_{(D^+ \cup D^-, A \cup S)}(X)$ is closed under modal rule of inference **NEC**,[33] by the set of defaults D^+.

---

[33] That is, $\psi \in \widehat{\mathbf{NM}}_S^{b,A}(X)$ implies $L\psi \in \widehat{\mathbf{NM}}_S^{b,A}(X)$.

However, (6) contradicts $w \models \Gamma^b_{(D^+ \cup D^-, A \cup S)}(X) \cup \{\psi\}$.

Since $\mathbf{Th}_b(\{w\})$ is both maximal and consistent, it belongs to U_b, see Definition 24. Therefore, by Theorem 10 and Remark 11, $\mathfrak{M}_b^{\mathbf{Th}_b(\{w\})} \models \Gamma^b_{(D^+ \cup D^-, A \cup S)}(X)$. Namely, satisfiability of $\Gamma^b_{(D^+ \cup D^-, A \cup S)}(X)$ by $\mathfrak{M}_b^{\mathbf{Th}_b(\{w\})}$ follows from the inclusion $\Gamma^b_{(D^+ \cup D^-, A \cup S)}(X) \subseteq \mathbf{Th}_b(\{w\})$ and the fact that, by the set of defaults D^+, the set of modal b-sentences $\Gamma^b_{(D^+ \cup D^-, A \cup S)}(X)$ is closed under **NEC**.

Since, by the set of defaults D^-,

$$A \cup S \cup \{M\varphi : \neg\varphi \notin X\} \subseteq \Gamma^b_{(D^+ \cup D^-, A \cup S)}(X),$$

it follows from (5) that also $\mathfrak{M}_b^{\mathbf{Th}_b(\{w\})} \models \widehat{\mathbf{NM}}_S^{b,A}(X)$. However, $\mathfrak{M}_b^{\mathbf{Th}_b(\{w\})} \not\models \varphi$ which contradicts our assumption $\varphi \in \widehat{\mathbf{NM}}_S^{b,A}(X)$. ∎

Now we are ready for the proof of Theorem 14: since, by Propositions 6 and 7, the operators $\mathbf{NM}_S^{b,A}$ and $\Gamma^b_{(D^+ \cup D^-, A \cup S)}$ have the same fixed points, the result follows from Theorem 12.

Example

Below we present an example of Zbar [2000] that, in contrast to Theorem 2, shows that b-clusters are not uniquely determined by the modal-free parts of their theories.

This example employs the following notation.

- The underlying language $\mathcal{L}$ consists of one unary predicate symbol P and infinitely many constant symbols $c_1, c_2, \ldots$;

- b (as usual) is an infinite set of new constant symbols;

- u' is a Herbrand b-interpretation defined by $u' = \{P(c_i)\}_{i=1,2,\ldots}$;

- U' is a set of all Herbrand b-interpretations u which include u', i.e., $U' = \{u : u' \subseteq u\}$;

- U'' is a set of Herbrand b-interpretations that results in removing u' from U', i.e., $U'' = U' \setminus \{u'\} = \{u : u' \subset u\}$; and

- $\mathfrak{M}' = \langle U', U' \times U', I' \rangle$ and $\mathfrak{M}'' = \langle U'', U'' \times U'', I'' \rangle$, where I' and I'' are the identity functions on U' and U'', respectively.

We shall prove that $\mathbf{Th}_b(\mathfrak{M}')$ and $\mathbf{Th}_b(\mathfrak{M}'')$ have the same ground parts, but, despite of this, their $\mathcal{L}$-theories are different. That is,

$$\mathbf{Th}_b(\mathfrak{M}') \cap \mathbf{GSt}_b = \mathbf{Th}_b(\mathfrak{M}'') \cap \mathbf{GSt}_b, \tag{7}$$

but $\mathbf{Th}(\mathfrak{M}') \neq \mathbf{Th}(\mathfrak{M}'')$.

Since $\mathbf{Th}_b(\mathfrak{M}') \cap \mathbf{GSt}_b = \mathbf{Th}_b(U')$ and $\mathbf{Th}_b(\mathfrak{M}'') \cap \mathbf{GSt}_b = \mathbf{Th}_b(U'')$, (7) is equivalent to $\mathbf{Th}_b(U') = \mathbf{Th}_b(U'')$. The inclusion $\mathbf{Th}_b(U') \subseteq \mathbf{Th}_b(U'')$ follows from $U'' \subset U'$ and for the proof of the converse inclusion it suffices to show that each ground b-sentence φ satisfied by u' is also satisfied by some $u \in U''$. Such a u results form u' by adding to it $P(c)$ for a new constant symbol $c \in b$ that does not appear in φ.

To show that $\boldsymbol{Th}_b(\mathfrak{M}') \neq \boldsymbol{Th}_b(\mathfrak{M}'')$, we shall prove that $\mathfrak{M}''$ satisfies the $\mathcal{L}$-sentence $\exists x(M \neg P(x) \wedge P(x))$, whereas $\mathfrak{M}'$ does not. Indeed, $\mathfrak{M}'' \models \exists x(M \neg P(x) \wedge P(x))$, because for every $u \in U''$ there exist $c \in b$ and $w \in U''$ such that $P(c) \in u$ and $P(c) \notin w$. On the other hand, $\mathfrak{M}' \not\models \exists x(M \neg P(x) \wedge P(x))$, because for every $t \in \boldsymbol{Tr}_b$, $(\mathfrak{M}', u') \models P(t)$ implies $t \in \{c_i\}_{i=1,2,\ldots}$ and $U' \models P(c_i)$ for all $i = 1, 2, \ldots$.

References

F.M. Donini, D. Nardi, and R. Rosati. Ground nonmonotonic modal logics. *Journal of Logic and Computation*, 7:523–548, 1997.

M.C. Fitting, W. Marek, and M. Truszczyński. The pure logic of necessitation. *Journal of Logic and Computation*, 2:349–373, 1992.

G. Gottlob. Translating default logic into standard autoepistemic logic. *Journal of the ACM*, 42:711–740, 1995.

R. Guerreiro and M. Casanova. An alternative semantics for default logic. Presented at *the 3rd International Workshop on Nonmonotonic Reasoning*, 1990.

G.E. Hughes and M.J. Cresswell. *An introduction to modal logic*. Methuen and Co., London, 1972.

G.E. Hughes and M.J. Cresswell. *A companion to modal logic*. Methuen and Co, London, 1984.

T. Janhunen. Representing autoepistemic introspection in terms of default rules. In W. Wahlster, editor, *Proceedings of the 12th European Conference on Artificial Intelligence – ECAI'96*, pages 70–74, Chichister, England, 1996. John Wiley & Sons.

T. Janhunen. On the intertranslatability of autoepistemic, default and priority logics, and parallel circumscription. In J. Dix, L.F. del Cerro, and U. Furbach, editors, *Proceedings of the 6th European Workshop on Logics in Artificial Intelligence – JELIA'98*, volume 1489 of *Lecture Notes in Computer Science*, pages 216–232, Berlin, 1998. Springer.

T. Janhunen. On the intertranslatability of non–monotonic logics. *Annals of Mathematics and Artificial Intelligence*, 27:79–128, 1999.

M. Kaminski. Embedding a default system into non-monotonic logics. *Fundamenta Informaticae*, XIV:345–353, 1991.

M. Kaminski. A comparative study of open default theories. *Artificial Intelligence*, 77:285–319, 1995.

M. Kaminski and B. Grimberg. First-order ground non-monotonic modal logic. *Fundamenta Informaticae*, 93:253–276, 2008.

M. Kaminski and G. Rey. First–order non-monotonic modal logics. *Fundamenta Informaticae*, 42:303–333, 2000.

M. Kaminski and G. Rey. Revisiting quantification in first-order autoepistemic logic. *ACM Transactions on Computational Logic*, 3:542–561, 2002.

M. Kaminski, J.A. Makowsky, and M. Tiomkin. Extensions for open default theories via the domain closure assumption. *Journal of Logic and Computation*, 8:169–187, 1998.

K. Konolige. Autoepistemic logic. In D.M. Gabbay, C.J. Hogger, and J.A. Robinson, editors, *Handbook of Logic in Artificial Intelligence and Logic Programming*, volume 3, pages 217–295, Oxford, 1994. Oxford University Press.

H.J. Levesque. All I know: A study in autoepistemic logic. *Artificial Intelligence*, 42:263–309, 1990.

V. Lifschitz. Computing circumscription. In *Proceedings of the 9th International Joint Conference on Artificial Intelligence*, pages 121–127, San Mateo, California, 1987. Morgan Kaufmann.

V. Lifschitz. On open defaults. In J.W. Lloyd, editor, *Computational Logic: Symposium Proceedings*, pages 80–95, Berlin, 1990. Springer–Verlag.

J.W. Lloyd. *Foundation of logic programming, second extended edition*. Springer–Verlag, Berlin, 1993.

V.W. Marek and M. Truszczyński. *Nonmonotonic Logic*. Springer–Verlag, Berlin, 1993.

J. McCarthy. Circumscription – a form of non-monotonic reasoning. *Artificial Intelligence*, 13:27–39, 1980.

D. McDermott. Non-monotonic logic II: Non-monotonic modal theories. *Journal of the ACM*, 29:33–57, 1982.

E. Mendelson. *Introduction to mathematical logic*. Chapman and Hall, London, 1997.

R.C. Moore. Semantical considerations on non-monotonic logics. *Artificial Intelligence*, 25:75–94, 1985.

R. Reiter. A logic for default reasoning. *Artificial Intelligence*, 13:81–132, 1980.

G. Schwarz. Minimal model semantics for nonmonotonic modal logics. In *Proceedings of the Seventh Annual IEEE Symposium on Logic in Computer Science*, pages 34–43, Los Alamitos, California, 1992. IEEE Computer Society Press.

K. Segerberg. An essay in classical modal logic. *Filosofiska Studier*, 13, 1971.

G.F. Shvarts. Autoepistemic modal logics. In R. Parikh, editor, *Proceedings of 3rd Conference on Theoretical Aspects of Reasoning about Knowledge*, pages 97–110, Los Altos, California, 1990. Morgan Kaufmann.

R.C. Stalnaker. A note on nonmonotonic modal logic. Unpublished manuscript, 1980.

M. Tiomkin and M. Kaminski. Nonmonotonic default modal logics. *Journal of the ACM*, 38:963–984, 1991.

M. Tiomkin and M. Kaminski. The modal logic of cluster-decomposable Kripke interpretations. *Notre Dame Journal of Formal Logic*, 48:511–520, 2007.

M. Truszczyński. Modal interpretation of default logic. In *Proceedings of the 12th International Joint Conference on Artificial Intelligence*, pages 393–398, San Mateo, California, 1991a. Morgan Kaufmann.

M. Truszczyński. Embedding default logic into non–monotonic logics. In W. Marek, A. Nerode, and V.S. Subrahmanian, editors, *Proceedings of the First International Workshop on Logic Programming and Nonmonotonic Reasoning*, pages 151–165, Cambridge, MA, 1991b. MIT Press.

Y. Zbar. Open default theories. Master's thesis, Department of Computer Science, Technion - Israel Institute of Technology, 2000.

Origins of Answer-Set Programming – Some Background And Two Personal Accounts

Victor W. Marek
Department of Computer Science
University of Kentucky
Lexington, KY 40506-0633, USA

Ilkka Niemelä
Department of Information and Computer Science
Aalto University, Finland

Mirosław Truszczyński
Department of Computer Science
University of Kentucky
Lexington, KY 40506-0633, USA

Abstract: We discuss the evolution of aspects of nonmonotonic reasoning towards the computational paradigm of answer-set programming (ASP). We give a general overview of the roots of ASP and follow up with the personal perspective on research developments that helped verbalize the main principles of ASP and differentiated it from the classical logic programming.

1 Introduction — Answer-Set Programming Now

Merely ten years since the term was first used and its meaning formally elaborated, answer-set programming has reached the status of a household name, at least in the logic programming and knowledge representation communities. In this paper, we present our personal perspective on influences and ideas — most of which can be traced back to research in knowledge representation, especially nonmonotonic reasoning, logic programming with negation, constraint satisfaction and satisfiability testing — that led to the two papers [Marek and Truszczyński, 1999, Niemelä, 1999] marking the beginning of answer-set programming as a computational paradigm.

Answer-set programming (*ASP*, for short) is a paradigm for declarative programming aimed at solving search problems and their optimization variants. Speaking informally, in ASP a search problem is modeled as a theory in some language of logic. This representation is designed so that once appended with an encoding of a particular instance of the problem, it results in a theory whose *models*, under the semantics of the formalism, correspond to solutions to the problem for this instance. The paradigm was first formulated in these terms by Marek and Truszczyński [1999] and Niemelä [1999].

The ASP paradigm is most widely used with the formalism of logic programming without function symbols, with programs interpreted by the *stable-model* semantics introduced by Gelfond and Lifschitz [1988]. Sometimes the syntax of programs is extended with the *strong* negation operator and disjunctions of literals are allowed in the heads of program rules. The semantics for such programs was also defined by Gelfond and Lifschitz [1991]. They proposed to use the term *answer sets* for sets of literals, by which programs in the extended syntax were to be interpreted. Ten years after the answer-set semantics was introduced, answer sets lent their name to the budding paradigm. However, there is more to answer-set programming than logic programming with the stable-model and answer-set semantics. Answer-set programming languages rooted directly in first-order logic, extending it in some simple intuitive ways to model definitions, have also been proposed over the years and have just matured to be computationally competitive with the original logic programming embodiments of the paradigm [Denecker, 1998, Denecker and Ternovska, 2008, East and Truszczyński, 2006].

Unlike Prolog-like logic programming, ASP is fully declarative. Neither the order of rules in a program nor the order of literals in rules have any effect on the semantics and only negligible (if any) effect on the computation. All ASP formalisms come with the functionality to model definitions and, most importantly, inductive definitions, in intuitive and concise ways. Further, there is a growing body of works that start addressing methods of modular program design [Dao-Tran, Eiter, Fink, and Krennwallner, 2009, Janhunen, Oikarinen, Tompits, and Woltran, 2009] and program development and debugging [Brain and Vos, 2005, Brummayer and Järvisalo, 2010]. These features facilitate modeling problems in ASP, and make ASP an approach accessible to non-experts.

Most importantly, though, ASP comes with fast software for processing answer-set programs. Processing of programs in ASP is most often done in two steps. The first step consists of *grounding* the program to its equivalent propositional version. In the second step, this propositional program is *solved* by a backtracking search algorithm that finds one or more of its answer sets (they represent solutions) or determines that no answer sets (solutions) exist. The current software tools employed in each step, commonly referred to as *grounders* and *solvers*, respectively, have already reached the level of performance that makes it possible to use them successfully with programs arising from problems of practical importance.

This effectiveness of answer-set programming tools is a result of a long, sustained and systematic effort of a large segment of the Knowledge Representation community, and can be attributed to a handful of crucial ideas, some of them creatively adapted to ASP from other fields. Specifically, *domain restriction* was essential to help control the size of ground programs. It was implemented in *lparse*, the first ASP grounder [Niemelä and Simons, 1996]. The *well-founded semantics* [Van Gelder et al., 1991] inspired strong propagation methods implemented in the first full-fledged ASP solver *smodels* [Niemelä and Simons, 1996]. *Program completion* [Clark, 1978] provided a bridge to satisfiability testing. For the class of tight programs [Erdem and Lifschitz, 2003], it allowed for a direct use of satisfiability testing software in ASP, the idea first implemented in an early version of the solver *cmodels*[1]. *Loop formulas* [Lin and

---

[1]http://www.cs.utexas.edu/users/tag/cmodels.html

Zhao, 2002] extended the connection to satisfiability testing to arbitrary programs. They gave rise to such successful ASP solvers as *assat* [Lin and Zhao, 2002], *pb-models* [Liu and Truszczyński, 2005] and later implementations of *cmodels* [Lierler and Maratea, 2004]. Database techniques for *query optimization* influenced the design of the grounder for the *dlv* system[2] [Leone, Pfeifer, Faber, Eiter, Gottlob, Perri, and Scarcello, 2006]. Important advances of satisfiability testing including the data structure of *watched literals*, *restarts*, and *conflict-clause learning* were incorporated into the ASP solver *clasp*[3], at present the front-runner among ASP solvers and the winner of one track of the 2009 SAT competition. Some of the credit for the advent of high-performance ASP tools is due to the initiative to hold ASP grounder and solver contests. The two editions of the contest so far [Gebser, Liu, Namasivayam, Neumann, Schaub, and Truszczyński, 2007, Denecker, Vennekens, Bond, Gebser, and Truszczynski, 2009] focused on modeling and on solver performance, and introduced a necessary competitive element into the process.

The modeling features of ASP and computational performance of ASP software find the most important reflection in a growing range of successful applications of ASP. They include molecular biology [Gebser, Guziolowski, Ivanchev, Schaub, Siegel, Thiele, and Veber, 2010a, Gebser, König, Schaub, Thiele, and Veber, 2010b], decision support system for space shuttle controllers [Balduccini, Gelfond, and Nogueira, 2006], phylogenetic systematics [Erdem, 2011], automated music composition [Boenn, Brain, Vos, and Fitch, 2011], product configuration [Soininen and Niemelä, 1998, Tiihonen, Soininen, Niemelä, and Sulonen, 2003, Finkel and O'Sullivan, 2011] and repair of web-service workflows [Friedrich, Fugini, Mussi, Pernici, and Tagni, 2010].

And so, ASP is now a declarative programming paradigm built on top of a solid theoretical foundation, with features that facilitate its use in modeling, with software supporting effective computation, and with a growing list of successful applications to its credit. How did it all come about? This paper is an attempt to reconstruct our personal journey to ASP.

2 Knowledge Representation Roots of Answer-Set Programming

One of the key questions for knowledge representation is how to model commonsense knowledge and how to automate commonsense reasoning. The question does not seem particularly relevant to ASP understood, as it now commonly is, as a general purpose computational paradigm for solving search problems. But in fact, knowledge representation research was essential. First, it recognized and emphasized the importance of principled modeling of commonsense and domain knowledge. The impact of the modeling aspect of knowledge representation and reasoning is distinctly visible in the current implementations of ASP. They support high level programming that separates modeling problem specifications from problem instances, provide intuitive means to model aggregates, and offer direct means to model defaults and inductive definitions. Second, knowledge representation research, and especially nonmonotonic reasoning

---

[2]www.dbai.tuwien.ac.at/proj/dlv/
[3]www.cs.uni-potsdam.de/clasp/

research, provided the theoretical basis for ASP formalisms: the answer-set semantics of programs can be traced back to the semantics of default logic and autoepistemic logic, the semantics of the logic FO(ID) [Denecker, 2000, Denecker and Ternovska, 2008] has it roots in the well-founded semantics of nonmonotonic provability operators.

In this section we discuss the development of those ideas in knowledge representation that eventually took shape of answer-set programming. In their celebrated 1969 paper, McCarthy and Hayes wrote

> [...] intelligence has two parts, which we shall call the epistemological and the heuristic. The epistemological part is the representation of the world in such a form that the solution of problems follows from the facts expressed in the representation. The heuristic part is the mechanism that on the basis of the information solves the problem and decides what to do.

With this paragraph McCarthy and Hayes ushered knowledge representation and reasoning into artificial intelligence and moved it to one of the most prominent positions in the field. Indeed, what they referred to as the *epistemological part* is now understood as knowledge representation, while the *heuristic part* has evolved into broadly understood automated reasoning — a search for proofs or models.

The question how to do knowledge representation and reasoning quickly reached the forefront of artificial intelligence research. McCarthy suggested first-order logic as the formalism for knowledge representation. The reasons behind the proposal were quite appealing. First-order logic is "descriptively universal" and proved itself as the formal language of mathematics. Moreover, key reasoning tasks in first-order logic could be automated, assuming one adopted appropriate restrictions to escape semi-decidability of first-order logic in its general form.

However, there is no free lunch and it turned out that first-order logic could not be just taken off the shelf and used for knowledge representation with no extra effort required. The problem is that domain knowledge is rarely complete. More often than not, information available to us has gaps. And the same is true for artificial agents we would like to function autonomously on our behalf. Reasoning with incomplete knowledge is inherently *defeasible*. Depending on how the world turns out to be (or depending on how the gaps in our knowledge are closed), some conclusions reached earlier may have to be withdrawn. The monotonicity of first-order logic consequence relation is at odds with the *nonmonotonicity* of defeasible reasoning and makes modeling defeasible reasoning in first-order logic difficult.

To be effective even when available information is incomplete, humans often develop and use *defaults*, that is, rules that typically work but in some exceptional situations should not be used. We are good at learning defaults and recognizing situations in which they should not be used. In everyday life, it is thanks to defaults that we are not bogged down in the *qualification problem* [McCarthy, 1977], that is, normally we do not check that every possible precondition for an action holds before we take it. And we naturally take advantage of the *frame axiom* [McCarthy and Hayes, 1969] when reasoning, that is, we take it that things remain as they are unless they are changed by an action. Moreover, we do so avoiding the difficulties posed by the *ramification problem* [Finger, 1987], which is concerned with side effects of actions. However, first-order logic conspicuously lacks defaults in its syntactic repertoire nor does it provide an obvious way to simulate them. It is not at all surprising, given that defaults have a defeasible flavor about them. Not being aware that a situation is "exceptional" one may apply a default but later be forced to withdraw the conclusion upon finding out the assumption of "non-exceptionality" was wrong.

Yet another problem for the use of first-order logic in knowledge representation comes from the need to model definitions, most notably the inductive ones. The way humans represent definitions has an aspect of defeasibility that is related to the *closed-world assumption*. Indeed, we often define a concept by specifying all its known instantiations. We understand such a definition as meaning also that *nothing else* is an instance of the concept, even though we rarely if ever say it explicitly. But the main problem with definitions lies elsewhere. Definitions often are *inductive* and their correct meaning is captured by the notion of a least fixpoint. First-order logic cannot express the notion of a least fixpoint and so does not provide a way to specify inductive definitions.

These problems did not go unrecognized and in late 1970s researchers were seeking ways to address them. Some proposals called for extensions of first-order logic by explicit means to model defaults while other argued that the language can stay the same but the semantics had to change. In 1980, the Artificial Intelligence Journal published a double issue dedicated to nonmonotonic reasoning, a form of reasoning based on but departing in major ways from that in first-order logic. The issue contained three papers by McCarthy [1980], Reiter [1980], and McDermott and Doyle [1980] that launched the field of nonmonotonic reasoning.

McCarthy's proposal to bend the language of first-order logic to the needs of knowledge representation was to adjust the semantics of first-order logic and to base the entailment relation among sentences in first-order logic on *minimal* models only [McCarthy, 1980]. He called the resulting formalism *circumscription* and demonstrated how circumscription could be used in several settings where first-order logic failed to work well. Reiter [1980] extended the syntax of first-order logic by *defaults*, inference rules with exceptions, and described formally reasoning with defaults. Reiter was predominantly interested in reasoning with *normal* defaults but his default logic was much more general. Finally, McDermott and Doyle proposed a logic based on the language of modal logic which, as they suggested, was also an attempt to model reasoning with defaults. This last paper was found to suffer from minor technical problems. Two years later, McDermott [1982] published another paper which corrected and extended the earlier one.

These three papers demonstrated that shortcomings of first-order logic in modeling incomplete knowledge and supporting reasoning from these representations could be addressed without giving up on the logic entirely but by adjusting it. They sparked a flurry of research activity directed at understanding and formalizing nonmonotonic reasoning. One of the most important and lasting outcomes of those efforts was the autoepistemic logic proposed by Moore [1984, 1985]. Papers by Moore can be regarded as closing the first phase of the nonmonotonic reasoning as a field of study.

Identifying nonmonotonic reasoning as a phenomenon deserving an in-depth study was a major milestone in logic, philosophy and artificial intelligence. The prospect of understanding and automating reasoning with incomplete information, of the type we humans are so good at, excited these research communities and attracted many re-searchers to the field. Accordingly, the first 10-12 years of nonmonotonic reasoning research brought many fundamental results and established solid theoretical founda-tions for circumscription [McCarthy, 1980, Lifschitz, 1988], default logic [Reiter and Criscuolo, 1981, Hanks and McDermott, 1986, Marek and Truszczyński, 1989, Pearl, 1990], autoepistemic logic [Moore, 1985, Niemelä, 1988, Marek and Truszczyński, 1991, Shvarts, 1990, Schwarz, 1991] and modal nonmonotonic logics in the style of McDermott and Doyle [Marek, Shvarts, and Truszczyński, 1993, Schwarz, 1992, Schwarz and Truszczyński, 1992]. Researchers made progress in clarifying the rela-tionship between these formalisms [Konolige, 1988, 1989, Marek and Truszczyński, 1989b, Bidoit and Froidevaux, 1991, Truszczyński, 1991]. Computational aspects re-ceived much attention, too. First complexity results appeared in late 1980s and early 1990s [Cadoli and Lenzerini, 1990, Marek and Truszczyński, 1991, Kautz and Sel-man, 1989, Gottlob, 1992, Stillman, 1992] and early, still naive at that time, imple-mentations of automated reasoning with nonmonotonic logics were developed around the same time [Etherington, 1987, Niemelä and Tuominen, 1986, 1987, Ginsberg, 1989]. Several research monographs were published in late 1980s and early 1990s systematizing that phase of nonmonotonic reasoning research and making it accessi-ble to outside communities [Besnard, 1989, Brewka, 1991, Marek and Truszczyński, 1993].

Expectations brought up by the advent of nonmonotonic reasoning formalisms were high. It was thought that nonmonotonic logics would facilitate concise and elab-oration tolerant representations of knowledge, and that through the use of defeasible inference rules like defaults it would support fast reasoning. However, around the time of the first Knowledge Representation and Reasoning Conference, KR 1989 in Toronto, concerns started to surface in discussions and papers.

First, there was the issue of multiple belief sets, depending on the logic used rep-resented as extensions or expansions. A prevalent interpretation of the problem was that multiple belief sets provided the basis for *skeptical* and *brave* modes of reasoning. Skeptical reasoning meant considering as consequences a reasoner was sanctioned to draw only those formulas that were in every belief set. Brave reasoning required a non-deterministic commitment to one of the possible belief sets with all its elements becoming consequences of the underlying theory (in the nonmonotonic logic at hand). The first approach was easy to understand and accept at the intuitive level. But as a reasoning mechanism it was rather weak as in general it supported few non-trivial in-ferences. The second approach was underspecified — it provided no guidelines on how to select a belief set, and it was not at all obvious how if at all humans perform such a selection. Both skeptical and brave reasoning suffered from the fact that there were no practical problems lying around that could offer some direction as to how to proceed with any of these two approaches.

Second, none of the main nonmonotonic logics seemed to provide a good for-malization of the notion of a default or of a defeasible consequence relation. This was quite a surprising and in the same time worrisome observation. Nonmonotonic reasoning brought attention to the concept of default and soon researchers raised the question of how to reason *about* defaults rather than with defaults [Pearl, 1990, Kraus,

Lehmann, and Magidor, 1990, Lehmann and Magidor, 1992]. A somewhat different version of the same question asked about defeasible consequence relations, whether they can be characterized in terms of intuitively acceptable axioms, and whether they have semantic characterizations [Gabbay, 1989, Makinson, 1989]. Despite the success of circumscription, default and autoepistemic logics in addressing several problems of knowledge representation, it was not clear if or how they could contribute to the questions above. In fact, it still remains an open problem whether any deep connection between these logics and the studies of abstract nonmonotonic inference relations exists.

Next, the complexity results obtained at about same time [Marek and Truszczyński, 1991, Eiter and Gottlob, 1993a, Gottlob, 1992, Stillman, 1992, Eiter and Gottlob, 1993b, 1995] were viewed as negative. They dispelled any hope of higher computational efficiency of nonmonotonic reasoning. Even under the restriction to the propositional case, basic reasoning tasks turned out to be as complex as and in some cases even more complex (assuming polynomial hierarchy does not collapse) than reasoning in propositional logic. Even more discouraging results were obtained for the general language.

Finally, the questions of applications and implementations was becoming more and more urgent. There were no practical artificial intelligence applications under development at that time that required nonmonotonic reasoning. Nonmonotonic logics continued to be extensively studied and discussed at AI and KR conferences, but the belief that they can have practical impact was diminishing. There was a growing feeling that they might amount to not much more but a theoretical exercise. Complexity results notwithstanding, the ultimate test of whether an approach is practical can only come from experiments, as the worst-case complexity is one thing but real life is another. But there was little work on implementations and one of the main reasons was lack of test cases whose hardness one could control. Researchers continued to analyze "by hand" small examples arguing about correctness of their default or autoepistemic logic representations. These toy examples were appropriate for the task of understanding basic reasoning patterns. But they were simply too easy to provide any meaningful insights into automated reasoning algorithms and their performance.

And so the early 1990s saw a growing sentiment that in order to prove itself, to make any lasting impact on the theory and practice of knowledge representation and, more generally, on artificial intelligence, practical and efficient systems for nonmonotonic reasoning had to be developed and their usefulness in a broad range of applications demonstrated. Despite of all the doom and gloom of that time, there were reasons for optimism, too. The theoretical understanding of nonmonotonic logics reached the level when development of sophisticated computational methods became possible. Complexity results were disappointing but the community recognized that they concerned the worst case setting only. Human experience tells us that there are good reasons to think that real life does not give rise to worst-case instances too often, in fact, that it rarely does. Thus, through experiments and the focus on reasoning with structured theories one could hope to obtain efficiency sufficient for practical applications. Moreover, it was highly likely that once implemented systems started showing up, they would excite the community, demonstrate the potential of nonmonotonic logics, and spawn competition which would result in improvements of algorithms and performance advances.

It is interesting to note that many of the objections and criticisms aimed at non-monotonic reasoning were instrumental in helping to identify key aspects of answer-set programming. Default logic did not provide an acceptable formalization of reasoning about defaults but inspired the answer-set semantics of logic programs [Bidoit and Froidevaux, 1987, Gelfond and Lifschitz, 1988, 1991] and helped to solve a long-standing problem of how to interpret negation in logic programming. Answer-set programming, which adopted the syntax of logic programs, as well as the answer-set semantics, can be regarded as an implementation of a significant fragment of default logic. The lack of obvious test cases for experimentation with implementations forced researchers to seek them outside of artificial intelligence and led them to the area of graph problems. This experience showed that the phenomenon of multiple belief sets can be turned from a bug to a feature, when researchers realized that it allows one to model *arbitrary* search problems, with extensions, expansions or answer sets, depending on the logic used, representing problem solutions [Cadoli, Eiter, and Gottlob, 1997, Marek and Remmel, 2003].

However important, knowledge representation was not the only source of inspiration for ASP. Influences of research in several other areas of computer science, such as databases, logic programming and satisfiability, are also easily identifiable and must be mentioned, if only briefly. One of the key themes in research in logic programming in the 1970s and 1980s was the quest for the meaning of the negation operator. Standard logic programming is built around the idea of a single intended Herbrand model. A program represents the declarative knowledge about the domain of a problem to solve. Some elements of the model, more accurately, ground terms the model determines, represent solutions to the problem. All works well for Horn programs, with the least Herbrand model of a Horn program as the natural choice for the intended model. But the negation operator, being ingrained in the way humans describe knowledge, cannot be avoided. The logic programming community recognized this and the negation was an element of Prolog, an implementation of logic programming, right from the very beginning. And so, the question arose for a declarative (as opposed to the procedural) account of its semantics.

Subsequent studies identified a non-classical nature of the negation operator. This nonmonotonic aspect of the negation operator in logic programming was also a complicating factor in the effort to find a single intended model of logic programs with negation. It became clear that to succeed one either had to restrict the class of programs or to move to the three-valued settings. The first line of research resulted in an important class of stratified programs [Apt, Blair, and Walker, 1988], the second one led Fitting [1985] and Kunen [1987] to the *Kripke-Kleene model* and, later on, Van Gelder, Ross, and Schlipf [1991] to the well-founded model.

In the hindsight, the connection to knowledge representation and nonmonotonic reasoning should have been quite evident. However, the knowledge representation and logic programming communities had little overlap at the time. And so it was not before the work by Bidoit and Froidevaux [1987] and Gelfond [1987] that the connection was made explicit and then exploited. That work demonstrated that intuitive constraints on an intended model cannot be reconciled with the requirement of its uniqueness. In other words, with negation in the syntax, we must accept the reality of multiple intended models. The connection between logic programming and knowledge representation, especially, default and autoepistemic logics was important. On the one hand, it showed that logic programming can provide syntax for an interesting non-trivial fragment of these logics, and drew attention of researchers attempting implementations of nonmonotonic reasoning systems. On the other hand, it led to the notion of a stable model of a logic program with negation. It also reinforced the importance of the key question how to adapt the phenomenon of multiple intended models for problems solving.

The work in databases provided a link between query languages and logic programming. One of the outcomes of this work was DATALOG, a fragment of logic programming without function symbols, proposed as a query language. The database research resulted in important theoretical studies concerning complexity, expressive power and connection of DATALOG to the SQL query language [Cadoli et al., 1997]. DATALOG was implemented, for instance as a part of DB2 database management system. DATALOG introduced an important distinction between extensional and intentional database components. Extensional database is the collection of tables that are stored in the database, the corresponding relation names known as extensional predicate symbols. The intensional database is a collection of intentional tables defined by DATALOG queries. In time this distinction was adopted by answer-set programming as a way to separate problem specification from data. The database community also considered extensions of DATALOG with the negation connective. Because of the semantics of the resulting language, multiplicity of answers in DATALOG$^\neg$ was a problem, as it was in a more general setting of arbitrary programs with negation. Therefore, DATALOG$^\neg$ never turned into a practical database query language (although, its stratified version could very well be used to this end). However, it was certainly an interesting fragment of logic programming. And even though its expressive power was much lower than that of general programs,[4] there was hope that fast tools to process DATALOG$^\neg$ can be developed. Jumping ahead, we note here that it was DATALOG$^\neg$ that was eventually adopted as the basic language of answer-set programming.

3 Towards Answer-Set Programming at the University of Kentucky

Having outlined some of the key ideas behind the emergence of answer-set programming, we now move on to a more personal account of research ideas that eventually resulted in the formulation of the answer-set programming paradigm. In this section, Victor Marek and Mirek Truszczynski, discuss the evolution of their understanding of nonmonotonic logics and how they could be used for computation that led to their paper *Stable logic programming — an alternative logic programming paradigm* [Marek and Truszczyński, 1999]. A closely intertwined story of Ilkka Niemelä, follows in the subsequent section. As the two accounts are strongly personal and necessarily quite

---

[4]It has to be noted though that the expressive power of general programs with function symbols and negation goes well beyond what could be accepted as computable under all reasonable semantics [Schlipf, 1995, Marek, Nerode, and Remmel, 1994].

subjective, for the most part they are given in the first person. And so, in this section "we" and us refers to Victor and Mirek, just as "I" in the next one to Ilkka.

In mid 1980s, one of us, Victor, started to study nonmonotonic logics following a suggestion from Witold Lipski, his former Ph.D. student and close collaborator. Lipski drew Victor's attention to Reiter's papers on closed-world assumption and default logic [Reiter, 1978, 1980]. In 1984, Victor attended the first Nonmonotonic Reasoning Workshop at Mohonk, NY, and came back convinced about the importance of problems that were discussed there. In the following year, he attracted Mirek to the program of the study of mathematical foundations of nonmonotonic reasoning.

In 1988 Michael Gelfond visited us in Lexington and in his presentation talked about the use of autoepistemic logic [Moore, 1985] to provide a semantics to logic programs. At the time we were already studying autoepistemic logic, inspired by talks Victor attended at Mohonk and by Moore's paper on autoepistemic logic in the Artificial Intelligence Journal [Moore, 1985]. We knew by then that stable sets of formulas of modal logic, introduced by Stalnaker [1980] and shown to be essential for autoepistemic logic, can be constructed by an iterated inductive definition from their modal-free part [Marek, 1989]. We also realized the importance of a simple normal form for autoepistemic theories introduced by Konolige [1988].

Thus, we were excited to see that logic programs can be understood as some simple autoepistemic theories thanks to Gelfond's interpretation [Gelfond, 1987]. Soon thereafter, we also realized that logic programs could be interpreted also as default logic theories and that the meaning of logic programs induced on them by default logic extensions is the same as that induced by autoepistemic expansions [Marek and Truszczyński, 1989b]. It is important to note that default logic was first used to assign the meaning to logic programs by Bidoit and Froidevaux [1987], but we did not know about their work at the time. Bidoit and Froidevaux effectively defined the stable model semantics for logic programs. They did so indirectly and with explicit references to default extensions. The direct definition of stable models in logic programming terms came about one year later in the celebrated paper by Gelfond and Lifschitz [1988].

What became apparent to us soon after Gelfond's visit was that despite both autoepistemic expansions and default extensions inducing the same semantics on logic programs, it was just serendipidity and not the result of the inherent equivalence of the two logics. In fact, we noticed that there was a deep mismatch between Moore's autoepistemic logic with the semantics of expansions and Reiter's default logic with the semantics of extensions. In the same time, we discovered a form of default logic, to be more precise, an alternative semantics of default logic, which was the perfect match for that of expansions for autoepistemic logic [Marek and Truszczyński, 1989]. This research culminated about 15 years later with a paper we co-authored with Marc Denecker that provided a definitive account of the relationship between default and autoepistemic logics [Denecker, Marek, and Truszczyński, 2003] and resolved problems and flaws of an earlier attempt at explaining the relationship due to Konolige [1988]. Another paper in this volume [Denecker, Marek, and Truszczynski, 2011] discusses the informal basis for that work and summarizes all the key results.

The relationship between default and autoepistemic logic was of only marginal importance for the later emergence of answer-set programming. But another result inspired by Gelfond's visit turned out to be essential. In our study of autoepistemic logic we wanted to establish the complexity of the existence of expansions. We obtained the result by showing that the problem of the existence of a stable model of a logic program is NP-complete and, by doing so, we obtained the same complexity for the problem of the existence of expansions of autoepistemic theories of some simple form but still rich enough to capture logic programs under Gelfond's interpretation [Marek and Truszczyński, 1991].

The result for autoepistemic logic did not turn out to be particularly significant as the class of autoepistemic theories it pertained to was narrow. And it was soon supplanted by a general result due to Gottlob [1992], who proved the existence of the expansions problem to be Σ_2^P-complete. But it was an entirely different matter with the complexity result concerning the existence of stable models of programs!

First, our proof reduced a combinatorial problem, that of the existence of a kernel in a directed graph, to the existence of stable model of a suitably defined program. This was a strong indication that stable semantics may, in principle, lead to a general purpose formalism for solving combinatorial and, more generally, search problems. Of course we did not fully realize it at the time. Second, it was quite clear to us, especially after the first KR conference in Toronto in May 1989 that the success of nonmonotonic logics can come only with implementations. Many participants of the conference (we recall David Poole and Matt Ginsberg being especially vocal) called for working systems. Since by then we understood the complexity of stable-model computation, we asked two University of Kentucky students Elizabeth and Eric Freeman to design and implement an algorithm to compute stable models of propositional programs. They succeeded albeit with limits — the implementation could process programs with about 20 variables only. Still, theirs was most likely the first working implementation of stable-model computation. Unfortunately, with the M.S. degrees under their belts, Eric and Elizabeth left the University of Kentucky.

For about three years after this first dab into implementing reasoning systems based on a nonmonotonic logic, our attention was focused on more theoretical studies and on the work on a monograph on mathematical foundations of nonmonotonic reasoning based on the paradigm of context-dependent reasoning. However, the matter of implementations had constantly been on the backs of our minds and in 1992 we decided to give the matter another try. As we felt we understood default logic well and as it was commonly viewed as the nonmonotonic logic of the future, in 1992 we started the project, Default Reasoning System DeReS. We aimed at implementing reasoning in the unrestricted language of propositional default logic. We also started a side project to DeReS, the TheoryBase project, aimed at developing a software system generating default theories to be used for testing DeReS. The time was right as two promising students, Pawel Cholewinski and Artur Mikitiuk, joined the University of Kentucky to pursue doctorate degrees in computer science.

As is common in such circumstances, we were looking for a sponsor of this research and found one in the US Army Research Office (US ARO), which was willing to support this work. A colleague of ours, Jurek Jaromczyk, also at the University of Kentucky, coined the term DeReS, a pun on an old polish word "deresz" presently rarely used and meaning a stallion, quite appropriate for the project to be conducted in Lexington, "the world capital of the horse." In the proposal to US ARO we promised to investigate basic reasoning problems of default logic:

1. Computing of extensions

2. Skeptical reasoning with default theories — testing if a formula belongs to all extensions of an input default theory

3. Brave reasoning with default theories — testing if a formula belongs to some extension of an input default theory.

The basic computational device was backtracking search for a basis of an extension of a finite default theory (D, W). This was based on two observations due to Reiter: that while default extensions of a finite default theory are infinite, they are finitely generated; and that the generators are all formulas of W and the consequent formulas of some defaults from D. We also employed ideas such as relaxed stratification of defaults [Cholewiński, 1995, Lifschitz and Turner, 1994] for pruning the search space and relevance graphs for simplifying provability.

We also thought it was important to have the nonmonotonic reasoning community accept the challenge of developing implementations of automated nonmonotonic reasoning. Our proposal to US ARO contained a request for funding of a retreat dedicated to knowledge representation, nonmonotonic reasoning and logic programming. The key goals for the retreat were:

1. To stimulate applications of nonmonotonic formalisms and implementations of automated reasoning systems based on nonmonotonic logics

2. To promote the project to create a public domain library of benchmark problems in nonmonotonic reasoning.

We held the workshop in Shakertown, KY, in October 1994. Over 30 leading researchers in nonmonotonic reasoning participated in talks and we presented there early prototypes of DeReS and TheoryBase. Importantly, we heard then for the first time from Ilkka Niemelä about the work on systems to perform nonmonotonic reasoning in the language of logic programs in his group at the Helsinki University of Technology. The meeting helped to elevate the importance of implementations of nonmonotonic reasoning systems and their applications. It evidenced first advances in the area of implementations, as well as in the area of benchmarks, essential as so far most problems considered as benchmarks were toy problems such as "Tweety" and "Nixon Diamond."

The DeReS system was not designed with any specific applications in mind. At the time we believed that, since default logic could model several aspects of commonsense reasoning, once DeReS became available, many artificial intelligence and knowledge representation researchers would use it in their work. And we simply regarded broadly understood knowledge representation problems as the main application area for DeReS.

Working on DeReS immediately brought up to our attention the question of testing and performance evaluation. In the summer of 1988, Mirek attended a meeting on combinatorics where Donald Knuth talked about the problem of testing graph algorithms and his proposal how to do it right. Knuth was of the opinion that testing algorithms on randomly generated graphs is insufficient and, in fact, often irrelevant. Graphs arising in real-life settings rarely resemble graphs generated at random from some probabilistic model. To address the problem, Knuth developed a software system, Stanford GraphBase, providing a mechanism for creating collections of graphs that could be then used in projects developing graph algorithms. Graphs produced by the Stanford GraphBase were mostly generated from real-life objects such as maps, dictionaries, novels and images. Some were based on rather obscure sources such as sporting events in Australia. The documentation was superb (the book by Knuth on the Stanford GraphBase is still available). The Stanford GraphBase was free and its use was not restricted. From our perspective, two aspects were essential. First, the Stanford GraphBase provided a unique identifier to every graph it created and so experiments could be described in a way allowing others to repeat them literally and perform comparisons on identical sets of graphs. Second, the Stanford GraphBase supported creating families of examples similar but increasing in size, thus allowing to test scalability of algorithms being developed.

In retrospect, the moment we started talking about testing our implementations of default logic was the defining moment on our path towards the answer-set programming paradigm. Based on our complexity result concerning the existence of stable models and its implication for default logic, we knew that all NP-complete graph problems could be reduced to the problem of the existence of extensions. The reductions expressed instances of graph problems as default theories. Thus, in order to get a family of default theories, similar but growing in size, we needed to select an NP-complete problem on graphs (say, the hamiltonian cycle problem), generate a family of graphs, and generate for each graph in the family the corresponding default theory. These theories could be used to test algorithms for computing extensions. This realization gave rise to the TheoryBase, a software system generating default theories based on reductions of graph problems to the existence of the extension problem and developed on top of the Stanford GraphBase, which served as the source of graphs. The TheoryBase provided default theories based on six well-known graph problems: the existence of k-colorings, Hamiltonian cycles, kernels, independent sets of size at least k, and vertex covers of size at most k. As the Stanford GraphBase provided an unlimited supply of graphs, the TheoryBase offered an unlimited supply of default theories.

We will recall here the TheoryBase encoding of the existence of a k-coloring problem as it shows that already then some fundamental aspects of the methodology of representing search problems as default theories started to emerge. Let $G = (V, E)$ be an undirected graph with the set of vertices $V = \{v_1, \ldots, v_n\}$. Let $C = \{c_1, \ldots, c_k\}$ be a set of colors. To express the property that vertex v is colored with c, we introduced propositional atoms $clrd(v, c)$. For each vertex $v_i, i = 1, \ldots, n$, and for each color $c_j, j = 1, \ldots, k$, we defined the default rule

$$color(v_i, c_j) = \frac{: \neg clr(v_i, c_1), \ldots, \neg clr(v_i, c_{j-1}), \neg clr(v_i, c_{j+1}), \ldots, \neg clr(v_i, c_k)}{clr(v_i, c_j)}.$$

The set of default rules $\{color(v_i, c_j) : j = 1, \ldots, k\}$ models a constraint that vertex v_i obtains exactly one color. The default theory $(D_0, \emptyset)$, where

$$D_0 = \{color(v_i, c_j) : i = 1, \ldots, n, \ j = 1, \ldots, k\},$$

has k^n extensions corresponding to all possible colorings (not necessarily *proper*) of the vertices of G. Thus, the default theory $(D_0, \emptyset)$ defines the basic space of objects within which we need to search for solutions. In the present-day answer-set programming implementations choice or cardinality rules, which offer much more concise representations, are used for that purpose. Next, our TheoryBase encoding imposed constraints to eliminate those colorings that are not proper. To this end, we used additional default rules, which we called *killing* defaults, and which now are typically modeled by logic program rules with the empty head. To describe them we used a new propositional variable F and defined

$$local(e, c) = \frac{clrd(x, c) \wedge clrd(y, c) : \neg F}{F},$$

for each edge $e = (x, y)$ of the graph and for each color c. Each default $local(e, c)$ "kills" all color assignments which give color c to both ends of edge e. It is easy to check (and it also follows from now well-known more general results) that defaults of the form $local(e, c)$ "kill" all non-proper colorings and leave precisely those that are proper. This two-step modeling methodology, in which we first define the space of objects that contains all solutions, and then impose constraints to weed away those that fail some problem specifications, constitutes the main way by which search problems are modeled in ASP.

The key lesson for us from the TheoryBase project was that combinatorial problems can be represented as default theories and that constructing these representations is easy. It was then for the first time that we sensed that programs finding extensions of default theories could be used as general purpose problem solving tools. It also lead us, in our internal discussions to thinking about "second-order" flavor of default logic, given the way it was used for computation. Indeed, in all theories we developed for the TheoryBase, extensions rather than their single elements represented solutions. In other words, the main reasoning task did not seem to be that of skeptical or brave reasoning (does a formula follow skeptically or bravely from a default theory) but computing *entire* extensions. We talked about this second-order flavor when presenting our paper on DeReS at the KR conference in 1996 [Cholewiński, Marek, and Truszczyński, 1996]. At that time, we knew we were closing in on a new declarative problem-solving paradigm based on nonmonotonic logics.

A problem for us was, however, a fairly poor performance of DeReS. The default extensions are closed under consequence. This means that processing of default theories requires testing provability of prerequisites and justifications of defaults. This turned out to be a major problem affecting the processing time of our implementations. It is not surprising at all in view of the complexity results of Gottlob [1992] and Stillman [1992]. Specifically, existence of extensions is a Σ_2^P-complete problem.

There is, of course, an easy case of provability when all formulas in a default theory are conjunctions of literals only. Now the problem with the provability of premises disappears. However, DeReS organized its search for solutions by looking for sets of generating defaults, inheriting this approach from the case of general default theories, rather than for literals generating an extension. And that was still a problem. There are typically many more rules in a default theory than atoms in the language.

At the International Joint Conference and Symposium on Logic Programming in

1996, Ilkka and his student Patrik Simons presented the first report on their *smodels* system [Niemelä and Simons, 1996]. But it seems fair to say that only a similar presentation and a demo Ilkka gave at the Logic Programming and Non-Monotonic Reasoning Conference in 1997, in Dagstuhl, made the community really take notice. The *lparse/smodels* constituted a major conceptual breakthrough and handled nicely all the traps DeReS did not avoid. First, *lparse/smodels* focused on the right fragment of default logic, logic programming with the stable-model semantics. Next, it organized search for a stable model by looking for atoms that form it. Finally, it supported programs with variables and separated, as was the standard in logic programming and databases, a program (a problem specification) from an extensional database (an instance of the problem).

The work by Niemelä had us focus our thinking about nonmonotonic logics as computational devices on the narrower but all-important case of logic programs. We formulated our ideas about the second-order flavor of problem solving with nonmonotonic logics and contrasted them with the traditional Prolog-style interpretation of logic programming. We stated our initial thoughts on the methodology of problem solving that exploited our ideas of modeling combinatorial problems that we used in the TheoryBase project, as well as the notion of program-data separation that came from the database community and was, as we just mentioned, already used in our field by Niemelä. These ideas formed the backbone of our paper on an alternative way logic programming could be used for solving search problems [Marek and Truszczyński, 1999].

4   Towards Answer-Set Programming at the Helsinki University of Technology

In this section Ilkka Niemelä discusses the developments at the Helsinki University of Technology that led to the paper *Logic Programs with Stable Model Semantics as a Constraint Programming Paradigm* [Niemelä, 1999]. Similarly as in the previous section, the account is very personal and quite subjective. Hence, in this section "I" refers to Ilkka.

I got exposed to nonmonotonic reasoning when I joined the group of Professor Leo Ojala at the Helsinki University of Technology in 1985. The group was studying specification and verification techniques of distributed systems. One of the themes was specification of distributed systems using modal, in particular, temporal and dynamic logics. The group had got interested in the solutions of the *frame problem* based on nonmonotonic logics when looking for compact and computationally efficient logic-based specification techniques for distributed systems. My role as a new research assistant in the group was to examine autoepistemic logic by Moore, nonmonotonic modal logics by McDermott and Doyle, and default logic by Reiter from this perspective.

There was a need for tool support and together with a doctoral student Heikki Tuominen we developed a system that we called the Helsinki Logic Machine, "an experimental reasoning system designed to provide assistance needed for application oriented research in logic" [Niemelä and Tuominen, 1986, 1987]. The system included tools for theorem proving, model synthesis, model checking, formula manipulation for modal, temporal, epistemic, deontic, dynamic, and *nonmonotonic* logics. It was written in Quintus Prolog and contained implementations, for instance, for Reiter's default logic, McDermott and Doyle style nonmonotonic modal logic, and autoepistemic logic *in the propositional case* based on the literature and some own work [Etherington, 1987, McDermott and Doyle, 1980, Niemelä, 1988]. While nonmonotonic reasoning was a side-track in the Helsinki Logic Machine, it seems that it was one of the earliest working nonmonotonic reasoning systems although we were not very well aware of this at the time.

The work and, in particular, the difficulties in developing efficient tools led me to further investigations to gain a deeper understanding of algorithmic issues and related complexity questions [Niemelä, 1988, Niemelä, 1988, 1990, Niemelä, 1992]. Similar questions were studied by others and in the early 90s results explaining the algorithmic difficulties started emerging. These results showed that key reasoning tasks in major nonmonotonic logics are complete for the second level of the polynomial hierarchy [Cadoli and Lenzerini, 1990, Gottlob, 1992, Stillman, 1992, Niemelä, 1992]. This indicated that these nonmonotonic logics have *two orthogonal sources of complexity* that we called classical reasoning and conflict resolution. Orthogonality means that even if we assume that classical reasoning can be done efficiently, nonmonotonic reasoning still remains NP-hard (unless the polynomial hierarchy collapses).

These results made me to focus more on conflict resolution to develop techniques for pruning the search space of potential expansions/extensions. One approach was to develop compact characterizations of expansions/extensions capturing their key ingredients. For autoepistemic logic I developed such a characterization based on the idea that expansions can be captured in terms of the modal subformulas in the premises and classical reasoning and exploited the idea in a decision procedure for autoepistemic logic [Niemelä, 1988]. Together with Jussi Rintanen we also showed that if one limits the theory in such a way that conflict resolution is easy by requiring stratification, then efficient reasoning is possible by further restrictions affecting the other source of complexity [Niemelä and Rintanen, 1992].

The characterization based on modal subformulas generalizes also to default logic where extensions can be captured using justifications in the rules and leads to an interesting way of organizing the search for expansions/extensions as a *binary search tree* very similar to that in the DPLL algorithm for SAT [Niemelä, 1994, 1995]. Further pruning techniques can be integrated to cut substantial parts of the potential search space for expansions/extensions and exploit, for instance, stratified parts of the rule set. My initial but very unsystematic experimentation gave promising results.

In 1994 encouraged and challenged by the Shakertown Workshop organized by Victor and Mirek, I decided to restrict to a simple subclass of default theories, that is, logic programs with the stable model semantics. For this subclass classical reasoning is essentially limited to Horn clauses and can be done efficiently in linear time using techniques proposed by Dowling and Gallier in the 1980s [Dowling and Gallier, 1984]. I had no particular application in mind. The goal was to study whether the conflict resolution techniques I had developed for autoepistemic and default logic would scale up so that it would be possible to handle *very large sets of rules* which meant at that time thousands or even tens of thousands of rules.

At that time Patrik Simons joined my group and started working on a C++ implementation of the general algorithm tailored to logic programs. Patrik had excellent insights into the key implementation issues from very early on and the first version was released in 1995 [Niemelä and Simons, 1995]. The C++ implementation was called smodels and it computed stable models for ground normal programs. It gave surprising good results immediately and could handle programs with a few thousand ground rules. Challenge benchmarks were combinatorial problems, mainly colorability and Hamiltonian cycles, an idea that I learnt from Mirek and Victor in Shakertown. For such hard problems the performance of smodels was substantially better than state-of-the-art tools such as the SLG system [Chen and Warren, 1996].

When developing benchmarks for evaluating novel algorithmic ideas and implementation techniques we soon realized that working with ground programs is too cumbersome. In practice, for producing large enough interesting ground programs for benchmarking we needed to write separate programs in some other language to generate ground logic programs. This took considerable time for each benchmark family and was quite inflexible and error-prone. We realized that in order to attract users and to be able to attack real applications we needed to support logic program rules with variables.

For handling rules with variables we decided to employ a two level architecture. The first phase was concerned with *grounding*, a process to generate a set of ground instances of the rules in the program so that stable models are preserved. Actual stable-model computation was taking place in the second *model search* phase on the program grounded in the previous one. The idea was to have a separation of concern, that is, be able to exploit advanced database and other such techniques in the first phase and novel search and pruning techniques in the other in such a way that both steps could be developed relatively independently. We released the first such system in 1996 [Niemelä and Simons, 1996].

This was a major step forward in attracting users and getting closer to applications. Such a system supporting rules with variables enabled *compact and modular* encodings of problems without any further host language. It was now also possible to separate the problem specification and the data providing the instance to be solved.

Working with the system and studying potential applications made me realize that logic programming with the stable model semantics is very different from traditional logic programming implemented in various Prolog systems. These systems are answering queries by SLD resolution and producing answer substitutions as results. But we were using logic programs more like in a constraint programming approach where rules are seen as constraints on a solution set (stable model) of the program and where a solution is not an answer substitution but a stable model, that is, a valuation that satisfies all the rules. This is like in constraint satisfaction problems where a solution is a variable assignment satisfying all the constraints. I wrote down these ideas in a paper *Logic Programs with Stable Model Semantics as a Constraint Programming Paradigm* which was first presented in a workshop on Computational Aspects of Nonmonotonic Reasoning in 1998 [Niemelä, 1998] and then appeared as an extended journal version in 1999 [Niemelä, 1999]. The paper emphasized, in particular, the knowledge representation advantages of logic programs as a constraint satisfaction framework:

"Logic programming with the stable model semantics is put forward as

an interesting constraint programming paradigm. It is shown that the
paradigm embeds classical logical satisfiability but seems to provide a
more expressive framework from a *knowledge representation* point of
view."

In 1998 we put more and more emphasis on potential applications and, in particular, on product configuration. This made us realize that a more efficient grounder
supporting an extended modeling language is needed. At that point another student,
Tommi Syrjänen, with excellent implementation skills and insight on language design,
joined the group and work on a new grounder, lparse, started. The goal was to en-
force a tighter typing of the variables in the rules to facilitate the application of more
advanced database techniques for grounding and the integration of built-in predicates
and functions, for instance, for arithmetic.

We also realized that for many applications normal logic programs were inadequate not allowing compact and intuitive encodings. This led to the introduction of
new language constructs: (i) choice rules for encoding choices instead of recursive
odd loops needed in normal programs and (ii) cardinality and weight constraints for
typical conditions needed in many practical applications [Soininen and Niemelä, 1998,
Niemelä et al., 1999]. In order to fully exploit the extensions computationally Patrik
Simons developed techniques to provide built-in support for them also in the model
search phase in the version 2 of smodels [Simons, 1999].

So in 1999 when Vladimir Lifschitz coined the term answer-set programming, the
system that we had with lparse as the grounder and smodels version 2 as the
model search engine offered quite promising performance. For example, for propositional satisfiability the performance of smodels compared nicely to the best SAT
solvers at that time (before more efficient conflict driven clause learning solvers like
zchaff emerged). Moreover, very interesting serious application work started. For
example, at the Helsinki University of Technology we cooperated with the product
data management group on automated product configuration which eventually led to
a spin-off company Variantum (http://www.variantum.com/). Moreover, in
Vienna the dlv project for handling disjunctive programs had started a couple years
earlier and had already made promising progress.

5   Conclusions

Now, more than 12 years since ASP became a recognizable paradigm of search problem solving, we see that the efforts of researchers in various domains: artificial intelligence, knowledge representation, nonmonotonic reasoning, satisfiability and others
resulted in a programming formalism that is being used in a variety of areas, but prin-
cipally in those where the modelers face the issues of defaults, frame axioms and
other nonmonotonic phenomena. The experience of ASP programmers shows that
these phenomena can be naturally incorporated into the practice of modeling real-life
problems.

We believe ASP is here to stay. It provides a venue for problem modeling, problem description and problem solving. This does not mean that the process of developing ASP is finished. Certainly new extensions of ASP will emerge in the future.

Additional desiderata include: software engineering tools for testing correctness of implementation, integrated development environments and other tools that will speed up the process of the use of ASP in normal programming practice. Better grounders and better solvers able to work with incremental grounding only will certainly emerge. Similarly, new application domains will surface and bring new generations of investigators and, more importantly, users for ASP.

Acknowledgments

The work of the second author was partially supported by the Academy of Finland (project 122399). The work of the third author was partially supported by the NSF grant IIS-0913459.

References

K. Apt, H. A. Blair, and A. Walker. Towards a theory of declarative knowledge. In J. Minker, editor, *Foundations of deductive databases and logic programming*, pages 89–142. Morgan Kaufmann, 1988.

M. Balduccini, M. Gelfond, and M. Nogueira. Answer set based design of knowledge systems. *Annals of Mathematics and Artificial Intelligence*, 47(1-2):183–219, 2006.

P. Besnard. *An Introduction to Default Logic*. Springer, Berlin, 1989.

N. Bidoit and C. Froidevaux. Minimalism subsumes default logic and circumscription. In *Proceedings of IEEE Symposium on Logic in Computer Science, LICS 1987*, pages 89–97. IEEE Press, 1987.

N. Bidoit and C. Froidevaux. General logical databases and programs: default logic semantics and stratification. *Information and Computation*, 91(1):15–54, 1991.

G. Boenn, M. Brain, M. De Vos, and J. Fitch. Automatic music composition using answer set programming. *Theory and Practice of Logic Programming*, 11(2-3):397-427, 2011.

M. Brain and M. De Vos. Debugging logic programs under the answer set semantics. In *Answer Set Programming, Advances in Theory and Implementation, Proceedings of the 3rd International ASP'05 Workshop*, CEUR Workshop Proceedings 142, 2005.

G. Brewka. *Nonmonotonic Reasoning: Logical Foundations of Commonsense*, volume 12 of *Cambridge Tracts in Theoretical Computer Science*. Cambridge University Press, Cambridge, UK, 1991.

R. Brummayer and M. Järvisalo. Testing and debugging techniques for answer set solver development. *Theory and Practice of Logic Programming*, 10(4-6):741–758, 2010.

M. Cadoli and M. Lenzerini. The complexity of closed world reasoning and circumscription. In *Proceedings of the 8th National Conference on Artificial Intelligence*, pages 550–555. MIT Press, 1990.

M. Cadoli, T. Eiter, and G. Gottlob. Default logic as a query language. *IEEE Transactions on Knowledge and Data Engineering*, 9(3):448–463, 1997.

W. Chen and D.S. Warren. Computation of stable models and its integration with logical query processing. *IEEE Transactions on Knowledge and Data Engineering*, 8(5):742–757, 1996.

P. Cholewiński. Stratified default theories. In L. Pacholski and J. Tiuryn, editors, *Computer Science Logic*, volume 933 of *LNCS*, pages 456–470. Springer, 1995

P. Cholewiński, W. Marek, and M. Truszczyński. Default reasoning system DeReS. In *Proceedings of the 5th International Conference on Principles of Knowledge Representation and Reasoning, KR 1996*, pages 518–528. Morgan Kaufmann, 1996.

K.L. Clark. Negation as failure. In H. Gallaire and J. Minker, editors, *Logic and data bases*, pages 293–322. Plenum Press, New York-London, 1978.

M. Dao-Tran, T. Eiter, M. Fink, and T. Krennwallner. Modular nonmonotonic logic programming revisited. In P. M. Hill and D. S. Warren, editors, *Proceedings of the 25th International Conference on Logic Programming, ICLP 2009*, volume 5649 of *LNCS*, pages 145–159. Springer, 2009.

M. Denecker. The well-founded semantics is the principle of inductive definition. In J. Dix, L. Fariñas del Cerro, and U. Furbach, editors, *Proceedings of the 6th European Workshop on Logics in Artificial Intelligence, JELIA 1998*, volume 1489 of LNCS, pages 1–16. Springer, 1998.

M. Denecker. Extending classical logic with inductive definitions. In J. W. Lloyd, V. Dahl, U. Furbach, M. Kerber, K.-K. Lau, C. Palamidessi, L. Moniz Pereira, Y. Sagiv, P. J. Stuckey, editors, *Computational Logic*, volume 1861 of *LNCS*, pages 703–717. Springer, 2000.

M. Denecker and E. Ternovska. A logic for non-monotone inductive definitions. *ACM Transactions on Computational Logic*, 9(2), 2008.

M. Denecker, V.W. Marek, and M. Truszczyński. Uniform semantic treatment of default and autoepistemic logics. *Artificial Intelligence Journal*, 143:79–122, 2003.

M. Denecker, J. Vennekens, S. Bond, M. Gebser, and M. Truszczynski. The second answer set programming competition. In E. Erdem, F. Lin, and T. Schaub, editors, *Proceedings of the 10th International Conference on Logic Programming and Nonmonotonic Reasoning, LPNMR 2009*, volume 5753 of *LNCS*, pages 637–654. Springer, 2009.

M. Denecker, V.W. Marek, and M. Truszczynski. Reiter's Default Logic Is a Logic of Autoepistemic Reasoning And a Good One, Too. This volume.

W.F. Dowling and J.H. Gallier. Linear-time algorithms for testing the satisfiability of propositional Horn formulae. *Journal of Logic Programming*, 1(3):267–284, 1984.

D. East and M. Truszczyński. Predicate-calculus based logics for modeling and solving search problems. *ACM Transactions on Computational Logic*, 7:38–83, 2006.

T. Eiter and G. Gottlob. Propositional circumscription and extended closed world reasoning are π_2^p-complete. *Theoretical Computer Science*, 114(2):231–245, 1993a.

T. Eiter and G. Gottlob. Complexity results for disjunctive logic programming and application to nonmonotonic logics. In D. Miller, editor, *Proceedings of the 1993 International Symposium on Logic Programming*, pages 266–278. MIT Press, 1993b.

T. Eiter and G. Gottlob. On the computational cost of disjunctive logic programming: propositional case. *Annals of Mathematics and Artificial Intelligence*, 15(3-4):289–323, 1995.

E. Erdem. Applications of answer set programming in phylogenetic systematics. In M. Balduccini and T.C. Son, editors, *Essays in Honor of Michael Gelfond*. Springer, 2011.

E. Erdem and V. Lifschitz. Tight logic programs. *Theory and Practice of Logic Programming*, 3(4-5):499–518, 2003.

D. W. Etherington. Formalizing nonmonotonic reasoning systems. *Artificial Intelligence*, 31 (1):41–85, 1987.

J. J. Finger. *Exploiting Constraints in Design Synthesis*. PhD thesis, Stanford University, 1987.

R. A. Finkel and B. O'Sullivan. Reasoning about conditional constraint specifications. *Artificial Intelligence for Engineering Design, Analysis and Manufacturing*, 25(2):163-174, 2011.

M. C. Fitting. A Kripke-Kleene semantics for logic programs. *Journal of Logic Programming*, 2(4):295–312, 1985.

G. Friedrich, M. Fugini, E. Mussi, B. Pernici, and G. Tagni. Exception handling for repair in service-based processes. *IEEE Transactions on Software Engineering*, 36(2):198–215, 2010.

D.M. Gabbay. Theoretical foundations for non-monotonic reasoning in expert systems. In *Proceedings of the NATO Advanced Study Institute on Logics and Models of Concurrent Systems*, pages 439–457. Springer, 1989.

M. Gebser, L. Liu, G. Namasivayam, A. Neumann, T. Schaub, and M. Truszczyński. The first answer set programming system competition. In C. Baral, G. Brewka, and J. Schlipf, editors, *Proceedings of the 9th International Conference on Logic Programming and Nonmonotonic Reasoning, LPNMR 2007*, volume 4483 of *LNCS*, pages 3–17. Springer, 2007.

M. Gebser, C. Guziolowski, M. Ivanchev, T. Schaub, A. Siegel, S. Thiele, and P. Veber. Repair and prediction (under inconsistency) in large biological networks with answer set programming. In F. Lin, U. Sattler and M. Truszczynski, editors, *Proceedings of the 12th International Conference on Principles of Knowledge Representation and Reasoning, KR 2010*, pages 497–507. AAAI Press, 2010a.

M. Gebser, A. König, T. Schaub, S. Thiele, and P. Veber. The BioASP library: ASP solutions for systems biology. In *Proceedings of the 22nd IEEE International Conference on Tools with Artificial Intelligence, ICTAI 2010*, pages 383–389, 2010b.

M. Gelfond. On stratified autoepistemic theories. In *Proceedings of AAAI 1987*, pages 207–211. Morgan Kaufmann, 1987.

M. Gelfond and V. Lifschitz. The stable model semantics for logic programming. In R. A. Kowalski and K. A. Bowen, editors, *Proceedings of the International Joint Conference and Symposium on Logic Programming*, pages 1070–1080. MIT Press, 1988.

M. Gelfond and V. Lifschitz. Classical negation in logic programs and disjunctive databases. *New Generation Computing*, 9:365–385, 1991.

M.L. Ginsberg. A circumscriptive theorem prover. In M. Reinfrank, J. de Kleer, M. L. Ginsberg, and E. Sandewall, editors, *Proceedings of the 2nd International Workshop on Non-Monotonic Reasoning*, volume 346 of *LNCS*, pages 100–114. Springer, 1989.

G. Gottlob. Complexity results for nonmonotonic logics. *Journal of Logic and Computation*, 2 (3):397–425, 1992.

S. Hanks and D. McDermott. Default reasoning, nonmonotonic logics and frame problem. In *Proceedings of AAAI 1986*, pages 328–333. Morgan Kaufmann, 1986.

T. Janhunen, E. Oikarinen, H. Tompits, and S. Woltran. Modularity aspects of disjunctive stable models. *Journal of Artificial Intelligence Research*, 35:813–857, 2009.

H.A. Kautz and B. Selman. Hard problems for simple default logics. In R. J. Brachman, H. J. Levesque, and R. Reiter, editors, *Proceedings of the 1st International Conference on Principles of Knowledge Representation and Reasoning, KR 1989*, pages 189–197. Morgan Kaufmann, 1989.

K. Konolige. On the relation between default and autoepistemic logic. *Artificial Intelligence*, 35(3):343–382, 1988.

K. Konolige. Errata: On the relation between default and autoepistemic logic. *Artificial Intelligence*, 41(1):115, 1989.

S. Kraus, D. Lehmann, and M. Magidor. Nonmonotonic reasoning, preferential models and cumulative logics. *Artificial Intelligence*, 44:167–207, 1990.

K. Kunen. Negation in logic programming. *Journal of Logic Programming*, 4(4):289–308, 1987.

D. Lehmann and M. Magidor. What does a conditional knowledge base entail? *Artificial Intelligence*, 55:1–60, 1992.

N. Leone, G. Pfeifer, W. Faber, T. Eiter, G. Gottlob, S. Perri, and F. Scarcello. The DLV system for knowledge representation and reasoning. *ACM Transactions on Computational Logic*, 7(3):499–562, 2006.

Y. Lierler and M. Maratea. Cmodels-2: SAT-based answer set solver enhanced to non-tight programs. In V. Lifschitz, I. Niemelä, editors, *Proceedings of the 7th International Conference on Logic Programming and Nonmonotonic Reasoning, LPNMR 2004*, volume 2923 of *LNCS*, pages 346–350. Springer, 2004.

V. Lifschitz. Circumscriptive theories: a logic-based framework for knowledge representation. *Journal of Philosophical Logic*, 17(4):391–441, 1988.

V. Lifschitz and H. Turner. Splitting a logic program. In P. Van Hentenryck, editor, *Proceedings of the 11th International Conference on Logic Programming, ICLP 1994*, pages 23–37. MIT Press, 1994.

F. Lin and Y. Zhao. ASSAT: Computing answer sets of a logic program by SAT solvers. In *Proceedings of the 18th National Conference on Artificial Intelligence, AAAI 2002*, pages 112–117. AAAI Press, 2002.

L. Liu and M. Truszczyński. Pbmodels - software to compute stable models by pseudoboolean solvers. In C. Baral, G. Greco, N. Leone, G. Terracina *Proceedings of the 8th International Conference on Logic Programming and Nonmonotonic Reasoning, LPNMR 2005*, volume 3662 of *LCNS*, pages 410–415. Springer, 2005.

D. Makinson. General theory of cumulative inference. In M. Reinfrank, J. de Kleer, M. L. Ginsberg, and E. Sandewall, editors, *Proceedings of the 2nd International Workshop on Non-Monotonic Reasoning*, volume 346 of *LNCS*, pages 1–18. Springer, 1989.

V.W. Marek and J.B. Remmel. On the expressibility of stable logic programming. *Theory and Practice of Logic Programming*, 3:551–567, 2003.

V.W. Marek and M. Truszczyński. Stable models and an alternative logic programming paradigm. In K.R. Apt, W. Marek, M. Truszczyński, and D.S. Warren, editors, *The Logic Programming Paradigm: a 25-Year Perspective*, pages 375–398. Springer, 1999.

W. Marek. Stable theories in autoepistemic logic. *Fundamenta Informaticae*, 12(2):243–254, 1989.

W. Marek and M. Truszczyński. Relating autoepistemic and default logics. In R. J. Brachman, H. J. Levesque, and R. Reiter, editors, *Proceedings of the 1st International Conference on Principles of Knowledge Representation and Reasoning, KR 1989*, pages 276–288. Morgan Kaufmann, 1989.

W. Marek and M. Truszczyński. Stable semantics for logic programs and default theories. In E.Lusk and R. Overbeek, editors, *Proceedings of the North American Conference on Logic Programming*, pages 243–256. MIT Press, 1989b.

W. Marek and M. Truszczyński. Autoepistemic logic. *Journal of the ACM*, 38(3):588–619, 1991.

W. Marek and M. Truszczyński. *Nonmonotonic Logic; Context-Dependent Reasoning*. Springer, 1993.

W. Marek, G.F. Shvarts, and M. Truszczyński. Modal nonmonotonic logics: ranges, characterization, computation. *Journal of the ACM*, 40(4):963–990, 1993.

W. Marek, A. Nerode, and J. B. Remmel. The stable models of predicate logic programs. *Journal of Logic Programming*, 21(3):129–154, 1994.

J. McCarthy. Epistemological problems of Artificial Intelligence. In *Proceedings of the 5th International Joint Conference on Artificial Intelligence*, pages 1038–1044, 1977.

J. McCarthy. Circumscription — a form of non-monotonic reasoning. *Artificial Intelligence*, 13(1-2):27–39, 1980.

J. McCarthy and P. Hayes. Some philosophical problems from the standpoint of artificial intelligence. In B. Meltzer and D. Michie, editors, *Machine Intelligence 4*, pages 463–502. Edinburgh University Press, 1969.

D. McDermott. Nonmonotonic logic II: nonmonotonic modal theories. *Journal of the ACM*, 29 (1):33–57, 1982.

D. McDermott and J. Doyle. Nonmonotonic logic I. *Artificial Intelligence*, 13(1-2):41–72, 1980.

R. C. Moore. Possible-world semantics for autoepistemic logic. In R. Reiter, editor, *Proceedings of the Workshop on Non-Monotonic Reasoning*, pages 344–354, 1984. Reprinted in: M. Ginsberg, editor, *Readings on Nonmonotonic Reasoning*, pages 137–142, Morgan Kaufmann, 1990.

R.C. Moore. Semantical considerations on nonmonotonic logic. *Artificial Intelligence*, 25(1): 75–94, 1985.

I. Niemelä. On the complexity of the decision problem in propositional nonmonotonic logic. In E. Börger, H. Kleine Büning, M. M. Richter, editors, *Proceedings of the 2nd Workshop on Computer Science Logic, CSL 1988*, volume 385 of *LNCS*, pages 226–239. Springer, 1988.

I. Niemelä. Decision procedure for autoepistemic logic. In R. Overbeek E. Lusk, editor, *Proceedings of the 9th International Conference on Automated Deduction*, volume 310 of *LNCS*, pages 675–684. Springer, 1988.

I. Niemelä. Towards automatic autoepistemic reasoning. In J. van Eijck, editor, *Proceedings of the European Workshop on Logics in Artificial Intelligence, JELIA 1990*, volume 478 of *LNCS*, pages 428–443. Springer, 1990.

I. Niemelä. On the decidability and complexity of autoepistemic reasoning. *Fundamenta Informaticae*, 17(1-2):117–155, 1992.

I. Niemelä. A decision method for nonmonotonic reasoning based on autoepistemic reasoning. In J. Doyle, E. Sandewall and P. Torasso, *Proceedings of the 4th International Conference on Principles of Knowledge Representation and Reasoning*, pages 473–484. Morgan Kaufmann, 1994.

I. Niemelä. Towards efficient default reasoning. In C. S. Mellish, editor, *Proceedings of IJCAI 1995*, pages 312–318. Morgan Kaufmann, 1995.

I. Niemelä. Logic programs with stable model semantics as a constraint programming paradigm. In *Proceedings of the Workshop on Computational Aspects of Nonmonotonic Reasoning*, pages 72–79. Helsinki University of Technology, Digital Systems Laboratory, Research Report A52, May 1998.

I. Niemelä. Logic programming with stable model semantics as a constraint programming paradigm. *Annals of Mathematics and Artificial Intelligence*, 25(3-4):241–273, 1999.

I. Niemelä and J. Rintanen. On the impact of stratification on the complexity of nonmonotonic reasoning. In B. Nebel, C. Rich, and W. R. Swartout, editors, *Proceedings of the 3rd International Conference on Principles of Knowledge Representation and Reasoning*, pages 627–638. Morgan Kaufmann, 1992.

I. Niemelä and P. Simons. Evaluating an algorithm for default reasoning. In *Working Notes of the IJCAI'95 Workshop on Applications and Implementations of Nonmonotonic Reasoning Systems, Montreal, Canada*, pages 66–72, Montreal, Canada, August 1995.

I. Niemelä and P. Simons. Efficient implementation of the well-founded and stable model semantics. In M. Maher, editor, *Proceedings of the Joint International Conference and Symposium on Logic Programming*, pages 289–303. MIT Press, 1996.

I. Niemelä and H. Tuominen. A system for logical expertise. In *Proceedings of the Finnish Artificial Intelligence Symposium, Volume 2*, pages 44–53, Espoo, Finland, August 1986. Finnish Society of Information Processing Science, 1986.

I. Niemelä and H. Tuominen. Helsinki Logic Machine: a system for logical expertise. Technical report B1, Helsinki University of Technology, Digital Systems Laboratory, Espoo, Finland, December 1987.

I. Niemelä, P. Simons, and T. Soininen. Stable model semantics of weight constraint rules. In M. Gelfond, N. Leone, and G. Pfeifer, editors, *Proceedings of the 5th International Conference on Logic Programming and Nonmonotonic Reasoning, LPNMR 1999*, volume 1730 of *LNCS*, pages 317–331. Springer, 1999.

J. Pearl. System Z: A natural ordering of defaults with tractable applications to nonmonotonic reasoning. In R. Parikh, editor, *Proceedings of the 3rd Conference on Theoretical Aspects of Reasoning about Knowledge, TARK 1990*, pages 121–135. Morgan Kaufmann, 1990.

R. Reiter. On closed world data bases. In H. Gallaire and J. Minker, editors, *Logic and data bases*, pages 55–76. Plenum Press, 1978.

R. Reiter. A logic for default reasoning. *Artificial Intelligence*, 13(1-2):81–132, 1980.

R. Reiter and G. Criscuolo. On interacting defaults. In P. J. Hayes, editor, *Proceedings of IJCAI 1981*, pages 270–276. William Kaufman, 1981.

J. Schlipf. The expressive powers of the logic programming semantics. *Journal of Computer and System Sciences*, 51(1):64–86, 1995.

G.F. Schwarz. Autoepistemic logic of knowledge. In A. Nerode, W. Marek, and V.S. Subrahmanian, editors, *Proceedings of the 1st International Workshop on Logic Programming and Nonmonotonic Reasoning, LPNMR 1991*, pages 260–274. MIT Press, 1991.

G.F. Schwarz. Minimal model semantics for nonmonotonic modal logics. In *Proceedings of LICS 1992*, pages 34–43. IEEE Computer Society, 1992.

G.F. Schwarz and M. Truszczyński. Modal logic **S4F** and the minimal knowledge paradigm. In Y. Moses, editor, *Proceedings of TARK 1992*, pages 34–43. Morgan Kaufmann, 1992.

G.F. Shvarts. Autoepistemic modal logics. In R. Parikh, editor, *Proceedings of TARK 1990*, pages 97–109, Morgan Kaufmann, 1990.

P. Simons. Extending the stable model semantics with more expressive rules. In M. Gelfond, N. Leone, and G. Pfeifer, editors, *Proceedings of the 5th International Conference on Logic Programming and Nonmonotonic Reasoning, LPNMR 1999*, volume 1730 of *LNCS*, pages 305–316. Springer, 1999.

T. Soininen and I. Niemelä. Developing a declarative rule language for applications in product configuration. In G. Gupta, editor, *Proceedings of the 1st International Workshop on Practical Aspects of Declarative Languages, PADL 1999*, volume 1551 of *LNCS*, pages 305–319. Springer, 1998.

R.C. Stalnaker. A note on nonmonotonic modal logic. *Artificial Intelligence*, 64(2):183–196, 1993; broadly available as unpublished manuscript, 1980.

J. Stillman. The complexity of propositional default logics. In W. R. Swartout, editor, *Proceedings of the 10th National Conference on Artificial Intelligence, AAAI 1992*, pages 794–800, MIT Press, 1992.

J. Tiihonen, T. Soininen, I. Niemelä, and R. Sulonen. A practical tool for mass-customizing configurable products. In *Proceedings of the 14th International Conference on Engineering Design*, pages 1290–1299, 2003.

M. Truszczyński. Modal interpretations of default logic. In R. Reiter and J. Mylopoulos, editors, *Proceedings of IJCAI 1991*, pages 393–398. Morgan Kaufmann, 1991.

A. Van Gelder, K.A. Ross, and J.S. Schlipf. The well-founded semantics for general logic programs. *Journal of the ACM*, 38(3):620–650, 1991.

Extensions of Answer Set Programming

Victor W. Marek
Department of Computer Science
University of Kentucky
Lexington, KY 40506-0633, USA

Jeffrey B. Remmel
Departments of Computer Science and Mathematics
University of California
La Jolla, CA 92093, USA

Abstract: We discuss a number of possible extensions of Answer Set Programming. The four formalisms we investigate are:
1. logic programs where the negative parts of the bodies in clauses can be replaced by arbitrary constraints which we call Arbitrary Constraint Logic Programming (ACLP),
2. logic programs where we are allowed arbitrary set constraint (SC) atoms,
3. logic programs where atoms represent sets from some fixed set X and the one-step provability operator is composed with a monotone idempotent operator on 2^X which we call Set-based Logic Programming (SBLP), and
4. logic programming where the clauses (rules) have embedded algorithms which we call Hybrid Answer Set Programming (H-ASP).

1 Introduction

Past research has demonstrated that logic programming with the answer-set semantics, known as *answer-set programming* or *ASP*, for short, is an expressive knowledge-representation formalism [Marek and Truszczynski, 1999, Niemelä, 1999, Lifschitz, 1999, Gelfond and Leone, 2002, Baral, 2003, Marek and Remmel, 2004]. The availability of the non-classical negation operator *not* allows the user to model incomplete information, frame axioms, and default assumptions such as normality assumptions and the closed-world assumption (CWA). Modeling these concepts in classical propositional logic is less direct [Gelfond and Leone, 2002] and typically requires much larger representations. In addition, current implementations of ASP support aggregate operations over finite sets or, more generally, constraints over finite sets.

   A fundamental methodological principle behind ASP, which was identified in the papers of Marek and Truszczynski [1999] and Niemelä [1999], is that to model a problem, one designs a program so that its answer sets *encode* or *represent* problem solutions. This is in contrast with the traditional way automated reasoning is used in knowledge representation, which relies on proof-theoretic methods of resolution with unification. Niemelä [1999] has argued that logic programming with the stable-model semantics should be thought of as a language for representing constraint satisfaction problems. Thought of from this point of view, ASP systems are ideal logic-based systems to reason about a variety of types of data and integrate quantitative and quantitative reasoning. ASP systems allow the users to describe solutions by giving a series of constraints and letting an ASP solver search for solutions. Marek and Remmel [2002] show that all FNP problems as defined by Bellare and Goldwasser [1994] can be solved by the ASP based on stable model semantics of logic programs.

In this paper, we shall consider several ways to extend answer set programming. One of our main motivations for considering extension of answer set programming is that current *solvers* such as *cmodels* [Babovich and Lifschitz, 2002], *smodels* [Simons, Niemelä, and Soininen, 2002], *assat* [Lin and Zhao, 2002], *clasp* [Gebser, Kaufmann, Neumann, and Schaub, 2007], *dlv* [Leone, Pfeifer, Faber, Eiter, Gottlob, Perri, and Scarcello, 2006], *pbmodels* [Liu and Tuszczynski, 2005] and *aspps* [East and Truszczynski, 2006, East, Iakhiaev, Mikitiuk, and Truszczynski, 2006] have no systematic way to reason about infinite sets. Of course, there is one obvious way that we can use to reason about infinite sets in logic programming, namely, by adding function symbols to the language. However, adding function symbols to the language has significant drawbacks, especially with regard to complexity. For example, finding the least model of a finite Horn program with no function symbols can be done in linear time [Dowling and Gallier, 1984] while the least model of a finite predicate logic Horn program with function symbols can be an arbitrary recursively enumerable set [Smullyan, 1968]. If we consider logic programs with negation, Marek and Truszczynski [1991] showed that the question of whether a finite propositional logic program has a stable model is NP-complete. However Marek, Nerode, and Remmel [1992] showed that the question of whether a finite predicate logic program with function symbols possesses a stable model is Σ_1^1 complete. Similarly, the stable models of logic programs that contain function symbols can be quite complex. Starting with Apt and Blair [1991] and continuing with Blair, Marek, and Schlipf [1995] and Marek, Nerode, and Remmel [1992], a number of results showed that the stable models of logic programs that allow function symbols can be exceedingly complex, even in the case where the program has a unique stable model. For example, Marek, Nerode, and Remmel [1992] showed that there exist finite predicate logic programs which have stable models but which have no hyperarithmetic stable models. Thus there is no hope to have general processing engines that will handle normal logic programs with function symbols.

These complexity results for logic programs with function symbols may seem quite negative, but they had a positive effect in the long run in that they forced researchers and designers mainly to limit themselves to cases where programs can be actually processed. The effect was that processing programs (*solvers*), as listed above had to focus on finite programs that do not admit function symbols[1]. At this point, none of the existing solvers have good ways to deal with infinite sets.

Why do we need to reason about infinite sets? Clearly, if one wants to reason about regions in Euclidean space or time intervals, then one is implicitly reasoning

---

[1]The *dlv* system does allow for some limited use of functions symbols, with the idea which is common in Computer Science that it is programmer's responsibility to write programs that the system can process.

about infinite sets although one often can get by with finite descriptions or approximations. However, we believe that another source of need for reasoning about infinite sets
comes from the interactions with the Internet that are required in modern applications
which give rise to the need to reason about sets and databases which are extremely
large and/or are constantly changing and evolving. We claim that infinite sets offer an
effective way to address problems involving large finite sets that do not have a clear
structure and may change rapidly. Such finite sets often do not have concise representations, and manipulating them based on their explicit enumerations is impractical.
On the other hand, infinite approximations to these large finite sets, if chosen appropriately, may have structure that makes concise finite representations feasible and
possibly allow for effective reasoning and processing. For instance, a large database
of documents and the set of WWW pages are examples of very large sets, interesting
subsets of which can be thought of conceptually as infinite sets, e.g., "all documents
containing a given string". Often such sets can be described as regular languages and
hence have a finite description. In addition, a set of localities that might be affected
by a tornado or the scope of a battlefield provide examples of finite sets that change
rapidly. Thus it may be more convenient to find approximations such as polygons cov-
ering the affected areas that lend themselves to easy manipulation. In each case, while
the finite sets of interest may have no small representations, the infinite sets used as
approximations do - a feature that can be exploited in automated reasoning.

For yet another example, consider the problem of controlling an unmanned under-
water vehicle V. Given parameters such as time, position, velocity and direction of
motion, as well as the model of the environment in which the vehicle moves, we can
describe constraints on various subsets of the set of possible trajectories of the vehicle that maintain the vehicle V in stable condition. Here, not only is each trajectory
infinite, but the set of trajectories that keep V stable may also be infinite. The many-
dimensional space describing the vehicle status and other features of the vehicle is a
region X included in R^n, and the desired regions, where the vehicle needs to be can
also be treated as subsets of the same R^n. There are, potentially many such regions.
One reasonable way to describing them is by means of constraints on subsets of R^n.

The designers of the ASP solvers have also focused on the issues of both improving the processing of the logic programs, i.e. finding more efficient ways to search for
a stable models, and improving the use of logic programs as a programming language.
The latter task consists of extending the constructs available to the programmer to
make programming easier and more readable. The extensions of ASP that we shall
talk about in this paper can be viewed as part of this latter task.

The basic idea behind all the extensions that we shall discuss in this paper is to
carefully consider the definition of the stable model or answer set semantics via the
Gelfond-Lifschitz transform and to consider ways in which that general mechanism
can be extended. To make our ideas more precise, we shall briefly review the definition
of stable models for propositional and predicate logic programs.

A (propositional) *logic programming clause* is an expression of the form

$$C = p \leftarrow q_1, \ldots, q_m, \text{ not } r_1, \ldots, \text{not } r_n \qquad (1)$$

where $p, q_1, \ldots, q_m, r_1, \ldots, r_n$ are atoms from a fixed set of atoms At. The atom p in the clause above is called the *head* of C ($head(C)$), and the expression

$$q_1, \ldots, q_m, not\ r_1, \ldots, not\ r_n,$$

with ',' interpreted as the conjunction, is called the *body* of C ($body(C)$). The set $\{q_1, \ldots, q_n\}$ is called the *positive part of the body* of C ($posBody(C)$) and the set $\{r_1, \ldots, r_m\}$ is called the *negative part of the body* of C ($negBody(C)$). Given any set $M \subseteq At$ and atom a, we say that M satisfies a (*not a*), written $M \models a$ ($M \models not\ a$), if $a \in M$ ($a \notin M$). We say that M satisfies C, written $M \models C$, if whenever M satisfies the body of C, then M satisfies the head of C. A normal logic program P is set of clauses of the form of (1). We say that $M \subseteq At$ is a model of P, written $M \models P$, if M satisfies every clause in C.

A (propositional) Horn clause is a logic programming clause of the form

$$H = p \leftarrow q_1, \ldots, q_m \tag{2}$$

where $p, q_1, \ldots, q_m \in At$. Thus in a Horn clause, the negative part of its body is empty. A Horn program P is a set of Horn clauses. Each Horn program P has a least model under the inclusion relation, LM_P, which can defined using the one-step provability operator T_P. For any set A, let 2^A denote the set of all subsets of A. The one-step provability operator $T_P : 2^A \rightarrow 2^A$ associated with the Horn program P [van Emden and Kowalski, 1976] is defined by setting

$$T_P(M) = \{p : \exists C \in P(p = head(C) \wedge M \models body(C))\} \tag{3}$$

for any $M \in 2^A$. We define $T_P^n(M)$ by induction by setting $T_P^1(M) = T_P(M)$ and $T_P^{n+1}(M) = T_P(T_P^n(M))$. Then the least model LM_P can be computed as

$$LM_P = T_P(\emptyset)^\omega = \bigcup_{n \geq 0} T_P^n(\emptyset).$$

If P is a normal logic program and $M \subseteq A$, then the Gelfond-Lifschitz transform of P with respect to M [Gelfond and Lifschitz, 1988] is the Horn program $GL_P(M)$ which results by eliminating those clauses C of the form (1) such that $r_i \in M$ for some i and replacing C by $p \leftarrow q_1, \ldots, q_n$ otherwise. We then say that M is a *stable model* or an *answer set* for P if M equals the least model of $GL_P(M)$.

We should note that the operator T_P makes perfectly good sense for any normal logic program [Apt and van Emden, 1982]. The fixpoints of the operator T_P are called *supported models* of P. One can prove that every answer set of P is a supported model. Supported models of P can be shown to coincide with models of the completion of P, $comp(P)$ [Clark, 1977]. As $comp(P)$ is a propositional theory, one can use SAT solvers to compute its models and so, the supported models of P. By pruning those supported models that are not answer sets, one can also compute answer sets by means of SAT solvers. This possibility was successfully used in systems such as *assat* [Lin and Zhao, 2002] and *cmodels* [Babovich and Lifschitz, 2002]. Moreover *assat* and *cmodels* implement pruning by expanding the input program with the so-called *loop formulas* [Lin and Zhao, 2002]. The process can be viewed as a version of clause learning used in SAT solvers. Recent solvers which are improvements on *assat* and *cmodels* such as *clasp* are very efficient.

One can extend the notion of stable models to predicate logic programs as follows. A (predicate) *logic programming clause* is an expression of the form

$$C = p \leftarrow q_1, \ldots, q_m, \, not \, r_1, \ldots, not \, r_n \qquad (4)$$

where $p, q_1, \ldots, q_m, r_1, \ldots, r_n$ are atoms from some fixed first order language $\mathcal{L}$. As in the case of propositional logic clauses, the atom p in the clause above is called the *head* of C ($head(C)$), and the expression $q_1, \ldots, q_m, not \, r_1, \ldots, not \, r_n$, with ',' interpreted as the conjunction, is called the *body* of C ($body(C)$). The set $\{q_1, \ldots, q_n\}$ is called the *positive body* of C ($posBody(C)$) and the set $\{r_1, \ldots, r_m\}$ is called the negative body of C ($negBody(C)$). A ground instance of the clause C is a substitution instance of C where we have uniformly replaced the free variables in C with ground terms, i.e. terms with no free variables, so that resulting substitution instance has no free variables. A predicate logic program P is a collection of clauses of the form (4). We then let $ground(P)$ denote the set of all ground instances of clauses in P. Thus $ground(P)$ can be thought of as propositional logic program. We then say that a collection of ground atoms M, i.e. a subset of atoms of $\mathcal{L}$ with no free variables, is a *stable model* or an *answer set* of P if and only if M is a stable model of $ground(P)$.

In this paper, in the subsequent sections, we shall consider four different extensions of the basic stable model paradigm described above.

Extension 1. *Arbitrary Constraint Logic Programming.*
Our first extension is to follow the paper of Marek et al. [1995] and consider logic programs with arbitrary constraints. Notice that in the definition of stable model of P, the negative bodies of the clauses of P only play a role in determining which Horn clauses end up in $GLP(M)$. Thus the idea of Marek, Nerode, and Remmel [1995] is to replace these negative bodies by arbitrary constraints so that we end up with clauses of the form

$$p \longleftarrow q_1, \ldots, q_n : \Psi. \qquad (5)$$

Here Ψ is *any* type of constraint such that given $M \subseteq At$, we can decide whether M satisfies Ψ. Thus Ψ does not even have to be in the original language of the program and it could even express an infinite constraint such as the ones studied by Marek, Nerode, and Remmel [1997]. Thus replacing negative bodies by arbitrary constraints provides a rich way to reason about all sorts of infinite constraints in ASP which we call Arbitrary Constraint Logic Programming (ACLP).

Extension 2. *Adding set constraint atoms to logic programming.*
A powerful extension of Answer Set Programming stems out of the work of Simons, Niemelä, and Soininen [2002]. The idea was to use as building blocks of programs not only atoms and negated atoms, but expressions of the form kXl where X is a finite set of literals, and k, l are nonnegative integers, smaller or equal than the size of X. The interpretation of such constraint is "at least k but not more than l of literals from X are true in a putative model M".

Later Marek and Remmel [2004] introduced set constraint atoms of the form $\langle X, \mathcal{F} \rangle$ where X is a set and $\mathcal{F}$ is a finite set of subsets of X. Research of several authors [Gelfond and Leone, 2002, Marek, Niemelä, and Truszczynski, 2008, Liu and

Truszczynski, 2005] led to significant progress in understanding such constraints. On the concrete level, arbitrary set constraints atoms include weight constraints, SQL constraints, parity constraints, and other kinds of common constraints, and, on the abstract level, include monotone, antimonotone, and convex constraints. Adding arbitrary set constraint atoms to logic programming is a natural mechanism that allows the user to reason about large variety of constraints in ASP solvers and SAT solvers.

Set constraint atoms $\langle X, \mathcal{F} \rangle$ where X is an infinite set and $\mathcal{F}$ is a finite subset of 2^X can also be used to reason about infinite sets. For example, Cenzer, Remmel, and Marek [2005] studied constraints of the form $\langle X, \mathcal{F} \rangle$ where X is an infinite recursive set and $\mathcal{F}$ is a finite set of indices for certain recursive or recursively enumerable subsets of X.

We shall also briefly outline the work of Brik and Remmel [2011a] that shows how one can use arbitrary set constraint atoms to reason about preferences in ASP. The ability to express preferences and to reason about them effectively has many important applications to problems in planning and negotiations. Brik and Remmel [2011a] have shown that set constraint atoms can be a very convenient and compact way to express a wide variety of such preferences. That is, suppose that we are given a set constraint atom $\langle X, \mathcal{F} \rangle$ and weight function $wt : 2^X \rightarrow \mathbf{Q}$ where $\mathbf{Q}$ is the set of rational numbers. Then the idea is that the weight function wt is defined in such a way so that we prefer those $F \in \mathcal{F}$ which have the smallest weight. For example, suppose that Dr. X is buying a car and the dealer offers several option packages such as you can have a red car with an automatic transmission with a high end CD player or you can have a blue car with standard transmission and a standard CD player. Suppose that the blue car costs \$25,000 and the red car costs \$35,000. Let B stand for blue, R stand for red, A stand for automatic transmission, S stand for standard transmission, HCD stand for high end CD player, and SCD stand for standard CD player. Suppose that the prices of the cars can be \$20,000, \$25,000, \$30,000, \$35,000. Then we let $X = \{B, R, A, S, HCD, SCD, 20000, 25000, 30000, 35000\}$ We can then view the set $F_1 = \{B, S, SCD, 20000\}$, $F_2 = \{R, A, HCD, 35000\}$, and $F_3 = \{B, S, HCD, 25000\}$ as option packages available from the car dealer. While an individual may prefer red cars to blue cars, standard transmissions to automatic transmission, and high end CD players to standard CD players and to get the car at the lowest possible price, there may be no such package as $\{R, S, HCD, 20000\}$. Thus the buyer has to choose from one the three packages F_1, F_2, or F_3 so that we may have a set constraint atom $\langle X, \{F_1, F_2, F_3\} \rangle$. Now we can insist that a model M satisfies the set constraint $\langle X, \mathcal{F} \rangle$ by adding a clause of the form

$$\langle X, \mathcal{F} \rangle \leftarrow . \tag{6}$$

One can use an auxiliary weighting function to expresses preference in this case. For example, we might define $wt(F_1) = 2$, $wt(F_2) = 1.5$ and $wt(F_3) = 1$. Of course, the buyer's spouse may have a different set of preferences so that we might want to create two copies of the $\langle X, \{F_1, F_2, F_3\} \rangle$, one for the husband and one for the wife. Thus we might want to consider programs which have several clauses of the form (6). This leads to a natural weighting on models M of the program defined to be the sum of $wt(M \cap X)$ for all such clauses. The idea is that lower weighted models satisfy more of the preferences incorporated by clauses of the form (6).

Often times in such situations, it is impossible to meet all individual preferences. This can lead to programs that do not have stable models. One way to handle this problem is to specify hard preferences, those that have to be satisfied, and soft preferences, those that do not necessarily have to be satisfied. Another approach is to look for subsets of preferences which can be satisfied and this naturally leads one to search for maximal subprograms of a program P which do have stable models. To date, none of the ASP solvers have the ability to find such maximal subprograms. However, there is an algorithm called the forward chaining algorithm developed by Marek, Nerode, and Remmel [1994] which does allow one to find such maximal subprograms when the original program does not have a stable model. Recently, Brik and Remmel [2010] have combined the forward chaining algorithm with the Metropolis algorithm [Metropolis et al., 1953] to produce a novel Monte Carlo type algorithm to find such maximal subprograms.

Extension 3. *Set-based logic programming.*
Blair, Marek, and Remmel [2001] observed that the ASP formalism can be significantly extended by allowing atoms to represent sets in some fixed universe X. That is, instead having the intended underlying universe be the Herbrand base of the program, one replaces the underlying Herbrand universe by some fixed space X and has the atoms of the program specify subsets of X, i.e. elements of the set 2^X, the set of all subsets of X.

If we reflect for a moment on the basic aspects of logic programming with an Herbrand model interpretation, a slight change in our point of view shows that interpreting atoms as subsets of the Herbrand base is quite natural. In normal logic programming, we determine the truth value of an atom p in an Herbrand interpretation I by declaring $I \models p$ if and only if $p \in I$. However, this is equivalent to defining the sense, $[\![p]\!]$, of a ground atom p to be the set $\{p\}$ and declaring that $I \models p$ if and only if $[\![p]\!] \subseteq I$. By this simple move, we have permitted ourselves to interpret the sense of an atom as a subset, rather than the literal atom itself.

This given, Blair et al. [2001] developed a system that they called *spatial logic programming*. They showed that it is a natural step to take the *sense* $[\![p]\!]$ of a ground atom p to be a fixed assigned subset of some nonempty set X and to define a $I \subseteq X$ to be a model of p, written $I \models P$, if and only if $[\![p]\!] \subseteq I$. This type of model theoretic semantics makes available, in a natural way, multiple truth values, intensional constructs, and interpreted relationships among the elements and subsets of X. Observe that the assignment $[\![\cdot]\!]$ of a *sense* to ground atoms is intrinsically intensional. Interpreted relationships among the elements and subsets of X allow the programs that use this approach to serve as front-ends for existing systems and still have a seamless model-theoretic semantics for the system as a whole.

Blair, Marek, and Remmel [2008], showed that if the underlying space X has structure such as a topology or an algebraic structure such as a group, ring, field, or vector space, then a number of natural options present themselves. For example, if we are dealing with a topological space, one can compose the one step consequence operator T_P with an operator that produces topological closures of sets or interiors of sets. In such a situation, one ensures that the the extended T_P operator always produces closed sets or always produces open sets. Similarly, if the underlying space

X is a vector space, one might insist that the extended T_P operator always produces a subspace of X or a subset of X which is convex. Notice that each of the operators: *closure, interior, span* and *convex-closure* is a *monotone idempotent operator*. That is, an operator $op : 2^X \rightarrow 2^X$ is an monotone operator if $I \subseteq J \Rightarrow op(I) \subseteq op(J)$ for all $I \subseteq J \subseteq X$ and is an idempotent operator if $op(op(I)) = op(I)$ for all $I \subseteq X$. We call such an operator a *miop* (pronounced "my op").

Unlike the situation in Extensions 1 and 2, there is a variety of options for how to interpret negation in spatial logic programming. In normal logic programming, a model M satisfies *not p* if $p \notin M$. From the set-based point of view when p is interpreted as a singleton $\{p\}$, this would be equivalent to saying that M satisfies *not p* if (i) $\{p\} \cap M = \emptyset$, or (equivalently) (ii) $\{p\} \not\subseteq M$. When the sense of p is a set with more than one element it is easy to see that saying that M satisfies *not p* if $[\![p]\!] \cap M = \emptyset$ which we call strong negation is different from saying that M satisfies *not p* if $[\![p]\!] \not\subseteq M$ which we call weak negation. There are thus two natural interpretations of the negation symbol. Again, when the underlying space has structure, one can get even more subsidiary types of negation by taking M to satisfy *not p* if $cl([\![p]\!]) \cap M = cl(\emptyset)$, or by taking M to satisfy *not p* if $cl([\![p]\!]) \not\subseteq M$ where cl is some natural miop. By composing the one-step provability operator with a miop, one naturally produces only those stable models which have desired properties such a being closed or being a subspace of a vector space. The familiar T_P operator corresponds to the case where the underlying miop operator is the simplest possible monotone idempotent operator, namely, the identity.

Blair et al. [2008] called the extension of spatial logic programming with miops *set-based logic programming* (SBLP). Set-based logic programming provides yet another powerful way to reason about infinite sets as one is allowed to have the sense $[\![a]\!]$ of an atom a be an infinite subset of X. Indeed, Marek and Remmel [2009] showed that one can effectively reason about infinite sets in SBLP provided that infinite sets have an indexing scheme with certain decision properties. For example, if the sense of all atoms are regular languages over some fixed finite alphabet Σ and $X = \Sigma^*$, then Marek and Remmel [2009] proved that the stable models of a finite SBLP program P are always regular languages over Σ and that one can effectively decide whether a given regular language $L \subseteq \Sigma^*$ is a stable model of P.

Extension 4. *Hybrid Answer Set Programming.* Brik and Remmel [2011b] introduced an extension of the ASP formalism in which one can reason about continuous trajectories which they called Hybrid Answer Set Programming (H-ASP). This extension is different than the other three in that the notion of a clause is greatly extended. To motivate this extension, consider the following situation where James Bond wants to take his Aston-Martin from point A to point B where the underlying trajectory his divided up into three regions: Region I which consists of ice and snow on a mountain, Region II which consists of lake, and Region III which consists of desert. With a push of button, Bond's Aston-Martin can change its configuration so that it can run on snow and ice, run as a boat, or run as a high performance car. This situation is pictured in Figure 1 where the rectangle in Region I is some building which must be avoided, the circles in Region II are some islands that must be avoided, and the hexagon is region III is some fort which must be avoided. We imagine that Bond makes certain decisions at regular intervals of length Δ as to what to do depending on his position $x(k\Delta)$, his velocity $v(k\Delta)$, his acceleration $a(k\Delta)$ and other requirements such as surface conditions, wind velocity and other logical conditions such as "I am being chased" or "I am at a minimum safe distance from an obstacle." In Figure 1, we have indicated Bond's position's at times $0, \Delta, 2\Delta, \ldots, 11\Delta$ by placing the $k\Delta$ at the position he has reached at time $k\Delta$.

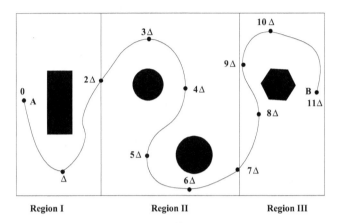

Figure 1: Picture of Bond's trajectory.

Brik and Remmel [2011b] discuss two systems of hybrid ASP. We shall briefly describe their simplified system of hybrid ASP in this introduction which they called *basic hybrid ASP* (BH-ASP) and discuss a more extended version of hybrid ASP in Section 4. In basic hybrid ASP, one specifies a parameter space S and a set of atoms At. The intended universe of an BH-ASP program is $At \times S$. That is, one thinks of the position and situation at time $k\Delta$ as being specified by a sequence of parameters $\mathbf{x}(k\Delta) = (x_1(k\Delta), x_2(k\Delta), \ldots, x_n(k\Delta)) \in S$ that specify such things as position, velocity, acceleration, etc. which are needed to compute the next position and the data base $M(k\Delta)_{\mathbf{x}(k\Delta)}$ of atoms a in At such that $(a, \mathbf{x}(k\Delta))$ are true at time $k\Delta$.

There are two types of clauses in a BH-ASP program.

1. *Stationary clauses* which are of the form

$$a \leftarrow a_1, \ldots, a_n, not\, b_1, \ldots, not\, b_m : O$$

where $a, a_1, \ldots, a_n, b_1, \ldots, b_m \in At$, $O \subseteq S$. The idea is that if $\mathbf{x}(k\Delta) \in O$, $a_i \in M(k\Delta)_{\mathbf{x}(k\Delta)}$ for $i = 1, \ldots, n$, and $b_j \notin M(k\Delta)_{\mathbf{x}(k\Delta)}$ for $j = 1, \ldots, m$, then $a \in M(k\Delta)_{\mathbf{x}(k\Delta)}$.

2. *Advancing clauses* which are of the form

$$a \leftarrow a_1, \ldots, a_n, not\, b_1, \ldots, not\, b_m : A, O$$

where $a, a_1, \ldots, a_n, b_1, \ldots, b_m \in A$, $O \subseteq S$, and A is an algorithm. The idea here is that if $\mathbf{x}(k\Delta) \in O$, $a_i \in M(k\Delta)_{\mathbf{x}(k\Delta)}$ for $i = 1, \ldots, n$, and $b_j \notin M(k\Delta)_{\mathbf{x}(k\Delta)}$ for $j = 1, \ldots, m$, then we can apply the algorithm A to the set of parameters $\mathbf{x}(k\Delta)$ to compute the set of parameters $\mathbf{x}((k+1)\Delta)$ at the next time step and the clause specifies that $(a, \mathbf{x}((k+1)\Delta))$ holds.

Here for advancing clauses, we envision that algorithm A could require that one solve a differential or integral equation to get the next set of parameters or it could require solving some system of linear equations or some linear programming problem to get the next set of parameters, etc.. From this point of view, we can think of an advancing clause as input-output device. Of course, classical logic rules also can be thought of as input-output devices, but one rarely thinks in this sort of terms.

The outline of the rest of this paper is as follows. In Section 2 we shall discuss ACLP of Extension 1. In Section 3, we shall discuss various extensions of ASP that use set constraint atoms. In Section 4, we shall discuss set-based logic programming and how it can be used to reason about certain classes of infinite set effectively. In Section 5, we shall briefly introduce basic hybrid ASP and its more general extension called Hybrid ASP (H-ASP) as described by Brik and Remmel [2011a]. Finally, in Section 6 we discuss conclusions and further research.

2   Arbitrary Constraint Logic Programming

In this section we discuss the variation of programs obtained by treating the negative part of the body of a clause as a constraint on applicability of a clause. That is, we shall give a detailed description of Arbitrary Constraint Logic Programming (ACLP) as described in Extension 1 of the introduction.

The basic Gelfond-Lifschitz transform mechanism of Answer Set Programming can be expressed as follows. The negative literals in the body of a clause serve as a "semaphore". Namely, when we guess a set of atoms M (a putative answer set), the negative part of the clause C tells us if the Horn part of C can be used in the computation or not. To make this intuition a bit more precise, given a clause

$$C = p \leftarrow q_1, \ldots, q_n, not\ r_1, \ldots, not\ r_n, \qquad (7)$$

let $h(C)$ be the Horn clause $p \leftarrow q_1, \ldots, q_n$. Then $P_M = \{h(C) : M \models \neg negBody(C)\}$. Thus $\neg negBody(C)$ is a constraint on usability of $h(C)$. There is no reason why such constraints should be restricted to be only conjunctions of negative literals. Motivated by this observation, we introduce the concept of ACLP clause and of ACLP program. An ACLP clause is a string C:

$$p \leftarrow q_1, \ldots, q_m : \Phi_C$$

where $p, q_1, \ldots, q_m$ are atoms, and Φ_C is a formula of some language $\mathcal{L}$ for which we have a satisfaction relation $\models$ which allows us to test if $M \models \Phi_C$. An ACLP program is a set of ACLP clauses. Now, the idea is to generalize the "semaphore" as defined above. Namely, we first guess a set M of atoms and then we test if the constraint Φ_C is satisfied by M or not. If it is satisfied, $h(C)$ is placed in P_M, otherwise it is eliminated. Then we compute the least model of P_M and check if it coincides with M. In such situation, we call M a *constraint answer set* for P.

The simplest case is when Φ_C consists of conjunctions of negative literals only. In that case, we get nothing new. Indeed, let us assign to a clause C of the form (7), a constraint clause:

$$C' = p \leftarrow q_1, \ldots, q_n : \neg r_1, \ldots, \neg r_n$$

and $P' = \{C' : C \in P\}$. Then one can prove the following proposition.

Proposition 1 *A set of atoms M is an answer set for P if and only if M is a constraint answer set for P'.*

We note that the notion of answer set for constraint programs includes the notion of a supported model via the following construction. Given a clause C we assign to C a constraint clause C'' as follows

$$C'' = p \leftarrow: q_1, \ldots, q_n, \neg r_1, \ldots, \neg r_n$$

and set $P' = \{C'' : C \in P\}$. Then one can prove the following proposition.

Proposition 2 *A set of atoms M is a supported model for P if and only if M is a constraint answer set for P''.*

As long as the constraints Φ are taken from the propositional language generated by the set of atoms of the program P, the expressive power of the concept of constraint answer set does not increase. We have the following fact.

Proposition 3 *Let $\mathcal{PC}$ be class of constraint programs where all the constraints are propositional formulas. Then the existence problem for constraint answer sets of programs in $\mathcal{PC}$ is an NP-complete problem.*

Pollett and Remmel [1997] studied a class of constraint programs where the constraints were quantified Boolean formulas over the set of atoms occurring in the program. That is, Pollett and Remmel consider programs whose clauses are of the form

$$p \leftarrow a_1, \ldots, a_n : B_1(\mathbf{b}_1), \ldots, B_m(\mathbf{b}_m) \tag{$*$}$$

where $p, a_1, \ldots, a_n$ are propositional variables and $B_i, 1 \leq i \leq m$ is a quantified Boolean formula and $\mathbf{b}_i$'s represents the free propositional variables in each B_i.

Let Σ_k^q denote the set of quantified Boolean formulas with at most k-alternations of quantifier type and whose outermost quantifier is an $\exists$. Similarly, let Π_k^q denote the set of quantified Boolean formulas with at most k-alternations of quantifier type and whose outermost quantifier is an $\forall$. In both cases, unless we say we are dealing with only sentences, we assume our formulas have free variables. Lastly, we write QBF_k to denote Boolean combinations of these two classes. In the $k = 0$ case, all of the above classes are the same. They each define the class of propositional formulas. We recall that the problem of determining whether a Σ_k^q-sentence is true is Σ_k^P-complete and the problem of determining whether a Π_k^q-sentence is true is Π_k^P-complete. The complexity of testing checking whether or not an assignment satisfies QBF_k formula is in Δ_{k+1}^P.

We now define LP_k as the class of *finite* arbitrary constraint logic programs whose constraints are all in QBF_k and $LP_\infty = \bigcup_{k \geq 0} LP_k$.

It turns out that Proposition 3 generalizes in a straightforward way. That is, Pollett and Remmel proved the following theorem.

Theorem 1   *1. The problem of determining whether an LP_k program has an answer set is Σ_{k+1}^P-complete.*

2. *The problem of determining whether a finite LP_∞ program has an answer set is PSPACE-complete.*
3. *The problem of deciding whether a given variable a is in an answer set of an LP_k program is Σ_{k+1}^P-complete.*
4. *The problem of deciding whether a given variable a is in an answer set of an LP_∞ program is PSPACE-complete.*

Pollett and Remmel pointed out that there are several ways to generalize the notion of logic programming with quantified Boolean constraints that fits the paradigm of ACLP programs. For example, rather than take our atoms of quantified Boolean formulas to be just propositional variables, we could let them be propositional variables and expressions of the form $a_1 a_2 \ldots a_n \in A$. That is, checking if the concatenation of some string propositional variables is in an oracle A. Given a variable assignment ν, we say $\bar{\nu}(a_1 a_2 \ldots a_n \in A) = 1$ if and only if the string $\nu(a_1)\nu(a_2)\ldots\nu(a_n)$ is in the set A. Thus, there is a well defined semantics for such formulas. We can define the classes $\Sigma_k^q(A)$, $\Pi_k^q(A)$, and $QBF_k(A)$ and use them in our logic programming theories. Hence, we can define $LP_k(A)$ to be those finite logic programs with $QBF_k(A)$ constraints. Then Pollett and Remmel proved the following generalization of Theorem 1.

Theorem 2   1. *The problem of deciding whether an $LP_k(A)$ program has an answer set is $\Sigma_{k+1}^P(A)$-complete.*
2. *The problem of deciding whether an $LP_\infty(A)$ program has an answer set is PSPACE(A)-complete.*
3. *The problem of deciding whether a given variable a is in an answer set of an $LP_k(A)$ program is $\Sigma_{k+1}^P(A)$-complete.*
4. *The problem of deciding whether a given variable a is in an answer set of an $LP_\infty(A)$ program is PSPACE(A)-complete.*

Going beyond propositional logic and quantified Boolean formulas leads to interesting but not investigated class of constraint programs. Specifically, for integers $j \geq 2$ and $0 \leq i < j$, we define a new formula $i\,mod\,j$ and stipulate, for a finite set of atoms M, $M \models i\,mod\,j$ if $|M| \equiv i \bmod j$. It should be clear that due to the localization properties of propositional logic and quantified Boolean formulas, the formulas $i \bmod j$ are not definable in the languages defined above. But once we defined satisfaction for expressions of the form $i\,mod\,j$, we have immediately satisfaction relation for the language of propositional formulas with atoms of the form $i\,mod\,j$. Let us call such constraints *mod-constraints*. We can then consider programs that use mod-constraints. Parity constraints as considered by Marek et al. [1995] are from this language. Complexity issues for the constraint programs in this language have not been studied. We illustrate the programs with mod-constraints with a simple example.

Example 1 Let P be a mod-constraint program consisting of the following clauses.
$p \leftarrow q, u, w, \quad r \leftarrow s, v, \quad u \leftarrow r, \quad v \leftarrow r, \quad s \leftarrow: 2mod3, \quad t \leftarrow: 2mod3,$
$q \leftarrow: 1mod3, \quad w \leftarrow: 1mod3, \quad u \leftarrow: 1mod3.$
We then check that $M_1 = \{p, s, t, u\}$ is a mod-constraint answer set for P. Indeed, the reduct of P by M_1 yields the program consisting of seven clauses: $p \leftarrow q, u, w,$
$r \leftarrow s, v, \quad u \leftarrow r, \quad v \leftarrow r, \quad q \leftarrow, \quad w \leftarrow, \quad u \leftarrow,$ with M_1 as its least model.

Likewise, we leave to the reader the task of testing that $M_2 = \{s, t\}$ is another mod-constraint answer set for P. $\quad\square$

The family of answer sets for a constraint program does not need to form an antichain. Minimal answer set of constraint programs have not been studied.

3   Logic Programming with Set Constraint Atoms

In this section we define a number of generalizations of cardinality and weight constraints.

One can think about propositional atoms as very simple constraints on assignments. Specifically, an atom p is just a requirement that the intended model M satisfies p. Likewise, a clause

$$C = p \leftarrow q_1, \ldots, q_m, not\ r_1, \ldots not\ r_n$$

propagates constraints and can be informally interpreted as a constraint on the intended model M. That is, once M satisfies the body of C, it has to satisfy the head of C as well. This point of view has been proposed by Niemelä [1999] and Marek and Truszczynski [1999] with the idea that the program specifies constraints of the problem at hand, while the answer sets encode intended solutions of the problem.

Once such point of view is adopted, it is natural to ask whether the ASP mechanism could be adopted to propagation of more complex constraints. Simons, Niemelä, and Soininen [2002] have shown that, indeed, one can adopt ASP formalism to a situation where the constraints are more complex. Specifically, Simons et al. [2002] describe a construction dealing with two specific types of constraints: cardinality constraints and a generalization, weight constraints. These constraints, under the name of pseudo-Boolean constraints are used in logic design and also in combinatorial optimization. Next, we describe the case of cardinality constraints which is the simplest case of pseudo-Boolean constraints where all the weights on the atoms are equal to 1. A cardinality constraint is a string C of the form kXl where X is a finite set of atoms and $k \leq l \leq |X|$. When $k = 0$ we drop it from the description of C, and similarly we drop l when it is $|X|$. When M is a set of atoms, we write $M \models kXl$ if $k \leq |X \cap M| \leq l$. We observe that $M \models p$ if and only if $M \models 1\{p\}$, and $M \models not\ p$ if and only if $M \models \{p\}0$. Thus cardinality constraints generalize literals. The satisfaction relation $\models$ can be extended to the language treating cardinality constraints as new atoms. We can also write program clauses where the heads of clauses and elements of bodies are cardinality constraints. The notion of a model of such clause generalizes the usual notion of a model of a program clause. Specifically, a set of atoms M satisfies a clause

$$kXl \leftarrow k_1 Y_1 l_1, \ldots k_m Y_m l_m \qquad (8)$$

if for some $1 \leq j \leq m$, $M \not\models k_j Y_j l_j$, or if $M \models kXl$. A set M is a model of a program if it is a model of each clause of the program. The cardinality constraints considered here concern the sets of atoms; the original definition of Simons et al. [2002] used literals, not only atoms.

The notion of an answer set of a cardinality constraint program P involves a significant modification of the usual Gelfond-Lifschitz transform (GL-transform). We will call it the NSS-transform. A number of steps are performed. As we shall see, some of these steps are different from the steps used to define the GL-transform. Let us guess M, a putative answer set. First, we require that M is a model of P. Next, one eliminates all clauses C in P of the form of (8) such that $M \not\models body(C)$. The final step transforms each remaining clause C of form (8) of the program P in two ways.

(a) First, one eliminates the upper bounds of all constraints in the body of C. Let us call the resulting clause C'.

(b) Second, one replaces the clause C' by all clauses of the form $p \leftarrow body(C')$ such that $p \in M \cap X$.

The NSS-transform of P is the set of clauses produced from P by this process. That program, P'' has two key properties. First, the heads of clauses of P'' are atoms. Second, for every clause C'' of the program P'', the collection of sets N that satisfy the body of the clause C'' is upper-closed. That is, if $N \models body(C'')$ and $N \subseteq N'$, then $N' \models body(C'')$. This implies that the one-step provability operator associated with P'' and M is monotone. It is also continuous, but, since in this section we shall limit ourselves to finite cardinality constraints programs, we shall not discuss this issue further. Thus by Knaster-Tarski theorem, this operator $T_{P,M}$ possesses a least fixpoint. If that fixpoint coincides with M, we call M an answer set of P. Observe that, by the construction, an answer set must be a model of P.

We observe that for normal logic programs, the NSS-transform of a program eliminates more clauses than the GL-transform. That is, the GL-transform tests only the negative part of the clauses, not the entire body. Yet, for normal programs, the fixpoints obtained via GL-transform and via NSS-transform are the same. The reason is that if a clause C has the property that $posBody(C) \setminus M \neq \emptyset$ and C survives the GL-test, then C will fire only if an atom $q_i \in posBody(C) \setminus M$ is computed. This guarantees, however, that M is different from the fixpoint and, hence, is not an answer set [Marek and Truszczynski, 1998].

A very similar construction can be done for weight constraints where atoms are weighted and the bounds k and l are on cumulative weight of $M \cap X$. We should note that if one allows weight functions which admit negative values, then the results are not always intuitive and alternative approaches have been proposed [Liu et al., 2007, Faber et al., 2011].

Recall that we defined a set constraint to be a pair $\langle X, \mathcal{F} \rangle$ where X is a finite set and $\mathcal{F} \subseteq 2^X$. A *set constraint* (SC) *clause* is a string of the form

$$\langle X, \mathcal{F} \rangle \leftarrow \langle Y_1, \mathcal{G}_1 \rangle, \ldots, \langle Y_n, \mathcal{G}_n \rangle.$$

A set constraint program is a finite set of SC clauses.

We note that there are interesting constraints that are not cardinality constraints nor weight constraints. An example of one such constraint is a constraint analogous to $1mod3$ discussed above, specifically, $M \models \langle X, 1mod3 \rangle$ if $|M \cap X| \equiv 1mod3$. Many other natural constraints can be defined as set-constraints. In fact generalized quantifiers over finite sets of atoms (we refer to the book by Libkin [2004]) can be expressed this way.

To see how the NSS-transform can be utilized for SC programs, we first need to define the satisfaction relation like we did in case of cardinality constraints. Let M be a set of atoms and $K = \langle X, \mathcal{F} \rangle$ be a set constraint atom. We say that $M \models K$ if $X \cap$

$M \in \mathcal{F}$. This is an abstract version of the satisfaction relation defined above. Let us notice that going to the abstract version of the constraint may significantly increase the size of the representation. For example, the cardinality constraint $1\{p, q, r\}2$ becomes $\langle\{p, q, r\}, \{\{p\}, \{q\}, \{r\}, \{p, q\}, \{p, r\}, \{q, r\}\}\rangle$.

We now show how the NSS-transform can be adopted for SC programs. The following observation is easy.

Proposition 4 *For every set X and a family $\mathcal{F}$ of subsets of X there exists a $\subseteq$-least family $\mathcal{G}$ of subsets of X such that*
1. *$\mathcal{F} \subseteq \mathcal{G}$*
2. *$\mathcal{G}$ is upper-closed that is, whenever $A \in \mathcal{G}$ and $A \subseteq B \subseteq X$ then $B \in \mathcal{G}$.*

Since the family $\mathcal{G}$ is $\subseteq$-least, it is unique. Hence we call the unique family $\mathcal{G}$ whose existence is established by Proposition 4, the closure of $\mathcal{F}$ and denote it $\overline{\mathcal{F}}$. It is easy to see that when $\langle X, \mathcal{F} \rangle$ is equivalent to the cardinality constraint kXl, then $\langle X, \overline{\mathcal{F}} \rangle$ is equivalent to the cardinality constraint kX.

At a cost of a possible large representation, we can describe answer sets for programs that include arbitrary set-constraints. Again the process of defining a stable model for SC programs is based on some form of "Horn" programs, GL-reduction, and least fixpoints of the one-step provability operators for Horn programs.

We will call an SC-clause *Horn* if
1. the head of that clause is a single atom (recall that atoms are represented as set constraints) and
2. whenever $\langle X_i, \mathcal{F}_i \rangle$ appears in the body, then $\mathcal{F}_i$ is an upper closed family of subsets of X_i.

A set-constraint Horn program P is an SC-program which consists entirely of Horn clauses. There is a natural one-step provability operator associated to an SC-Horn program P, $T_P : 2^X \to 2^X$ where X is the underlying set of atoms of the program. Specifically, $T_P(S)$ consists of all p such that there is clause

$$C = p \leftarrow \langle X_1, \mathcal{F}_1 \rangle, \ldots, \langle X_m, \mathcal{F}_m \rangle \in P$$

such that S satisfies the body of C. Our definitions ensure that T_P is a monotone operator and hence each SC-Horn program P has a least model M^P. M^P can be computed in a manner analogous to the computation of the least model of a definite Horn program as $T_P^\omega(\emptyset)$. The NSS transform $\mathbf{NSS}_M(P)$ of the set-constraint program P for a given set of atoms M which is a model of P is defined as follows. First eliminate all clauses with bodies not satisfied by M. Next, for each remaining clause

$$\langle X, \mathcal{F} \rangle \leftarrow \langle X_1, \mathcal{F}_1 \rangle, \ldots, \langle X_m, \mathcal{F}_m \rangle$$

and each $p \in M \cap X$, put the clause

$$p \leftarrow \langle X_1, \overline{\mathcal{F}}_1 \rangle, \ldots, \langle X_m, \overline{\mathcal{F}}_m \rangle$$

into $\mathbf{NSS}_M(P)$. Clearly the resulting program $\mathbf{NSS}_M(P)$ is an SC-Horn program and hence has a least model $M^{\mathbf{NSS}_M(P)}$. M is a stable model of P if M is a model of P and $M = M^{\mathbf{NSS}_M(P)}$. It can be shown that this construction corresponds to the same notion of Gelfond-Lifschitz stable models when we restrict ourselves to ordinary logic programs.

We note that there are other semantics available for SC programs. For example, Son, Pontelli, and Tu [2007], observed that the stable models for an SC programs may be included one in another. That is, consider the following SC program P consisting of four clauses: $a \leftarrow, \quad b \leftarrow, \quad c \leftarrow q$, and $q \leftarrow \langle \{a, b, c\}, \{\{a, b, c\}\} \rangle$. One can easily check that there are two stable models $M_1 = \{a, b, c, q\}$ and $M_2 = \{a, b\}$. Son et al. [2007] define an alternative semantics for SC constraints that does not allow for nested stable models.

We end this section with a few remarks on how set constraint atoms can allow one to reason about infinite sets and preferences.

3.1   Using set constraint atoms to reason about infinite sets.

Cenzer, Remmel, and Marek [2005] suggested a way to use set constraint atoms to reason about infinite sets. The basic idea is as follows. First we allow X to be an infinite recursive set and assume that we have a particular coding scheme for some family of subsets of X. Let $\mathcal{F}$ be a finite family of such codes. We will write C_e for the set with the code e. Then we can write two types of constraints. One constraint $\langle X, \mathcal{F} \rangle^{\subseteq}$ has the meaning that the putative set of integers M satisfies $\langle X, \mathcal{F} \rangle^{\subseteq}$ if and only if $M \cap X \supseteq C_e$ for some $e \in \mathcal{F}$. Similarly, we shall also consider constraints of the form $\langle X, \mathcal{F} \rangle^{=}$ where we say that M satisfies $\langle X, \mathcal{F} \rangle^{=}$ if and only if $M \cap X = C_e$ for some $e \in \mathcal{F}$. Observe that constraints of the form $\langle X, \mathcal{F} \rangle^{\subseteq}$ behave like atoms p in that they are preserved when the set grows while constraints of the form $\langle X, \mathcal{F} \rangle^{=}$ behave like constraints *not* p in that they are not always preserved as the set grows.

Now, it is clear that once we introduce these types of constraint schemes, we can consider various coding schemes for the set of indices. For example, Cenzer et al. [2005] used three such schemes: explicit indices of finite sets, recursive indices of recursive sets and recursively enumerable (r.e.) indices of recursively enumerable (r.e.) sets. They then defined extended set constraint clause C to be a clause of the form

$$\langle X, \mathcal{A} \rangle^{*} \leftarrow \langle Y_1, \mathcal{B}_1 \rangle^{\subseteq}, \ldots, \langle Y_k, \mathcal{B}_k \rangle^{\subseteq}, \langle Z_1, \mathcal{C}_1 \rangle^{=}, \ldots, \langle Z_l, \mathcal{C}_l \rangle^{=},$$

where $*$ is either $=$ or $\subseteq$.

Formally, Cenzer et al. [2005] defined three types of indices indices (i.e. codes) for certain subsets of the natural numbers ω.

1. **Explicit indices of finite sets**. For each finite set $F \subseteq \omega$, we define the explicit index of F as follows. The explicit index of the empty set is 0 and the explicit index of $\{x_1 < \cdots < x_n\}$ is $2^{x_1} + \cdots + 2^{x_n}$. We shall let F_n denote the finite set whose index is n.

2. **Recursive indices of recursive sets**. Let $\phi_0, \phi_1, \ldots$, be an effective list of all partial recursive functions. By a recursive index of a recursive set R, we mean an e such that ϕ_e is the characteristic function of R. If ϕ_e is a total $\{0, 1\}$-valued function, then R_e will denote the set $\{x \in \omega : \phi_e(x) = 1\}$.

3. **R.E. indices of r.e. sets**. By a r.e. index of a r.e. set W, we mean an e such that W equals the domain of ϕ_e, that is, $W_e = \{x \in \omega\} : \phi_e(x) \text{ converges}\}$.

Then for any subset $M \subseteq \omega$, we shall say that *M is a model of* $\langle X, \mathcal{F} \rangle^=$, written $M \models \langle X, \mathcal{F} \rangle^=$, if there exists an $e \in \mathcal{F}$ such that $M \cap X$ equals that set with index e. Similarly, we shall say that *M is a model of* $\langle X, \mathcal{F} \rangle^\subseteq$, written $M \models \langle X, \mathcal{F} \rangle^\subseteq$, if there exists an $e \in \mathcal{F}$ such that $M \cap X$ contains the set with index e.

Based on these three different types of indices, Cenzer et al. [2005] considered three different types of constraints, which in turn, leads to three categories of programs.

(A) **Finite constraints**. Here we assume that we are given an explicit index x of a finite set X and a finite family $\mathcal{F}$ of explicit indices of finite subsets of X. We identify the finite constraints $\langle X, \mathcal{F} \rangle^=$ and $\langle X, \mathcal{F} \rangle^\subseteq$ with their codes, $ind(0, 0, x, n)$ and $ind(0, 1, x, n)$ respectively where $\mathcal{F} = F_n$, that is, the finite set with explicit index n. Here the first coordinate 0 tells us that the constraint is finite, the second coordinate is 0 or 1 depending on whether the constraint is $\langle X, \mathcal{F} \rangle^=$ or $\langle X, \mathcal{F} \rangle^\subseteq$, and the third and fourth coordinates are the codes of X and $\mathcal{F}$ respectively.

(B) **Recursive constraints**. Here we assume that we are given a recursive index x of a recursive set X and a finite family $\mathcal{R}$ of recursive indices of recursive subsets of X. Again we shall identify the recursive constraints $\langle X, \mathcal{R} \rangle^=$ and $\langle X, \mathcal{R} \rangle^\subseteq$ with their codes, $ind(1, 0, x, n)$ and $ind(1, 1, x, n)$ respectively, where $\mathcal{R} = F_n$. Here the first coordinate 1 tells us that the constraint is recursive, the second coordinate is 0 or 1 depending on whether the constraint is $\langle X, \mathcal{R} \rangle^=$ or $\langle X, \mathcal{R} \rangle^\subseteq$, and the third and fourth coordinates are the codes of X and $\mathcal{R}$ respectively.

(C) **R.E. constraints**. Here we are given a r.e. index x of a r.e. set X and a *finite* family $\mathcal{W}$ of r.e. indices of r.e. subsets of X. Again we identify the finite constraints $\langle X, \mathcal{W} \rangle^=$ and $\langle X, \mathcal{W} \rangle^\subseteq$ with their codes, $ind(2, 0, x, n)$ and $ind(2, 1, x, n)$ respectively, where $\mathcal{W} = F_n$. The first coordinate 2 tells us that the constraint is r.e., the second coordinate is 0 or 1 depending on whether the constraint is $\langle X, \mathcal{W} \rangle^=$ or $\langle X, \mathcal{W} \rangle^\subseteq$, and the third and fourth coordinates are the codes of X and $\mathcal{W}$.

An *extended set constraint* (ESC) *clause* is defined to be a clause of the form

$$\langle X, \mathcal{A} \rangle^* \leftarrow \langle Y_1, \mathcal{B}_1 \rangle^\subseteq, \ldots, \langle Y_k, \mathcal{B}_k \rangle^\subseteq, \langle Z_1, \mathcal{C}_1 \rangle^=, \ldots, \langle Z_l, \mathcal{C}_l \rangle^= \qquad (9)$$

where $*$ is either $=$ or $\subseteq$. We shall refer to $\langle X, \mathcal{A} \rangle^*$ as the head of C, written $head(C)$, and $\langle Y_1, \mathcal{B}_1 \rangle^\subseteq, \ldots, \langle Y_k, \mathcal{B}_k \rangle^\subseteq, \langle Z_1, \mathcal{C}_1 \rangle^=, \ldots, \langle Z_l, \mathcal{C}_l \rangle^=$ as the body of C, written $body(C)$. Here either k, or l, or both may be 0. M is said to be a model of C if either M does not model every constraint in $body(C)$ or $M \models head(C)$. *An extended set constraint* (ESC) *program* P is a set of clauses of the form of (1).

A (ESC) *Horn program* P is a set of clauses of the form

$$\langle X, \mathcal{A} \rangle^\subseteq \leftarrow \langle Y_1, \mathcal{B}_1 \rangle^\subseteq, \ldots, \langle Y_k, \mathcal{B}_k \rangle^\subseteq. \qquad (10)$$

where $\mathcal{A}$ is a singleton, that is $\mathcal{A}$ consists of a single index. We define the *one-step provability operator*, $T_P : 2^\omega \to 2^\omega$, so that for any $S \subseteq \omega$, $T_P(S)$ is the union of the set of all D_e such that there exists a clause $C \in P$ such that $S \models body(C)$, $head(C) = \langle X, \mathcal{A} \rangle^\subseteq$ and $A = \{e\}$ where $D_e = F_e$ if $head(C)$ is a finite constraint, $D_e = R_e$ if $head(C)$ is a recursive constraint, and D_e is W_e if $head(C)$ is an r.e. constraint. It is easy to see that T_P is a monotone operator and hence there is a least fixpoint of T_P which we denote by N^P. Moreover it is easy to check that N^P is a model of P.

If P is an ESC Horn program in which the body of every clause consists of *finite* constraints, then one can easily prove that the least fixpoint of T_P is reached in ω-steps, that is, $N^P = T_P^\omega(\emptyset)$. However, if we allow clauses whose bodies contain either recursive or r.e. constraints, then we can no longer guarantee that we reach the least fixpoint of T_P in ω steps. Here is an example.

Example 2 Let e_n be the explicit index of the set $\{n\}$ for all $n \geq 0$, let w be a recursive index of ω and f be a recursive index of the set of even numbers E. Consider the following program.

$$\langle\{0\}, \{e_0\}\rangle^\subseteq \quad \leftarrow$$
$$\langle\{2x+2\}, \{e_{2x+2}\}\rangle^\subseteq \quad \leftarrow \quad \langle\{2x\}, \{e_{2x}\}\rangle^\subseteq \text{ (for every number } x)$$
$$\langle\omega, \{w\}\rangle^\subseteq \quad \leftarrow \quad \langle E, \{f\}\rangle^\subseteq$$

Clearly ω is the least model of P but it takes $\omega + 1$ steps to reach the fixpoint. That is, it is easy to check that $T_P^\omega = E$ and that $T_P^{\omega+1} = \omega$. □

Once we have a notion of ESC Horn program, we are in a position to define the analogue of stable models for ESC programs.

Definition 1 *Suppose that M is a model of an ESB program P.*
1. *We define the analogue of the NSS-transform by saying that $\mathbf{NSS}_M(C)$, where $C \in P$ is a clause of the form (1), is nil if M does not satisfy the body of C. If M does satisfy the body of C, then since M is model of P, it must also be a model of the head of C, $\langle X, \mathcal{A}\rangle^*$ where $*$ is either $=$ or $\subseteq$. If $* =\subseteq$, there must be an explicit (recursive, r.e.) index in $\mathcal{A}$, of such that either $M \cap X$ contains the set with index e and for each such e, we add the clause*

$$\langle X, \{e\}\rangle^\subseteq \leftarrow \langle Y_1, \mathcal{B}_1\rangle^\subseteq, \ldots, \langle Y_k, \mathcal{B}_k\rangle^\subseteq, \langle Z_1, \mathcal{C}_1\rangle^\subseteq, \ldots, \langle Z_l, \mathcal{C}_l\rangle^\subseteq. \quad (11)$$

Similarly, if $*$ is $=$, there must be an index e such that $M \cap X$ is the set coded by e and again for each such e, we add the clause

$$\langle X, \{e\}\rangle^\subseteq \leftarrow \langle Y_1, \mathcal{B}_1\rangle^\subseteq, \ldots, \langle Y_k, \mathcal{B}_k\rangle^\subseteq, \langle Z_1, \mathcal{C}_1\rangle^\subseteq, \ldots, \langle Z_l, \mathcal{C}_l\rangle^\subseteq. \quad (12)$$

Then $\mathbf{NSS}_M(P) = \{\mathbf{NSS}_M(C) : C \in P\}$ will be an ESB Horn program.
2. *We then say that M is a stable model of P if M is a model of P and M equals the least model of $\mathbf{NSS}_M(P)$.*

Cenzer et al. [2005] explored the complexity of the least models of recursive ESC Horn programs and, more generally, of ESC programs restricted to the three categories of constraints considered above.

3.2   Using set constraint atoms to reason about preferences

In this subsection, we briefly describe how we can use set constraint atoms to describe preferences based on ideas in a forthcoming paper by Brik and Remmel [2011b]. The basic idea is to consider triples of the form $\langle X, \mathcal{F}, wt\rangle$ or $\langle X, \mathcal{F}, \preccurlyeq\rangle$ where
1. X is a finite set of atoms,

2. $\mathcal{F} \subseteq 2^X$,
3. $wt : \mathcal{F} \to [0, \infty) \subseteq \mathbf{R}$,
4. $\preccurlyeq$ is a partial order on $\mathcal{F}$.

We call triples of the form $\langle X, \mathcal{F}, wt \rangle$ *weight preference set constraint atoms* and triples of the form $\langle X, \mathcal{F}, \preccurlyeq \rangle$ *partially ordered preference set constraint atoms*. We say that a set of atoms M is satisfies $\langle X, \mathcal{F}, wt \rangle$ or $\langle X, \mathcal{F}, \preccurlyeq \rangle$ if and only if M satisfies $\langle X, \mathcal{F} \rangle$.

Now suppose that we have an SC program P which in addition has a finite set of clauses T of the form

$$\langle X_i, \mathcal{F}_i, wt_i \rangle \leftarrow$$

$i \in \{1, \ldots, n\}$. Now suppose that M is a stable model of $P \cup T$. Then we can define the weight of the model M as

$$W(M) = \sum_{i=1}^{n} wt_i(X_i \cap M).$$

As described in the introduction, we can use the weight functions to describe our preferences for what we want $M \cap X_i$ to be by saying that for $F_1, F_2 \in \mathcal{F}_i$, F_1 is preferred over F_2 if $wt_i(F_1) < wt_i(F_2)$. Then we say that a stable model M_1 of $P \cup T$ is preferred over the stable model M_2 of $P \cup T$ if $W(M_1) < W(M_2)$. Thus the introduction of weight preference set constraint atoms can lead to a natural weighting of stable models which can be used to model preferences.

Similarly, suppose that we have an SC program P which in addition has a finite set of clauses T of the form

$$\langle X_i, \mathcal{F}_i, \preccurlyeq_i \rangle \leftarrow$$

for $i \in \{1, \ldots, n\}$. Now suppose that we are given two stable models M_1 and M_2 of $P \cup T$. Then we say that $M_1 \preccurlyeq M_2$ if and only if $M_1 \cap X_i \preccurlyeq_i M_2 \cap X_i$ for $i = 1, \ldots, n$. Thus the introduction of partial order preference set constraint atoms can lead to a natural partial order on stable models which can be used to model preferences.

4 Set-Based Logic Programming

We start this section with a review the basic definitions of set-based logic programming as introduced by Blair et al. [2008]. The syntax of set-based logic programs will essentially be the syntax of DATALOG programs with negation. We will then briefly discuss some results of Marek and Remmel [2009] on conditions which ensure that we can effectively process set-based logic programs.

Following Blair et al. [2008], we define a **set-based augmented first-order language (set-based language**, for short) $\mathcal{L}$ as a triple $(L, X, \llbracket \cdot \rrbracket)$, where
(1) L is a language for first-order predicate logic (without function symbols other than constants),
(2) X is a nonempty (possibly infinite) set, called the **interpretation space**, and
(3) $\llbracket \cdot \rrbracket$ is a mapping from the ground atoms of L to the power set of X, called the *sense assignment*. If p is an atom, then $\llbracket p \rrbracket$ is called the *sense* of p.

Intuitively, one can treat the set of atoms A of $\mathcal{L}$ as a set of descriptions or codes of subsets of X. For example, if $X = \Sigma^*$ where Σ is a finite alphabet, then a description might be a regular expression for a language $A \subseteq X$ or a deterministic finite automaton (DFA) that accepts L. If $X = \mathbf{R}^n$, then a convex polygon of X can be described by the finite set of extreme points of X. We shall see later that the properties we need to effectively process set-based logic programs is that our set of atoms or descriptions come with algorithms which allow us to decide things like whether for any given atoms A and B, $[\![A]\!] \subseteq [\![B]\!]$ or $[\![A]\!] \cap [\![B]\!] = \emptyset$ holds and how to find an atom or code C such that $[\![C]\!] = [\![A]\!] \cup [\![B]\!]$.

For the rest of this section, we shall fix a set X and a first order language $\mathcal{L}$ with no function symbols except constants. We let HB_L denote the Herbrand base of L, i.e. the set of atoms of L. We omit the subscript L when the context is clear. We let 2^X be the power set of X. Given $[\![\cdot]\!] : \mathrm{HB}_L \longrightarrow 2^X$, an *interpretation* I of the set-based language $\mathcal{L} = (L, X, [\![\cdot]\!])$ is a subset of X.

A set-based logic programming clause is a clause of the form

$$\mathcal{C} = A \leftarrow B_1, \ldots, B_n, not\, C_1, \ldots, not\, C_m. \tag{13}$$

where A, B_i, and C_j are atoms for $i = 1, \ldots, n$ and $j = 1, \ldots, m$. We let $head(\mathcal{C}) = A$, $Body(\mathcal{C}) = B_1, \ldots, B_n, not\, C_1, \ldots, not\, C_m$, and $posBody(\mathcal{C}) = \{B_1, \ldots, B_m\}$, and $negBody(\mathcal{C}) = \{C_1, \ldots, C_m\}$. A set-based program is a set of clauses of the form (13) and a set-based Horn program is a set of clauses of the form (13) which contain no occurrences of *not*.

A second component of a set-based logic program is one or more monotonic idempotent operators $O : 2^X \rightarrow 2^X$ that are associated with the program. Recall that an operator $O : 2^X \rightarrow 2^X$ is *monotonic* if for all $Y \subseteq Z \subseteq X$, we have $O(Y) \subseteq O(Z)$ and is *idempotent* if for all $Y \subseteq X$, $O(O(Y)) = O(Y)$. We call a monotonic idempotent operator a *miop*. We say that a set Y is *closed* with respect to miop O if and only if $Y = O(Y)$.

For example, suppose that the interpretation space X is either $\mathbf{R}^n$ or $\mathbf{Q}^n$ where $\mathbf{R}$ is the reals and $\mathbf{Q}$ is the rationals. Then, X is a topological vector space under the usual topology so that we have a number of natural miop operators:

1. $op_{id}(A) = A$, i.e. the identity map is simplest miop operator,
2. $op_c(A) = \bar{A}$ where $\bar{A}$ is the smallest closed set containing A,
3. $op_{int}(A) = int(A)$ where $int(A)$ is the largest open set included in A,
4. $op_{convex}(A) = K(A)$ where $K(A)$ is the convex closure of A, i.e. the smallest set $K \subseteq X$ such that $A \subseteq K$ and whenever $x_1, \ldots, x_n \in K$ and $\alpha_1, \ldots, \alpha_n$ are elements of the underlying field ($\mathbf{R}$ or $\mathbf{Q}$) such that $\sum_{i=1}^{n} \alpha_i = 1$, then $\sum_{i=1}^{n} \alpha_i x_i$ is in K, and
5. $op_{subsp}(A) = (A)^*$ where $(A)^*$ is the subspace of X generated by A.

We should note that (5) is a prototypical example if we start with an *algebraic* structure. That is, in such cases, we can let $op_{substr}(A) = (A)^*$ where $(A)^*$ is the substructure of X generated by A. Examples of such miops include the following:

(a) if X is a group, we can let $op_{subgrp}(A) = (A)^*$ where $(A)^*$ is the subgroup of X generated by A,

(b) if X is a ring, we can let $op_{subrg}(A) = (A)^*$ where $(A)^*$ is the subring of X generated by A,

(c) if X is a field, we can let $op_{subfld}(A) = (A)^*$ where $(A)^*$ is the subfield of X generated by A,

(d) if X is a Boolean algebra, we can let $op_{subalg}(A) = (A)^*$ where $(A)^*$ is the subalgebra of X generated by A or we can let $op_{ideal}(A) = Id(A)$ where $Id(A)$ is the ideal of X generated by A, and

(e) if $(X, \leq_X)$ is a partially ordered set, we can let $op_{uideal}(A) = Uid(A)$ where $Uid(A)$ is the upper order ideal of X, that is, the least subset S of X containing A such that whenever $x \in S$ and $x \leq_X y$, then $y \in S$.

For simplicity, for the rest of this section, we shall assume that all our miops O have the additional property that $Y \subseteq O(Y)$ for all $Y \in 2^X$. Now suppose that we are given a miop $op^+ : 2^X \to 2^X$ and Horn set-based logic program P over X. Blair et al. [2008] generalized the one-step provability operator to set-based logic programs relative to a miop operator op^+ as follows. First, for any atom A and $I \subseteq X$, we say that $I \models_{[\cdot], op^+} A$ if and only if $op^+([\![A]\!]) \subseteq I$. Then, given a set-based logic program P, let P' be the set of ground instances of a clauses in P and let

$$T_{P,op^+}(I) = op^+(I_1)$$

where $I_1 = \bigcup\{[\![A]\!] : A \leftarrow A_1, \ldots, A_n \in P' \text{ \& } I \models_{[\cdot], op^+} A_i, i = 1, \ldots, n\}$. We then say that a *supported model relative to* op^+ of P is a fixpoint of T_{P,op^+}.

We iterate T_{P,op^+} according to the following.

$$
\begin{aligned}
T^0_{P,op^+}(I) &= I \\
T^{\alpha+1}_{P,op^+}(I) &= T_{P,op^+}(T^\alpha_{P,op^+}(I)) \\
T^\lambda_{P,op^+}(I) &= op^+(\bigcup_{\alpha < \lambda} \{T^\alpha_{P,op^+}(I)\}), \lambda \text{ limit}
\end{aligned}
$$

It is easy to see that if P is a set-based Horn program and op^+ is a miop, then T_{P,op^+} is monotonic. Blair et al. [2008] proved the following.

Theorem 3 *Given a miop op^+, the least model of a Horn set-based logic program P exists and is closed under op^+, is supported relative op^+, and is given by $\mathbf{T}^\alpha_{P,op^+}(\emptyset)$ for the least ordinal α at which a fixpoint is obtained.*

We note, however, that if the underlying universe X universe of a set-based logic program is infinite, then, unlike the situation with ordinary Horn programs, T_{P,op^+} will not in general be continuous even in the case where $op^+(A) = A$ for all $A \subseteq X$. That is, consider the following example which was given by Blair et al. [2008].

Example 3 Assume that op^+ is the identity operator on 2^X. Let $\mathcal{L} = (L, X, [\![\cdot]\!])$ where L has four unary predicate symbols: p, q, r and s, and countably many constants $e_0, e_1, \ldots, $. X is the set $\omega \cup \{\omega\}$. $[\![\cdot]\!]$ is specified by $[\![q(e_n)]\!] = \{0, \ldots, n\}$, $[\![p(e_n)]\!] = \{0, \ldots, n+1\}$, $[\![r(e_n)]\!] = \omega$, and $[\![s(e_n)]\!] = \{\omega\}$. The set-based program P consists of the following three clauses:
$q(e_0) \leftarrow$
$p(X) \leftarrow q(X)$ and
$s(e_0) \leftarrow r(e_0)$.
It is then easy to see that after ω iterations of the T_P operator starting from the empty set, $r(e_0)$ becomes satisfied. One more iteration is required to reach an interpretation that satisfies $s(e_0)$ which is the least fixpoint of T_P. $\square$

Next, we consider how we should deal with negation in the setting of miop operators. Suppose that we have a miop operator op^- on the space X. We do not require that op^- is the same as the miop op^+, but it may be. Our goal is to define two different satisfaction relations for negative literals relative to the miop operator op^- which are called strong and weak negation by Blair et al. [2008] [2].

Definition 2 Suppose that P is a set-based logic program over X and op^+ and op^- are miops on X and $a \in \{s, w\}$.

(I) Given any atom A and set $J \subseteq X$, we say

$\qquad J \models^a_{[\![\cdot]\!], op^+, op^-} A$ if and only if $op^+([\![A]\!]) \subseteq J$.

$(II)_s$ (Strong negation) Given any atom A and set $J \subseteq X$, we say

$\qquad J \models^s_{[\![\cdot]\!], op^+, op^-}$ *not* A if and only if $op^-([\![A]\!]) \cap J \subseteq op^-(\emptyset)$.

$(II)_w$ (Weak negation) Given any atom A and set $J \subseteq X$, we say

$\qquad J \models^w_{[\![\cdot]\!], op^+, op^-}$ *not* A if and only if $op^-([\![A]\!]) \not\subseteq J$.

This given, we can naturally define two analogues of the Gelfond-Lifschitz transform and two analogues of stable models depending on whether we want to use strong or weak negation to definition the satisfaction of *not A*.

Definition 3 Given a set $J \subseteq X$, we define the *strong Gelfond-Lifschitz transform*, $GL^s_{J, [\![\cdot]\!], op^+, op^-}(P)$, of a program P with respect to miops op^+ and op^- on 2^X, in two steps. First, we consider all clauses in P,

$$C = A \leftarrow B_1, \ldots, B_n, not\ C_1, \ldots, not\ C_m \qquad (14)$$

where $A, B_1, \ldots, B_n, C_1, \ldots, C_m$ are atoms. If for some i, it is *not* the case that $J \models^s_{[\![\cdot]\!], op^+, op^-}$ *not* C_i, then we eliminate clause C. Otherwise we replace C by the Horn clause

$$A \leftarrow B_1, \ldots, B_n. \qquad (15)$$

Then, $GL^s_{J, [\![\cdot]\!], op^+, \mathcal{R}}(P)$ consists of the set of all Horn clauses produced by this two step process.

We define the *weak Gelfond-Lifschitz transform*, $GL^w_{J, [\![\cdot]\!], op^+, op^-}(P)$, of a program P with respect to miops op^+ and op^- on 2^X in a similar manner except that we use $\models^w_{[\![\cdot]\!], op^+, op^-}$ in place of $\models^s_{[\![\cdot]\!], op^+, op^-}$ in the definition.

Notice that since $GL^a_{J, [\![\cdot]\!], op^+, op^-}(P)$ is a Horn set-based logic program for either $a = s$ or $a = w$, the least model of $GL^a_{J, [\![\cdot]\!], op^+, op^-}(P)$ relative to op^+ is defined. We then define the a-stable model semantics for a set-based logic program P over X relative to the miops op^+ and op^- on X for $a \in \{s, w\}$ as follows.

Definition 4 J is an a-*stable* model of P *relative* to op^+ and op^- if and only if J is the least fixpoint of $T_{GL^a_{J, [\![\cdot]\!], op^+, op^-}(P), op^+}$.

---

[2]Lifschitz [1994] observed that different modalities, thus different operators, can be used to evaluate positive and negative part of bodies of clauses of normal programs.

Next we give a simple example to show that there is a difference between s-stable and w-stable models.

Example 4 Suppose that the space $X = \mathbf{R}^2$ is the real plane. Our program will have two atoms $\{a, b\}, \{c, d\}$ where a, b, c and d are reals. We let $[a, b]$ and $[c, d]$ denote the line segments connecting a to b and c to d respectively. We let the sense of the these atoms be the corresponding subsets, i.e. we let $[\![\{a, b\}]\!] = \{a, b\}$ and $[\![\{c, d\}]\!] = \{c, d\}$. We let $op^+ = op^- = op_{convex}$. Then consider the following program $\mathcal{P}$.

(1) $\{a, b\} \leftarrow not \{c, d\}$
(2) $\{c, d\} \leftarrow not \{a, b\}$

There are four possible candidates for stable models in this case, namely (i) $\emptyset$, (ii) $[a, b]$, (iii) $[c, d]$, and (iv) $op_{convex}\{a, b, c, d\}$. Let us recall that $op_{convex}(X)$ is the convex closure of X which, depending on a, b, c, and d may be either a quadrilateral, triangle, or a line segment.

If we are considering s-stable models where $J \models^s_{[\![\cdot]\!], op^+, op^-} not\ C$ if and only if $op^-(C) \cap J = op^-(\emptyset) = \emptyset$, then the only case where there are s-stable models if $[a, b]$ and $[c, d]$ are disjoint in which (ii) case and (iii) are s-stable models.

If we are considering w-stable models where $J \models^w_{[\![\cdot]\!], op^+, op^-} not\ C$ if and only if $op^-(C) \not\subseteq J$, then there are no w-stable models if $[a, b] = [c, d]$, (ii) is a w-stable model if $[a, b] \not\subseteq [c, d]$, (iii) is w-stable model if $[c, d] \not\subseteq [a, b]$ and (ii) and (iii) are w-stable models if neither $[a, b] \subseteq [c, d]$ nor $[c, d] \subseteq [a, b]$. □

It is still the case that the a-stable models of a set-based logic program P form an antichain for $a \in \{s, w\}$. That is, Blair et al. [2008] proved the following result.

Theorem 4 *Suppose that P is a set-based logic program over X, op^+ and op^- are miops on X, and $a \in \{s, w\}$. If M and N are a-stable models of P and $M \subseteq N$, then $M = N$.*

We end this section by considering the question of what conditions are required if one is to effectively process a finite set-based logic program where the sense of the underlying atoms are allowed be infinite sets. This question was considered by Marek and Remmel [2009, 2011]. The idea was to start with a finite set-based logic program P and let $\mathcal{S}_P$ denote set of fixpoints over all finite unions of sets represented by the atoms of a finite set-based logic program P of the miops associated with P. Here the elements of $\mathcal{S}_P$ may be finite or infinite. Marek and Remmel [2011] showed that if there is a way of associating codes $c(A)$ to the elements of $A \in \mathcal{S}_P$ such that there are effective procedures which, given codes $c(A)$ and $c(B)$ for elements of $A, B \in \mathcal{S}_P$, will

(i) decide if $A \subseteq B$,
(ii) decide if $A \cap B = \emptyset$, and
(iii) produce of the codes of closures of $A \cup B$ and $A \cap B$ under miop operators associated with P,

then we can effectively decide whether a code $c(A)$ is the code of a weak or strong stable model of P.

There are several examples where conditions (i), (ii), and (iii) can be realized.
(1) Let $X = \omega$ and assume that the atoms are codes for finite sets. As above we let the

code of the finite set $\{x_1, \ldots, x_n\}$ be $\sum_i 2^{x_i}$ and the code of the empty set be 0. If the miops are just the identity operators, then clearly conditions (i)-(iii) are satisfied. Thus the scheme proposed above can be realized for programs using such codes.

(2) Another example consists of the finite dimensional subspaces of the space $\mathbf{Q}^n$. Such subspace can be coded by any of its bases. The miop in this case is the subspace generated by a given set of vectors. Clearly, given two bases B_1 and B_2 for subspaces S_1 and S_2 of $\mathbf{Q}^n$, respectively, we can generate effectively from B_1 and B_2 a basis for the least space containing the union of S_1 and S_2. We can test if one space is included in another, and see if there is any vector different from the 0-vector in their intersection. Again we can use set-based logic programs to effectively reason about such subspaces relative to either weak or strong stable models.

(3) The third example is one where one naturally wants to use non-trivial miops. Namely, the space is $\mathbf{Q}^2$ and the collection $\mathcal{X}$ consists of convex polygons in $\mathbf{Q}^2$ determined by lines with rational slopes. The codes are sets of the extreme points of polygons. The miop is cl_{convex}. There are effective procedures for the computation of the code of the closure of the union of two polygons, as well as for testing inclusion and disjointness. Thus, we can reason about such polygons, and compute weak and strong stable models for programs with atoms being codes for convex polygons.

(4) The fourth example of a situation where we can reason about infinite sets that are regular languages. Here, the codes are the regular expressions for the language or a DFA which accepts the language. The sense function assigns to the code the regular set it describes.

We shall expand on Example 4 to illustrate how conditions (i), (ii), and (iii) can naturally be satisfied. It is well known that given two deterministic finite automata (DFA) A_1 and A_2, one can effectively decide whether the languages $L(A_1)$ and $L(A_2)$ accepted by A_1 and A_2, respectively, satisfy $A_1 \subseteq A_2$, $A_1 = A_2$, or $A_1 \cap A_2 = \emptyset$. Similarly, one can effectively construct DFA's which accept $L(A_1) \cup L(A_2)$, $L(A_1) \cap L(A_2)$, and $\Sigma^* - L(A_1)$.

We say that a miop $op : 2^{\Sigma^*} \rightarrow 2^{\Sigma^*}$ is *effectively automata-preserving* if for any DFA M whose underlying alphabet of symbols is Σ, we can effectively construct a DFA N whose underlying alphabet of symbols is Σ such that $L(N) = op(L(M))$. For example, suppose that $\Sigma = \{0, 1, \ldots, m\}$. Then, the following are effectively automata-preserving operators.

1. If N is a DFA whose underlying set of symbols is Σ, then we can define $op : 2^{\Sigma^*} \rightarrow 2^{\Sigma^*}$ by setting $op(S) = S \cup L(N)$ for any $S \subseteq \Sigma^*$. Clearly if $S = L(M)$ for some DFA M whose underlying set of symbols is Σ, then $op(L(M)) = L(M \cup N)$ so op is effectively automata-preserving.

2. If N is a DFA whose underlying set of symbols is Σ, then we can define $op : 2^{\Sigma^*} \rightarrow 2^{\Sigma^*}$ by setting $op(S) = S \cap L(N)$ for any $S \subseteq \Sigma^*$. Clearly if $S = L(M)$ for some DFA M whose underlying set of symbols is Σ, then $op(L(M)) = L(M \cap N)$ so op is effectively automata-preserving.

3. If T is any subset of Σ, we can let $op(S) = S(T^*)$. Again op will be an effectively automata-preserving miop since if M is DFA whose underlying set of symbols is Σ, then let N be NFA constructed from M by adding loops on all the accepting states labeled with letters from T. It is easy to see that N accepts $L(M)T^*$ and then one can use the standard construction to find a DFA N' such

that $L(N') = L(N)$. Notice that in the special case where T equals Σ, we can think of op as constructing the upper ideal of S in Σ^* relative to the partial order $\sqsubseteq$ where for any words $u, v \in \Sigma^*$, $u \sqsubseteq v$ if and only if u is prefix of v, i.e. v is of the form uw for some $w \in \Sigma^*$. For any poset $(P, \leq_P)$, we say that a set $U \subseteq P$ is an *upper ideal* in P, if whenever $x \leq_P y$ and $x \in P$, then $y \in P$. Clearly, for the poset $(\Sigma^*, \sqsubseteq)$, $op(S)$ is the upper ideal of $(\Sigma^*, \sqsubseteq)$ generated by S.

4. Let $\mathcal{P} = (\Sigma, \leq)$ be a partially-ordered set. For any $w, w' \in \Sigma^*$, we say that w' is a factor of w if there are words $u, v \in \Sigma^*$ with $w = uw'v$. Define the *generalized factor order* on P^* by letting $u \leq w$ if there is a factor w' of w having the same length as u such that $u \leq w'$, where the comparison of u and w' is done componentwise using the partial order in $\mathcal{P}$. Again we can show that if $op(S)$ is the upper ideal generated by S the generalized factor order relative to P^*, then op is an effectively automata-preserving miop. That is, if we start with a DFA $M = (Q, \Sigma, \delta, s, F)$, then we can modify M to an NFA that accepts $op(L(M))$ as follows. Think of M as a digraph with edges labeled by elements of Σ in the usual manner. First, we add a new start state s_0. There are loops from s_0 labeled with all letters in Σ. There is also a λ-transition from s_0 to the old start state s. We then modify the transitions in M so that if there is an edge from state q to q' labeled with symbol r, then we add an edge from q to q' with any symbol s such that $r \leq s$. Finally we add loops to all accepting states such that labeled with all letters in in Σ.

Marek and Remmel [2011] proved the following theorem.

Theorem 5 *Suppose that P is a finite set-based logic program over $\mathcal{L} = (L, X, [\![\cdot]\!])$ where $X = \Sigma^*$ for some finite alphabet Σ and $op^+ : 2^{\Sigma^*} \to 2^{\Sigma^*}$ and $op^- : 2^{\Sigma^*} \to 2^{\Sigma^*}$ are effectively automata-preserving miops. Moreover, assume that for any atom A which appears in Q, $[\![A]\!]$ is a regular language whose underlying set of symbols is Σ. Then:*

1. *every weak (strong) stable model of P is a regular language and*
2. *for any regular language L whose underlying set of symbols is Σ, we can effectively decide whether L is a weak or strong stable model of P.*

5 Hybrid ASP

In this section, we discuss Hybrid ASP programs and stable models as defined by Brik and Remmel [2011b].

The goal of Hybrid ASP is to allow the user to reason about dynamical systems that exhibit both discrete and continuous aspects. The unique feature of Hybrid ASP is that Hybrid ASP rules can be thought of as general input-output devices. In particular, Hybrid ASP programs allow the user to include ASP type rules that act as controls for when to apply a given algorithm to advance the system to the next position.

Modern computational models and simulations such as the model of dog's heart described by Kerckhoffs et al. [2007] rely on existing PDE solvers and ODE solvers to determine the values of parameters. Such simulations proceed by invoking appropriate algorithms to advance a system to the next state, which is often distanced by a short time interval into the future from the current state. In this way, a simulation of continuously changing parameters is achieved, although the simulation itself is a discrete system. The parameter passing mechanisms and the logic for making decisions regarding what algorithms to invoke and when are part of the ad-hoc control algorithm. Thus the laws of a system are implicit in the ad-hoc control software.

On the other hand, action languages [Gelfond and Lifschitz, 1998] which are also used to model dynamical systems allow the users to describe the laws of a system explicitly. Initially action languages did not allow simulation of the continuously changing parameters, which severely limited applicability of such languages. Recently, Chintabathina [2010] introduced an action language H where he proposed an elegant approach to modeling continuously changing parameters. That is, a program in H describes a state transition diagram of a system where each state models a time interval in which the parameter dynamics is a known function of time. However, the implementation of H discussed by Chintabathina [2010] cannot use PDE solvers nor ODE solvers.

Hybrid ASP is an extension of ASP that allows users to combine the strength of the ad-hoc approaches, i.e. the use of numerical methods to faithfully simulate physical processes, and the expressive power of ASP which provides the ability to elegantly model laws of a system. Hybrid ASP provides mechanisms to express the laws of the modeled system via hybrid ASP rules which can control execution of algorithms relevant for simulation.

We should note that any given dynamical system may have a single trajectory or have multiple trajectories if the system is non-deterministic. For instance, in our James Bond example given in the introduction, our agent may have two possible trajectories which would get him to his desired destination as pictured in Figure 2.

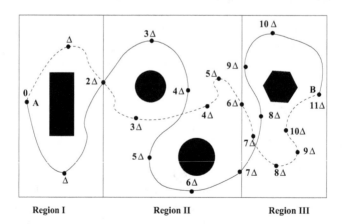

Figure 2: Multiple trajectories.

We shall start out this section by describing a simplified version of Hybrid ASP program which Brik and Remmel called Basic Hybrid ASP (BH-ASP) programs.

5.1 Basic Hybrid ASP

A BH-ASP program P will have an underling parameter space S. For example, in our secret agent example, imagine that we allow James Bond to make decision every Δ seconds where $\Delta > 0$. Then one can think of describing the position and situation at time $k\Delta$ by a sequence of parameters

$$\mathbf{x}(k\Delta) = (x_0(k\Delta), x_1(k\Delta), x_2(k\Delta), \ldots, x_m(k\Delta))$$

that specify both continuous parameters such as time, position, velocity, and acceleration as well as discrete parameters such as is the car configured as an all-terrain vehicle or as a boat. In a BH-ASP program, we shall always think of the parameter x_0 as specifying time and the range of x_0 is $\{k\Delta : k = 0, \ldots, n\}$ for some fixed n or of the form $\{k\Delta : k \in \omega\}$. In particular, for finite BH-ASP programs, we shall assume that the range of x_0 is $\{k\Delta : k = 0, \ldots, n\}$ for some fixed n and $\Delta > 0$. Thus we shall always write an element of S in the form $\mathbf{x} = (k\Delta, x_1(k\Delta), \ldots, x_m(k\Delta))$ for some k. We refer to the elements of S as *generalized positions*. A BH-ASP program will also have an underlying set of atoms At. Then the underlying universe of the program will be $At \times S$.

Suppose that $M \subseteq At \times S$. Then we let $\widehat{M} = \{\mathbf{x} : (\exists a \in At)((a, \mathbf{x}) \in M)\}$. We will say that M satisfies $(a, \mathbf{x}) \in At \times S$, written $M \models (a, \mathbf{x})$, if $(a, \mathbf{x}) \in M$. For any element $(k\Delta, x_1, \ldots, x_m) \in S$, we let $W_M((k\Delta, x_1, \ldots, x_m)) = \{a \in At : (a, (k\Delta, x_1, \ldots, x_m)) \in M\}$ and we shall refer to $W_M(k\Delta, x_1, \ldots, x_m)$ as the *world of M at the generalized position* $(k\Delta, x_1, \ldots, x_m)$. We say that M is a **single trajectory model** if for each $k \in \{0, \ldots, n\}$, there is exactly one generalized position of the form $(k\Delta, x_1, \ldots, x_m)$ in $\widehat{M}$. If M is a single trajectory model, then we let $(k\Delta, x_1(k\Delta), \ldots, x_m(k\Delta))$ be the unique element of the form $(k\Delta, x_1, \ldots, x_m)$ in $\widehat{M}$ and we can write M as a disjoint union

$$M = \bigsqcup_{k=0}^{n} W_M(k\Delta, x_1(k\Delta), \ldots, x_m(k\Delta)) \times \{(k\Delta, x_1(k\Delta), \ldots, x_m(k\Delta))\}.$$

We will say that M is a **multiple trajectory model** if for each $k \in \{0, \ldots, n\}$, there is at least one generalized positions of the form $(k\Delta, x_1, \ldots, x_m)$ in $\widehat{M}$ and for some $0 \leq k_0 \leq n$, there are at least two generalized position of the form $(k\Delta, x_1, \ldots, x_m)$ in $\widehat{M}$. The reason for introducing multiple trajectory models is that we may want to reason about all possible trajectories of our secret agent rather than just reasoning about a single trajectory. If we drop the requirement that for each $k\Delta$, there is a generalized position $(k\Delta, x_1, \ldots, x_m) \in \widehat{M}$ in the definition of single trajectory or multiple trajectory models, we get what we call *partial single trajectory* and *partial multiple trajectory* models.

BH-ASP programs consist of collections of the following two types of clauses.

Stationary clauses are of the form

$$a \leftarrow a_1, \ldots, a_s, not\ b_1, \ldots, not\ b_t : O \tag{16}$$

where $a, a_1, ..., a_n, b_1, ..., b_m \in At$, O is a set of generalized positions in the parameter space S. The idea is that if for a generalized position $\mathbf{p} \in O$, if $(a_i, \mathbf{p})$ holds for $i = 1, ..., s$ and $(b_j, \mathbf{p})$ does not hold for $j = 1, ..., t$, then $(a, \mathbf{p})$ holds. Thus stationary clauses are typical normal logic programming clauses relative to a fixed world $W_M(\mathbf{p})$.

Advancing clauses are of the form

$$a \leftarrow a_1, ..., a_s, not\ b_1, ..., not\ b_t : A, O \qquad (17)$$

where $a, a_1, ..., a_n, b_1, ..., b_m \in At$, O is a set of generalized positions in the parameter space S and A is an algorithm such that for any generalized position $\mathbf{p} \in O$, $A(\mathbf{p})$ is defined and is an element of S. Here A can be any sort of algorithm which might be the result of solving a differential or integral equation, solving a set of linear equations or linear programming equations, running a program or automaton, etc. The idea is that if for a generalized position $\mathbf{p} \in O$, if $(a_i, \mathbf{p})$ holds for $i = 1, ..., s$ and $(b_j, \mathbf{p})$ does not hold for $j = 1, ..., t$, then $(a, A(\mathbf{p}))$ holds. We will require that for all $\mathbf{p} \in O$, $A(\mathbf{p})$ always produces the same output. In a BH-ASP program, we will always assume that if $\mathbf{p} = (k\Delta, x_1, \ldots, x_m)$, then $A(\mathbf{p})$ is of the form $((k+1)\Delta, y_1, \ldots, y_m)$ for some $y_1, \ldots, y_m$. Thus advancing clauses are like input-output devices in that the algorithm A allows certain elements a to hold at the next generalized position.

In both advancing clauses and stationary clauses, we shall refer to the set O as the *constraint set* of the clause. The idea here is that O allows one to use a single clause to specify clauses that can be used at a variety of generalized positions. We shall refer to the algorithm A in advancing clause as the *advancing algorithm* of the clause.

A *BH-ASP Horn program* H is a collection of clauses of the form (16) and (17) such that there are no occurrences of *not* in any of its clauses. A *consistent BH-ASP Horn program* G is a BH-ASP Horn program such that if (A, O) and (A', O') appear in H, then $A \upharpoonright_{O \cap O'} = A' \upharpoonright_{O \cap O'}$, where for an algorithm B and set of positions K, $B \upharpoonright_K$ is the restriction of the algorithm to the domain K.

Next we introduce the one-step provability operator for BH-ASP Horn programs. Let I be an initial generalized position in S, M be a subset of $At \times S$, and P be a basic hybrid Horn program. Then we define $T_{P,I}(M)$ to be the union of M and the set of all atoms $(a, \mathbf{p})$ such that either

1. there exists $C = a \leftarrow a_1, ..., a_n : O \in P$ and $\mathbf{p} \in (\widehat{M} \cup \{I\}) \cap O$ such that $(a_i, \mathbf{p}) \in M$ for $i = 1, ..., n$ or
2. there exists $C = a \leftarrow a_1, ..., a_n : A, O \in P$ and $\mathbf{q} \in (\widehat{M} \cup \{I\}) \cap O$ such that $(a_i, \mathbf{q}) \in M$ for $i = 1, ..., n$ and $\mathbf{p} = A(\mathbf{q})$.

It is easy to see that $T_{P,I}$ is a continuous monotone operator so that the least fixpoint of $T_{P,I}$ is

$$T_{P,I}(\emptyset)^\omega = \bigcup_{n \geq 0} T_{P,I}^n(\emptyset).$$

We shall define two types of stable models: partial multiple trajectory stable models and partial single trajectory stable models. Let P be a BH-ASP program over the parameter set S and sets of atoms At. Suppose that $I = (0, x_1, \ldots, x_m) \in S$ is an initial condition and $M \subseteq At \times S$ is a partial multiple trajectory model. Then the

Gelfond-Lifschitz reduct of P *with respect to* M *and* I is denoted by $P^{M,I}$ and is defined by the following procedure.

1. Eliminate from P all clauses $C = a \leftarrow a_1, ..., a_n, not\ b_1, ..., not\ b_m : A, O$ such that $(\forall \mathbf{p} \in (\widehat{M} \cup \{I\}) \cap O)((\exists i)((b_i, \mathbf{p}) \in M)$ or $A(\mathbf{p}) \notin \widehat{M})$.

2. If a clause C is not eliminated in step (1), then replace it by the clause $a \leftarrow a_1, ..., a_n : A, O'$ where O' equals the set of all $\mathbf{p}$ such that $\mathbf{p} \in (\widehat{M} \cup \{I\}) \cap O$ and $(b_i, \mathbf{p}) \notin M$ for $i = 1, \ldots, m$ and $A(\mathbf{p}) \in \widehat{M}$.

3. Eliminate from P all clauses $C = a \leftarrow a_1, ..., a_n, not\ b_1, ..., not\ b_m : O$ such that $(\forall \mathbf{p} \in (\widehat{M} \cup \{I\}) \cap O)(\exists i)((b_i, \mathbf{p}) \in M)$.

4. If a clause C in (3) is not eliminated, then replace it by the clause

$$a \leftarrow a_1, ..., a_n : O'$$

where $O' = \left\{ \mathbf{p} | \mathbf{p} \in (\widehat{M} \cup \{I\}) \cap O \text{ and } \forall i = 1, ..., m\ ((b_i, \mathbf{p}) \notin M) \right\}$.

Clearly, $P^{M,I}$ is always a BH-ASP Horn program. If M is a partial single trajectory model, then it is easy to see that it must be the case that $P^{M,I}$ is a consistent BH-ASP Horn program. Then we say that M *is a stable model of* P *with initial condition* I if $T_{P^{M,I},I}(\emptyset)^\omega = M$. If, in addition, M is a partial single trajectory model so that $P^{M,I}$ is a consistent hybrid Horn program, then we say that M is a *partial single trajectory stable model of* P *with initial condition* I.

5.2 Hybrid ASP programs.

There are several features which are not available in BH-ASP programs that one would like to have in an ASP system that can reason about dynamic systems exhibiting a mixture of continuous and discrete phenomena. For example, here is a partial list of features that one would like to have.

1. The restriction that in BH-ASP programs, advancing clauses always have conclusions that represent information that occur at a fixed time interval Δ later than the current time is inconvenient. For example, in our James Bond example, the dynamics changes as we go from the mountain to the lake and as we go from the lake to the desert, but the time of such transitions may not be a multiple of Δ.

2. It is often useful to specify that certain invariance properties hold over a set of times or generalized positions so that it may be desirable to have clauses whose hypothesis refer to two or more different generalized positions.

3. In a BH-ASP program, every algorithm is required to produce the values for all the parameters in a generalized position. As a number of parameters grows such a requirement could become a serious drawback. In many cases, it would be more convenient if an algorithm was allowed to specify values of only some of the parameters, letting other parameters be "unspecified" and possibly allow unspecified values to be assigned by algorithms associated with other clauses.

4. There is the issue of how to deal with imprecise computations. That is, if our algorithm is to solve a partial differential equation numerically, we may not be able to get exact answers but only produce an answer that lies distance ϵ from

the exact answer. Similarly, we may want to use randomized algorithms. For this reason, we might want to allow our algorithms to be set-valued rather than specify functions.

To deal with such issues, Brik and Remmel [2011b] introduced an extension of BH-ASP which they called Hybrid ASP (H-ASP for short). As before, we start with a parameter space S consisting of tuples of parameters $\mathbf{p} = (v_t, v_1, ..., v_k)$ and a set of atoms At. If $\mathbf{p} = (v_t, v_1, ..., v_k) \in S$, we will assume that v_t is always the time parameter and we let $t(\mathbf{p})$ denote v_t and $v_i(\mathbf{p})$ denote v_i for $i = 1, \ldots, k$. The universe U of a hybrid ASP program will equal $At \times S$.

Given $M \subseteq At \times S$, $B = a_1, \ldots, a_n, not\ b_1, ..., not\ b_m$, and $\mathbf{p} \in S$, we say that M satisfies B at the generalized position $\mathbf{p}$, written $M \models (B, \mathbf{p})$, if $(a_i, \mathbf{p}) \in M$ for $i = 1, \ldots, n$, and $(b_j, \mathbf{p}) \notin M$ for $j = 1, \ldots, m$. Notice that if B is empty then $M \models (B, \mathbf{p})$ holds. We let $posBody(B) = a_1, \ldots, a_n$ and $negBody(B) = not\ b_1, ..., not\ b_m$. Let $\widehat{M} = \{\mathbf{x} : (\exists a \in At)((a, \mathbf{x}) \in M)\}$.

There are two types of clauses in H-ASP programs.
Extended stationary clauses are of the form

$$a \leftarrow B_1, B_2, \ldots, B_r : H, O \qquad (18)$$

where each B_i is of the form $a_1^{(i)}, \ldots, a_{n_i}^{(i)}, not\ b_1^{(i)}, ..., not\ b_{m_i}^{(i)}$ where $a_1^{(i)}, \ldots a_{n_i}^{(i)}, b_1^{(i)}, ..., b_{m_i}^{(i)}$ are atoms, a is an atom, $O \subseteq S^r$ is such that if $(\mathbf{p}_1, \ldots, \mathbf{p}_r) \in O$, then $t(\mathbf{p}_1) < \cdots < t(\mathbf{p}_r)$, and H is a Boolean valued algorithm. Here and in subsequent clauses, we allow n_i or m_i to be equal to 0 for any given i. Moreover, if $n_i = m_i = 0$, then B_i is empty and we automatically assume that B_i is satisfied by any $M \subseteq At \times S$.

The idea is that if $(\mathbf{p}_1, \ldots, \mathbf{p}_r) \in O$ and for each i, B_i is satisfied at the general position $\mathbf{p}_i$, and $H(\mathbf{p}_1, \ldots, \mathbf{p}_r)$ is true, then $(a, \mathbf{q})$ holds. Thus extended stationary clauses in H-ASP are similar to stationary clauses in BH-ASP except that we allow the clause to refer to generalized positions that occur at multiple times up to and including the time $t(\mathbf{p}_r)$ where we require the pair $(a, \mathbf{p}_r)$ to hold, and we allow a user to specify an additional constraint on the tuples of positions via an algorithm H. For example, H could involve such non-logical conditions as that the generalized position satisfies some system of linear equations or that there exist clauses in the program that could be used to advance position $\mathbf{p}_i$ to position $\mathbf{p}_{i+1}$ for all $i = 1, \ldots, r-1$. We shall refer to O as the *constraint set* of the clause and the algorithm H as the *Boolean algorithm* of the clause.
Extended advancing clauses are of the form

$$a \leftarrow B_1, B_2, \ldots, B_r : A, O \qquad (19)$$

where A is an algorithm and each B_i is of the form $a_1^{(i)}, \ldots, a_{n_i}^{(i)}, not\ b_1^{(i)}, ..., not\ b_{m_i}^{(i)}$ where $a_1^{(i)}, \ldots a_{n_i}^{(i)}, b_1^{(i)}, ..., b_{m_i}^{(i)}$ are atoms, a is atom, and $O \subseteq S^r$ is such that if $(\mathbf{p}_1, \ldots, \mathbf{p}_r) \in O$, then $t(\mathbf{p}_1) < \ldots < t(\mathbf{p}_r)$ and for all $\mathbf{q} \in A(\mathbf{p}_1, \ldots, \mathbf{p}_r)$, $t(\mathbf{q}) > t(\mathbf{p}_r)$.

The idea is that if $(\mathbf{p}_1, \ldots, \mathbf{p}_r) \in O$ and for each i, B_i is satisfied at the general position $\mathbf{p}_i$, then the algorithm A can be applied to $(\mathbf{p}_1, \ldots, \mathbf{p}_r)$ to produce a set of generalized positions O' such that if $\mathbf{q} \in O'$, then $t(\mathbf{q}) > t(\mathbf{p}_r)$ and $(a, \mathbf{q})$ holds. Thus advancing clauses in H-ASP are similar to advancing clauses in BH-ASP except that we allow a clause to refer to generalized positions that occur at multiple times up to and including the time $t(\mathbf{p}_r)$ and our algorithm A is set-valued rather than single valued. As before, we shall refer to O as the *constraint set* of the clause and the algorithm A as the *advancing algorithm* of the clause.

An H-ASP program is a collection of clauses of the form (18) and (19). A *H-ASP Horn program* is a H-ASP program which does not contain any occurrences of *not* . A *consistent H-ASP Horn program* P is an H-ASP program such that if whenever two pairs of an advancing algorithm and a constraint set, (A, O) and (A', O'), appear in P and $O, O' \subseteq S^r$, then $A \restriction_{O \cap O'} = A' \restriction_{O \cap O'}$.

Let P be a H-ASP Horn program and $I \in S$ be an initial condition such that $t(I) = 0$. Then the one-step provability operator $T_{P,I}$ is defined so that given $M \subseteq At \times S$, $T_{P,I}(M)$ consists of M together with the set of all $(a, J) \in At \times S$ such that

1. there exists an extended stationary clause $C = a \leftarrow B_1, B_2, \ldots, B_r : H, O$ and
   $(\mathbf{p}_1, \ldots, \mathbf{p}_r) \in O \cap \left(\widehat{M} \cup \{I\} \right)^r$ such that $H(\mathbf{p}_1, \ldots, \mathbf{p}_r) = 1$ and
   $(a, J) = (a, \mathbf{p}_r)$ and $M \models (B_i, \mathbf{p}_i)$ for $i = 1, \ldots, r$ or
2. there exists an advancing clause $C = a \leftarrow B_1, B_2, \ldots, B_r : A, O$ and
   $(\mathbf{p}_1, \ldots, \mathbf{p}_r) \in O \cap \left(\widehat{M} \cup \{I\} \right)^r$ such that $J \in A(\mathbf{p}_1, \ldots, \mathbf{p}_r)$ and
   $M \models (B_i, \mathbf{p}_i)$ for $i = 1, \ldots, r$.

It is easy to see that for all H-ASP Horn programs P and initial conditions $I \in S$ such that $t(I) = 0$, $T_{P,I}$ is a continuous monotone operator so that the least model of P relative to the initial condition I is given by $T_{P,I}^\omega(\emptyset)$.

We can then define the stable model semantics for general H-ASP programs as follows. Suppose that we are given a hybrid ASP program P, over a set of atoms At and a parameter space S, a set $M \subseteq At \times S$, and an initial condition $I \in S$ such that $t(I) = 0$. Then we form the Gelfond-Lifschitz reduct of P over M and I, $P^{M,I}$ as follows.

1. Eliminate from P all advancing clauses $C = a \leftarrow B_1, \ldots, B_r : A, O$ such that for all $(\mathbf{p}_1, \ldots, \mathbf{p}_r) \in O \cap \left(\widehat{M} \cup \{I\} \right)^r$ there is an i such that $M \not\models (negBody(B_i), \mathbf{p}_i)$ or $A(\mathbf{p}_1, \ldots, \mathbf{p}_r) \cap \widehat{M} = \emptyset$.
2. If the advancing clause $C = a \leftarrow B_1, \ldots, B_r : A, O$ is not eliminated by (1), then replace it by $a \leftarrow B_1^+, \ldots, B_r^+ : A^+, O^+$ where for each i, $B_i^+ = posBody(B_i)$, O^+ is equal to the set of all $(\mathbf{p}_1, \ldots, \mathbf{p}_r)$ in $O \cap \left(\widehat{M} \cup \{I\} \right)^r$ such that $M \models (negBody(B_i), \mathbf{p}_i)$ for $i = 1, \ldots, r$ and $A(\mathbf{p}_1, \ldots, \mathbf{p}_r) \cap \widehat{M} \neq \emptyset$, and A^+ is defined so that the domain of $A+$ is O^+ and $A^+(\mathbf{p}_1, \ldots, \mathbf{p}_r)$ is $A(\mathbf{p}_1, \ldots, \mathbf{p}_r) \cap \widehat{M}$ for all $(\mathbf{p}_1, \ldots, \mathbf{p}_r) \in O^+$.
3. Eliminate from P all extended stationary clauses $C = \leftarrow B_1, \ldots, B_r : H, O$ such that for all $(\mathbf{p}_1, \ldots, \mathbf{p}_r) \in O \cap \left(\widehat{M} \cup \{I\} \right)^r$, either there is an i such that $M \not\models (negBody(B_i), \mathbf{p}_i)$ or $H(\mathbf{p}_1, \ldots, \mathbf{p}_r) = 0$.
4. If the extended stationary clause $C = a \leftarrow B_1, \ldots, B_r : H, O$ is not eliminated by (3), then replace it by $a \leftarrow B_1^+, \ldots, B_r^+ : H, O^+$ where for each i, $B_i^+ = posBody(B_i)$, O^+ is equal to the set of all $(\mathbf{p}_1, \ldots, \mathbf{p}_r)$ in $O \cap \left(\widehat{M} \cup \{I\} \right)^r$

such that $M \models (negBody(B_i), \mathbf{p}_i)$ for $i = 1, \ldots, r$ and $H(\mathbf{p}_1, \ldots, \mathbf{p}_r) = 1$.
Let H^+ be the restriction of H to O^+.
We then say that M is a *general stable model of P with initial condition I* if $T_{PM,I} (\emptyset)^\omega$
$= M$.

We believe that the point of view of thinking of rules as general input-output devices has the potential for many new applications of ASP techniques. Thus we believe that one should view the work on H-ASP programs [Brik and Remmel, 2011b] as a first step for further work that will lead to both theoretical tools used for the modeling and analysis of dynamic systems and for computer applications that simulate dynamical systems. There is considerable work to be done in developing a theory of such programs which would be similar to the theory that has been developed for ASP programs. For example, a careful analysis of the complexity of the stable models of a H-ASP programs as a function to the complexity of the advancing and Boolean algorithms in the program needs to be done. One should explore extended sets of rules that allows for partial parameter passing or allow different rules to instantiate disjoint sets of parameters for the next time step. We need to develop extensions of ASP solvers that can process Hybrid ASP programs. That is, in action languages like H, the goal is to compile an H program into a variant ASP program that can be processed with current variant ASP solvers. The existence of Hybrid ASP solvers would allow us to develop Hybrid ASP type extensions of action languages like H that could be compiled to Hybrid ASP programs which, in turn, would be processed by Hybrid ASP solvers.

6   Conclusions

While there are several declarative formalisms that deal with finite-domain constraint-satisfaction, two of these, ASP and Satisfiability (SAT) [Biere, Heule, van Maaren, and Walsh (eds), 2009] are *logic-based*. However, the motivations of these two technologies are different. ASP is extensively discussed above and is based on generalizations of Horn logic and knowledge representation. On the other hand, SAT has its roots in the theory of computation and has significant applications in electronic design automation. However, at least up until now, the SAT community has not payed much attention to the issue of constraint representation. The tools available for a programmer to prepare the input clausal theory for the solver as well as tools for decoding the results returned by the solver have traditionally been very limited. By contrast, ASP has its roots in knowledge representation as understood by the artificial intelligence community. Thus researchers in ASP have been much more sensitive to the issue of proper representation of constraints and providing a bigger repertoire of tools that could support the programmer. Example of such tools include *grounders* that support the use of variables and pseudo-Boolean constraints. The work of the dlv designers shows that solvers can also be tightly coupled with traditional database systems.

It is only natural to ask whether there are further steps that can be taken to increase the applicability of ASP and its underlying logic and universal algebra mechanisms. As ASP solvers such as *clasp* [Gebser et al., 2007] have recently become fast enough to compete with SAT solvers, it is worth to ask whether the knowledge representation tools available to the ASP programmer can be further extended. Put slightly differently, one should ask if the mechanism of context-dependent reasoning as discovered by Gelfond and Lifschitz can be applied to a richer class of applications that go well beyond finite-domain constraint satisfaction. As shown in a number of our papers cited above, the mechanism of fixpoint computation, based on the abstract form of the Knaster-Tarski fixpoint theorem, can be extended to much richer programming environments by properly interpreting the Gelfond-Lifschitz construction of stable models. Conceptually, this is akin to Satisfiability-Modulo-Theories. This observation has been made in a variety of forms by a number of other authors, especially, by Niemelä [2009] and Mellarkod et al. [2008]. We have presented four such extensions in this paper. However, we should note that all of these extensions are compatible in that one can incorporate all the features of these extensions into a single system. In other words, by choosing appropriate libraries for processing various classes of constraints, one could use an abstract Gelfond-Lifschitz mechanism for stable model computation as a single processing paradigm. A step in this direction was made by Lierler [2008] who proposed such abstract mechanism. The next step is to develop efficient solvers for such extensions so that one can extend the range of applications of ASP systems. This is a highly complex task, but one that we think is worth the effort.

We hope our review of these four extensions of ASP will motivate other researchers in ASP to investigate new extensions of ASP. This is a topic that has interested us over the last 15 years and we feel that there is still much more work to be done on the theory of such extensions, the implementations of such extensions, and the applications of such extensions.

Acknowledgements

The authors acknowledge very helpful suggestions of the anonymous referee, as well as discussions with Alex Brik, Howard Blair, and Mirek Truszczyński.

References

K. R. Apt and H. A. Blair. Arithmetic classification of perfect models of stratified programs. *Fundam. Inform.*, 14(3):339–343, 1991.

K. R. Apt and M. H. van Emden. Contributions to the theory of logic programming. *J. ACM*, 29(3):841–862, 1982.

Y. Babovich and V. Lifschitz. Cmodels package. http://www.cs.utexas.edu/users/tag/cmodels.html, 2002.

C. Baral. *Knowledge Representation, Reasoning and Declarative Problem Solving*. Cambridge University Press, 2003.

M. Bellare and S. Goldwasser. The complexity of decision versus search. *SIAM J. Comput.*, 23 (1):97–119, 1994.

A. Biere, M. Heule, H. van Maaren, and T. Walsh (eds), editors. *Handbook of Satisfiability*. IOS Press, 2009.

H. A. Blair, V. W. Marek, and J. B. Remmel. Spatial logic programming. In *Proceedings of SCI 2001*, 2001.

H. A. Blair, V. W. Marek, and J. B. Remmel. Set based logic programming. *Ann. Math. Artif. Intell.*, 52(1):81–105, 2008.

H. A. Blair, V. W. Marek, and J. S. Schlipf. The expressiveness of locally stratified programs. *Ann. Math. Artif. Intell.*, 15(2):209–229, 1995.

A. Brik and J.B. Remmel. Computing stable models of logic programs using the Metropolis algorithm. Proceedings of Answer Set Programming and Other Computing Paradigms (ASPOCP 2011), 2011.

A. Brik and J.B. Remmel. Expressing preferences in asp using set constraint atoms. In preparation, 2011a.

A. Brik and J.B. Remmel. Hybrid ASP. Technical Communications of the 27th International Conference on Logic Programming, *Leibniz International Proceedings in Informatics*, 2011.

D. A. Cenzer, J. B. Remmel, and V. W. Marek. Logic programming with infinite sets. *Ann. Math. Artif. Intell.*, 44(4):309–339, 2005.

S. Chintabathina. *Towards Answer Set Programming Based Architectures for Intelligent Agents*. PhD thesis, Texas Tech University, 2010.

K. L. Clark. Negation as failure. In H. Gallaire and J. Minker, editors, *Logic and Data Bases*, pages 293–322. Plenum Publishers, New York, 1978.

W. F. Dowling and J. H. Gallier. Linear-time algorithms for testing the satisfiability of propositional horn formulae. *J. Log. Program.*, 1(3):267–284, 1984.

D. East and M. Truszczynski. Predicate-calculus-based logics for modeling and solving search problems. *ACM Trans. Comput. Log.*, 7(1):38–83, 2006.

D. East, M. Iakhiaev, A. Mikitiuk, and M. Truszczynski. Tools for modeling and solving search problems. *AI Commun.*, 19(4):301–312, 2006.

W. Faber, G. Pfeifer, and N. Leone. Semantics and complexity of recursive aggregates in answer set programming. *Artif. Intell.*, 175(1):278–298, 2011.

M. Gebser, B. Kaufmann, A. Neumann, and T. Schaub. *clasp* : A conflict-driven answer set solver. In C. Baral, G. Brewka, and J. S. Schlipf, editors, *Logic Programming and Nonmonotonic Reasoning, 9th International Conference, LPNMR 2007, Proceedings*, volume 4483 of *Lecture Notes in Computer Science*, pages 260–265. Springer, 2007.

M. Gelfond and N. Leone. Logic programming and knowledge representation - the A-prolog perspective. *Artif. Intell.*, 138(1-2), 2002.

M. Gelfond and V. Lifschitz. The stable model semantics for logic programming. In *ICLP/SLP*, pages 1070–1080, 1988.

M. Gelfond and V. Lifschitz. Action languages. *Electron. Trans. Artif. Intell.*, 2:193–210, 1998.

R. Kerckhoffs, M. Neal, Q. Gu, J.B. Bassingthwaighte, J.H. Omens, and A.D. McCulloch. Coupling of a 3d finite element model of cardiac ventricular mechanics to lumped systems models of the systemic and pulmonic circulation. *Ann. Biomed. Eng.*, 35:1–18, 2007.

N. Leone, G. Pfeifer, W. Faber, T. Eiter, G. Gottlob, S. Perri, and F. Scarcello. The dlv system for knowledge representation and reasoning. *ACM Trans. Comput. Log.*, 7(3):499–562, 2006.

L. Libkin. *Elements of Finite Model Theory*. Springer, 2004.

Y. Lierler. Abstract answer set solvers. In M. G. de la Banda and E. Pontelli, editors, *Logic Programming, 24th International Conference, ICLP 2008, Proceedings*, volume 5366 of *Lecture Notes in Computer Science*, pages 377–391. Springer, 2008.

V. Lifschitz. Minimal belief and negation as failure. *Artif. Intell.*, 70(1-2):53–72, 1994.

V. Lifschitz. Action languages, answer sets, and planning. In *The Logic Programming Paradigm: a 25-Year Perspective*, pages 357–373. Springer, 1999.

F. Lin and Y. Zhao. Assat: Computing answer sets of a logic program by sat solvers. In *AAAI/IAAI*, pages 112–118, 2002.

L. Liu and M. Truszczynski. Properties of programs with monotone and convex constraints. In M. M. Veloso and S. Kambhampati, editors, *Proceedings, The Twentieth National Conference on Artificial Intelligence and the Seventeenth Innovative Applications of Artificial Intelligence Conference*, pages 701–706. AAAI Press / MIT Press, 2005.

L. Liu and M. Truszczynski: Pbmodels - Software to Compute Stable Models by Pseudoboolean Solvers. In C. Baral, G. Greco, N. Leone, and G. Terracina eds. *Logic Programming and Nonmonotonic Reasoning, 8th International Conference, Proceedings*, volume 3662 of *Lecture Notes in Computer Science*, Springer, pages 410-415, 2005.

L. Liu, E. Pontelli, T. C. Son, and M. Truszczynski. Logic programs with abstract constraint atoms: The role of computations. In V. Dahl and I. Niemelä, editors, *Logic Programming, 23rd International Conference, ICLP 2007, Proceedings*, volume 4670 of *Lecture Notes in Computer Science*, pages 286–301. Springer, 2007.

V. W. Marek and M. Truszczynski. Autoepistemic logic. *J. ACM*, 38(3):588–619, 1991.

V. W. Marek and M. Truszczynski. Revision programming. *Theor. Comput. Sci.*, 190(2):241–277, 1998.

V. W. Marek and M. Truszczynski. Stable Logic Programming, an alternative logic programming paradigm. In *The Logic Programming Paradigm: a 25-Year Perspective*, pages 375–398. Springer, 1999.

V. W. Marek, A. Nerode, and J. B. Remmel. How complicated is the set of stable models of a recursive logic program? *Ann. Pure Appl. Logic*, 56(1-3), 1992.

V. W. Marek, A. Nerode, and J. B. Remmel. A context for belief revision: Forward chaining - normal nonmonotonic rule systems. *Ann. Pure Appl. Logic*, 67(1-3):269–323, 1994.

V. W. Marek, A. Nerode, and J. B. Remmel. On logical constraints in logic programming. In V. W. Marek and A. Nerode, editors, *Logic Programming and Nonmonotonic Reasoning, Third International Conference, LPNMR'95, Proceedings*, volume 928 of *Lecture Notes in Computer Science*, pages 43–56. Springer, 1995.

V. W. Marek, A. Nerode, and J. B. Remmel. Nonmonotonic rule systems with recursive sets of restraints. *Arch. Math. Log.*, 36(4-5):339–384, 1997.

V. W. Marek and J. B. Remmel. On logic programs with cardinality constraints. In S. Benferhat and E. Giunchiglia, editors, *9th International Workshop on Non-Monotonic Reasoning (NMR 2002), Proceedings*, pages 219–228, 2002.

V. W. Marek and J. B. Remmel. Set constraints in logic programming. In V. Lifschitz and I. Niemelä, editors, *Logic Programming and Nonmonotonic Reasoning, 7th International Conference, LPNMR 2004, Proceedings*, volume 2923 of *Lecture Notes in Computer Science*, pages 167–179. Springer, 2004.

V. W. Marek and J. B. Remmel. Automata and answer set programming. In S. N. Artëmov and A. Nerode, editors, *Logical Foundations of Computer Science, International Symposium, LFCS 2009, Proceedings*, volume 5407 of *Lecture Notes in Computer Science*, pages 323–337, Springer, 2009.

V. W. Marek and J. B. Remmel. Effectively reasoning about infinite sets in answer set programming. In M. Balduccini and T.C. Son, editors, *Logic Programming, Knowledge Representation, and Nonmonotonic Reasoning: Essays in Honor of Michael Gelfond*, volume 6565 of *Lecture Notes in Computer Science*, pages 131–147, Springer, 2011.

V. W. Marek, I. Niemelä, and M. Truszczynski. Logic programs with monotone abstract constraint atoms. *TPLP*, 8(2):167–199, 2008.

V. S. Mellarkod, M. Gelfond, and Y. Zhang. Integrating answer set programming and constraint logic programming. *Ann. Math. Artif. Intell.*, 53(1-4):251–287, 2008.

N. Metropolis, A. W. Rosenbluth, M. N. Rosenbluth, A. H. Teller, and E. Teller. Equation of state calculations by fast computing machines. *J. Chem. Phys*, 21:1087–1092, 1953.

I. Niemelä. Logic programs with stable model semantics as a constraint programming paradigm. *Ann. Math. Artif. Intell.*, 25(3-4):241–273, 1999.

I. Niemelä. Integrating answer set programming and satisfiability modulo theories. In E. Erdem, F. Lin, and T. Schaub, editors, *Logic Programming and Nonmonotonic Reasoning, 10th International Conference, LPNMR 2009, Proceedings*, volume 5753 of *Lecture Notes in Computer Science*, page 3. Springer, 2009.

Ch. Pollet and J. B. Remmel. Non-monotonic reasoning with quantified boolean constraints. In J. Dix, U. Furbach, and A. Nerode, editors, *Logic Programming and Nonmonotonic Reasoning, 4th International Conference, LPNMR'97, Proceedings*, volume 1265 of *Lecture Notes in Computer Science*, pages 18–39. Springer, 1997.

P. Simons, I. Niemelä, and T. Soininen. Extending and implementing the stable model semantics. *Artif. Intell.*, 138(1-2):181–234, 2002.

R.M. Smullyan. *First-order logic.* Springer, 1968.

T. C. Son, E. Pontelli, and P. H. Tu. Answer sets for logic programs with arbitrary abstract constraint atoms. *J. Artif. Intell. Res. (JAIR)*, 29:353–389, 2007.

M. H. van Emden and R. A. Kowalski. The semantics of predicate logic as a programming language. *J. ACM*, 23(4):733–742, 1976.

Reminiscences on the Anniversary of 30 Years of Nonmonotonic Reasoning

Jack Minker
Department of Computer Science and
Institute for Advanced Computer Studies
University of Maryland
College Park, Maryland 20742, USA

Abstract: I review some key early events and individuals whose research led to the start of nonmonotonic reasoning in 1980. Research in three fields of artificial intelligence was significant: deductive databases, logic programming and reasoning with incomplete information. I focus on those events that related to my work. I discuss how I became involved in nonmonotonic reasoning, remember individuals whose work influenced me, some contributions I and my students made and workshops and conferences in which I participated. I finally discuss some observations on the contributions made in nonmonotonic reasoning since the start of the field in 1980.

1 Introduction

The field of nonmonotonic reasoning is considered to have started in 1980, stimulated by three papers that appeared in the Artificial Intelligence Journal. I was fortunate to have participated and contributed to this field since its inception. It is probably true that many contributions had to have been made in any scientific field before its purported starting date. This applies to nonmonotonic reasoning. Having worked in artificial intelligence and databases for over 40 years, in this paper I review some of the developments that occurred before 1980 that influenced future work in this field, including my own contributions.

The field of nonmonotonic reasoning was built on the shoulders of two researchers who were unable to attend the celebration: John McCarthy and the late Raymond Reiter. Their work lives on in their papers and references to them that were published as long as more than 50 years ago in the case of McCarthy and more than 35 years ago in the case of Reiter. It is unusual to have papers referenced 5 years after their publication and yet these giants in our field are the few exceptions. They both influenced my work and undoubtedly the work of many participants at the 30th celebration of nonmonotonic reasoning. It is appropriate that the proceedings of invited papers are dedicated to John McCarthy.

2 Beginning Experiences - Pre Logic Programming

My efforts leading to my work in deductive databases (DDB), logic programming (LP) and nonmonotonic reasoning (NMR) began in 1967 when I joined the faculty at the University of Maryland after working in industry for 17 years.

At the time I joined the Maryland faculty, I had no real direction to my research. In looking around for a topic, it struck me that work I had done at the RCA Corporation on deduction in databases would be an appropriate topic for research. I had used ad-hoc techniques to perform deduction, and the work could hardly have been worthy of being considered a general approach to the problem.

In surveying the literature, I came across the work of McCarthy. I believe that almost anything in AI starts with McCarthy. He named the field and wrote seminal articles. In his 1959 paper on common sense reasoning [McCarthy, 1959], he defined and proposed a solution to a planning problem, which for many years was the "Oldest Problem in AI": At home, your car is in the garage, and you want to get to the airport. How do you make a plan to get there? McCarthy proposed an approach, which was not precisely accurate. Yehoshua Bar-Hillel, a philosopher and mathematician, provided a critique of the approach primarily because the formal commonsense reasoning was oversimplified and that a proper formalization would be more complex (reported in the paper by Bar-Hillel, McCarthy, and Selfridge [1990]). McCarthy recognized the correctness of the critique. It has taken 41 years before an acceptable solution was found using techniques from nonmonotonic reasoning by Lifschitz, McCain, Remolina, and Tacchella [2000]. Lifschitz et al. [2000] also provide an interesting analysis of Bar Hillel's objections to McCarthy's solution.

McCarthy's 1963 paper [McCarthy, 1963a] was another milestone. He defined the situation calculus and the importance to represent actions and causal laws for realistic problems in AI. Cordell Green was the first person to use the situation calculus for planning [Green, 1969b], however his approach did not address the frame problem, defined below. There was the belief in the AI community that a logic-based approach could not solve the frame problem. As noted by Fiora Pirri and Reiter, this critique was premature [Pirri and Reiter, 2000]. Reiter in his 2001 book, *Knowledge in Action*, demonstrates the importance of the situation calculus for robotics and dynamic domains.

McCarthy and Patrick Hayes's paper on philosophical problems and frame axioms was another milestone on the road to nonmonotonic reasoning [McCarthy and Hayes, 1969]. The frame problem was initially formulated as the problem of expressing a dynamical domain in logic without explicitly specifying which conditions are not affected by an action. According to Sandewall[1], the ramification problem, concerned with indirect consequences of an action, was part of the original definition of the frame problem. Lifschitz et al. [2000] have said,

> The discovery of the frame problem and the invention of nonmonotonic formalisms that are capable of solving it may be the most important significant events so far in the history of research in reasoning about actions. A large part of this story is described in Shanahan's book [Shanahan, 1997]

---

[1]Discussion with Erik Sandewall at NonMon@30.

Early work on the frame problem was done by Bertram Raphael [1971], who surveyed and critiqued proposed solutions. Sandewall [1972] proposed a solution based on a default operator, *Unless* that had some difficulties. Additional solutions to the frame problem may be found in the books by Reiter [2001] and Shanahan [1997].

I learned about Alan Robinson's work in theorem proving [Robinson, 1965] after reading Green's paper on planning and his paper with Raphael on a system, called Question Answering 3 [Green and Raphael, 1968] that they developed using Robinson's resolution method for deduction. That 1968 paper was the first work in deductive databases using a logic-based approach, rather than ad-hoc techniques. Robinson's seminal paper was very influential and important for NMR and logic-based AI. Green and Raphael's 1968 paper made a deep impression on me. The work generalized ad-hoc techniques used earlier by Raphael [1968] in his thesis on the SIR system and by other researchers. As Green and Raphael noted,

> SIR used a different subroutine to answer each type of question, and when a new relation was added to the system, not only was a new subroutine required to deal with that relation but also changes throughout the system were usually necessary to handle the interaction of the new relation with the previous relations. This programming difficulty was the basic obstacle in enlarging SIR.

Their paper marks the start of the field of deductive databases. However, it took another 10 years before the field began to be recognized, as discussed in Section 4.1.3.

After learning about the application of the Robinson resolution principle to the field of Question/Answering Systems, I read Robinson's landmark paper and was impressed by its clarity and the deep results with respect to automating deduction in theorem provers. It was clear to me at that time that work that I would do in deductive databases should be based on using resolution. It was also clear that I had to become involved in automated theorem proving.

It was fortuitous that in 1969 the U.S. Army was interested in finding out more about the use of deduction in databases. The Auerbach Corporation, with which I had been associated before I went to the University of Maryland, received a contract in 1969 based on a proposal that I had written to investigate deductive systems in the United States. Among the systems I was anxious to investigate was QA3. In my visit to the Stanford Research Institute (SRI) (now SRI International), I was able to learn details about Green and Raphael's work and co-authored a report on it [Minker and Sable, 1970]. The report also described other approaches to deduction in databases. In addition to work on QA3, Green [1969b], as described above, using the situation calculus, expressed state transformations in logic. This work was incorporated in SHAKEY the robot, developed at SRI. He also introduced the ANSWER predicate which permits answers to be extracted from a theorem prover. It seems to me that Green, in this paper and in his thesis [Green, 1969c], essentially had the idea of using a theorem prover to execute a computer program. However, it is not clear that he realized that a subset of clause form, Horn clauses, were sufficient to write computer programs. The restriction to Horn logic is an important factor in making logic programming competitive with LISP and conventional programming languages.

2.1 *Machine Intelligence* Books

In the 1960s and 1970s, the series of books, *Machine Intelligence*, edited by Michie and subsequently together with Meltzer were important sources of information on all aspects of artificial intelligence, especially with respect to the field of automated theorem proving. Logic programming as a formal discipline is a direct outgrowth of automated theorem proving. Volumes 3 [Michie, 1968], 4 [Meltzer and Michie, 1969], 5 [Meltzer and Michie, 1970], 6 [Meltzer and Michie, 1971] and 7 [Meltzer and Michie, 1972] were especially important to me. As discussed later, logic programming is the major tool used to implement nonmonotonic reasoning theories.

Volumes 3 and 4 contained papers by Meltzer [1968a,b], Robinson [1968, 1970b], David Luckham [1968], Dag Prawitz [1970], Loveland [1969a], Kowalski and Hayes [1968], and Robinson and Larry Wos [1970], all devoted to automated theorem proving and papers by Jared Darlington [1968, 1970] and Green [1969a,b] on deductive information retrieval. In addition to Green and Raphael, Darlington was one of the early researchers to recognize the importance of automated theorem proving for databases. Loveland described his work on Model Elimination (ME); Kowalski and Hayes described their work on the use of semantic trees in theorem proving; and the technique of paramodulation in automated theorem proving was described by Wos.

Volume 5 included, in particular, papers by Robinson [1970a], J.C. Reynolds [1970], Gordon Plotkin [1970], Meltzer [1970], Kowalski and Hayes [1970], Robin Popplestone [1970], and J.R. Allen and Luckham [1970]. Popplestone originated the idea that generalizations and least generalizations of literals existed and would be useful when performing induction. Plotkin introduced the idea of generalization of a clause for use in induction. These are important ideas with respect to machine learning. As stated by Meltzer [1970], induction and deduction are inverses of one another. The paper by Allen and Luckham influenced my ideas to develop a general theorem proving system with many inference systems and search strategies to test their effectiveness on alternative problem sets. Kowalski's paper was also instrumental in this regard. He described search strategies for theorem proving. In particular, he introduced the idea of a proof procedure, defined as:

proof procedure = inference system + search strategy

The concept of a proof procedure is, I believe, the counterpart to Kowalski's definition of a program, namely:

algorithm = logic + control

the fundamental paradigm of logic programming.

Volume 6 included papers by Robinson [1971], Kuehner [1971], Sergei Maslov [1971], Plotkin [1971] and Elcock, Foster, Gray, McGregor, and Murray [1971]. Robinson discussed the unification computation; Kuehner discussed the relation between resolution and Maslov's inverse method; the great Soviet logician, Maslov, discussed his work on proof-strategies for methods of the resolution type; Plotkin expanded on his earlier work on inductive generalization; and Elcock and his co-authors discussed their declarative language ABSET, based on sets (see Section 2.2 for further discussion on Elcock's work).

Volume 7 included papers by Robert Boyer and J Moore [1972], Sandewall [1972], Fikes, Hart, and Nilsson [1972], Patrick Winston [1972], and Harry Barrow and G.F. Crawford [1972]. Boyer and Moore discussed their work on structure sharing in theorem proving, which is also important in logic programming; Kuehner discussed several specialized resolution systems. Work on robotics was discussed by: Fikes et al., on new directions in robot problem solving; Winston on the MIT robot; and Barrow and Crawford on the Mark 1.5 Edinburgh facility.

2.2   1970 NATO Workshop on Artificial Intelligence

In the summer of 1970, an important NATO sponsored conference was held on artificial intelligence. Nick Findler and Meltzer were the organizers. I applied for permission to attend the Workshop and for support. Findler was kind enough to agree to both my attendance and financial support.

The Workshop was held in Menaggio, a lovely resort area on Lake Como, Italy. It was held at the Grand Hotel Victoria, a delightful place, conducive to a relaxed atmosphere and work. I remember waking to the sound of the rooster's crow and watching the nuns play soccer in their sneakers and habits in the garden of the nunnery behind the hotel. There was also an evening get together after dinner, with wine flowing.

The Workshop had a major influence in convincing me to work in the area of applying theorem proving to work in databases. Many key individuals who were known for their work in AI and others who would subsequently be known for their work in that area attended the Workshop. Included among the attendees and participants were Robinson, Meltzer, Michie, Elcock, Sandewall, Findler, Raphael, the late Max Clowes, Jacques Pitrart, Jacques Cohen and others. Most of the AI group at Edinburgh, with the exception of Kowalski, attended the Workshop.

Although there were many excellent lectures at the Workshop, all of which broadened my background in AI, lectures by Robinson, Meltzer, Raphael and Elcock most influenced me. Robinson is a spell-binding speaker. It was a delight to hear his talk on building deduction machines [Robinson, 1971].

The talk by Meltzer [1971] was influenced by Bob Kowalski's thesis [Kowalski, 1970]. I first heard about Kowalski's definition of a proof procedure, discussed above, from Meltzer.

As noted above, Raphael surveyed and critiqued several approaches to solving the frame problem.

The talk by Elcock [1971] was on his work on the declarative languages ABSYS (standing for *Ab*erdeen *Sys*tem) and ABSET (the name because the primitive concepts of the language are founded on the elementary theory of sets). Elcock [1990] argues cogently that he has a claim to having developed the first logic programming language. I recall that Robinson was impressed with Elcock's work, pointed out the significance of programming in a declarative language and referred to the work as the first implementation of a declarative language. The paper by Elcock [1990] describes the major features in ABSYS implemented in 1967: SLD resolution, solving equations, a fair computation rule, negation as failure, delay mechanisms, aggregation operators and constraint solving. Although he incorporated the above in his system, it was not until later that others, not necessarily influenced by Elcock's work, such as Hill who formalized SLD resolution [Hill, 1974], Jean-Louis Lassez and Michael Maher who formalized the fair computation [Lassez and Maher, 1984], and Keith Clark who developed negation-as-failure [Clark, 1978], independently developed results on the techniques implemented in ABSYS. The use of negation-as-failure makes ABSYS the first nonmonotonic declarative language.

3   Logic Programming Beginnings

In 1961, McCarthy wrote a paper, "A Mathematical Theory of Computation", that he expanded and published in a book [McCarthy, 1963b]. The paper "...tried to lay a basis as to how computations are built up from elementary operations and also of how data spaces are built up." He also noted that the formalism differed from that in the theory of computability where the emphasis was on cases of proving statements within the systems. He hoped that the "...the relationship between computation and mathematical logic will be as fruitful in the next century as that between analysis and physics in the last."

The relationship between computation and logic started to evolve faster than he anticipated. In a little known abstract in the 1966 Communications of the ACM, Louis Hodes proposed the idea of using logic as a programming language [Hodes, 1966]. The relationship between computation and deduction was proposed more concretely by Hayes [1973], namely, that computation is controlled deduction. The paper by Hayes was very influential.

The papers by Elcock and by Green were steps towards logic programming. Although the programming language Planner [Hewitt, 1969], developed by Carl Hewitt, was a procedural language, it was also based on logic. Implications could be interpreted as Planner procedures. Planner also incorporated the default rule THNOT that made it nonmonotonic, which was another important step towards nonmonotonic reasoning.

The Prolog language and implementation was started in 1970 in Marseille, France, by Alain Colmerauer and his group (Colmerauer and Roussel [1993] give an account of these events). Kowalski [1988], at the invitation of Colmerauer, visited Marseille for a few days in the summer of 1971. The group which consisted of Bob Pasero, Philippe Roussel, and Colmerauer, was developing a natural language question-answering system. Roussel was interested in using the SL-resolution theorem prover developed by Kowalski and Kuehner [1971] for the deductive component of their question-answering system. His work during this visit led to discussions with Colmerauer about using logic to represent grammar and using resolution to parse sentences. In the summer of 1972, Roussel and Colmerauer designed and implemented the first Prolog system in Algol-W as an adaptation of Roussel's existing SL-resolution theorem prover. In 1972, Colmerauer and his group wrote the first major program in Prolog [Colmerauer, Kanoui, Pasero, and Roussel, 1973].

The semantics for Prolog is due to Kowalski, as his work was used in Prolog. However, at about the same time, both Loveland and Reiter developed inference systems that were essentially the same as Kowalski and Kuehner's SL-resolution. Loveland [1970] and independently, Luckham [1970], developed linear resolution. Loveland also developed model elimination [Loveland, 1969a,b]. The original papers on linear resolution solved n conjunctive subgoals in all $n!$ ways, but SL-resolution and Model Elimination demonstrated how to avoid the problem of dealing with $n!$ subproblems. SL-resolution can be viewed as an amalgamation of linear resolution and model elimination. Independently, Reiter [1971], in his paper on clause ordering, discovered the same ordering restriction on linear resolution.

As Loveland[2] wrote to me,

> ME and SL-resolution were much more powerful than needed for Prolog. What was actually used was the facts that deduction "clauses" could be ordered, were linear (as in linear resolution) and factoring was not needed. That was all shown in ME. (Actually, I recall that SL-resolution had factoring, a point where it differed from the original ME.) Both ME and SL-resolution were complete for first-order logic so had much more horsepower than needed. For Horn clause logic it is very easy to show that, starting with a negative clause (as Prolog does), all the properties you want, including linearity, come easily. So, the roles of ME and SL-resolution are historic 'only'.

Colmerauer and Roussel first used SL-resolution in Prolog and then realized that all that machinery was not needed to handle their language processing needs.

While Kowalski, Kuehner, Loveland and Reiter contributed to the semantics of Horn clause logic programming, it is recognized that Kowalski, Colmerauer and Roussel are the founders of Prolog. Kowalski is one of the founders of Prolog for more than his historic role: he played a key role in its development, its foundations and expansions of use. In addition to his paper [Kowalski, 1974] and his paper with Maarten van Emden [van Emden and Kowalski, 1976], his book, "Logic for Problem Solving" [Kowalski, 1979], was influential for researchers and students. He is one of the principal founders of the field of logic programming.

The incorporation of the *not* operator, the *cut* operator and other constructs makes Prolog more of a logic programming tool than strictly a Horn logic engine. The *not* operator makes it a nonmonotonic logic programming language. The semantics for the *not* operator may be viewed as given by Clark's [1978] work on *negation-as-failure* and by Reiter's [1978a] closed world assumption.

Finally, the development of Edinburgh Prolog in 1977 by Warren, Pereira, and Pereira [1977] demonstrated that the performance of Edinburgh Prolog was competitive with LISP.

4 Early Work at Maryland in Theorem Proving and Deductive Databases

Following the 1970 NATO Workshop, together with my students, we developed a theorem proving system in which we incorporated Kowalski's Σ^* algorithm as the search strategy and many inference systems, including binary resolution and factoring, input

---

[2] E-mail message from Don Loveland to Jack Minker, August 28, 2010

resolution, P1-deduction, SL-resolution and others. Gerry Wilson, one of my Ph.D. students and I experimented with the system on a wide range of problems that Wos and his group had developed and used on their theorem prover. Our paper [Wilson and Minker, 1976] gives the results of that experimental study. Our objective was to try to find which proof procedure worked best on which problems. We were unable to come to general conclusions, but found that if one had a good search strategy, even binary resolution and factoring would work well. When we compared the time to execute a problem in the best proof procedure system of our system against the same problem executed by Wos and his group, their timing results were almost uniformly faster. There were several reasons for this: their program was written in assembly language, and they had paid more careful attention to program efficiency and data structures than had we. This is important when one is implementing a system for general use. Indeed, the effectiveness of Prolog is due to the careful attention that is paid to low level implementation details such as structure sharing and tail recursion. The lack of an occur check in Prolog results in a fast unification algorithm at the expense of completeness. Current work by Torsten Schaub and his colleagues [Gebser, Kaufmann, and Schaub, 2009] on their answer set programming system CLASP, shows the importance of careful implementations to use with large systems.

Wilson and I worked out a method to keep track of proofs when arbitrarily selecting literals in a proof [Wilson and Minker, 1976]. It introduced nesting into the proof process so that one knew where the parents came from. This was used in MRPPS 3.0 system [Minker, 1978].

Guy Zanon, a student from Hervé Gallaire's group in Toulouse, France observed that the way in which nesting was being done allowed one to introduce arbitrary literal selection into non-Horn theorem proving based on linear resolution. We showed our inference system was complete and sound [Minker and Zanon, 1982] and named it LUST resolution (Linear resolution with Unrestricted Selection function based on Trees), influenced by Hill's LUSH resolution for Horn theories [Hill, 1974]. Subsequently, Arcot Rajasekar and I renamed LUST resolution to be SLI resolution (Linear Resolution with Selection Function for Indefinite Clauses), to be consistent with van Emden [1978] who renamed Hill's LUSH resolution to SLD resolution (Linear Resolution with Selection Function Based on Definite Clauses). The paper by van Emden was also one of the early papers on logic databases. According to Thomas Eiter[3], his group was considering the use of SLI for their disjunctive Datalog system, DLV system. Gerald Pfeifer, a student in Eiter's group, implemented SLI resolution for his MS degree [Pfeifer, 1996]. At the recommendation of Nicole Leone they decided upon a bottom up operation for DLV. This was a better choice for disjunctive databases.

4.1   External Influences in the 1970s

Aside from my research, there were several important events for me in the 1970s: my meeting Ray Reiter in 1973; attending IFIPS 1974 in Stockholm, Sweden, where I met Jean-Marie Nicolas and Bob Kowalski; my visits to Toulouse, where I first met Gallaire and together with Nicolas organized the 1976 workshop on logic and

---

[3]Discussion with Thomas Eiter at NonMon@30, October 2010

databases in Toulouse; and reading Loveland's book [Loveland, 1978] on theorem proving.

4.1.1 Meeting Ray Reiter in 1973

In 1973, I vacationed with my family in Canada on my way to IJCAI73, held at Stanford University. We visited Vancouver. I knew of Reiter's work in theorem proving and decided to stop by the University of British Columbia to see if Ray was there. Fortunately he was in his office and we had a productive discussion about our work. We became close friends until the day he died. I learned about his work [Reiter, 1971], which he called "clause-ordered linear resolution strategy".

I first learned about Reiter's work in databases around 1976. He spent his sabbatical leave at Bolt, Beranek and Newman (BBN), from where he sent me a document subsequently published as a BBN technical report "An approach to deductive question-answering" [Reiter, 1977]. I was impressed by the technical report and told him he should publish it as a monograph. However, in his modest way, he said he did not believe it was worthy of a monograph. Thirty-three years after he wrote the report, I still believe I was right; however, I benefited from his modesty as described below.

4.1.2 IFIPS 1974, Stockholm, Sweden

I attended IFIPS 1974 in Stockholm, Sweden, where I first met Nicholas and Kowalski. Both played important roles in my career.

Nicolas spoke about "Question-Answering and Automatic Deduction in SYN-TEX", a deductive database system using a logic-based approach he had developed with Jean-Claude Syre [Nicolas and Syre, 1974]. As I was working in deductive databases, I decided to visit Nicolas when I went on sabbatical leave in 1975 to learn more about his work.

I was introduced to logic programming by listening to Kowalski who spoke "On Predicate Logic as a Programming Language" [Kowalski, 1974]. After his talk, I told Bob my experience with theorem proving led me to be skeptical it would be as efficient as conventional programming languages. Bob said the reason was that logic programming was not a full theorem prover since it dealt only with Horn clauses. After reading his paper, I realized he was correct. As I learned from van Emden, although logic programming is based on Robinson's resolution principle, Alan remained faithful to programming in LISP.

4.1.3 Visit With Gallaire and Nicolas in Toulouse

In 1975 I was on sabbatical leave and spent a week in Toulouse with Gallaire and Nicolas. I had met Nicolas at IFIPS, but I had not met Gallaire. I lectured about my work in Deductive Databases (DDB) and learned more about their work. They asked what I thought about organizing a workshop on deductive databases. I thought their idea was excellent. Although the field of DDBs arose out of the work of Green and Raphael, there was no recognition of the importance of deduction in databases in the database community. We discussed possible participants and Ray was at the top of my list. It was the first workshop organized to focus on the use of logic for databases.

Nicolas did the major work in organizing the workshop with assistance from Gallaire, some of his colleagues in Toulouse and me.

The workshop was important not only for deductive databases, but for the publication of the first two theoretical papers on nonmonotonic reasoning: Reiter's article on the closed world assumption and Clark's paper on completion and negation-as-failure provided the semantics for negation in logic programming; and was an important milestone in nonmonotonic reasoning.

The workshop, held in November 1977, was very productive. It demonstrated for the first time the intimate connection between logic and databases. A paper by Nicolas and Gallaire [1978] discussed theory vs. interpretation; Reiter presented two papers, one on his Closed World Assumption (CWA) [Reiter, 1978a] and the other on deductive question-answering on relational databases [Reiter, 1978c]; Clark presented his completion theory and the concept of negation-as-failure [Clark, 1978]; Kowalski presented his work on logic for data description [Kowalski, 1978]; Nicolas and Kioumars Yazdanian described techniques for integrity checking in deductive databases [Nicolas and Yazdanian, 1978]. In addition, there were several papers that discussed knowledge representation and deduction: Minker discussed an experimental relational database system based on logic, MRPPS3.0 Minker [1978]; Chang discussed a deductive query language, DEDUCE 2 [Chang, 1978]; Charles Kellogg, Philip Klahr and Larry Travis described a prototype system, Deductively Augmented Data Management (DADM) for relational databases [Kellogg et al., 1978]; and Ivan Futo, Ferenec Darvas and Péter Szeredi described the application of Prolog to the development of Question-Answering and Data Management Systems [Futo et al., 1978]. It was not until the book, *Logic and Databases* [Gallaire and Minker, 1978], that I edited with Gallaire, where these papers appeared, that this field came to be recognized as important. Following the first work by Green and Raphael, I believe the book represents the start of the field of deductive databases. Although Nicolas chose not to be one of the co-authors, he was the major organizer of the workshop that led to the book, and is an important contributor to the resurgence of the field of DDBs.

4.2   First Formal Results on Nonmonotonic Reasoning

The Toulouse workshop and the book were also important to nonmonotonic reasoning, due to the papers by Reiter [1978a] and Clark [1978]. Reiter's was on the *closed world assumption* (CWA), and Clark's was on *negation-as-failure*. Following my visit to Toulouse, I spent a few days visiting Clark and Kowalski at Imperial College in London. Keith discussed his completion theory and negation-as-failure. I was impressed with his work and pleased that he would be at our workshop in Toulouse.

Reiter's paper, "On Closed World Databases", discussed the evaluation of questions under two different assumptions referred to as open and closed world assumptions. The open world assumption corresponds to the first order approach to query evaluation: Given a database DB and a query Q, the only answers to Q are those obtained from proofs of Q, given DB as hypotheses. Under the Reiter CWA, certain answers are admitted as result of failure to find a proof. The default rule used is that if no proof of a ground literal exists, then the negation of that literal is assumed *true*. The database assumed is a Horn database under the *unique names axiom*, which does not permit aliases of names, and the *domain closure axiom*, which requires that the only constants that apply are those specifically listed in DB. Reiter showed that the CWA evaluation of an arbitrary query may be reduced to open world evaluation of atomic queries. He also showed that the CWA can lead to inconsistencies, but no such inconsistencies apply to Horn DBs. He further proved that for Horn DBs under the CWA purely negative clauses were not needed for deductive retrieval and function instead as integrity constraints. However, as shown by Sharma Chakravarthy, John Grant and Minker, integrity constraints can be valuable for retrieval [Chakravarthy, Grant, and Minker, 1986].

Clark's paper, "Negation as Failure", discussed Horn theories augmented with a special rule, *negation-as-failure* whereby $\neg P$ can be inferred if every possible proof of *P fails*. He showed that the negation-as-failure rule allows one to conclude negated facts that could be inferred from the axioms of the completed data base, a data base of relational definitions and equality schemas that are assumed to be implicitly given by the data base clauses. Clark's completion of a database, $comp(R)$, is defined as follows. Let

$$R(t_1, \ldots, t_n) \leftarrow A_1, \ldots, A_m$$

be a database clause about R. If $=$ is the equality relation and $x_1, \ldots, x_n$ are variables not appearing in the clause, it is equivalent to the clause

$$R(x_1, \ldots, x_n) \leftarrow x_1 = t_1, x_2 = t_2, \ldots, x_n = t_n, A_1, \ldots, A_m$$

If $y_1, \ldots, y_p$ are the variables in the original rule for relation R, this is equivalent to

$$R(x_1, \ldots, x_n) \leftarrow (\exists y_1, \ldots, y_p)[x_1 = t_1, x_2 = t_2, \ldots, x_n = t_n, A_1, \ldots, A_m]$$

If there are exactly k clauses, $k > 0$, in the database for the relation R, let

$$R(x_1, \ldots, x_n) \leftarrow E_1$$
$$\ldots$$
$$R(x_1, \ldots, x_n) \leftarrow E_k$$

be the k general forms of these clauses. Each of the E_i will be an existentially quantified conjunction of literals. The *definition of R*, implicitly given by the database is

$$(\forall x_1, \ldots, x_n)[R(x_1, \ldots, x_n) \leftrightarrow E_1 \vee E_2 \vee \ldots \vee E_k]$$

The if-half of the definition is the general form of clauses grouped together as a single implication. The only-if is the *completion law* for R.

Clark proved that the query evaluation process will find every answer that is a logical consequence of the completed database.

The results by Clark and Reiter provide the semantics for negation in logic programming languages.

4.3 Loveland's Book on Theorem Proving

While recovering from knee surgery in 1979, I spent time reading Don Loveland's book [Loveland, 1978]. The book provides a rigorous presentation of theoretical work performed in theorem proving up to 1978. It includes proofs of theorems, algorithms and historical developments. It should be required reading by all those interested in deductive methods.

5 Beginnings of a Theory of Logic Programming

The late 1970s and 1980s was an exciting time for logic programming. The theoretical foundations of the field were under constant assault in an attempt to provide meaning to logic programs and to extensions to the basic concept of Horn clause logic programming as first specified by Kowalski [1974]. The first theoretical developments were by van Emden and Kowalski [1976] in which they detailed model theoretic, fixpoint, and operational semantics of logic programs as a programming language. They demonstrated that the fixpoint semantics corresponded to model theory, while operational semantics corresponded to proof theory. They provided a formal semantics of a definite clause logic formula, viewed as a statement in a programming language. Their use of the least model and least fixpoint constructions, as well as the procedural interpretation, have become standard tools in logic programming theory.

Krzysztof Apt and van Emden [1982] built upon the theoretical treatment of van Emden and Kowalski. They followed van Emden's renaming of LUSH resolution to be SLD resolution and the renaming became the standard. They also characterized the finite failure set of an atom relative to a program in terms of a fixpoint operator.

Lassez and Maher [1984] then proved that the finite failure set is characterized by the difference between the Herbrand base and the largest fixpoint of the operator described by Apt and van Emden. The result by Lassez and Maher extended the results of Apt and van Emden which only guarantee the existence of one finitely failed SLD tree while others may be infinite. Using the concept of fairness, which specifies that every clause generated must be expanded before any infinite path is taken, they identified the computation rules which guarantee finitely failed SLD trees leading to a strong soundness and completeness result on finite failure.

Clark [1978] tied these results to the completion of a logic program. He showed the soundness of the negation-as-failure rule for any Horn logic program, P, augmented by $comp(P)$ and equality axioms. He showed that every goal G with a finitely failed SLD-tree is a logical consequence of $comp(P)$ and equality axioms (actually his results extend to general programs using safe computation rules, i.e. rules that select only ground negative literals). Joxan Jaffar, Lassez and John Lloyd [1983] proved the completeness result and showed that if a goal G is a logical consequence of $comp(P)$, then there is a finitely failed SLD-tree for G.

Shepherdson [1988] provides a comprehensive description of negation in logic programs. The excellent research monograph by John Lloyd, first published in 1984 [Lloyd, 1984] and then updated in 1987 [Lloyd, 1987], pulled together the theoretical developments in logic programming up to the time of the publication date.

6 First Step Towards Disjunctive DBs and Disjunctive Logic Programs

I had been intrigued by Reiter's CWA when I first learned about it. I thought that it would be of interest to extend his work to non-Horn theories. I had been chair of the Department of Computer Science at the University of Maryland. I had devoted much of my time to starting the new department. Around 1980 the department was in good shape, was rated 13th in the country and I decided that it was time to emphasize my research and stepped down as chair.

I developed a procedural definition of negation in disjunctive theories. It stated that "one can conclude $\neg p$, if one cannot prove p and there is no positive clause K such that one can prove $p \vee K$, but cannot prove K." In January 1981 I was invited to lecture at Simón Bolívar University in Caracas, Venezuela. Roussel, who had developed the first Prolog interpreter with Colmerauer, and whom I had met at the first Toulouse Workshop had a faculty position at Simón Bolívar. I discussed what I had been working on with him, and we came up with the idea of a model theoretic approach to handling negation. Unlike Horn clause theories, there may be no unique minimal model for a disjunctive theory. For instance, the theory $p \vee q$ possesses two minimal models, $\{p\}$ and $\{q\}$.

Using the concept of minimal models, one can conclude the negation of an atom if it does not appear in any minimal model. Thus, in the above example one cannot prove p and one cannot prove q. Hence the definition of the minimal models approach was not inconsistent with the original theory. The problem was then to demonstrate the equivalence of the proof procedure method and the minimal models approach. As I was only in Caracas for a few days, Roussel and I did not have sufficient time to try to prove this result.

I returned to Maryland and started to work on the problem. As I obtained results, I sent them along to Roussel. Alas, I never received a response from him. Finally I proved the equivalence as well as some other results, and named the approach the *generalized closed world assumption* (GCWA). Since I did not hear from Roussel and his contribution was the concept of minimal models that we had come up with in Caracas, I decided to publish the paper without his co-authorship [Minker, 1982] and acknowledged his contribution to the work. The paper provided a procedural and a declarative semantics for negation in disjunctive logic programs.

I believe the GCWA was the first result in disjunctive theories. Following this work, there were alternative definitions of negation for disjunctive theories by Ross and Topor [1988a]; by Michael Gelfond and Halina Przymusinska [1986]; and by Larry Henschen and Adnan Yahya [1985]. I did not pursue disjunctive theories until six years later.

7 Nonmonotonic Theories

NMR is considered to have started in 1980 with the publication of a seminal issue of the Artificial Intelligence Journal, devoted exclusively to nonmonotonic reasoning and developing the initial theories. As noted by Daniel Bobrow in his "Editor's Preface" to the journal, the approaches to nonmonotonic reasoning were characterized broadly by different ways. McCarthy [1980] formalized the theory of *circumscription* that he introduced earlier [McCarthy, 1977]; Reiter [1980] introduced his theory of *default reasoning*; and Drew McDermott and Jon Doyle [1980] used *modal logic* to handle non-monotonicity.

The three approaches to NMR have led to a large literature in the past 30 years. Building upon this work, numerous results have been obtained. I briefly review some of the early work in this field.

Matthew Ginsberg captured most of the significant developments in nonmonotonic logic up to approximately 1987 in his book *Readings in Nonmonotonic Reasoning* [Ginsberg, 1987]. This is a major source document for work up to that date. It consists of the original articles on the subject of nonmonotonic reasoning. Ginsberg ties the work together nicely with comments interspersed throughout the book.

The founders of the field of NMR are the authors of the seminal AIJ articles, McCarthy, Reiter, McDermott and Doyle. As discussed below, Moore, in 1985, wrote a paper on autoepistemic reasoning [Moore, 1985] that has been very influential. I believe that Moore should be added as a founder of NMR.

7.1   Circumscription

McCarthy's 1980 paper on circumscription [McCarthy, 1980] is the starting point of this work. Circumscription deals with the minimization of predicates subject to restrictions expressed by predicate formulas. If A is a sentence of a first-order language containing a predicate symbol $P(x_1, \ldots, x_n)$, written $P(\overline{x})$, then the result of replacing all occurrences of P in A by the predicate expression Φ is written as $A(\Phi)$. The *circumscription* of P in $A(P)$ is the sentence schema

$$A(\Phi) \wedge \forall \overline{x}(\Phi(\overline{x}) \supset P(\overline{x})) \supset \forall \overline{x}(P(\overline{x}) \supset \Phi(\overline{x}))$$

The formula states that the only tuples $\overline{x}$ that satisfy P are those that have to - assuming the sentence A. McCarthy shows how a slight generalization allows circumscribing several predicates jointly.

Lifschitz [1986] modified circumscription so that instead of being a single minimality condition, it becomes an "infinite conjunction" of "local" minimality conditions; each condition expresses the impossibility of changing the value of a predicate from *true* to *false* at one point. This is referred to as *pointwise circumscription*. Lifschitz [1985a] then defines *prioritized circumscription* which provides for priorities between predicates and *parallel circumscription*. Benjamin Grosof [1991] generalized prioritized circumscription to a partial order of priorities.

David Etherington, Robert Mercer and Reiter [1985] discussed the adequacy of circumscription. Lifschitz [1985b] addressed the problem of computing circumscription. He noted that circumscription is difficult to implement because its definition involves a second-order quantifier. He introduced metamathematical results that, in some cases, allowed circumscription to be replaced by an equivalent first-order formula. Etherington, Mercer and Reiter established results about the consistency of circumscription, showing that predicate circumscription cannot account for some kinds of default reasoning, and also provided no information about equality predicates. Donald Perlis [1986] showed inadequacies of circumscription to deal with counterexamples.

There have been other results about circumscription. Perlis and Minker [1986] developed completeness results for some cases of circumscription. McCarthy's original paper [McCarthy, 1980] discussed only the soundness of circumscription.

7.2 Default Reasoning

Reiter's 1978 article [Reiter, 1978b] surveyed default reasoning in AI: the paper discussed default assignments to variables, the Reiter CWA, frame default for causal worlds, exceptions as defaults, and negation in AI programming languages. This led to his seminal article in the AIJ on default reasoning [Reiter, 1980].

A default is a rule of the form $\frac{\alpha:\beta}{\gamma}$, whose meaning is intended to state: 'if α is *true*, and it is consistent to assume that β is *true*, then conclude that γ is *true*.

Default rules act as mappings from some incomplete theory to a more complete *extension* of the theory. An extension is a maximal set of conclusions that can be drawn from the default theory. Reiter defines a *default theory* to be a pair (D, W), where D is a set of closed default rules and W is a set of first-order sentences. Extensions are defined by a fixpoint construction. For any set of first-order sentences S, define $\Gamma(S)$ to be the smallest set U satisfying the following three properties:

1. $W \subseteq U$.
2. U is closed under first-order logical consequence.
3. If $\frac{\alpha:\beta}{\gamma}$ is a default rule of D and $\alpha \in U$ and $S \not\vdash \neg\beta S$, then $\gamma \in U$.

Then E is defined to be an *extension* of the default theory (D, W) iff $\Gamma(E) = E$, that is, E is a fixpoint of the operator Γ.

A theory consisting of general default rules does not always have an extension. A subclass of theories admitting only default rules called *normal defaults*, and of the form, $\frac{\alpha:\beta}{\beta}$ have an extension. Reiter developed a complete proof theory for normal defaults and shows how it interfaces with a top down resolution theorem prover.

Reiter and Giovanni Criscuolo [1981] show that default rules may be normal when viewed in isolation, however, they can interact in ways that lead to derivations of anomalous default assumptions. Non-normal default rules are required to deal with default interactions. Handling non-normal default rules is computationally more complex than dealing with normal default rules.

Gelfond, Lifschitz, Przymusinska and Mirek Truszczynski [1991] generalized Reiter's default logic to handle disjunctive information. The generalization is due to difficulties with disjunctive information that arise because there may be multiple extensions - one containing a sentence α and another a sentence β - and the theory with a simple extension $\alpha \vee \beta$.

A *disjunctive default* is an expression of the form

$$\frac{\alpha:\beta_1,\ldots,\beta_m}{\gamma_1|\ldots|\gamma_n}$$

where $\alpha, \beta_1, \ldots, \beta_m, \gamma_1, \ldots, \gamma_n (m, n \geq 0)$ are quantifier free formulas. Formula α is the prerequisite of the default, $\beta_1, \ldots, \beta_m$ are justifications, and $\gamma_1, \ldots, \gamma_n$ are its consequents. A *disjunctive default theory* (DDT) is a set of disjunctive defaults.

Gelfond et al. [1991] show that one cannot simulate a DDT with a default theory in polynomial time (under standard complexity theory assumptions).

Chitta Baral and VS Subrahmanian [1992] show that even though all default theories do not necessarily have extensions, two-fold iteration of Reiter's operator Γ, Γ^2 always has fixpoints over the power lattice. Such fixpoints indicate that when extensions do not exist, it is due to oscillations. They show that considering these oscillations leads to a semantics for all default theories including those that do not have extensions. They also define a well-founded semantics for default theories.

Etherington and Reiter [1983] use default logic to formalize NETL-like [Fahlman, 1980] inheritance hierarchies. They provided the first attempt at a semantics for such hierarchies; a provably correct inference algorithm for acyclic networks; a guarantee that the acyclic networks have extensions; and a provably correct quasi-parallel inference algorithm for such networks.

Bart Selman [1989] in his excellent thesis explores three default reasoning formalisms. He obtains the first characterization of tractable forms of default reasoning. He also gives a high-level characterization of the main factors contributing to the intractability of the general reasoning. He considers the following formalisms: *model-preference defaults*, *default reasoning*, and *path-based defeasible inheritance*. Horty, Thomason, and Touretzky [1987] provide references to other related work. In model-preference defaults, the preference ordering on models is defined by statements of the form "a model where α holds is preferable to a model where β holds." He proves that only systems with quite limited expressible power lead to tractable reasoning, e.g., $\mathcal{D}H$ and $\mathcal{D}H_a^+$, containing respectively Horn defaults and acyclic specificity-ordered Horn defaults.

Selman and Henry Kautz [1990] and Selman [1989] consider the complexity of various forms of default reasoning. They consider *unary, disjunction-free ordered, ordered unary, disjunction-free normal, Horn*, and *normal unary* theories. These theories form a hierarchy and Kautz and Selman show that to find an extension in disjunction-free and unary theories is NP-hard, whereas it is $\mathcal{O}(n^3)$ for the remaining theories. They also show that the complexity of determining if a given literal p appears in any extension of a Horn default theory or a normal unary theory is $\mathcal{O}(n)$, where n is the number of occurrences of literals in the theory, but otherwise is NP-hard. In skeptical theories where one wants to determine if a literal is in all extensions, they show that for a normal unary theory the time-complexity is $\mathcal{O}(n^3)$, where n is the number of occurrences of literals in the theory. For other theories it is co-NP-complete.

Path-based inheritance reasoning as defined by David Touretzky is NP-hard, even when restricted to acyclic unambiguous networks. Horty et al. [1987] identify tractable forms of defeasible reasoning. Kautz and Selman [1989] and Selman [1989] show that the tractability of an inheritance theory depends upon the kinds of chaining involved in the path construction. Selman [1989] further shows that the standard upward chaining notion of path construction leads to tractable algorithms; those definitions based on a double chaining notion lead to exponential algorithms.

7.3   Modal Theories

McDermott and Doyle [1980] introduced a modal operator M into nonmonotonic logics. If p is a sentence, then Mp denotes the sentence in modal logic whose intended meaning is "p is consistent with what is known," or "maybe p." The semantics of the M operator have not been clearly described and the formalism is weak, as noted by Moore [1985].

Moore [1985] developed *autoepistemic logic (AEL)*, which builds the work by Mc-Dermott and Doyle. Instead of "possibly" modal operator M, he used a "necessarily" modal operator L. Intuitively, Lp is to be read as "I know p." Moore constructs non-monotonic logic as a model of an ideally rational agent's reasoning about his own beliefs. He defines a semantics for which he shows that autoepistemic logic is sound and complete. Moore shows the relationship between autoepistemic logic and Kripke's approach to modal logic [Moore, 1984].

Hector Levesque [1990] generalized Moore's notion of a stable expansion [Moore, 1985] to the full first-order case. He provides a semantic account of stable expansions in terms of a second modal operator $\mathcal{O}$, where $\mathcal{O}(w)$ is read as "w is all that is believed." He characterizes stable expansions as: $\mathcal{O}(w)$ is *true* exactly when all formulas that are believed form a stable expansion of w.

7.4   Relationships Among Nonmonotonic Theories

The theoretical papers on NMR were followed by a number of relationships that were developed amongst them.

Perlis [1988] and Lifschitz [1989] have developed variants of circumscription analogous to autoepistemic logic.

Although default logic and autoepistemic logic are seemingly different, and motivated by slightly different concerns, Kurt Konolige [1988] showed that autoepistemic logic can be strengthened so that there is an equivalence between the propositional form of both logics. He gave an effective translation of default logic into autoepistemic logic and shows there is a reverse translation; every set of sentences in autoepistemic logic can be effectively rewritten as a default theory. Unfortunately, Konolige's work contained substantial and not easily correctable mistakes.

Marek and Truszczynski [1989a] revised and extended the work by Konolige. They take a syntactic approach to investigating the relationships between autoepistemic and default logics. Together with Marc Denecker, they show that the interpretation of defaults as modal formulas proposed by Konolige allows them to represent all semantics for default logic in terms of the corresponding semantics for autoepistemic logic [Denecker, Marek, and Truszczynski, 2003]. They also demonstrate, the semantics of Moore and Reiter are given by different operators and occupy different locations in their corresponding families of semantics.

Truszczynski [1991] develops a natural modal interpretation of defaults. He shows that under this interpretation there are whole families of modal nonmonotonic logics that accurately represent default reasoning. He applies the method to logic programs and obtains results that relate stable models to expansions in several nonmonotonic logics. His results show that there is no single modal logic for describing default reasoning and that there exist a whole range of modal logics that can be used in the embedding as a "host" logic.

Marek and Subrahmanian [1992] show the relationship between supported models of normal programs and expansions of autoepistemic theories.

Przymusinski [1991] notes several drawbacks of autoepistemic logic, notably some "reasonable" theories are inconsistent in AEL; even for consistent theories AEL does not always lead to expected semantics; it insists upon completely deciding all of our beliefs; and it does not offer flexibility in terms of selecting application-dependent formalisms on which to base our beliefs. He shows that autoepistemic logics of closed beliefs of Moore coincides with autoepistemic logic of closed beliefs in which the negative introspection operator used is Reiter's CWA. He then extends autoepistemic logic to *generalized autoepistemic logic (GAEL)*, which uses Minker's generalized closed world assumption as the basis for the negative introspection operator and demonstrates how other forms of AEL may be achieved.

7.5   Logic Programming and Nonmonotonic Theories

As important as it was to show that autoepistemic logic, circumscription and default logic were related to one another under certain conditions, their relationship to logic programming was established. This was significant as normal logic programming systems could be used to implement parts of these three theories.

Reiter [1982] was the first to relate circumscription to logic programming. Minker [1985] introduced *protected circumscription* and demonstrated how one can compute in this theory with logic programs. Subrahmanian and Lu [1988] extended the concept of protected circumscription. Gelfond, Przymusinska, and Przymusinski [1989] relate a propositional form of circumscription to stratified theories. Minker, Lobo, and Rajasekar [1991] complement the work by Gelfond et al. [1989] by developing a procedure to compute circumscription in stratified disjunctive logic programs.

The use of logic programming is a natural computation vehicle for a large class of circumscription problems as they both deal with the concept of minimal models. In general, it will be difficult to compute in circumscriptive theories, except for those that are equivalent to *normal theories* which allow a single atom in the head of the clause and literals in the body of a clause. Schlipf [1986] has shown that circumscription is Π_2^1-complete. In disjunctive theories, the computational complexity increases so that unless the theory is *near-Horn*, in which there are $log(n)$ non-Horn formulas, it will be computationally complex.

Three papers, two by Gelfond and Lifschitz and one by Lifschitz, showed that stable model semantics was a unifying theme for logic programming and NMR theories. Gelfond and Lifschitz [1988] developed stable model semantics and proved that it can be obtained from Moore's semantics of autoepistemic logic under a suitably chosen translation. Lifschitz [1989] showed that autoepistemic logic, stable models and introspective circumscription provided three equivalent descriptions of the meaning of propositional logic programs. Lifschitz also noted that default logic and autoepistemic logic provided more expressive possibilities that apparently have no counterpart in circumscription. Gelfond and Lifschitz [1991] extended normal logic programs to include classical negation $\neg$ and negation-as-failure *not*. They based the semantics of these programs on stable model semantics. They extended disjunctive databases in the same way. They referred to the use of stable models in such logic programs as answer set semantics. The answer set semantics is important in that it has an elegant and simple definition and handles negation for both extended logic programs and extended disjunctive logic programs.

Marek and Truszczynski [1989a] showed that normal logic programs translate to autoepistemic logic and default logic. Also Przymusinski [1988b] developed relationships between logic programming and nonmonotonic reasoning. He extended autoepistemic logic to generalized autoepistemic logic and related Reiter's CWA to autoepistemic logic and Minker's generalized CWA to generalized autoepistemic logic.

In the above, I have provided a brief discussion of the theories of nonmonotonic reasoning, some of their relationships, and the importance of logic programming as the preferred implementation tool for them. The important book by Marek and Truszczynski, Nonmonotonic Logic: Context-Dependent Reasoning [Marek and Truszczynski, 1993], provides a thorough analysis of formalisms of nonmonotonic reasoning. They show how they are related to one another and that they provide the formal foundations for logic programming.

8 Towards a Semantics for Disjunctive Databases and Disjunctive Logic Programming

In 1986 I believed that it was time to have another workshop devoted to logic and databases since it was 10 years since the last such workshop was held in Toulouse. The *Workshop on the Foundations of Deductive Databases and Logic Programming*, was sponsored by the National Science Foundation, the Institute for Advanced Computer Studies at the University of Maryland, and the Department of Computer Science. I subsequently edited a book consisting of papers drawn from the workshop [Minker, 1988]. The workshop brought together almost all of the leading researchers in deductive databases and logic programming. I believed it essential to bring these two groups together as the two topics were intimately related. Although every paper in the book contains important results, I focus only on a few articles as they relate to my subsequent work in developing a foundation for disjunctive logic programs with two of my students, Rajasekar and Lobo and the subsequent use of the work in extending nonmonotonic theories.

The papers that most influenced me were by Apt, Howard Blair and Adrian Walker, who developed a fixpoint for *stratified logic programs* [Apt et al., 1988]; Allen Van Gelder, who had also developed the concept of a stratified logic program [Van Gelder, 1988]; John Sheperdson, who surveyed all work on negation and contributed new results on negation in logic programming [Shepherdson, 1988]; Przymusinski, who had developed the idea of a *perfect model* and a locally stratified logic program [Przymusinski, 1988a]; and Lifschitz, who had shown the relationship of stratified logic programs to circumscription [Lifschitz, 1988]. Earlier work on stratification was done by Ashok Chandra and David Harel [1985]. The work on stratification extended the correct use of negation to a wider class of programs than that in Prolog.

Rodney Topor and I had a brief discussion during a break period at the workshop. He asked if I were pursuing research in disjunctive theories and remarked about the difficulty in dealing with the GCWA. We did not discuss this topic further as I was busy tending to the workshop.

Following the workshop, I addressed the problem of finding a fixpoint semantics
for disjunctive databases and disjunctive logic programming. I came to realize that
there would have to be a change in perspective because we were now dealing with
disjunctive logic programs, not Horn or generalized Horn programs. I realized that
the T_P operator developed by van Emden and Kowalski [1976] mapped from subsets
of the Herbrand Base to subsets of the Herbrand Base; that is, from sets of atoms to
sets of atoms. However, what was needed was to map from sets of positive disjuncts
to sets of positive disjuncts. This gave rise to what we called the *extended Herbrand
base*, which consists of the set of all positive disjuncts that can be constructed from the
Herbrand base. It then became obvious how to modify the van Emden and Kowalski
[1976] T_P-operator to apply to disjunctive logic programs. Basically, we used hyper-
resolution starting with ground clauses from the program - in much the same way as
van Emden and Kowalski do with their T_P-operator. We are mapping from sets of
positive clauses in the extended Herbrand base to sets of positive clauses also in the
extended Herbrand base [Minker and Rajasekar, 1990].

Once I realized this, it was straightforward to both me and Rajasekar, with whom
I discussed my observation, as to how to proceed. The proof that the T_P-operator
reaches a fixpoint is identical to that given in Lloyd [1987]. The fixpoint yields all of
the positive disjuncts that can be derived from the program. If one were to form a set
consisting of an atom from each disjunct in the fixpoint, one would obtain a model of
the program from which minimal models may be found. We were also able to show,
using the fixpoint operator and ideas borrowed from Jaffar et al. [1983] how one can
find those atoms that may be assumed false and showed that this gave the same result
as the GCWA.

We also addressed the problem of devising a procedural method for finding answers to negated atoms [Minker and Rajasekar, 1988]. For this we used SLI resolution. To handle negation by failure, we modified SLI to achieve an inference system
SLINF, the counter-part to negation-as-finite-failure. In the presence of Horn theories,
SLINF reduces to SLDNF. Hence, there now exists a procedural, a model theoretic,
and a fixpoint semantics for disjunctive logic programs. Our paper [Minker and Rajasekar, 1990] provides an account of these results.

While we were writing our results for a technical report, I received a copy of
a paper from Rodney Topor. It was the paper he co-authored with Kenneth Ross
[Ross and Topor, 1988b]. The abstract indicated that they had developed a fixpoint
operator for disjunctive programs. After a moment of panic, I scanned the paper and
found that his work differed from ours. They had developed a new idea, called the
disjunctive database rule (DDR), to compute answers to negated queries in disjunctive
databases. The computation of negation is much simpler than computing negation in
the GCWA. I suspect that Rodney must have had this thought in mind when we spoke
at the workshop.

Having found a fixpoint operator for disjunctive logic programs, we wanted to find
analogues of results developed for logic programs. In particular, we were interested
in the analogue of Clark's completion axiom [Clark, 1978], an alternative form of
negation to the GCWA [Minker, 1982], and stratified disjunctive logic programs.

Rajasekar and I, together with Lobo, considered the idea of weakening the GCWA
in disjunctive theories. In the GCWA, to infer a ground literal, $\neg A$, from a disjunctive

logic program P, *all positive clauses* K such that $A \vee K$ is provable from P must satisfy the condition that K is also provable from P. A weaker form of the GCWA is to infer a negative ground literal $\neg A$ when, for *some positive clause* K, $A \vee K$ is derivable from a program P. We termed this the *Weak Generalized Closed World Assumption* (WGCWA). It is much easier to compute answers to negated atoms using the WGCWA than the GCWA. We showed that the WGCWA is equivalent to Ross and Topor's Disjunctive Database Rule. In a discussion that Rajasekar had with Topor at the Logic Programming Conference in 1988, Topor stated that he liked the name WGCWA instead of the DDR. He also said that he preferred the name disjunctive logic programs to what we had been referring to as non-Horn logic programming. When Rajasekar told me this, I realized that Topor was correct and we have since adopted the terminology, disjunctive databases and disjunctive logic programs. This terminology is now standard.

Lobo, Minker, and Rajasekar observed that it was straightforward to extend Clark's completion axiom to apply to disjunctive logic programs - one simply replaces each disjunctive clause in the program with a set of Horn clauses. The set of Horn clauses consists of an atom from the disjunct as the head of the new clause, and the body remains the same as the body of the disjunctive clause. This is repeated for all atoms in the head of the clause. Now, for each clause in this set, the direction of the arrow is reversed. This gave us the analogue of Clark's *only if* part of the completion axiom. We were able to show that the completion of a disjunctive program P is consistent and that there is a finitely failed fair SLINF tree if and only if $\neg A$ is a logical consequence of the $comp(P)$ [Lobo, Rajasekar, and Minker, 1988].

To round out the current work in this area, Rajasekar and I extended the work by Apt et al. [1988] and Van Gelder [1988] to stratified disjunctive logic programs [Minker and Rajasekar, 1990]. As with stratification in logic programs, recursion through negation is not permitted. We apply the nonmonotonic fixpoint theory developed by Apt, Blair and Walker to develop a fixpoint semantics for stratified disjunctive programs. We also develop an iterative definition for negation, called the *Generalized Closed World Assumption for Stratified Programs* (GCWAS), and show that our semantics captures this definition. We show that the least state characterized by the fixpoint semantics corresponds to a stable-state defined in a manner similar to stable models of Gelfond and Lifschitz [1988].

The work summarized above extends logic programming theory to apply to disjunctive logic programs. Indefinite answers and answers to negated atoms can be found in these logic programs. There are theories where it is not possible to stratify the program or where it is possible only to partially stratify the program. In such theories users may want to have disjunctive answers rather than "unknown". The results that we have found apply to such theories - one does, of course, pay for such a convenience. Namely, it is more complex to solve problems in disjunctive logic programs, than it is to solve problems in logic programs.

At the same time as my students and I had been working on disjunctive logic programs, other significant developments had been taking place on theoretical aspects of logic programming. Gelfond and Lifschitz [1988] developed the concept of stable model semantics. Gelfond et al. [1989] extended the CWA to stratified programs and developed the concept of an Iterated Closed World Assumption (ICWA), defined by iteration on the level of predicates. Marek [1989] and Marek and Truszczynski [1989b] made contributions to autoepistemic reasoning and its relationship to logic programming and other forms of default reasoning. Van Gelder, Ross, and Schlipf [1988] developed the concept of the well-founded model that gives semantics to all logic programs including those that are not stratified. They develop a three-valued logic to handle this case. Przymusinski [1989] extended this work. Baral, Lobo, and I extended well-founded semantics both to general Horn programs and to disjunctive logic programs. Our objective was to provide a more intuitive semantics than one achieves with well-founded semantics. For example, given the normal program, $\{p \leftarrow a, p \leftarrow b, a \leftarrow \neg b, b \leftarrow \neg a\}$, the well-founded semantics yields unknown for all atoms, while in our semantics we obtain that p is true, while a and b are unknown.

9   Answer Set Programming and Nonmonotonic Reasoning

As noted earlier, ASP has become the standard generally used in knowledge representation and reasoning. In the following I describe its influence.

9.1   First Logic Programming and Nonmonotonic Research Workshop 1991

The start of conferences on logic programming and nonmonotonic reasoning began in 1991 in Washington, DC. The workshop was organized by Anil Nerode, Marek and VS Subrhamanian. Workshops on this topic have become important to our field since that time. I was invited to give the banquet address. I provided an overview of the accomplishments in the field and discussed several different semantics that had been proposed for normal LPs, by me and my students, and by others, including stable model semantics by Gelfond and Lifschitz [1988]. Others were proposed for disjunctive logic programs and Gelfond and Lifschitz had discussed stable model semantics [Gelfond and Lifschitz, 1991] for disjunctive theories, including logical and default negation in 1991. I published an updated version of the talk in Journal of Logic Programming [Minker, 1993].

Following my talk, I was asked which semantics was the most useful. I replied, it would depend upon what the user wanted with respect to the answer to the problem. In the disjunctive case, one might like a disjunctive answer or an answer according to the well-founded semantics. A Russian-sounding voice from the back of the room modestly shouted out, *stable model semantics*. Startled, I responded, cautiously and said, "time will tell". Nineteen years later, I must admit, what I have realized for almost all these years and have acknowledged in other lectures is that Vladimir was almost correct, it is answer set programming (ASP). Since Gelfond and Lifschitz renamed stable model semantics to be ASP, it has become the preferred semantics for logic programming with negation.

9.2 Answer Set Programming and Systems

In the 19 years since it was proposed, ASP programming has been turned into a veritable industry. Almost every conference in AI contains one or more papers using ASP. This includes theoretical research-work, including linguistic extensions, semantic properties, evaluation algorithms, and optimization techniques. There have been expansions of the core language, including probabilities, handling aggregate functions, function symbols, complex terms (allowing the explicit usage of lists and sets), consistency-restoring rules, ontology handling constructs, nested expressions, and causality. The paper "Sixty Years of Stable Models" by David Pearce gives a fascinating discussion of stable models [Pearce, 2008]. Pearce re-examines there the conceptual foundations underlying the approach of stable models and relates them to various developments in logic around the middle of the last century.

As formulated by Marek and Truszczyński [1999] and Niemelä [1999] Answer Set Programming (ASP) is an approach to knowledge representation and reasoning in which a problem is formalized in a logical language so that the models of a logic program, capture solutions to the problem. The models of the program are computed in terms of a dedicated search engine, called an answer set solver. A full-fledged ASP system provides a programmer with a rule-based input language in which problems are encoded. The front-end of the system consists of a parser for this language and the outcome is an intermediate representation of the problem in a simplified language directly supported by the search engine. The search of models, i.e., variable assignments potentially fulfilling additional criteria, is then performed using the respective answer set solver. Solvers may carry out optional compilation steps, possibly giving rise to additional intermediate representations of the problem. For a description of some intermediate representations, see the paper by Janhunen [2007].

9.3 Dagstuhl Workshops on LPNMR and ASP

Workshops have been held at Dagstuhl, Germany on many scientific topics. The Dagstuhl workshops have become important for LP and NMR in the last 15 years.

In 1996, I was invited to talk at a Dagstuhl workshop. I spoke on deductive databases. According to Ulrich Furbach, I proposed that the LP community should develop a database of information about logic programming system implementations and applications. I have no remembrance of having made the suggestion, but Furbach took the initiative at the University of Koblenz to develop a web site that listed 32 systems and applications for LPNMR. The web page was last updated in 2000. The initial work done at the University of Koblenz has been replaced by the Dagstuhl Initiative discussed, below.

At the Fourth International Conference on Logic Programming and Nonmonotonic Reasoning (LPNMR '97), Dagstuhl, Germany 28-31 July 1997, Robert Milnikel wrote a conference report [Milnikel, 1997] in which he stated that the

> LPNMR reflected a new stage in the development of this cross-disciplinary field, with demonstrations of ten implemented systems joining the submitted papers, invited talks and panel discussions.

This was the first major conference to feature systems. It demonstrated the importance of implementations to enhance research and applications in LPNMR. Ten systems were demonstrated or discussed.

At the 2002 Dagstuhl conference on LPNMR, it was recognized that there was a need for systematic benchmarking of ASP systems that had been developed. This inaugurated the Dagstuhl Initiative [Borchert, Anger, Schaub, and Truszczynski, 2004] to create a web-based benchmark archive, *asparagus* at the University of Potsdam. The web page consists of a depository of test problems and a uniform platform on which to test ASP solvers on problems from the depository. Test problems and application problems are continually added to the depository. The Dagstuhl Initiative is important for several reasons: an international board has been appointed to supervise the effort and to assure its continuation. Yearly competitions have taken place with existing ASP solvers. This brings together the principal implementors in the field and has stimulated discussions as to how to improve their systems. This has led to major improvements with ASP solvers. The results of the contests are published and made available to the NMR community.

Efficient ASP systems include DLV, developed by Nicola Leone, Pfeifer, Faber, Eiter, Gottlob, Perri, and Scarcello [2006]; smodels built by Ilkka Niemelä, Patrik Simons and Tommi Syrjänen [2000] (with some contributions of Timo Soininen [Simons, Niemelä, and Soininen, 2002]); and CLASP, developed by Gebser, Kaufmann, and Schaub [2009], as well as others. As one measure of systems, it is of interest to note that CLASP has outperformed all other systems in the past competitions. This does not account for the features in the language to support the user. In 2007, I spoke at LPNMR07 and noted that there were few applications in NMR. I noted that if NMR were to be recognized as being of value, there would have to be more applications of the systems to real problems. Since then, there have been numerous applications reported in bioinformatics, database integration, diagnosis, hardware design, planning, security protocols and high-level control of the NASA space shuttle. However, there has not been, what I referred to as, a "killer application" that has captured the attention of the AI, the database or user communities. A "killer application" would be one such as the work by Nogueira, Balduccini, Gelfond, Watson, and Barry [2001] with the NASA space shuttle on the use of a nonmonotonic system for space applications. The work, although valuable and used for training of space missions, has not made an impact. Were it used on an actual space mission, it might have attracted the attention of the various communities. A "killer application" need not be a large application.

10   Logic-Based Artificial Intelligence Workshop (1999) and Book (2000)

In early 1998, John McCarthy phoned and asked if I would organize a workshop with him on Logic-Based AI. If McCarthy makes such a request, it is not possible to reject the idea. The workshop was supported by the NSF and the University of Maryland Institute for Advanced Computer Studies. We had sufficient funds to invite and to support researchers throughout the world and students. McCarthy and I discussed those who we would invite to lecture. We also decided that it was time to feature implementations of nonmonotonic systems as well as research contributions. At the time, we did not know about the 1997 Dagstuhl conference at which some NMR systems were demonstrated. We believed that demonstrating NMR systems was an integral part of the conference since, without systems and applications, NMR would ultimately become sterile.

Twelve systems were demonstrated during the evenings. Only one of the systems, XSB, based on well-founded semantics (WFS) by David Warren et. al. [Sagonas, Swift, and Warren, 1994] and his group was commercially available. A second system, DLV, was based on answer set semantics (ASP) by Eiter and his group. A third system, smodels, by Niemelä and Simons featured both ASP and WFS. The last two systems subsequently became and remain commercially available. All twelve systems demonstrated the field of NMR had matured beyond theoretical papers to systems used to solve significant problems. The diversity of topics demonstrated covered: agents, causal theories, deductive databases, dynamic domains, inductive logic programming, planning, and robotics.

Following the workshop I edited a book [Minker, 2000], Logic-Based Artificial Intelligence, based on reviewed and expanded papers from the Workshop. I focus on a few papers that are important for nonmonotonic reasoning. There were other papers of significance.

In addition to demonstrating their comprehensive systems on ASP, two papers described experiments and improvements in their systems. Eiter, Faber, Leone, and Pfeifer [2000] discussed enhancements to DLV, a disjunctive system for databases and logic programming. They discussed how problems can be solved in a natural way using a *"Guess & Check"* paradigm where solutions are guessed and verified by parts of the program. Experiments indicated that DLV solved some complex problems. DLV is probably the first comprehensive system that handles disjunction. Niemelä and Simons [2000] used ASP in smodels for normal logic programs and extended the system to have cardinality and weight constraints. They summarize the extensions in the system, illustrate the level of performance and compare it to state-of-the-art satisfiability checkers. Since then, these systems have been extended with many powerful features. For instance, Leone, Pfeifer, Faber, Eiter, Gottlob, Perri, and Scarcello [2006] discuss recent extensions of DLV, and Tomi Janhunen and Niemelä [Janhunen and Niemelä, 2004] built a system GnT developed on top of smodels to handle disjunctive logic programs.

Several papers of importance to nonmonotonic reasoning in the book are as follows. Kautz and Selman Kautz and Selman [2000] discussed the success of satisfiability planning. They describe experiments with SATPLAN, possibly the fastest planning for domain-independent planning. They present a way to encode domain knowledge in a declarative, algorithm-independent manner. They show the same heuristic knowledge can be used by entirely different search engines and present experimental results.

Two papers were presented on the important subject of dynamic systems which deal with time and actions over time among many other things. Pirri and Reiter [2000] wrote about natural actions in the situation calculus intimately related to dynamic systems. They describe the theory and implementation of a deductive planner in the situation calculus for domains with two kinds of actions: "free will" actions on the part of agents with the ability to perform or withhold their actions; and natural actions whose occurrence times are predictable in advance, in which case they must occur at those times unless something happens to prevent them. They "...suggest that the rejection of logic and the situation calculus for planning was perhaps a bit too hasty." Baral and Gelfond [2000] discussed an architecture for intelligent agents based on the use of A-Prolog, a language based on ASP. They represent agent's knowledge about the domain and to formulate the agent's reasoning tasks. They demonstrate their methodology of constructing such programs.

As noted earlier, an important contribution to nonmonotonic reasoning was McCarthy's 1959 paper [McCarthy, 1959], in which he defined what had become "the oldest planning problem in artificial intelligence". Lifschitz, McCain, Remolina, and Tacchella [2000] provided the first solution to the problem. The reason it took so long to provide an acceptable solution was due to the need for a combination of results to be developed in nonmonotonic reasoning. These were: Reiter's default logic, a modification of the completion semantics of negation-as-failure in logic programming by Clark as modified by McCain and Turner, the formalization of Reiter's closed world assumption in causal logic, which can be viewed as the use of circumscription by McCarthy; Reiter's solution of the frame problem in the situation calculus; and action languages defined by Gelfond and Lifschitz.

11 Significant Developments and Summary

I have been fortunate to have been working in the field of AI for over 40 years and for over 30 years in nonmonotonic reasoning. I have discussed some of my contributions at this commemorative workshop. I am reluctant to make predictions on future work, since my record is not great in the prediction field. Instead, I briefly discuss some of the great strides that have been made in nonmonotonic reasoning since its start in 1980.

1. The development of theoretical basis for nonmonotonic reasoning by: McCarthy, Reiter, Doyle and McDermott, and Moore. This laid the foundation for further developments.

2. The extension of normal logic programming to include disjunction in the head of clauses and to develop a semantics by Minker, Rajasekar and Lobo; and the extension to include both default and logical negation in the body of normal and disjunctive logic programs by Gelfond and Lifschitz. This allowed more sophisticated knowledge representation and reasoning systems to be developed.

3. The demonstration that there were relationships among classes of the various nonmonotonic theories: autoepistemic logic, circumscription and default reasoning and that logic programming and its extensions can be used to implement nonmonotonic theories. The book by Marek and Truszczynski [1993] provides a rigorous approach to these topics and should be read by all those interested in nonmonotonic reasoning.

4. The development of Answer Set Programming by Lifschitz and Gelfond and the recognition that it is the preferred tool for use with logic programs and nonmonotonic theories.

5. The development of sophisticated and mature ASP systems such as CLASP [Gebser et al., 2009], and commercial systems such as DLV [Leone et al., 2006] and smodels [Niemelä et al., 2000, Niemelä and Simons, 2000, Simons et al., 2002] has allowed many applications to be developed. The book by Baral, *Knowledge Representation, Reasoning and Declarative Problem Solving* [Baral, 2003] has been important for application development.

6. The Dagstuhl Initiative that instituted a test bed for nonmonotonic reasoning systems and for maintaining application and combinatorial problems for testing systems. The competition among different systems has served to advance tools available to users and to sharing insights by system developers for ways to speed up and enhance ASP systems.

7. The development of abductive reasoning within logic programming by Kakas, Kowalski, and Toni [1992] and induction by Plotkin [1970] and by Popplestone [1970], were important. Their relationship to ASP programming was described by Lin and You [2002].

8. The solution of the frame problem by Reiter within the situation calculus and the book by Shanahan [1997] which discusses alternate solutions to the frame problem.

9. The recognition that the situation calculus is a major mechanism for implementing dynamic systems. The 2001 book, *Knowledge in Action*, by Reiter, is a major accomplishment. It should be required reading for anyone involved with research in dynamic systems. The book provides motivation, a rigorous approach using logic and based on the situation calculus, including theorems, examples, code, exercises and bibliographic remarks. The book is important for cognitive robotics, a field named by Reiter, and also for other real world problems.

10. The integration of Judea Pearl's [1988] work on causality and probabilistic reasoning with ASP as in the work of Baral, Gelfond, and Rushton [2009].

I have described some of the early work in nonmonotonic reasoning starting with the work by John McCarthy in 1959, before the field was defined. I described the developments that have influenced my efforts in the field particularly in disjunctive databases and disjunctive logic programming as they relate to nonmonotonic reasoning. Having had the opportunity to participate in research at the start of a new field has been exciting. The field is growing, is vibrant and has exciting young researchers. I am fortunate to have come along at the right time.

Almost all those named in this paper and others who may have been omitted, contributed to my education in nonmonotonic reasoning. I thank them for their research and for their friendship.

One of the great pleasures is to have had very capable and wonderful students, colleagues and visitors with whom I worked with in deductive databases, logic programming and nonmonotonic reasoning. I have not had time to discuss their contributions in this paper. A list of those with whom I worked is contained in the footnote[4] They provided me with much intellectual stimulus. I believe I learned much from them, just as I hope they learned something from me.

Acknowledgments I greatly appreciate Gerhard Brewka, Witek Marek, and Mirek Truszczynski for inviting me to talk at the NonMon@30 Conference, commemorating 30 years of the field of nonmonotonic reasoning. The paper expands on the remarks I made at the workshop. I am indebted to Maarten van Emden, Michael Gelfond and Donald Loveland who read a draft of the paper and made many useful suggestions. Views expressed in this paper are those of the author.

References

J.R. Allen and D. Luckham. An interactive theorem proving program. In Meltzer and Michie, editors, *Machine Intelligence 5*, pages 321–336. University Press, Edinburgh, 1970.

K.R. Apt and M.H. van Emden. Contributions to the Theory of Logic Programming. *J.ACM*, 29(3):841–862, 1982.

K.R. Apt, H.A. Blair, and A. Walker. Towards a Theory of Declarative Knowledge. In J. Minker, editor, *Foundations of Deductive Databases and Logic Programming*, pages 89–148. Morgan-Kaufmann Publishers, Washington, D.C., 1988.

Y. Bar-Hillel, J. McCarthy, and O. Selfridge. Discussion of the paper: Programs with common sense. In V. Lifschitz, editor, *Formalizing Common Sense: Papers by John McCarthy*, pages 17–20. Ablex, London, 1990.

C. Baral. *Knowledge representation, reasoning and declarative problem solving*. Cambridge University Press, 2003.

C. Baral and M. Gelfond. Reasoning agents in dynamic domains. In J. Minker, editor, *Logic-Based Artificial Intelligence*, chapter 12, pages 257–279. Kluwer Academic Publishers, Boston/Dordrecht/London, 2000.

C. Baral and V.S. Subrahmanian. Stable and Extension Class Theory for Logic Programs and Default Logics. *Journal of Automated Reasoning*, 8(3):345–366, 1992.

C. Baral, M. Gelfond, and N. Rushton. Probabilistic reasoning with answer sets. *Theory and Practice of Logic Programming*, 9:55–144, 2009.

H.G. Barrow and G.F. Crawford. The Mark 1.5 Edinburgh Robot Facility. In B. Meltzer and D. Michie, editors, *Machine Intelligence 7*, pages 465–480. American Elsevier Publishing Company, 1972.

P. Borchert, C. Anger, T. Schaub, and M. Truszczynski. Towards systematic benchmarking in Answer Set Programming: The dagstuhl initiative. In M. De Vos and T. Schaub, editors, *Proceedings of the Logic Programming and Nonmonotonic Reasoning Conference*, pages 3–7, 2004.

---

[4]Chitta Baral, Dan Fishman, Jose Alberto Fernandez, Terry Gaasterland, Gallaire, Parke Godfrey, Jarek Gryz, John Grant, John Horty, Madhur Kohli, Sarit Kraus, Jorge Lobo, Jean-Marie Nicolas, Shekhar Pradhan, Arcot Rajasekar, Carolina Ruiz, Dietmar Seiple, VS Subrahmanian, Gerry Wilson, Adnan Yahya and Guy Zanon.

R.S. Boyer and J S. Moore. The Sharing of Structure in Theorem-Proving Programs. In B. Meltzer and D. Michie, editors, *Machine Intelligence 7*, pages 101–116. American Elsevier Publishing Company, 1972.

U.S. Chakravarthy, J. Grant, and J. Minker. Foundations of semantic query optimization for deductive databases. In J. Minker, editor, *Proc. Workshop on Foundations of Deductive Databases and Logic Programming*, pages 67–101, Washington, D.C., August 1986.

A. Chandra and D. Harel. Horn clause queries and generalizations. *Journal of Logic Programming*, 2(1):1–15, April 1985.

C.L. Chang. Deduce 2: Further investigations of deduction in relational databases. In H. Gallaire and J. Minker, editors, *Logic and Databases*, pages 201–236. Plenum Publishers, New York, 1978.

K.L. Clark. Negation as Failure. In H. Gallaire and J. Minker, editors, *Logic and Data Bases*, pages 293–322. Plenum Publishers, New York, 1978.

A. Colmerauer and P. Roussel. The birth of Prolog. In *Proceedings of HOPL-II The second ACM SIGPLAN conference on History of programming languages*, 1993.

A. Colmerauer, H. Kanoui, R. Pasero, and P. Roussel. Un systeme de communication homme-machine en francais. Technical report, Groupe de Intelligence Artificielle Universitae de Aix-Marseille II, Marseille, 1973.

J.L. Darlington. Automatic theorem proving with equality substitutions and mathematical induction. In D. Michie, editor, *Machine Intelligence 3*, pages 113–127. American Elsevier Publishing Company, New York, 1968.

J.L. Darlington. Theorem Proving and Information Retrieval. In B. Meltzer and D. Michie, editors, *Machine Intelligence 5*, pages 173–182. American Elsevier Publishing Company, New York, 1970.

M. Denecker, V.W. Marek, and M. Truszczynski. Uniform semantic treatment of default and autoepistemic logics. *Artificial Intelligence Journal*, 143:73–112, 2003.

T. Eiter, W. Faber, N. Leone, and G. Pfeifer. Declarative problem-solving in dlv. In J. Minker, editor, *Logic-Based Artificial Intelligence*, chapter 4, pages 79–103. Kluwer Academic Publishers, Boston/Dordrecht/London, 2000.

E.W. Elcock. Problem solving compilers. In N.V. Findler and B. Meltzer, editors, *Artificial Intelligence and Heuristic Programming*, pages 37–50. American Elsevier Publishing Company, Inc., 1971.

E.W. Elcock. Absys: The first logic programming language – a retrospective and a commentry. *Journal of Logic Programming*, 9(1):1–17, July 1990.

E.W. Elcock, J.M. Foster, P.D.M. Gray, J.J. McGregor, and A.M. Murray. ABSET: A Programming Language Based on Sets; Motivation and Examples. In B. Meltzer and D. Michie, editors, *Machine Intelligence 6*, pages 467–492. American Elsevier Publishing Company, 1971.

D. Etherington and R. Reiter. On inheritance hierarchies with exceptions. In *Proceedings AAAI 1983*, pages 104–108, 1983.

D. Etherington, R. Mercer, and R. Reiter. On the adequecy of predicate circumscription for closed world reasoning. *Computational Intelligence*, 1:33–57, 1985.

S.E. Fahlman. Design sketch for a million-element netl machine. In *Proceedings AAAI 1980*, pages 249–252, 1980.

R.E. Fikes, P.E. Hart, and N.J. Nilsson. Some New Directions in Robot Problem Solving. In B. Meltzer and D. Michie, editors, *Machine Intelligence 7*, pages 405–430. American Elsevier Publishing Company, 1972.

I. Futo, F. Darvas, and P. Szeredi. The application of prolog to the development of qa and dbm systems. In H. Gallaire J. Minker, editor, *Logic and Data Bases*, pages 347–375. Plenum Press, New York, 1978.

H. Gallaire and J. Minker, editors. *Logic and Databases*. Plenum Press, New York, April 1978.

M. Gebser, B. Kaufmann, and T. Schaub. The conflict-driven answer set solver clasp: Progress report. In *International Conference on Logic Programming and Nonmonotonic Reasoning*, pages 509–514, 2009.

A. Van Gelder. Negation as Failure Using Tight Derivations for General Logic Programs. In J. Minker, editor, *Foundations of Deductive Databases and Logic Programming*, pages 1149–176. Morgan-Kaufmann Publishers, 1988.

A. Van Gelder, K.A. Ross, and J.S. Schlipf. Unfounded Sets and Well-founded Semantics for General Logic Programs. In *Proc. 7th Symposium on Principles of Database Systems*, pages 221–230, 1988.

M. Gelfond and V. Lifschitz. The Stable Model Semantics for Logic Programming. In R.A. Kowalski and K.A. Bowen, editors, *Proc. 5th International Conference and Symposium on Logic Programming*, pages 1070–1080, Seattle, Washington, August 15-19 1988.

M. Gelfond and V. Lifschitz. Negation in logic programs and disjunctive databases. *New Generation Computing*, 9(3/4):365–386, 1991.

M. Gelfond and H. Przymusinska. Negation as failure: Careful closure procedure. *Artificial Intelligence*, 30(3):273–286, December 1986.

M. Gelfond, H. Przymusinska, and T.C. Przymusinski. On the Relationship between Circumscription and Negation as Failure. *Artificial Intelligence*, 38:75–94, 1989.

M. Gelfond, V. Lifschitz, H. Przymusinska, and M. Truszczynski. Disjunctive defaults. In *Proceedings of Principles of Knowledge Representation Conference, KR'91*, pages 230–237, 1991.

M. Ginsberg, editor. *Readings in Nonmonotonic Reasoning*. Morgan-Kaufmann Publishers, 1987.

C.C. Green. Theorem proving by resolution as a basis for question - answering systems. In B. Meltzer D. Michie, editor, *Machine Intelligence 4*, pages 183–205. Edinburgh University Press, New York, 1969a.

C.C. Green. Application of theorem proving to problem solving. In D.E. Walker L.M. Norton, editor, *Proc. International Conference on Artificial Intelligence*, pages 219–240, Washington D.C., 1969b.

C.C. Green. *The Application of Theorem Proving to Question-Answering Systems.* PhD thesis, Computer Science Department Stanford University, June 1969c.

C.C. Green and B. Raphael. Research in intelligent question answering systems. *Proc. ACM 23rd National Conference*, pages 169–181, 1968.

B. Grosof. Generalizing prioritization. In *Proceedings of Knowledge Representation 1991*, pages 289–300, 1991.

P.J. Hayes. Computation and deduction. In *Proceedings of the 2nd Symposium on Mathematical Foundations of Computer Science*, Czechoslovakia: Czechoslovakian Academy of Sciences, 1973.

C.E. Hewitt. PLANNER: A Language for Proving Theorems in Robots. In *First International Joint Conference in Artificial Intelligence*, pages 295–301, 1969.

R. Hill. Lush resolution and its completeness. Technical Report DCL Memo 78, Department of Artificial Intelligence, University of Edinburgh, August 1974.

L. Hodes. Programming languages, logic and cooperative games. *Presented at Symposium for Symbolic and Algebraic Manipulation*, 1966. An abstract of the unpublished paper appears in the August 1966 Issue of the Communications of the ACM.

J. Horty, R. Thomason, and D. Touretzky. A skeptical theory of inheritance in nonmonotonic semantic networks. Technical Report CMU-CS-87-175, Computer Secience Department, Carnegie Mellon University, 1987.

J. Jaffar, J.-L. Lassez, and J.W. Lloyd. Completeness of the negation as failure rule. In *Proceedings Eighth International Joint Conference on Artificial Intelligence*, pages 500–506, Karlsruhe, West Germany, August 1983.

T. Janhunen. Intermediate languages of asp systems and tools. In M. De Vos and T. Schaub, editors, *Proceedings of the 1st International Workshop on Software Engineering for Answer Set Programming*, Tempe, AZ, May 2007.

T. Janhunen and I. Niemelä. A solver for disjunctive logic programs. In *Lecture Notes in Computer Science, 2004 Logic Programming and Nonmonotonic Reasoning*, volume 2923, pages 331–335, 2004.

A.C. Kakas, R.A. Kowalski, and F. Toni. Abductive logic programming. *Journal of Logic and Computation*, 2(6):719–770, 1992.

H. Kautz and B. Selman. The tractability of path-based inheritance. In *International Joint Conference on Artificial Intelligence*, pages 1140–1145, 1989.

H. Kautz and B. Selman. Encoding domain knowledge for propositional planning. In J. Minker, editor, *Logic-Based Artificial Intelligence*, chapter 8, pages 169–186. Kluwer Academic Publishers, Boston/Dordrecht/London, 2000.

C. Kellogg, P. Klahr, and L. Travis. Deductive planning and pathfinding for relational data bases. In H. Gallaire and J. Minker, editors, *Logic and Databases*, pages 179–200. Plenum Press, New York, 1978.

K. Konolige. On the relation between default and autoepistemic logic. *Artificial Intelligence*, 35:343–382, 1988.

R.A. Kowalski. *Studies in the Completeness and Efficiency of Theorem-Proving by Resolution.* PhD thesis, University of Edinburgh, 1970.

R.A. Kowalski. Predicate logic as a programming language. *Proc. IFIPS 4*, pages 569–574, 1974.

R.A. Kowalski. Logic for data description. In H. Gallaire J. Minker, editor, *Logic and Data Bases*, pages 77–102. Plenum Press, New York, 1978.

R.A. Kowalski. *Logic for Problem Solving.* Elsevier North Holland, Inc., New York, 1979.

R.A. Kowalski. The early years of logic programming. *Communications of the ACM*, 31(1): 38–43, January 1988.

R.A. Kowalski and P.J. Hayes. Semantic trees in automatic theorem-proving. In B. Meltzer D. Michie, editor, *Machine Intelligence 4*, pages 87–101. Edinburgh University Press, 1968.

R.A Kowalski and P.J Hayes. Semantic Trees in Automatic Theorem-Proving. In B. Meltzer and D. Michie, editors, *Machine Intelligence 5*, pages 87–101. American Elsevier Publishing Company, 1970.

R.A. Kowalski and D. Kuehner. Linear Resolution with Selection Function. *Artificial Intelligence*, 2:227–260, 1971.

D. Kuehner. A Note on the Relation Between Resolution and Maslov's Inverse Method. In B. Meltzer and D. Michie, editors, *Machine Intelligence 6*, pages 73–76. American Elsevier Publishing Company, 1971.

J.-L. Lassez and M.J. Maher. Closure and fairness in the semantics of programming logic. *Theoretical Computer Science*, 29:167–184, 1984.

N. Leone, G. Pfeifer, W. Faber, T. Eiter, G. Gottlob, S. Perri, and F. Scarcello. The dlv system for knowledge representation and reasoning. *ACM Transactions on Computational Logic (TOCL)*, 7(3), 2006.

H.J. Levesque. All I know: A study in autoepistemic logic. *Artificial Intelligence Journal*, 42 (2-3):263–309, 1990.

V. Lifschitz. Computing circumscription. In *IJCAI 85*, pages 121–127, 1985a.

V. Lifschitz. Computing circumscription. *Proc. Ninth International Joint Conference on Artificial Intelligence*, pages 121–127, 1985b. Morgan Kaufman Publishers.

V. Lifschitz. Pointwise circumscription: A preliminary report. *Proceedings of the AAAI*, pages 406–410, 1986.

V. Lifschitz. On the declarative semantics of logic programs with negation. In J. Minker, editor, *Foundations of Deductive Databases and Logic Programming*, pages 177–192. Morgan Kaufmann Publishers, 1988.

V. Lifschitz. Between circumscription and autoepistemic logic. In *Proceedings of the Conference on Knowledge Representation 1989*, 1989.

V. Lifschitz, N. McCain, E. Remolina, and A. Tacchella. Getting to the airport: The oldest planning problem in ai. In J. Minker, editor, *Logic-Based Artificial Intelligence*, chapter 7, pages 157–165. Kluwer Academic Publishers, Boston/Dordrecht/London, 2000.

F. Lin and J.-H. You. Abduction in logic programming: a new definition and an abductive procedure based on rewriting. *Artificial Intelligence*, 140(1-2):180–190, September 2002.

J.W. Lloyd. *Foundations of Logic Programming*. Springer–Verlag, 1984.

J.W. Lloyd. *Foundations of Logic Programming*. Springer–Verlag, second edition, 1987.

J. Lobo, A. Rajasekar, and J. Minker. Weak Completion Theory for Non-Horn Programs. In R.A. Kowalski and K.A. Bowen, editors, *Proc. 5th International Conference and Symposium on Logic Programming*, pages 828–842, Seattle, Washington, August 15-19 1988. MIT Press.

D.W. Loveland. Theorem-provers combining model elimination and resolution. In B. Meltzer and D. Michie, editors, *Machine Intelligence 4*, pages 73–86. University Press, Edinburgh, 1969a.

D.W. Loveland. A simplified format for the model elimination procedure. *J. ACM*, 16:349–363, July 1969b.

D.W. Loveland. A linear format for resolution. In *Proceedings of the INRIA Symposium on Automatic Demonstration*, pages 147–162. Springer Verlag, New York, 1970.

D.W. Loveland. *Automated Theorem Proving: A Logical Basis*. North–Holland Publishing Co., 1978.

D. Luckham. Some Tree-Paring Strategies for Theorem Proving. In D. Michie, editor, *Machine Intelligence 3*, pages 95–112. American Elsevier Publishing Company, New York, 1968.

D. Luckham. Refinement theorems in resolution theory. In *Lecture Notes in Mathematics 125*, pages 163–190. Springer Verlag, Berlin, 1970.

V. Marek and V.S. Subrahmanian. The relationship between stable, supported, default and autoepistemic semantics for general logic programs. *Theor. Comput. Sci.*, 103(2):365–386, 1992.

V.W. Marek and M. Truszczynski. Relating autoepistemic and default logics. In *Proc. of the First International Conference on Principle of Knowledge Representation and Reasoning*, pages 276–278. Morgan-Kaufmann Publishers, San Mateo, California, 1989a.

V.W. Marek and M. Truszczynski. *Nonmonotonic Logic: Context-Dependent Reasoning (Artificial Intelligence)*. Springer-Verlag, December 1993.

W. Marek. Stable theories in autoepistemic logic. *Fundamenta Informaticae*, 12:243–254, 1989.

W. Marek and M. Truszczynski. Stable semantics for logic programs and default theories. In *Proceedings of NACLP 89*, 1989b.

V.W. Marek and M. Truszczyński. Stable models and an alternative logic programming paradigm. In K.R. Apt, W. Marek, M. Truszczyński, and D.S. Warren, editors, *The Logic Programming Paradigm: a 25-Year Perspective*, pages 375–398. Springer, Berlin, 1999.

S.Ju. Maslov. Proof-Search Strategies for Methods of the Resolution Type. In B. Meltzer and D. Michie, editors, *Machine Intelligence 6*, pages 77–90. American Elsevier Publishing Company, 1971.

J. McCarthy. Programs with common sense. *Mechanization of Thought Processes*, 1, 1959.

J. McCarthy. Situations, actions and causal laws. Technical report, Stanford Artificial Intelligence Project, Stanford University, 1963a. Memo 2.

J. McCarthy. A basis for a mathematical theory of computation. In *Computer Programming and Formal Systems*. North Holland, 1963b.

J. McCarthy. Epistemological problems in artificial intelligence. In *Proc. 5th International Conference on Artificial Intelligence*, pages 1038–1044, 1977.

J. McCarthy. Circumscription - a form of non-monotonic reasoning. *Artificial Intelligence*, 13 (1 and 2):27–39, 1980.

J. McCarthy and P.J. Hayes. Some philosophical problems from the standpoint of artificial intelligence. *Machine Intelligence 4*, 1969. (Reprinted with permission of the publisher and the authors).

D. McDermott and J. Doyle. Nonmonotonic Logic I. *Artificial Intelligence*, 13:41–72, 1980.

B. Meltzer. A New Look at Mathematics and Its Mechanization. In D. Michie, editor, *Machine Intelligence 3*, pages 63–70. American Elsevier Publishing Company, New York, 1968a.

B. Meltzer. Some Notes on Resolution Strategies. In D. Michie, editor, *Machine Intelligence 3*, pages 71–76. American Elsevier Publishing Company, New York, 1968b.

B. Meltzer. Power Amplification for Automatic Theorem-Provers. In B. Meltzer and D. Michie, editors, *Machine Intelligence 5*, pages 165–180. American Elsevier Publishing Company, New York, 1970.

B. Meltzer. Prologomena to a theory of efficiency of proof procedures. In N.V. Findler and B. Meltzer, editors, *Artificial Intelligence and Heuristic Programming*, pages 15–31. American Elsevier Publishing Company, Inc., 1971.

B. Meltzer and D. Michie, editors. *Machine Intelligence*, volume 4. American Elsevier Publishing Company, New York, 1969.

B. Meltzer and D. Michie, editors. *Machine Intelligence*, volume 5. American Elsevier Publishing Company, New York, 1970.

B. Meltzer and D. Michie, editors. *Machine Intelligence*, volume 6. American Elsevier Publishing Company, New York, 1971.

B. Meltzer and D. Michie, editors. *Machine Intelligence*, volume 7. American Elsevier Publishing Company, New York, 1972.

D. Michie, editor. *Machine Intelligence*, volume 3. American Elsevier Publishing Company, New York, 1968.

R.S. Milnikel. Fourth international conference on logic programming and nonmonotonic reasoning (lpnmr '97) dagstuhl, germany, 28-31 july 1997. *AI Communications*, 10(3/4):203–207, 1997.

J. Minker. An experimental relational data base system based on logic. In H. Gallaire and J. Minker, editors, *Logic and Data Bases*, pages 107–147. Plenum Press, New York, 1978.

J. Minker. On indefinite databases and the closed world assumption. In *Lecture Notes in Computer Science 138*, pages 292–308. Springer-Verlag, 1982.

J. Minker. Computing protected circumscription. *Journal of Logic Programming*, 2(4):235–249, 1985.

J. Minker, editor. *Foundations of Deductive Databases and Logic Programming*. Morgan Kaufmann Publishers, 1988.

J. Minker. An overview of nonmonotonic reasoning and logic programming. *Journal of Logic Programming*, 17(2, 3 and 4):95–126, November 1993.

J. Minker, editor. *Logic-Based Artificial Intelligence*. Kluwer Academic Publishers, Boston/Dordrecht/London, 2000.

J. Minker and A. Rajasekar. Procedural Interpretation of Non-Horn Logic Programs. In E. Lusk and R. Overbeek, editors, *Proc. 9th International Conference on Automated Deduction*, pages 278–293, Argonne, IL, 23-26, May 1988.

J. Minker and A. Rajasekar. A fixpoint semantics for disjunctive logic programs. *Journal of Logic Programming*, 9(1):45–74, July 1990.

J. Minker and J.D. Sable. Relational data system study. Technical report, Auerbach Corporation, July 1970.

J. Minker and G. Zanon. An Extension to Linear Resolution with Selection Function. *Information Processing Letters*, 14(3):191–194, June 1982.

J. Minker, J. Lobo, and A. Rajasekar. Circumscription and disjunctive logic programming. In V. Lifschitz, editor, *Artificial Intelligence and Mathematical Theory of Computation*, pages 281–304. Academic Press, 1991.

R. C. Moore. Possible-world semantics for autoepistemic logic. In *Proceedings of the Workshop on Non-Monotonic Reasoning*, pages 344–354, 1984. Reprinted in: M. L. Ginsberg, ed., *Readings on Nonmonotonic Reasoning*, pages 137–142, Morgan Kaufmann, 1990.

R.C. Moore. Semantical considerations on nonmonotonic logic. *Artificial Intelligence*, 25(1): 75–94, 1985.

J-M. Nicolas and H. Gallaire. Data base: Theory vs. interpretation. In H. Gallaire and J. Minker, editors, *Logic and Data Bases*. Plenum Press, New York, 1978.

J-M. Nicolas and J-C. Syre. Natural question-answering and automatic deduction in system syntex. *Proc. IFIP Congress 1974*, pages 595–599, 1974.

J-M. Nicolas and K. Yazdanian. Integrity checking in deductive databases. In H. Gallaire and J. Minker, editors, *Logic and Data Bases*, pages 325–599. Plenum Press, New York, 1978.

I. Niemelä. Logic programs with stable model semantics as a constraint programming paradigm. *Annals of Mathematics and Artificial Intelligence*, 25(3/4):241–273, 1999.

I. Niemelä and P. Simons. Extending the smodels system with cardinality and weight constraints. In J. Minker, editor, *Logic-Based Artificial Intelligence*, chapter 21, pages 491–521. Kluwer Academic Publishers, Boston/Dordrecht/London, 2000.

I. Niemelä, P. Simons and T. Syrjänen, Smodels: A system for Answer Set Programming. *CoRR*, cs.AI/0003033, http://arxiv.org/abs/cs.AI/0003033, 2000.

M. Nogueira, M. Balduccini, M. Gelfond, R.W., and M. Barry. An A-prolog decision support system for the space shuttle. answer set programming. In *PADL 2001*, 2001.

D. Pearce. Sixty years of stable models. In *International Conference on Logic Programming*, page 52, 2008.

J. Pearl. *Probabilistic Reasoning in Intelligent Systems: Networks of Plausible Inference*. Morgan-Kaufmann Publishers, San Mateo, California, 1988.

D. Perlis. On the consistency of commonsense reasoning. *Computational Intelligence*, 2:180–190, 1986.

D. Perlis. Autocircumscription. *Artificial Intelligence*, 36(2):223–236, 1988.

D. Perlis and J. Minker. Completeness results for circumscription. *Artificial Intelligence*, 28 (1):29–42, 1986.

G. Pfeifer. Disjunctive datalog–an implementation by resolution. Master's thesis, Institut für Informationssysteme, Technische Universität Wien, Wien, Österreich, 1996.

F. Pirri and R. Reiter. Natural actions in the situation calculus. In J. Minker, editor, *Logic-Based Artificial Intelligence*, chapter 10, pages 213–231. Kluwer Academic Publishers, Boston/Dordrecht/London, 2000.

G.D. Plotkin. A Note on Inductive Generalization. In B. Meltzer and D. Michie, editors, *Machine Intelligence 5*, pages 153–164. American Elsevier Publishing Company, New York, 1970.

G.D. Plotkin. A Further Note on Inductive Generalization. In B. Meltzer and D. Michie, editors, *Machine Intelligence 6*, pages 101–124. American Elsevier Publishing Company, 1971.

R.J. Popplestone. An Experiment in Automatic Induction. In B. Meltzer and D. Michie, editors, *Machine Intelligence 5*, pages 203–215. American Elsevier Publishing Company, 1970.

D. Prawitz. Advances and problems in mechanical theorem proving. In B. Meltzer and D. Michie, editors, *Machine Intelligence 5*, pages 59–72. American Elsevier Publishing Company, 1970.

T. C. Przymusinski. On the declarative semantics of deductive databases and logic programming. In J. Minker, editor, *Foundations of Deductive Databases and Logic Programming*, chapter 5, pages 193–216. Morgan-Kaufmann Publishers, Washington, D.C., 1988a.

T.C. Przymusinski. On the relationship between logic programming and non-monotonic reasoning. In *Proc. AAAI-88*, pages 444–448, 1988b.

T.C. Przymusinski. Every Logic Program has a Natural Stratification and an Iterated Fixed Point Model. In *"Proceedings of the 8th ACM SIGACT-SIGMOD-SIGART Symposium on Principle of Database Systems"*, pages 11–21, 1989.

T.C. Przymusinski. Autoepistemic logics of closed beliefs and logic programming. In *International Conference on Logic Programming and Nonmonotonic Reasoning*, 1991.

B. Raphael. Sir: A computer program for semantic information retrieval. In M. Minsky, editor, *Semantic Information Processing*, pages 33–134. The MIT Press, Cambridge, Massachusetts, 1968.

B. Raphael. The frame problem in problem solving systems. In N.V. Findler and B. Meltzer, editors, *Artificial Intelligence and Heuristic Programming*, pages 159–169. American Elsevier Publishing Company, Inc., 1971.

R. Reiter. Two results on ordering for resolution with merging and linear format. *J.ACM*, 18: 630–646, October 1971.

R. Reiter. An approach to deductive question-answering. Technical Report BBN Report No. 3649, Bolt, Beranek and Newman, Inc., Cambridge, 1977.

R. Reiter. On Closed World Data Bases. In H. Gallaire and J. Minker, editors, *Logic and Data Bases*, pages 55–76. Plenum Press, New York, 1978a.

R. Reiter. On reasoning by default. In *TINLAP '78 Proceedings of the 1978 workshop on Theoretical issues in natural language processing*, pages 210–218, 1978b.

R. Reiter. Deductive question-answering on relational data bases. In H. Gallaire and J. Minker, editors, *Logic and Data Bases*, pages 149–177. Plenum Press, New York, 1978c.

R. Reiter. A Logic for Default Reasoning. *Artificial Intelligence*, 13(1 and 2):81–132, April 1980.

R. Reiter. Circumscription implies predicate completion (sometimes). *Proc. Amer. Assoc. for Art. Intell. National Conference*, pages 418–420, 1982.

R. Reiter. *Knowledge In Action*. The MIT Press, Cambridge, Massachusetts, 2001.

R. Reiter and G. Criscuolo. On interacting defaults. *Proc. IJCAI 7*, pages 270–276, 1981.

J.C. Reynolds. Transformation Systems and the Algebraic Structure of Atomic Formulas. In B. Meltzer and D. Michie, editors, *Machine Intelligence 5*, pages 135–152. American Elsevier Publishing Company, New York, 1970.

G. Robinson and L. Wos. Paramodulation and Theorem-Proving in First-Order Theories With Equality. In B. Meltzer and D. Michie, editors, *Machine Intelligence 5*, pages 135–150. American Elsevier Publishing Company, New York, 1970.

J.A. Robinson. A machine-oriented logic based on the resolution principle. *J.ACM*, 12(1): 23–41, jan 1965.

J.A. Robinson. The Generalized Resolution Principle. In D. Michie, editor, *Machine Intelligence 3*, pages 77–94. American Elsevier Publishing Company, New York, 1968.

J.A. Robinson. A Note on Mechanizing Higher Order Logic. In B. Meltzer and D. Michie, editors, *Machine Intelligence 5*, pages 121–134. American Elsevier Publishing Company, New York, 1970a.

J.A. Robinson. Mechanizing Higher-Order Logic. In B. Meltzer and D. Michie, editors, *Machine Intelligence 5*, pages 151–170. American Elsevier Publishing Company, New York, 1970b.

J.A. Robinson. Building deduction machines. In N.V. Findler and B. Meltzer, editors, *Artificial Intelligence and Heuristic Programming*, pages 3–13. American Elsevier Publishing Company, Inc., 1971.

K.A. Ross and R.W. Topor. Inferring negative information from disjunctive databases. *Journal of Automated Reasoning*, 4(2):397–424, December 1988a.

K.A. Ross and R.W. Topor. Inferring Negative Information from Disjunctive Databases. *Journal of Automated Reasoning*, 4(2):397–424, December 1988b.

K. Sagonas, T. Swift, and D. S. Warren. XSB as an efficient deductive database engine. In *Proceedings of the SIGMOD 1994 Conference*, pages 442–453, 1994.

E. Sandewall. An approach to the frame problem, and its implementation. In B. Meltzer and D. Michie, editors, *Machine Intelligence 7*, pages 195–204. American Elsevier Publishing Company, 1972.

J.S. Schlipf. How uncomputable is general circumscription? In *Proceedings of the 1st IEEE Symposium on Logic in Computer Science*, Cambridge, MA, 1986.

B. Selman. *Tractable Default Reasoning*. PhD thesis, University of Toronto, 1989. Department of Computer Science.

B. Selman and H.A. Kautz. Model-preference default theories. *Artificial Intelligence*, 45(3): 287–322, October 1990.

M. Shanahan. *Solving the Frame Problem: A Mathematical Investigation of the Common Sense Law of Inertia*. The MIT Press, Cambridge, Massachusetts, 1997.

J.C. Shepherdson. Negation in Logic Programming. In J. Minker, editor, *Foundations of Deductive Databases and Logic Programming*, pages 19–88. Morgan-Kaufman Publishers, 1988.

P. Simons, I. Niemelä, and T. Soininen. Extending and implementing the stable model semantics. *Journal of Artificial Intelligence*, 138(1/2):181–234, 2002.

V.S. Subrahmanian and J. Lu. Protected Completions of First Order General Logic Programs. Technical report, Syracuse University, 1988.

M. Truszczynski. Modal interpretations of default logics. In *Proc. of the International Joint Conference on Artificial Intelligence '91*, pages 393–398. Morgan-Kaufmann Publishers, San Mateo, California, 1991.

M.H. van Emden. Computation and deductive information retrieval. In E. Neuhold, editor, *Formal Description of Programming Concepts*, pages 421–440. North Holland, 1978.

M.H. van Emden and R.A. Kowalski. The Semantics of Predicate Logic as a Programming Language. *J.ACM*, 23(4):733–742, 1976.

D.H.D.. Warren, L.M. Pereira, and F. Pereira. Prolog - the language and its implementation compared with lisp. In *Proc. Symp. on AI and Programming Languages SIGPLAN Notices 12 (8) SIGART Newsletter*, pages 109–115, August 1977.

G.A. Wilson and J. Minker. Resolution Refinements and Search Strategies – A Comparative Study. *IEEE Transactions on Computers*, C-25(8):782–800, 1976. Extended version of the paper appears as TR 470 of University of Maryland.

P.H. Winston. The MIT Robot. In B. Meltzer and D. Michie, editors, *Machine Intelligence 7*, pages 431–463. American Elsevier Publishing Company, 1972.

A. Yahya and L.J. Henschen. Deduction in Non-Horn Databases. *J. Automated Reasoning*, 1 (2):141–160, 1985.

The Gödel and the Splitting Translations[1]

David Pearce
Universidad Politécnica de Madrid
Madrid, Spain
david.pearce@upm.es

Levan Uridia
Universidad Rey Juan Carlos
Madrid, Spain
uridia@ia.urjc.es

Abstract: When the new research area of logic programming and non-monotonic reasoning emerged at the end of the 1980s, it focused notably on the study of mathematical relations between different non-monotonic formalisms, especially between the semantics of stable models and various non-monotonic modal logics. Given the many and varied embeddings of stable models into systems of modal logic, the modal interpretation of logic programming connectives and rules became a dominant view until well into the new Century. Recently, modal interpretations are once again receiving attention in the context of hybrid theories that combine reasoning with non-monotonic rules and ontologies or external knowledge bases.

This paper discusses how some familiar embeddings of stable models into modal logics can be derived from two translations that are very well-known in non-classical logic. They are, first, the translation used by Gödel in 1933 to embed Heyting's intuitionistic logic $\mathcal{H}$ into a modal provability logic equivalent to Lewis's S4; second, the splitting translation, known since the mid-1970s, that allows one to embed extensions of $S4$ into extensions of the non-reflexive logic, $wK4$. By composing the two translations one can obtain an adequate provability interpretation of $\mathcal{H}$ within the Gödel-Löb logic GL, the system shown by Solovay to capture precisely the provability predicate of Peano Arithmetic. These two translations and their composition not only apply to monotonic logics extending $\mathcal{H}$ and $S4$, they also apply in several relevant cases to non-monotonic logics built upon such extensions, including equilibrium logic, non-monotonic $S4F$ and autoepistemic logic. The embeddings obtained are not merely

[1]Based on an invited talk given by the first-named author at the conference on 30 Years of Non-Monotonic Reasoning held in Lexington, Kentucky, USA in October 2010. The paper draws on some joint ongoing work with Pedro Cabalar and Mirek Truszczyński. We are grateful for their permission to make use of this material. Additional thanks are due to Michael Gelfond and Larisa Maksimova for help with some historical matters. The first-named author is grateful to Misha Zakharyaschev for explaining some aspects of the Gödel and splitting translations a number of years ago. Both authors are indebted to the late Leo Esakia for his friendship and inspiration. This paper is dedicated to Victor Marek and Mirek Truszczyński who have done so much to turn the area of non-monotonic logic into a systematic discipline. Work reported here was partially supported by the MICINN projects TIN2009-14562-C05 and CSD2007-00022.

faithful and modular, they are based on fully recursive translations applicable to arbitrary logical formulas. Besides providing a uniform picture of some older results in non-monotonic logic, the translations yield a perspective from which some new logics of belief emerge in a natural way.

1 Introduction

When the new area of logic programming and non-monotonic reasoning emerged towards the end of the 1980s much attention was paid to the semantics of stable models and its relation to systems of non-monotonic modal logic. Several different formal embedding relations of stable model semantics into modal logics were discovered in papers by Gelfond [1987], Gelfond and Lifschitz [1988], Przymusinski [1991], Lifschitz and Schwarz [1993], Marek and Truszczyński [1993], Chen [1993] and others. In this manner the modal-epistemic interpretation of logic programming rules was born and became a dominant view in the 1990s and even well into the new Century.

These embedding relations are interesting for several reasons. For one thing, it is well-known that provability in formal systems can be given a modal interpretation and so the new negation as failure-to-prove in logic programming might perhaps turn out to be related in a natural way to modal concepts of this type. Secondly, logic programs had a special kind of syntax based on *rules*, while their modal translations were sets of ordinary logical formulas. So the embeddings seemingly related quite different kinds of syntactic objects. Thirdly, since stable models could be related to extensions in Reiter's default logic [Reiter, 1980, Gelfond and Lifschitz, 1991], these embeddings also provided a method to connect default and epistemic logics. Fourthly, while the translations were not arbitrary (cf. provability concepts of modality), they were largely *ad hoc*. That is to say, they were discovered and found to work in a practical sense, but were not based on a single underlying, systematic methodology. Indeed it seemed that as one moved from simpler to more complex types of program rules, their modal translations actually changed. A comprehensive theory to explain this was lacking. Moreover, while the successful embedding relations were *modular*, they were generally defined on complete rules, one at a time, rather than being built up recursively from the rule components.

The aim of this paper is to revisit the modal and epistemic interpretation of logic programs under stable model semantics and try to supply an overarching theory that links these different translations and explains in a certain sense why they work. What we shall try to show is how some of the familiar embeddings of stable models into modal logics can be seen as special cases of two translations that are very well-known in non-classical logic. They are, first, the translation used by Gödel in 1933 to embed Heyting's intuitionistic logic $\mathcal{H}$ into a modal provability logic equivalent to Lewis's $S4$; second, the splitting translation, known since the mid-1970s, that allows one to embed extensions of $S4$ into extensions of the non-reflexive logic, $wK4$.

We hope this approach will be of historical as well as didactical interest. But it may also prove useful beyond this. Recently, modal interpretations of logic programs are once again receiving attention in the context of hybrid theories that combine reasoning with non-monotonic rules and ontologies or external knowledge bases, often considered in the framework of the so-called Semantic Web. It seems plausible that at least some of these interpretations – since they involve translating answer set programs – will also be closely related to the Gödel and the splitting translations. Another feature of interest is that when we build up a picture of different logics, monotonic and non-monotonic, related by these translations, we actually discover some gaps in the picture that can be filled by 'new' logics that may be of interest in their own right. We will encounter an example of this later in the paper.

1.1 Plan of the paper

The rest of the paper will be organised around commutative diagrams of different logics and their inter-relations. We will start with the basic case that was dealt with by Gödel and by Tarski and McKinsey: that of intuitionistic logic and modal $S4$. As we proceed we will extend this diagram to include more logics, adding also the splitting translation. Eventually we will have a set of base logics, their extensions and also their non-monotonic versions. Once this diagram is complete we will return to some of the embeddings of stable model semantics that arose in the early years of logic programming and non-monotonic reasoning. We will see how to derive those embeddings using the two basic translations that we started with. We will conclude by discussing some possible extensions of this approach as well as some of its limitations.

2 Logical preliminaries

We assume familiarity with basic intuitionistic and modal propositional logic. What follows is a brief summary that serves mainly to fix notation and terminology. For more details the reader should consult the numerous introductions that can be found in handbook chapters and several textbooks, e.g. those by van Dalen [2004], Chellas [1980] and Chagrov and Zakharyaschev [1997]. We consider propositional languages equipped with an infinite set $Prop$ of propositional letters. In both cases the symbols $\vee, \wedge, \neg, \rightarrow$ denote the propositional connectives disjunction, conjunction, negation and implication, respectively.[2] We consider normal modal logics containing the additional necessity operator, L. We use $M\varphi$ as an abbreviation for the proposition $\neg L\neg\varphi$. The axioms are all classical tautologies plus a selection of the axioms listed below. Rules of inference are: modus ponens, substitution and necessitation.

$$K : L(p \rightarrow q) \rightarrow (Lp \rightarrow Lq)$$
$$4 : Lp \rightarrow LLp$$
$$w4 : Lp \wedge p \rightarrow LLp$$
$$5 : \neg L\neg Lp \rightarrow Lp$$
$$W5 : \neg L\neg Lp \rightarrow (p \rightarrow Lp)$$
$$D : \neg Lp \vee \neg L\neg p$$
$$T : Lp \rightarrow p$$
$$F : (p \wedge MLq) \rightarrow L(Mp \vee q)$$
$$f : p \wedge M(q \wedge L\neg p) \rightarrow L(q \vee Mq)$$

---

[2]Context should make it clear whether we are dealing with an intuitionistic connective or a classical one.

Well-known examples of modal logics we shall deal with are $S4$, containing the axioms K, T and 4, the logic $S5$ which extends $S4$ by adding axiom 5, and $KD45$ comprising precisely $K, D, 4$ and 5.

Intuitionistic propositional logic is determined by the usual intuitionistic axioms and the rules modus ponens and substitution (see e.g. the book by van Dalen [2004]). We denote this calculus by $\mathcal{H}$ for Heyting. *Super-intuitionistic logics* are obtained by adding further axioms to $\mathcal{H}$. Here we consider one such logic in particular, the logic of *here-and-there*, denoted by HT. This is obtained by adding to $\mathcal{H}$ the axiom of Hosoi [1966]:

$$\alpha \vee (\neg\beta \vee (\alpha \to \beta))$$

HT is the strongest extension of $\mathcal{H}$ that is properly contained in classical logic. It can be equivalently presented as a 3-valued logic and is sometimes known as Gödel's 3-valued logic.

Semantics. An (intuitionistic, Kripke) *frame* is a pair $\mathcal{F} = \langle W, \leq \rangle$, where W is a non-empty set and $\leq$ a partial ordering on W. A Kripke *model* $\mathcal{M}$ is a frame together with a valuation $V : Prop \times W \to \{0, 1\}$. V is extended recursively to all formulas by the usual truth conditions for intuitionistic Kripke semantics (van Dalen [2004] and other standard reference texts provide a detailed discussion of these concepts). In particular we have the following clause for implication

$$V(\alpha \to \beta, w) = 1 \quad \text{iff} \quad \text{for all } w' \geq w, \ V(\alpha, w') = 1 \Rightarrow V(\beta, w') = 1 \quad (1)$$

Given a model $\mathcal{M} = \langle W, \leq, V \rangle$ we also write $\mathcal{M}, w \models \varphi$ to denote $V(\varphi, w) = 1$. Similarly, an intuitionistic formula φ is said to be *true* in a model $\mathcal{M}$, denoted by $\mathcal{M} \models \varphi$, if $\mathcal{M}, w \models \varphi$ for all $w \in W$ (likewise for sets of formulas). A formula is valid on a class of frames if it is true in all models based on those frames. Heyting's calculus $\mathcal{H}$ is sound and complete with respect to this semantics, in particular the theorems of $\mathcal{H}$ are precisely the formulas valid on the class of all frames. For any super-intuitionistic logic $\mathcal{I}$ based on some class $\mathcal{K}$ of frames, we can define a concept of (local) *consequence* as follows: φ is a consequence of Σ, in symbols $\Sigma \models_{\mathcal{I}} \varphi$, if for any model $\mathcal{M} = \langle W, \leq, V \rangle$ where $\langle W, \leq \rangle$ is a frame in $\mathcal{K}$, and any point $x \in W$, $\mathcal{M}, x \models \varphi$ whenever $\mathcal{M}, x \models \Sigma$.

The logic HT is based on rooted frames having just two elements, h and t, with $h \leq t$. It is often convenient to represent an HT-model $\langle \{h, t\}, \leq V \rangle$ as an ordered pair of sets of atoms, $\langle H, T \rangle$, where $H = \{p \in Prop : V(p, h) = 1\}$ and similarly $T = \{p \in Prop : V(p, t) = 1\}$. By the usual persistence requirement it follows that $H \subseteq T$. Truth, validity and consequence for HT are defined in the usual way.

Similarly a *modal frame* $\mathcal{F} = \langle W, R \rangle$ comprises a set W together with a binary relation R on W. Again, modal models are frames equipped with a valuation $V : Prop \times W \to \{0, 1\}$, where V is extended recursively to all formulas by the usual truth conditions. In particular we have the standard condition for necessity:

$$V(L\varphi, w) = 1 \quad \text{iff} \quad V(\varphi, w') = 1 \quad \text{for all } w' \text{ s.t. } wRw'. \quad (2)$$

For modal logics the notions of truth, validity and consequence are defined analogously to the intuitionistic cases described above and we adopt similar notational conventions. For a normal modal logic $\mathcal{S}$ we denote its consequence relation by $Cn_{\mathcal{S}}$, i.e. $Cn_{\mathcal{S}}(I)$ is the set of consequences of a set of formulas I.

For Kripke structures, we will deal with properties of relations such as *weak-transitivity*, *quasi order* and classes defined by these properties. Below we give the definition of some of these properties and point out some important classes of frames.

Definition 1 *We say that a relation $R \subseteq W \times W$ is a quasi order if it is reflexive and transitive i.e. $(\forall x)(xRx)$ and $(\forall x, y, z)\ (xRy \wedge yRz \Rightarrow xRz)$.*

Quasi orders may contain sets where every two points are related to each other. These sets are called *clusters*. More formally $U \subseteq W$ in a Kripke frame (W, R) is called a cluster if $(\forall w, v \in U)(wRv)$. We say that a Kripke frame (W, R) is a cluster if W is a cluster in (W, R). A weaker version of transitivity, occurs in weakly-transitive frames.

Definition 2 *We say that a relation $R \subseteq W \times W$ is weakly-transitive if $(\forall x, y, z)$ $(xRy \wedge yRz \wedge x \neq z \Rightarrow xRz)$.*

Clearly every transitive relation is weakly-transitive as well. Moreover, the only difference between weakly-transitive and transitive frames is that weakly-transitive relations allow irreflexive points inside the sets where every two distinct points are related with each other. Such sets will be called *weak-clusters*. More precisely $U \subseteq W$ in a Kripke frame (W, R) is called a weak-cluster if $(\forall w, v \in U)(w \neq v \rightarrow wRv)$. It is easy to see that original definition of cluster given above enforces reflexivity, while weak-clusters may allow for irreflexive points. On the other hand if a weak-cluster is transitive (but not weakly-transitive) than reflexivity follows automatically and hence onae obtains a cluster. In the same way as reflexive weak clusters are clusters, it is easy to notice that quasi orders are obtained by adding the reflexivity condition to weakly-transitive relations.

An important class of frames are those known as *cluster-closed* Schwarz [1992]. As we will see in Section 4, such classes of frames are used to provide a minimal model semantics for non-monotonic logics.

Definition 3 *Let $\mathcal{N} = (N, S, U)$ be a Kripke model. A nonempty set $W \subseteq N$ is called a final cluster if:*
a) W is an upper cone i.e. $w \in W$ and wSv implies that $v \in W$,
b) W is cluster,
c) For every $v \in N - W$ and for every $w \in W$, vRw.

Observe that a quasi order may not always have a final cluster, for example if it contains an infinite ascending chain. However, if the quasi order has a maximum then it does have a final cluster.

Definition 4 *Let $\mathcal{N} = (N, S, U)$ be a Kripke model and let N_2 be its final cluster. Let $\mathcal{M} = (W, R, V)$ be a cluster. By cluster substitution of $\mathcal{M}$ in $\mathcal{N}$ we mean the model $< (N - N_2) \cup W, S', V' >$, where for each $w, v \in (N - N_2) \cup W, wS'v$ if and only if wSv or $v \in W$ and V' agrees with U on $(N - N_2)$ and agrees with V on W. In other words we substitute the cluster W instead of the weak-cluster N_2 into $\mathcal{N}$.*

Definition 5 *By the concatenation of two models (W, R, V) and (N, S, U) with $W \cap N = \varnothing$ we mean the model $(N \cup W, S \cup N \times W \cup R, U \cup V)$.*

We can now give define the notion of of cluster closed class of frames.

Definition 6 *Let C be a class of models. We say that C is cluster closed if C contains all clusters and for each $\mathcal{N} \in C$, at least one of the following two conditions holds: the concatenation of $\mathcal{N}$ and each cluster belongs to C, or $\mathcal{N}$ has a final cluster and for each $S5$-model $\mathcal{M}$, the cluster substitution of $\mathcal{M}$ in $\mathcal{N}$ belongs to C.*

Note that the classes of all quasi orders, all weakly-transitive relations, and of all clusters are cluster closed, while for example the collection of all weak-clusters is not. Although the latter contains the set of all clusters it is easy to check that it does not contain a final cluster and neither is it closed under concatenation. Nevertheless we shall see in Section 4 that the class of all weak-clusters can be applied to give a minimal model semantics for non-monotonic logics. In particular we can apply a concept of weak-cluster closed class, whose definition is easily obtained by substituting everywhere weak-cluster for cluster in Definitions 3,4,5 and 6.

2.1   The Gödel translation

Gödel [1933] provided an interpretation of $\mathcal{H}$ into a logical system equivalent to $S4$. His original translation of intuitionistic into modal formulas, denoted here by τ, is defined as follows:

$$
\begin{aligned}
\tau(p) &= Lp \\
\tau(\varphi \wedge \psi) &= \tau(\varphi) \wedge \tau(\psi) \\
\tau(\varphi \vee \psi) &= \tau(\varphi) \vee \tau(\psi) \\
\tau(\varphi \rightarrow \psi) &= L(\tau(\varphi) \rightarrow \tau(\psi)) \\
\tau(\neg\varphi) &= L\neg\tau(\varphi)
\end{aligned}
$$

If we add $\tau(\bot) = \bot$, then the translation of $\neg\varphi$ becomes derivable via the definition $\neg\varphi := \varphi \rightarrow \bot$. Gödel also noted that variations on this translation will also work; a frequently used variant is the translation that places an 'L' before every subformula[3]. Gödel essentially established that for an intuitionistic formula φ,

$$
\vdash_{\mathcal{H}} \varphi \Rightarrow \vdash_{S4} \tau(\varphi), \tag{3}
$$

(where $\vdash$ with subscripts denotes theoremhood) and he conjectured that the converse relation also holds, in other words that τ is a faithful interpretation or *embedding*. This conjecture was later proved by McKinsey and Tarski [1948].

Following Tarski [1939], McKinsey and Tarski [1944, 1946, 1948] also laid the foundations for the algebraic and topological study of intuitionistic and modal logics. The basic idea, recalled and developed in a recent paper by Leo Esakia [2004], is that from an arbitrary topological space X we can generate three different algebraic structures each giving rise to different logical systems[4]. In particular, by considering the algebra of open sets, $Op(X)$, one is led to the well-known Heyting algebra that also forms a semantical basis for intuitionistic logic. On the other hand by considering

---

[3]The version of the Gödel translation given here by τ is actually the one used by McKinsey and Tarski [1948] and is commonly found in the literature.

[4]For the basic notions of topology see e.g. the monograph by Kuratowski [1976] or any appropriate textbook.

the closure algebra, $(\mathcal{P}(X), \mathbf{c})$ one is led to the modal system $S4^5$. McKinsey and Tarski [1948] proved the topological completeness of $S4$ where M is interpreted as the closure operator $\mathbf{c}$.

By the 1970s logicians began the systematic study of relations between (the lattice of) extensions of $\mathcal{H}$ and (the lattice of) normal extensions of $S4$. Also, since (as Gödel [1933] already observed) $S4$-necessity does not correspond to provability in a formal system, interest grew in finding modal systems that directly model provability concepts. Around 1976 Esakia introduced the expression "modal companion" of a super-intuitonistic logic $\mathcal{I}$ to refer to those modal systems into which $\mathcal{I}$ can be embedded via the Gödel translation. Esakia [1976b, 1979] established that $\mathcal{H}$ has a strongest modal companion, the so-called Grzegorczyk logic, Grz, already known [Grzegorczyk, 1967] to be a modal companion of $\mathcal{H}$. Grz is the extension of $S4$ obtained by adding the schema

$$L(L(p \to Lp) \to p) \to p.$$

Esakia's result is part of a more general pattern: the Blok-Esakia Theorem establishes an isomorphism between the lattice of intermediate logics and the lattice of all normal extensions of Grz [Blok, 1976, Esakia, 1976b].

2.2 The splitting translation

In the same year, Solovay [1976] proved the arithmetical completeness of the so-called Gödel-Löb modal logic, GL, establishing a correspondence between derivability in GL and provability in the formal system of Peano Arithmetic, PA. GL results from $K4$ by adding the schema

$$L(Lp \to p) \to Lp.$$

At this point it is appropriate to turn to the splitting translation that will be denoted here by a superscript operator '$+$'. This is a translation from modal formulas to modal formulas that replaces each occurrence of L by L^+ where $L^+\varphi$ abbreviates

$$\varphi \wedge L\varphi,$$

other formulas being left unchanged. It appears that the splitting translation was independently discovered by several authors and its first main application is usually attributed to Kuznetsov and Muravitsky [1976], Goldblatt [1978] and Boolos [1980]. They established the following embedding of the reflexive logic Grz into the non-reflexive GL.

$$\vdash_{Grz} \varphi \Leftrightarrow \vdash_{GL} \varphi^+. \tag{4}$$

Not surprisingly, as Goldblatt [1978] showed, one can form the composition τ^+ of τ with $^+$ and combine (3) and (4) to yield

$$\vdash_{\mathcal{H}} \varphi \Leftrightarrow \vdash_{GL} \tau^+(\varphi), \tag{5}$$

---

[5]Recall that a Heyting algebra $(\mathbf{H}, \vee, \wedge, \to, \perp)$ is a distributive lattice with smallest element $\perp$ containing a binary operation $\to$ such that $x \le a \to b$ iff $a \wedge x \le b$. $(\mathbf{B}, \vee, \wedge, -, \mathbf{c})$ is a closure algebra if $(\mathbf{B}, \vee, \wedge, -)$ is a Boolean algebra and $\mathbf{c}$ is a closure operator satisfying: $a \le \mathbf{c}a, \mathbf{cc}a = \mathbf{c}a, \mathbf{c}(a \vee b) = \mathbf{c}a \vee \mathbf{c}b, \mathbf{c}\perp = \perp$.

(setting $\tau^+(\varphi) = (\tau(\varphi)^+)$) which using Solovay's result yields a provability interpretation of intuitionistic logic.

If we depict these relations in diagrammatic form, from the simple picture

$$\mathcal{H} \xrightarrow{\ \tau\ } S4$$

we have now reached the following situation

$$
\begin{array}{ccc}
 & Grz \xrightarrow{\ +\ } GL \\
 & \nearrow^{\tau} \quad \uparrow \\
\mathcal{H} & \xrightarrow{\ \tau\ } S4
\end{array}
$$

The splitting translation has two very natural interpretations. One of them is topological. One path from topology to logic is via what are known as *derivative algebras*, $(\mathcal{P}(X), der)$. These are Boolean algebras with a unary operation der representing topological derivation: if A is a subset of X then $der(A)$ is the set of all accumulation or limit points of A. Under this topological reading of modal logics Mp is interpreted as derivation. Since in topology closure is definable in terms of derivation, viz. for a point set A, $\mathbf{c}(A) = A \cup der(A)$, we obtain the following modal connection: Mp is identified with $p \vee Mp$. But this is precisely the splitting translation when applied to M, namely $M^+p = p \vee Mp$.

Esakia [2001] showed that the derivative algebra $(P(X), der)$ gives rise to the modal logic $wK4$, a slightly weaker version of the logic $K4$, that was first studied from a topological point of view by Esakia [1976a] (Esakia [2004] provides a detailed overview). The logic $wK4$ is obtained by adding the axiom schema $w4$ to K. It is known to be the weakest normal extension of K into which $S4$ embeds via the splitting translation. This means that we can now complete our previous picture by adding in a missing logic, namely the logic of all topological spaces interpreting M as the derivative operator:[6]

$$
\begin{array}{ccc}
 & Grz \xrightarrow{\ +\ } GL \\
 & \nearrow^{\tau} \quad \uparrow \qquad \uparrow \\
\mathcal{H} & \xrightarrow{\ \tau\ } S4 \xrightarrow{\ +\ } wK4
\end{array}
$$

The logics on the right of our picture are interesting from another perspective as well. While extensions of $S4$ may be considered good candidates for modelling epistemic reasoning about knowledge, extensions of $wK4$ are candidates to form logics of belief. In particular they may lack the strong T axiom $Lp \rightarrow p$. A standard doxastic logic, $KD45$, is one such extension of $wK4$. From this epistemic perspective the splitting translation is also very natural. After interpreting Lp in extensions of $S4$ as "the agent knows p" and in extensions of $wK4$ as "the agent believes p", the splitting translation gives us the interpretation of knowledge as 'truth plus belief'.

---

[6] Vertical arrows always denote extensions.

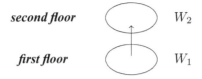

Figure 1: Reflexive frames

3 Monotonic embeddings

Up to this point we have remarked how the systems $\mathcal{H}$, $S4$ and $wK4$ represent three distinct paths to logic via topology, each logic captured in its own way by the class of all topological spaces. We also saw how the Gödel translation τ embeds $\mathcal{H}$ into $S4$ and how $wK4$ is the least logic into which $S4$ embeds via the splitting translation, $^{+}$.

Let us now turn to logics obtained from *minimal* topological spaces. A topological space is minimal if it has only three open sets. Considering the algebra of open sets we arrive at the 3-element Heyting algebra which captures precisely the logic HT of here-and-there. Taking as a starting point the closure algebra $(\mathcal{P}(X), \mathbf{c})$ we obtain the modal logic $S4F$ that adds the F axiom schema to $S4$. The semantics of $S4F$ was first studied by Segerberg [1971]. However Schwarz and Truszczyński [1992] and Schwarz and Truszczyński [1994] proposed a new approach to minimal knowledge and suggested that this concept is precisely captured by non-monotonic $S4F$. Schwarz and Truszczyński [1994] have shown that non-monotonic $S4F$ captures, under some intuitive encodings, several important approaches to knowledge representation. They include disjunctive logic programming under answer set semantics [Gelfond and Lifschitz, 1991], (disjunctive) default logic [Reiter, 1980, Gelfond et al., 1991], the logic of grounded knowledge [Lin and Shoham, 1990], the logic of minimal belief and negation as failure [Lifschitz, 1994] and the logic of minimal knowledge and belief [Schwarz and Truszczyński, 1994]. Recently, Truszczyński [2007] and Cabalar and Lorenzo [2004] have revived the study of $S4F$ in the context of a general approach to default reasoning.

Frames for $S4F$ have the form depicted in Figure 1. where W_1 and W_2 are clusters, all points are reflexive and every point in W_2 is accessible from every point in W_1. We call W_1, W_2 respectively the *first* and the *second floor* of the model. The former may be empty but the latter not. In these frames the acessibility relation is of course a preorder or quasi-order.

If the possibility operator 'M' is construed as topological derivation then, as Pearce and Uridia [2010] recently showed, as the logic corresponding to the class of minimal topological spaces one obtains $wK4f$, the system extending $wK4$ by adding the schema f. Pearce and Uridia [2010] also establish the soundness and completeness of $wK4f$ with respect to a class of modal frames. The frames for $wK4f$ are similar to those of $S4F$ except that we drop the condition of reflexivity on frames: in Figure 2 some points in W_1, W_2 may now be irreflexive (where i and r label this difference). Accessibility for these frames is only weakly-transitive [Pearce and Uridia, 2010].

Therefore by considering logics extending H, $S4$ and $wK4$ based on *minimal*

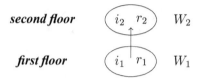

second floor $i_2 \quad r_2$ W_2

first floor $i_1 \quad r_1$ W_1

Figure 2: Reflexive and irreflexive points

topological spaces, we obtain respectively the systems HT, $S4F$ and $wK4f$. So our picture now looks like this:

$$min-topologies \qquad HT \qquad S4F \qquad wK4f$$

$$topologies \qquad H \xrightarrow{\;\tau\;} S4 \xrightarrow{\;+\;} wK4$$

Our main objective is to consider non-monotonic versions of the logics HT, $S4F$ and $wK4f$ and to show how the first of these, equilibrium logic, in the diagram below EL, embeds into the other two. In diagrammatic form we want to show

$$nonmonotonic \qquad EL \xrightarrow{\;\tau\;} S4F^* \xrightarrow{\;+\;} wK4f^*$$

where the starred version of a modal logic indicates its non-monotonic version. Rather than attempt to do this directly, we will build on the embedding relations that hold at the monotonic level and then show how these can be lifted to the non-monotonic cases. So we will deal with this picture (where dotted arrows denote non-monotonic extensions):

$$nonmonotonic \qquad EL \xrightarrow{\;\tau\;} S4F^* \xrightarrow{\;+\;} wK4f^*$$

$$min-topologies \qquad HT \xrightarrow{\;\tau\;} S4F \xrightarrow{\;+\;} wK4f$$

Actually the embeddings at the bottom of this picture can be established by standard methods. But since these are quite general and applicable to many logics, it may be a useful exercise to review them here. We will omit detailed proofs but give the main ideas and lemmas needed.

Let $\mathcal{F} = \langle W, R \rangle$ be a modal frame. We define the *reflexivization* of $\mathcal{F}$ to be the frame $\mathcal{F}^r = \langle W, R^r \rangle$ where

$$xR^r y \quad \text{iff} \quad x = y \quad \text{or} \quad xRy. \tag{6}$$

It is easy to see that if $\mathcal{F}$ is a one-step, weakly transitive frame for $wK4f$ then $\mathcal{F}^r$ is transitive and a frame for $S4F$. Given a model $\mathcal{M} = (\mathcal{F}, V)$, $\mathcal{M}^r$ is the model

$(\mathcal{F}^r, V)$.[7]

Lemma 1 *For every model $\mathcal{M}$, every point x in $\mathcal{M}$ and every formula φ,*

$$\mathcal{M}, x \models \varphi^+ \text{ iff } \mathcal{M}^r \models \varphi.$$

The proof is by induction on φ. From this "reflexivization" lemma we can deduce that for every frame $\mathcal{F}$ and formula φ,

$$\mathcal{F} \models \varphi^+ \text{ iff } \mathcal{F}^r \models \varphi. \tag{7}$$

On this basis we can relate the frames for $wK4f$ to those for $S4F$. Now consider the Gödel translation, τ. Corresponding to this there is a semantic map that relates quasi-ordered modal frames to partially ordered intuitionistic frames. We denote this map by ρ and define it as follows. Let $\mathcal{F}$ be a quasi-ordered frame and $\mathcal{M} = \langle \mathcal{F}, V \rangle$ a model based on it. The *skeleton* $\rho\mathcal{F}$ of $\mathcal{F}$ is obtained from $\mathcal{F}$ by collapsing clusters to single points, resulting in a partially-ordered frame. The corresponding model $\rho\mathcal{M} = \langle \rho\mathcal{F}, \rho V \rangle$, called the *skeleton* of $\mathcal{M}$, is defined by the intuitionistic valuation ρV which for each variable p, sets

$$\rho V(p) = \{C(x) : \mathcal{M}, x \models Lp\}, \tag{8}$$

where $C(x)$ is the cluster generated by x, i.e. it is a single point of $\rho\mathcal{F}$. □
   The following "skeleton lemma" is another well-known result that is proved by induction on formulas φ.

Lemma 2 *Let $\mathcal{M}$ be a modal model based on a quasi-ordered frame. Then for every point x in $\mathcal{M}$ and every modal-free formula φ,*

$$\rho\mathcal{M}, C(x) \models \varphi \text{ iff } \mathcal{M}, x \models \tau(\varphi).$$

Likewise we can deduce that for every quasi-ordered frame $\mathcal{F}$ and modal-free formula φ we have

$$\rho\mathcal{F} \models \varphi \text{ iff } \mathcal{F} \models \tau(\varphi). \tag{9}$$

We define the translation τ^+ by setting $\tau^+(\varphi) = (\tau(\varphi))^+$. Then our embedding theorem can be stated as follows.

Theorem 1 *The translation τ^+ is an embedding of HT into $wK4f$, i.e. for any intutionistic formula φ,*

$$\models_{HT} \varphi \text{ iff } \models_{wK4f} \tau^+(\varphi).$$

Proof. Suppose that $\not\models_{wK4f} \tau^+(\varphi)$. Then there is a $wK4f$ frame $\mathcal{F}$ such that $\mathcal{F} \not\models \tau^+(\varphi)$. Hence $\mathcal{F}^r \not\models \tau(\varphi)$ and so $\rho(\mathcal{F}^r) \not\models \varphi$. However $\rho(\mathcal{F}^r)$ is isomorphic to the HT-frame $\mathcal{H}$, meaning that $\not\models_{HT} \varphi$. For the other direction, if $\not\models_{HT} \varphi$ then $\mathcal{H} \not\models \varphi$. Hence $\mathcal{F}^r \not\models \tau(\varphi)$ and so $\mathcal{F} \not\models \tau^+(\varphi)$. □
   Evidently we can decompose τ^+ to deduce that τ embeds HT into $S4F$ and $^+$ embeds $S4F$ into $wK4f$.

---

[7]The following few steps and results are quite standard and well-known in modal logic. The first-named author learnt the method from Misha Zakharyaschev who several years ago supplied him a detailed account and proofs. The reader is referred to a standard reference for this topic, the paper by Chagrov and Zakharyaschev [1992], or to their mnograph [Chagrov and Zakharyaschev, 1997].

4   Lifting embeddings to the non-monotonic case

Now we turn to the non-monotonic versions of these logics. For the case of a normal modal logic S the standard way to define its non-monotonic version that we will denote by S^* is via a fixpoint condition that defines the so-called S-expansions.

Definition 7 *Let S be a normal modal logic with consequence relation Cn_S. Let I be a set of modal formulas. A set of formulas E is said to be an S-expansion of I if*

$$E = Cn_S(I \cup \{\neg L\varphi : \varphi \notin E\}).$$

Then S^* is the non-monotonic logic determined by truth in all S-expansions, i.e. we can define the non-monotonic entailment relation by $I \mathrel{\vicenter\triangleright}_{S^*} \varphi$ iff $\varphi \in E$ for each S-expansion E of I.

Similarly, the logic HT has a natural non-monotonic extension that we call *equilibrium logic, EL* [Pearce, 1997]. This can also be captured by a similar fixpoint condition Pearce [1999].

Definition 8 *Let X be a set of intuitionistic formulas. A set of formulas C is said to be a* completion *of X iff*

$$C = Cn_{HT}(X \cup \{\neg\varphi : \varphi \notin C\}).$$

Again equilibrium logic is determined by truth in all completions.[8]

For proving results about these logics it is often easier to work with equivalent characterisations using minimal or preferred models. For the case of EL this was the original definition of the logic in terms of special kinds of minimal HT-models called *equilibrium* models.

Definition 9 *Among HT-models we define the order $\trianglelefteq$ by: $\langle H, T\rangle \trianglelefteq \langle H', T'\rangle$ if $T = T'$ and $H \subseteq H'$. If the subset relation is strict, we write '$\lhd$'.*

Definition 10 *Let X be a set of intuitionistic formulas and $\mathcal{M} = \langle H, T\rangle$ a model of X. $\mathcal{M}$ is said to be an* equilibrium *model of X if it is minimal under $\trianglelefteq$ among models of X, and it is total, i.e. $H = T$.*

Equilibrium models of theories correspond to their completions, so we can define inference by

$$X \mathrel{\vicenter\triangleright} \varphi \quad \text{iff} \quad \mathcal{M} \models \varphi \text{ for each equilibrium model } \mathcal{M} \text{ of } X. \tag{10}$$

Certain kinds of non-monotonic modal logics are captured by a concept of minimal model that was introduced by Schwarz [1992].

Definition 11 *Let $\mathcal{N} = (N, S, U)$ be a two-floor S4F model as depicted in Figure 1. We say that $\mathcal{N}$ is preferred over an S5-model $\mathcal{M} = (W, R, V)$ if:*
a) There is a propositional (ie. modal-free) formula ψ such that $\mathcal{M} \models \psi$ and $\mathcal{N} \not\models \psi$,
b) (W, R) is the second floor of (N, S) and V equals to the restriction of U to the second floor. Briefly, $\mathcal{M}$ is the model which is obtained by deleting the first floor in $\mathcal{N}$.

---

[8]Pearce [1999] calls completions *negation-stable expansions*.

From this one obtains the notion of minimal model that is central for the semantics of non-monotonic modal logics.

Definition 12 *An S5-model* $\mathcal{M} = (W, R, V)$ *is called a* $\mathcal{K}$*-minimal model for the set of formulas* I *if* $\mathcal{M} \models I$ *and for every preferred model* $\mathcal{N} \in \mathcal{K}$ *we have* $\mathcal{N} \not\models I$.

Schwarz showed that for logics $\mathcal{S}$ such as $S4F$ that are cluster-closed, minimal models correspond to $\mathcal{S}$-expansions and so characterise the logic. For the case of $wK4f$ we cannot directly apply Schwarz's Theorem since the class $\mathcal{K}$ which characterises $wK4f$ is not cluster-closed. In particular some two-floor models in $\mathcal{K}$ may not have a cluster as a maximum. On the other hand every model in $\mathcal{K}$ has a maximal *weak-cluster* (that is a cluster where irreflexive points are allowed or more precisely it is a rooted, symmetric, weakly-transitive frame).

It turns out that Schwarz's Theorem can be extended to the weak-cluster closed class $\mathcal{K}$, so accordingly we extend Definitions 11 and 12 to include this class. We then obtain the following recent result of Pearce and Uridia [2010].

Theorem 2 *Let* $\mathcal{M} = (W, R, V)$ *be an S5-model, and* $T = \{\varphi : \mathcal{M} \models \varphi\}$. *Then* T *is a* $wK4f$*-expansion of* I *if and only if* $\mathcal{M}$ *is a* $\mathcal{K}$*-minimal model of* I.

We are now ready to prove that τ^+ embeds equilibrium logic into non-monotonic $wK4f$. Although we could prove this in one step, it is more convenient to deal with the τ and the $^+$ embeddings separately and then compose them at the end. So, first let's show how the reflexivization map ^{r} preserves the property of being a minimal model.

Lemma 3 *Let* I *be a set of modal formulas.* $\mathcal{M}$ *is a minimal* $wK4f$*-model of* I^+ *if and only if* $\mathcal{M}^r$ *is a minimal* $S4F$*-model of* I.

Proof. For the 'if' direction suppose that ^{r} is the mapping from $wK4f$ models to $S4F$ models that makes irreflexive points reflexive. Let I be a theory in $S4F$ and consider any minimal model $\mathcal{M}$ of I^+. Since $\mathcal{M}$ is an $S5$-model, $\mathcal{M}^r = \mathcal{M}$. Suppose it is not a minimal model of I. Then there is a preferred model $\mathcal{M}' < \mathcal{M}^r$ such that $\mathcal{M}' \models I$ but there is a modal-free formula α such that $\mathcal{M}^r$ verifies α but $\mathcal{M}'$ does not. In particular $\mathcal{M}'$ is a 2-floor model that we can represent as the pair $(\mathcal{N}, \mathcal{M}^r)$, where $\mathcal{N}$ is the first-floor cluster. Now consider $(\mathcal{N}, \mathcal{M})$ as a $wK4f$ model. Evidently $(\mathcal{N}, \mathcal{M})^r = \mathcal{M}'$. So, by the reflexivization lemma, for any formula φ,

$$\mathcal{M}' \models \varphi \quad \text{iff} \quad (\mathcal{N}, \mathcal{M})^r \models \varphi \quad \text{iff} \quad (\mathcal{N}, \mathcal{M}) \models \varphi^+.$$

Now since α is a modal-free formula, $\alpha^+ = \alpha$ and so $(\mathcal{N}, \mathcal{M}) \not\models \alpha^+$. On the other hand, $(\mathcal{N}, \mathcal{M}) \models I^+$. But this contradicts the assumption that $\mathcal{M}$ is a minimal model of I^+. It follows that $\mathcal{M}^r$ is a minimal model of I.

For the 'only if' direction the argument is quite similar, applying the Reflexivization Lemma. It is left to the reader. $\square$

We have established one half of the embedding. Now let us turn to the semantic mapping between $S4F$-models and HT models and show that it preserves the property of being a minimal model.

Lemma 4 *Let X be a set of HT-formulas. $\mathcal{M}$ is a minimal model of $\tau(X)$ if and only if $\rho\mathcal{M}$ (short for $\langle\rho\mathcal{M}, \rho\mathcal{M}\rangle$) is an equilibrium model of X.*

Proof. For the 'if' direction, suppose that $\mathcal{M}$ is a minimal model of $\tau(X)$. Recall that $\mathcal{M}$ is an $S5$-model and $\rho V(p) = \{C(x) : \mathcal{M}, x \models Lp\}$. Now suppose for the contradiction that $\rho\mathcal{M}$ is not an equilibrium model of X. Then there is an HT-model $\mathcal{N}$ of X such that $\mathcal{N} \lhd \rho\mathcal{M}$. This means that for some propositional variable p we have $\mathcal{N}, t \models p$ and $\mathcal{N}, h \not\models p$. Suppose $\mathcal{N}$ has the form $\mathcal{N} = \langle\rho\mathcal{F}', U\rangle$ where $\mathcal{F}' = \langle W, R\rangle$ is an $S4F$ frame. By setting for each p

$$V'(p) = \{x \in W : \mathcal{N}, C(x) \models p\}$$

one obtains a modal ($S4F$) model $\langle\mathcal{F}', V'\rangle$ whose skeleton is (isomorphic to) $\mathcal{N}$. By construction $\langle\mathcal{F}', V'\rangle$ has two clusters. Its second floor is equivalent to $\mathcal{M}$ and verifies Lp and hence p, while its first floor does not verify p. It follows that $\langle\mathcal{F}', V'\rangle$ is preferred over $\mathcal{M}$. Now since $\mathcal{M}$ is a minimal model of $\tau(X)$, $\langle\mathcal{F}', V'\rangle \not\models \tau(X)$. But applying the skeleton Lemma we have for any φ,

$$\langle\mathcal{F}', V'\rangle \models \tau(\varphi) \text{ iff } \rho\langle\mathcal{F}', V'\rangle \models \varphi \text{ iff } \mathcal{N} \models \varphi.$$

This shows that $\mathcal{N} \not\models X$ contradicting our assumption.

The 'only if' direction is similar and left to the reader. $\square$

Theorem 3 *τ^+ is an embedding of equilibrium logic into non-monotonic $wK4f$. In particular we have for any set X of HT-formulas, and formula φ*

$$X \hspace{1pt}\vdash\hspace{-6pt}\sim \varphi \text{ iff } \tau(X) \hspace{1pt}\vdash\hspace{-6pt}\sim_{S4F*} \tau(\varphi) \text{ iff } \tau^+(X) \hspace{1pt}\vdash\hspace{-6pt}\sim_{wK4f*} \tau^+(\varphi).$$

Proof. Apply Lemmas 3 and 4 together with Theorem 2. $\square$

To complete our picture of modal embeddings, let us turn to two well-known strengthenings of $S4F$ and $wK4f$ respectively. $SW5$ is the extension of $S4F$ whose frames consist of a single reflexive point that sees a cluster (in Figure 1, W_1 becomes a single point). Likewise $KD45$ is captured by frames that comprise a single irreflexive point that sees a cluster (in Figure 2, W_1 becomes an irreflexive point while W_2 now contains only reflexive points). The non-monotonic version of $KD45$ is well-known as being (equivalent to) autoepistemic logic. Since the frames for $SW5$ are similar, but reflexive, its non-monotonic version, $SW5^*$ has been called by Lifschitz and Schwarz [1993] and Marek and Truszczyński [1993] *reflexive* autoepistemic logic.

It is easy to see that our previous embeddings also hold for $SW5$ and $KD45$ together with their autoepistemic extensions (notice that these logics fall under the scope of Schwarz's theorem and so they can be characterised in terms of minimal models). The proofs are entirely analogous. Without stating these properties as theorems, we merely present them as the picture in Figure 3.

The close relation between autoepistemic and reflexive autoepistemic logic was already studied by Marek and Truszczyński [1993].

5   Deriving embeddings for logic programs

Our next task is to examine some of the familar embeddings of logic programs under stable model and answer set semantics to nonmonotonic modal logics. We consider

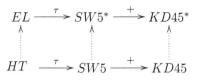

Figure 3

some of the most general ones here and show how to derive them from our main theorem.

Even before the birth of the stable model semantics for logic programs [Gelfond and Lifschitz, 1988] efforts were made to connect negation as failure with modal logic. Early steps were taken by Gabbay [1991] and Gelfond [1987].[9] Gabbay used the provability logic GL (and a specially adapted extension) to interpret negation-as-failure, while Gelfond provided a connection between SLDNF-resolution for stratified programs and provability in autoepistemic logic, a similar connection being maintained for arbitrary normal programs under the stable model semantics developed with Lifschitz around the same time [Gelfond and Lifschitz, 1988].

The concept of stable model or answer set was soon extended by Gelfond and Lifschitz [1991] to embrace more general kinds of logic programs, containing disjunctive rules and a second, strong negation operator, though it was not immediately apparent how these extensions could be related to non-monotonic modal systems. Answers were provided at the 2nd International LPNMR workshop held in Lisbon in June 1993. No fewer than three papers by five authors proposed similar embeddings of answer set semantics into autoepistemic logic [Marek and Truszczyński, 1993, Lifschitz and Schwarz, 1993, Chen, 1993], while two of these papers [Marek and Truszczyński, 1993, Lifschitz and Schwarz, 1993] also dealt with the relation to non-monotonic $SW5$, alias "reflexive" autoepistemic logic.

5.1   Answer sets and SW5 expansions

The main results of Marek and Truszczyński [1993] and Lifschitz and Schwarz [1993] concern disjunctive logic programs. They consist of rules α, that, if written as logical formulas, have the shape

$$p_1 \wedge \ldots \wedge p_m \wedge \neg p_{m+1} \wedge \ldots \wedge \neg p_n \rightarrow q_1 \vee \ldots \vee q_k \qquad (11)$$

where the p_i and q_j are atoms.[10] To establish modal embeddings they consider the following translation of a rule α of form (11), which we denote by $\sigma(\alpha)$:

$$Lp_1 \wedge \ldots \wedge Lp_m \wedge L\neg Lp_{m+1} \wedge \ldots \wedge L\neg Lp_n \rightarrow Lq_1 \vee \ldots \vee Lq_k \qquad (12)$$

A disjunctive logic program is a set of such rules and if Π is a disjunctive program let $\sigma(\Pi) := \{\sigma(\alpha) : \alpha \in \Pi\}$. The results of Lifschitz and Schwarz [1993] and Marek

---

[9]Gabbay's paper from 1991 was circulated in draft form from the mid-80s.
[10]They also include the case where the p_i, q_j can be atoms or their strong negations but let us postpone the topic of strong negation to a later section.

and Truszczyński [1993] are quite similar but they are presented and proved some-what differently. Lifschitz and Schwarz [1993] use the bi-modal system of minimal belief and negation-as-failure (as does Chen [1993]) while Marek and Truszczyński [1993] use the method of preferred models. We state the main property using their formulation.

Proposition 5 (Marek and Truszczyński [1993], Lifschitz and Schwarz [1993])
Let Π *be a logic program and* T *a set of atoms. Then* T *is an answer set for* Π *if and only if* $ST(T)$ *is an SW5-expansion of* $\sigma(\Pi)$.

If X is a set of modal-free formulas, $ST(X)$ stands for the unique stable theory E whose modal-free part is precisely $Cn(X)$, the classical propositional consequences of X.

It is well-known [Pearce, 1997, 2006] that equilibrium logic generalises stable model semantics to the full language of propositional logic. In particular, for any kind of logic program Π, a set T of atoms is a stable model of Π if and only if there is an equilibrium model $\mathcal{M}$ of Π (writing rules of Π as logical formulas) such that for any atom p, $p \in T$ iff $\mathcal{M} \models p$. If we represent HT-models as pairs $\langle H, T \rangle$ where H is the set if atoms verified 'here' and T is the set of atoms verified 'there', then T is an answer set of Π iff $\langle T, T \rangle$ is an equilibrium model of Π. It follows that all embeddings of equilibrium logic include as a special case corresponding embeddings of logic programs under stable models.

To derive Proposition 5, it suffices to note that equilibrium models and answer sets coincide and to show that the embedding σ is closely related to to τ. By inspection it is clear that for any program rule α of form (11),

$$\tau(\alpha) = L\sigma(\alpha). \tag{13}$$

Since answer sets correspond to equilibrium models, the embedding depicted in Figure 3 establishes the relation between answer sets of a disjunctive program Π and the reflexive autoepistemic expansions of $L\sigma(\Pi)$, where for any set of formulas Γ, we put $L\Gamma = \{L\varphi : \varphi \in \Gamma\}$. Although in extensions of $S4$, $\tau(\Pi)$ and $\sigma(\Pi)$ are not logically equivalent, it is easy to see that in $S4F$ or $SW5$ they have the same models, i.e. for any Γ,

$$\mathcal{M} \models \Gamma \Leftrightarrow \mathcal{M} \models L\Gamma \tag{14}$$

Consequently they have the same minimal models and hence the same $SW5$-expansions.[11]

To obtain Proposition 5 one may use the following simple lemma.

Lemma 5 *Let* $\mathcal{M}$ *be a minimal* $SW5$-*model and* φ *a modal-free formula. Then* $\mathcal{M} \models \varphi$ *if and only if* $\mathcal{M} \models \tau(\varphi)$.

This can be shown by induction on φ. Now notice that if T is an answer set of Π and hence $\langle T, T \rangle$ is an equilibrium model, then the formulas true in $\langle T, T \rangle$ are precisely the classical consequences $Cn(T)$ of T. Now we apply Lemmas 4 and 5. Then $\mathcal{M}$

---

[11] Alternatvely one can apply the definition of reflexive autoepistemic expansion Lifschitz and Schwarz [1993], Marek and Truszczyński [1993] to see immediately for that any set Γ, $L\Gamma$ and Γ have the same $SW5$-expansions.

is a minimal model of $\tau(\Pi)$ if and only if $\rho(\mathcal{M})$ is an equilibrium model of Π, and $\rho(\mathcal{M}) \models \varphi$ iff $\mathcal{M} \models \tau(\varphi)$ iff $\mathcal{M} \models \varphi$, for modal-free φ. Since the stable expansion is given by the formulas true in the minimal model, we have established that the modal-free formulas true in the stable expansion are precisely those in $Cn(T)$, where $\langle T, T \rangle$ is $\rho(\mathcal{M})$ and T is an answer set of Π. So T is answer set for Π if and only if $ST(T)$ is an SW5-expansion of $\tau(\Pi)$. Proposition 5 then follows from (14) and (13).

5.2 Autoepistemic logic

There have been several embeddings of stable model semantics into autoepistemic logic. For the case of disjunctive logic programs with rules of shape (11), the following translation γ was used by Lifschitz and Schwarz [1993], Marek and Truszczyński [1993] and Chen [1993]: for α of the form (11), $\gamma(\alpha)$ is the expression:

$$(p_1 \wedge Lp_1) \wedge \ldots \wedge (p_m \wedge Lp_m) \wedge \neg Lp_{m+1} \wedge \ldots \wedge \neg Lp_n \rightarrow$$
$$(q_1 \wedge Lq_1) \vee \ldots \vee (q_k \wedge Lq_k) \qquad (15)$$

The main result obtained by Marek and Truszczyński [1993] (and in a similar form also by Lifschitz and Schwarz [1993], Chen [1993]) can be stated thus:

Proposition 6 (Marek and Truszczyński [1993]**)** *Let Π be a logic program. A set of atoms T is a stable model of Π if and only if $ST(T)$ is an autoepistemic expansion of $\gamma(\Pi)$.*

To derive this result we will relate γ to our translation τ^+. We start by observing that for any formula α, since $\tau(\alpha) = L\sigma(\alpha)$, we have

$$\tau^+(\alpha) = \sigma^+(\alpha) \wedge L\sigma^+(\alpha), \qquad (16)$$

by applying $^+$ to both sides. Next we will show that in $KD45$, σ^+ and γ are equivalent. Given a program formula α, apply $^+$ to $\sigma(\alpha)$ to obtain:

$$(p_1 \wedge Lp_1) \wedge \ldots \wedge (p_m \wedge Lp_m) \wedge (\neg(p_{m+1} \wedge Lp_{m+1}) \wedge L(\neg(p_{m+1} \wedge Lp_{m+1})))$$
$$\wedge \ldots \wedge (\neg(p_n \wedge Lp_n) \wedge L(\neg(p_n \wedge Lp_n))) \rightarrow (q_1 \wedge Lq_1) \vee \ldots \vee (q_k \wedge Lq_k)$$

This can be simplified, in particular by noting that in $KD45$ the equivalence

$$\neg L\varphi \equiv L(\neg(\varphi \wedge L\varphi))$$

holds. Substituting for the RHS of this equivalence, the middle terms of $\sigma^+(\alpha)$ become

$$\neg(p_{m+i} \wedge Lp_{m+i}) \wedge \neg Lp_{m+i}$$

etc, which by propositional logic is equivalent to $\neg Lp_{m+i}$. So $\sigma^+(\alpha)$ becomes

$$(p_1 \wedge Lp_1) \wedge \ldots \wedge (p_m \wedge Lp_m) \wedge \neg Lp_{m+1} \wedge \ldots \wedge \neg Lp_n \rightarrow$$
$$(q_1 \wedge Lq_1) \vee \ldots \vee (q_k \wedge Lq_k)$$

which is $\gamma(\alpha)$ as required. So from (16) we can infer that the following equivalence holds in $KD45$:

$$\tau^+(\alpha) \equiv \gamma(\alpha) \wedge L\gamma(\alpha).$$

Now let Π be any theory and consider $\tau^+(\Pi) = \{\tau^+(\alpha) : \alpha \in \Pi\}$. Then this is equivalent to $\{\gamma(\alpha) \wedge L\gamma(\alpha) : \alpha \in \Pi\}$. This, in turn, is equivalent to the union of $\{\gamma(\alpha) : \alpha \in \Pi\}$ and $\{L\gamma(\alpha) : \alpha \in \Pi\}$, i.e. to $\gamma(\Pi) \cup L\gamma(\Pi)$, in the sense that the two sets of formulas have the same models.

We can now show that $\tau^+(\Pi)$ and $\gamma(\Pi)$ have the same autoepistemic expansions by noting that $\gamma(\Pi)$ has the same autoepistemic expansions as $\gamma(\Pi) \cup L\gamma(\Pi)$ in virtue of the follow simple lemma:

Lemma 6 *For any set of formulas Γ, in autoepistemic logic Γ and $\Gamma \cup L\Gamma$ have the same expansions.*

Proof. It suffices to recall that the autoepistemic expansions E of a set Γ are characterised by

$$E = Cn(\Gamma \cup \{L\varphi : \varphi \in E\} \cup \{\neg L\varphi : \varphi \notin E\},$$

where Cn is ordinary propositional consequence. So, for any set $X \supseteq \Gamma$,

$$Cn(\Gamma \cup \{L\varphi : \varphi \in X\}) = Cn(\Gamma \cup L\Gamma \cup \{L\varphi : \varphi \in X\})$$

which, applying some simple properties of Cn, proves the lemma.

This shows that $\tau^+(\Pi)$ and $\gamma(\Pi)$ have the same autoepistemic expansions and Proposition 6 follows by an analogous argument to the one we used to conclude Proposition 5.

6   Some other modal embeddings

Normal logic programs consist of non-disjunctive rules that correspond to formulas of shape (11) where $k = 1$. A starting point for the development of the stable model semantics was Gelfond's realisation that provability in logic programs could be brought into correspondence with derivability in autoepistemic logic [Gelfond, 1987]. In fact, as Marek and Truszczyński [1993] noted, Gelfond [1987] had already obtained a more general correspondence. Gelfond's translation (call it $g(\alpha)$) of a normal rule α:

$$p_1 \wedge \ldots \wedge p_m \wedge \neg p_{m+1} \wedge \ldots \wedge \neg p_n \to q \tag{17}$$

was simply

$$p_1 \wedge \ldots \wedge p_m \wedge \neg Lp_{m+1} \wedge \ldots \wedge \neg Lp_n \to q \tag{18}$$

and his result said that a set of atoms T is a stable model of a program Π if and only if there is a consistent autoepistemic expansion E of the translation $g(\Pi)$ of Π such that $T = E \cap At$, where At denotes the collection of all atoms in the language of E.[12]

---

[12] This result was obtained in an unpublished technical report. It was later extended to disjunctive programs by Przymusinski [1991].

It is easy to see that in general $g(\alpha)$ is equivalent neither to $\tau(\alpha)$ nor to $\gamma(\alpha)$ in $KD45$. Furthermore, although the stable models of Π and the consistent autoepistemic expansions of $g(\Pi)$ are in correspondence, the expansions of $g(\Pi)$ and of $\tau(\Pi)$ are not necessarily the same; g may determine different expansions than τ and γ. A further distinguishing feature is that using or τ and γ the translations of equivalent programs (or theories) are equivalent: if Π_1 and Π_2 have the same equilibrium models then their modal expansions are equivalent too (cf. Theorem 3). This is not guaranteed under the g embedding.

For normal programs with rules like (17) Marek and Truszczyński [1993] discuss a further embedding that we may label δ. It transforms (17) into

$$p_1 \wedge Lp_1 \wedge \ldots \wedge p_m \wedge Lp_m \wedge \neg Lp_{m+1} \wedge \ldots \wedge \neg Lp_n \to q.$$

In other words δ simplifies γ by not applying splitting to the head of the rule, q. However, δ and γ are fully equivalent in $KD45$ because for any normal α, $\vdash_{KD45} \delta(\alpha) \leftrightarrow \gamma(\alpha)$. Consequently these translations produce equivalent epistemic theories and thus the same expansions.

Let us also mention another embedding of $\mathcal{H}$ into $S4$, discussed by Egly [2000], who attributes it to Girard. This translation - call it τ' - is not equivalent to τ and leads to a different embedding relation. Formally, φ is derivable from Γ in $\mathcal{H}$ if and only if $\tau'(\varphi)$ is derivable from $L\tau'(\Gamma)$ in $S4$. A feature of this embedding τ' is that on disjunctive rules it agrees completely with the translation σ: for any such rule α, $\tau'(\alpha) = \sigma(\alpha)$. So if Π is a logic program, then $\tau(\Pi)$, $L\sigma(\Pi)$ and $L\tau'(\Pi)$ are all the same.

Finally, we should recall briefly two recent modal embeddings that were designed specifically to translate the logic HT and its non-monotonic extension, equilibrium logic. One of these is due to Lin and Zhou [2007] who embed equilibrium logic into the logic GK of knowledge and justified assumptions of Lin and Shoham [1990]. Since GK is a logic with two modal operators (both of which are employed in the translation), it lies somewhat outside the scope of our present discussion. For details of the somewhat complex translation and the results obtained, the reader is referred to the paper by Lin and Zhou [2007].

A second embedding of HT and equilibrium logic was proposed by Truszczyński [2007]. Since this mapping takes HT into $S4F$ and $SW5$ it falls within the scope of our present discussion; however we shall postpone a detailed analysis to a future work. We note merely that, like the embedding of Lin and Zhou [2007], Truszczyński's translation is composed of two parts. For any modal-free formula φ, the modal formula φ_{ML} is obtained by replacing each $a \in At \cup \{\bot\}$ in φ by MLa. Then, for any modal-free formula φ the corresponding modal formula φ_{mp} is defined inductively:

1. $p_{mp} = Lp$ for any atom p or $p = \bot$;

2. $(\varphi \wedge \psi)_{mp} = \varphi_{mp} \wedge \psi_{mp}$ and $(\varphi \vee \psi)_{mp} = \varphi_{mp} \vee \psi_{mp}$;

3. $(\varphi \to \psi)_{mp} = (\varphi_{mp} \to \psi_{mp}) \wedge (\varphi \to \psi)_{ML}$

One can show that for any propositional φ and set of formulas Γ,

$$\Gamma \models_{HT} \varphi \quad \text{iff} \quad \Gamma_{mp} \models_S \varphi_{mp}.$$

where $\mathcal{S}$ is $S4F$ or $SW5$; and this embedding lifts as well to the non-monotonic case. It seems likely that $\tau(\varphi)$ and φ_{mp} are closely related in $S4F$, but this matter will be studied in a future work.

7   Strong Negation

The concept of *strong negation* was introduced into logic by Nelson [1949] and later axiomatised by Vorob'ev [1952a,b]. We denote the strong negation operator by '$\sim$'. Nelson's logic $\mathcal{N}$ is obtained from $\mathcal{H}$ by adding the new negation $\sim$ together with the following axiom schemata (where '$\alpha \leftrightarrow \beta$' abbreviates $(\alpha \to \beta) \wedge (\beta \to \alpha)$):

N1. $\sim (\alpha \to \beta) \leftrightarrow \alpha \wedge \sim \beta$ **N2.** $\sim(\alpha \wedge \beta) \leftrightarrow \sim\alpha \vee \sim \beta$

N3. $\sim(\alpha \vee \beta) \leftrightarrow \sim\alpha \wedge \sim\beta$ **N4.** $\sim \sim\alpha \leftrightarrow \alpha$

N5. $\sim\neg\alpha \leftrightarrow \alpha$ **N6.** (for atomic α) $\sim\alpha \to \neg\alpha$

taken from the calculus of Vorob'ev [1952a,b]. These axioms can be added to any super-intuitionistic logic and the least strong negation extension of any such logic is always a conservative extension. Extensions by strong negation also preserve many metalogical properties of the base logic [Kracht, 1998].

We denote the least strong negation extension of HT by $\mathcal{N}_5$ since it is equivalent to a five-valued logic. Its models are like HT-models except that at each world a set of *literals* is verified, i.e. a set of atoms or their strong negations. Equilibrium models are defined by Pearce [1997, 2006] in an analogous fashion and provide a foundation for answer set semantics for logic programs containing strong negation as defined by Gelfond and Lifschitz [1991].[13]

Nelson's logic $\mathcal{N}$ can also be embedded in modal $S4$ by suitable extending the Gödel translation; Jaspers [1994] provides a detailed account. Briefly, the new translation is obtained by extending our previous τ by means of the following new clauses governing strong negation:

$$
\begin{aligned}
\tau(\sim p) &= L\sim p \text{ (for } p \text{ an atom)} \\
\tau(\sim(\varphi \wedge \psi)) &= \tau(\sim\varphi) \vee \tau(\sim\psi) \\
\tau(\sim(\varphi \vee \psi)) &= \tau(\sim\varphi) \wedge \tau(\sim\psi) \\
\tau(\sim(\varphi \to \psi)) &= \tau(\varphi) \wedge \tau(\sim\psi) \\
\tau(\sim\sim\varphi) &= \tau(\varphi) \\
\tau(\sim\neg\varphi) &= \tau(\varphi)
\end{aligned}
$$

If we denote the derivability relation for $\mathcal{N}$ by $\vdash_{\mathcal{N}}$, then we have the usual embedding relation (defined by Jaspers [1994]) given by

$$\vdash_{\mathcal{N}} \varphi \Leftrightarrow \vdash_{S4} \tau(\varphi).$$

All the embeddings that we have established for HT and equilibrium logic extend straightforwardly to their strong negation extensions, so our previous diagram now evolves to Figure 4. where the language of EL now contains strong negation. Simi-

---

[13] Strong negation is called *classical* negation by Gelfond and Lifschitz [1991].

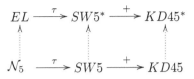

Figure 4

larly, by inspection, for logic programs Π with strong negation we have the same relation between the Gödel translation τ and the translation σ of Lifschitz and Schwarz [1993] and Marek and Truszczyński [1993]. Program rules α have the same shape as in (11) where now the p_i and q_j are literals. $\sigma(\alpha)$ is defined as before as in (12) and so as in the previous section we have $\tau(\alpha) = L\sigma(\alpha)$. Consequently it is a routine matter to derive the embedding results of Lifschitz and Schwarz [1993] and Marek and Truszczyński [1993] for programs with strong negation.

8 Concluding discussion

In this paper we have tried to bring together two lines of logical research that have proceeded largely independently in the past. On the one hand we have seen how the Gödel and the splitting translations, well-known in modal logic for many years, can be lifted from super-intuitionistic and (monotonic) modal logics to their non-monotonic extensions. This generalisation is quite straightforward given the minimal model characterisations of these logics. At the same time, we have also extended some familiar embeddings of normal and disjunctive logic programs under answer set semantics to embrace general theories in equilibrium logic, obtaining those previous results as a special case. Our embeddings are therefore not only modular and faithful but are based on fully recursive translations. As a by-product, our commutative diagrams suggest that the non-reflexive logic $wK4f$ and its non-monotonic extension may be interesting candidates for modelling a minimal notion of belief (Pearce and Uridia [2010] provide additional comments on the topic).

Our approach has some limitations, since we saw that there are embeddings of logic programs, such as the translation g of Gelfond [1987], that we cannot derive from τ^+. However, compared to the other translations we have encountered, g seems to be a rather special case. First, it is not clear whether it can be extended to more general kinds of program rules. Second, we saw that it gives rise to modal expansions different from those of other translations. Moreover, unlike those other translations it does not necessarily preserve the ordinary equivalence of programs. On balance, while g is historically significant in connecting logic programming with autoepistemic reasoning, it does not seem to provide us with a unifying picture of different non-monotonic logics.

On the other hand, there are also prospects for extending our methods to other logical systems that are currently of interest in AI and non-monotonic reasoning. One important area is the combination of reasoning with ontologies and classical knowledge bases together with non-monotonic rules. Recently, de Bruijn, Pearce, Polleres, and Valverde [2007b] have shown how the first-order version of equilibrium logic captures several versions of hybrid knowledge bases under answer set semantics, such as that of Rosati [2005a,b]. Fink and Pearce [2010] show that by extending the semantics of here-and-there logic one can also obtain a logical foundation for the dl-programs of Eiter, Lukasiewicz, Schindlauer, and Tompits. [2004]. It is well-known that the Gödel embedding extends to quantified versions of $\mathcal{H}$, $S4$ and their extensions (the result is discussed for instance by Egly [2000]). A topic for future work is to apply this extension to study embeddings of first-order equilibrium logic and non-ground logic programs. Already de Bruijn, Eiter, Tompits, and Polleres [2007a] have investigated embeddings of hybrid approaches based on ASP into a quantified version of autoepistemic logic. Several different embeddings are studied which are clearly related to the Gödel and splitting translations. The exact connection is an interesting open topic for future study.

Let us conclude with some words about the epistemic interpretation of logic programs. While default logic and non-monotonic modal logics were presented from the outset as full, general purpose logical systems, the semantics of stable models was designed for a special, restricted syntax of program rules. Embedding relations provided interpretations of the semantics into logical calculi but did not initially lead to a propositional logic for stable model semantics, in other words to a fully-fledged non-monotonic logic. Even when stable models began to be implemented and the idea of answer set programming emerged at the end of the 1990s, it was with a restricted syntax. This was soon augmented with additional, but non-logical constructions like weight constraints, cardinality constraints and aggregates. We have since learnt that all the main syntactic extensions of ASP rules, as well those additional programming devices, can be adequately represented within equilibrium logic which does offer a full system of non-monotonic reasoning. [14]

Since this logic can be embedded in non-monotonic modal systems, it can also be given an epistemic interpretation. At first sight this might seem to confirm the epistemic reading of program rules as still the most adequate interpretation. However, another, equally viable, picture emerges from our commuting diagrams. The logic of stable model semantics is a non-monotonic extension of super-intuitionistic logics that in turn express different degrees of constructive reasoning. Since Gödel's early results, we know that constructive logics can be given provability, modal or epistemic interpretations. This is an option, but not a necessity. We don't have to understand Heyting's and Kolmogorov's calculi as reflecting states of knowledge, that is simply a mathematical or philosophical possibility. We can also stay closer to the original ideas that these calculi provide a logic of mathematical constructions, or perhaps a logic of problems. In this case, if we want to add mechanisms for reasoning about knowledge, time or other modalities, we can do so directly. We do not need to proceed by translation into modal languages based on classical logic. The main lesson we can learn from the commuting diagrams is how apparently very different systems, based on quite distinct logical languages, can be seen with hindsight to be part of a single, common approach to non-monotonic reasoning.

---

[14] The paper by Ferraris [2008] and further references given there offer an extensive discussion.

References

W. J. Blok. *Varieties of interior algebras*. PhD thesis, University of Amsterdam, 1976.

G. Boolos. Provability in arithmetic and a schema of Grzegorczyk. *Fundamenta Mathematicae*, 106:41–45, 1980.

P. Cabalar and D. Lorenzo. New insights on the intuitionistic interpretation of default logic. In R. López de Mántaras and L. Saitta, editors, *Proceedings ECAI 2004*, pages 798–802, 2004.

A. Chagrov and M. Zakharyaschev. Modal companions of intermediate propositional logics. *Studia Logica*, 51:49–82, 1992.

A. Chagrov and M. Zakharyaschev. *Modal Logic*. Oxford Science Publications, 1997.

B. F. Chellas. *Modal Logic, an introduction*. CUP, 1980.

J. Chen. Minimal knowledge + negation as failure = only knowing (sometimes). In A. Nerode and L. M.Pereira, editors, *Proceedings Logic Programming and Nonmonotonic Reasoning*, pages 132–150, 1993.

J. de Bruijn, T. Eiter, H. Tompits, and A. Polleres. Embedding non-ground logic programs into autoepistemic logic for knowledge-base combination. In *Proceedings IJCAI2007*, pages 304 –309, 2007a.

J. de Bruijn, D. Pearce, A. Polleres, and A. Valverde. Quantified equilibrium logic and hybrid rules. In M. Marchiori *et al*, editor, *Proceedings RR2007*, Web Reasoning and Rule Systems, pages 58 –72. Springer LNCS, 2007b.

U. Egly. Properties of embeddings from **Int** to **S4**. In R. Dyckhoff, editor, *Proceedings Tableaux 2000*, LNAI 1847, pages 205–219, 2000.

T. Eiter, T. Lukasiewicz, R. Schindlauer, and H. Tompits. Combining answer set programming with description logics for the semantic web. In *Proceedings KR2004*, 2004.

L. Esakia. The modal logic of topological spaces. Preprint, 22pp, Georgian Academy of Sciences, 1976a.

L. Esakia. About modal 'companions' of superintuitionistic logics (abstract). In *The VII Logic Symposium, Kiev*, 1976b.

L. Esakia. On varieties of Grzegorczyk algebras(in russian). *Studies in Non-Classical Logic and Set Theory, Nauka, Moscow*, pages 257–287, 1979.

L. Esakia. Weak transitivity - restitution. *Logical Studies*, 8:244 –255, 2001.

L. Esakia. Intuitionistic logic and modality via topology. *Annals of Pure Applied Logic*, 127: 155–170, 2004.

P. Ferraris. Logic programs with propositional connectives and aggregates. *CoRR*, abs/0812.1462, 2008.

M. Fink and D. Pearce. A logical semantics for description logic programs. In *Proceedings JELIA 2010*, LNAI 6341, pages 156–168, 2010.

D. Gabbay. Modal provability foundations for negation by failure. In P. Schroeder-Heister, editor, *Extensions of Logic Programming*, LNAI 475. Springer, 1991.

M. Gelfond. On stratified autoepistemic theories. In *Proceedings AAAI-87*, pages 207–211, 1987.

M. Gelfond and V. Lifschitz. The stable model semantics for logic programming. In & K. Bowen R. Kowlaski, editor, *Logic Programming: proceedings 5th Int. Conf. & Symp.*, pages 1070–1080, 1988.

M. Gelfond and V. Lifschitz. Classical negation in logic programs and disjunctive databases. In *New Generation Computing*, volume 9, pages 365–385, 1991.

M. Gelfond, V. Lifschitz, H. Przymusinska, and M. Truszczyński. Disjunctive defaults. In *Proceedings KR91*. Cambridge, MA, 1991.

K. Gödel. Eine interpretation intuitionistischen aussagenkalkuls. *Ergebnisse eines mathematischen Kolloquiums 4*, pages 39–40, 1933.

R. Goldblatt. Arithmetical necessity, provability and intuitionistic logic. *Theoria*, 44:38–46, 1978.

A. Grzegorczyk. Some relational systems and the associated topological spaces. *Fund. Math.*, 60:223–231, 1967.

T. Hosoi. The axiomatization of the intermediate propositional systems S_n of Gödel. *Journal of the Faculty of Science of the University of Tokio*, 13:183–187, 1966.

J. Jaspers. *Calculi for Constructive Communication*. PhD thesis, ILLC Dissertation Series, Amsterdam, 1994.

M. Kracht. On extensions of intermediate logics by strong negation. *Journal of Philosophical Logic*, 27(1):49–73, 1998.

R. Kuratowski. *Topology*, volume 1. Academic Press, 1976.

A. V. Kuznetsov and A. Yu. Muravitsky. Provability logic (in russian). In *Proceedings IVth Soviet Union Conf. Math. Logic*, page 73, 1976.

V. Lifschitz. Minimal belief and negation as failure. *Artificial Intelligence*, 70:53–72, 1994.

V. Lifschitz and G. Schwarz. Extended logic programs as autoepistemic theories. In A. Nerode and L. M. Pereira, editors, *Proc. Logic Programming and Nonmonotonic Reasoning, 2nd Int. Workshop*, pages 101–114, 1993.

F. Lin and Y. Shoham. Epistemic semantics for fixed-points nonmonotonic logics. In Rohit Parikh, editor, *Theoretical Aspects of Reasoning about Knowledge: Proc, of the 3rd Conference*, pages 111–120, 1990.

F. Lin and Y. Zhou. From answer set logic programming to circumscription via logic of GK. In *Proceedings IJCAI2007*, pages 441–446, 2007.

V. Marek and M. Truszczyński. Reflexive autoepistemic logic and logic programming. In A. Nerode and L. M. Pereira, editors, *Proceedings Logic Programming and Nonmonotonic Reasoning, 2nd Int. Workshop*, pages 115–131. MIT Press, 1993.

J. McKinsey and A. Tarski. The algebra of topology. *Annals of Mathematics*, 45:141–191, 1944.

J. McKinsey and A. Tarski. On closed elements in closure algebras. *Annals of Mathematics*, 47:122–162, 1946.

J. McKinsey and A. Tarski. Some theorems about the sentential calculi of Lewis and Heyting. *Journal of Symbolic Logic*, 13:1–15, 1948.

D. Nelson. Constructible falsity. *Journal of Symbolic Logic*, 14:16–26, 1949.

D. Pearce. A new logical characterisation of stable models and answer sets. In *Non-Monotonic Extensions of Logic Programming, NMELP 96*, LNCS 1216, pages 57–70, 1997.

D. Pearce. From here to there: Stable negation in logic programming,. In D Gabbay & H Wansing, editor, *What is Negation?* Kluwer, 1999.

D. Pearce. Equilibrium logic. *Annals of Mathemathics and Artificial Intelligence*, 47(1-2):3–41, 2006.

D. Pearce and L. Uridia. Minimal knowledge and belief via minimal topology. In T. Janhunen and I. Niemela, editors, *JELIA*, 6341 LNCS, pages 273–285, 2010.

T. Przymusinski. Stable semantics for disjunctive programs. *New Generation Computing*, 9: 401–424, 1991.

R. Reiter. A logic for default reasoning. *Artificial Intelligence*, 13:81–132, 1980.

R. Rosati. On the decidability and complexity of integrating ontologies and rules. *Journal of Web Semantics*, 3(1), 2005a.

R. Rosati. Semantic and computational advantages of the safe integration of ontologies and rules. In *PPSWR2005*, 2005b.

G. Schwarz and M. Truszczyński. Modal logic S4F and the minimal knowledge paradigm. In *Proceedings TARK-IV*, pages 184–198, 1992.

G. Schwarz and M. Truszczyński. Minimal knowledge problem: A new approach. *Artificial Intelligence*, 67(1):113–141, 1994.

G. F. Schwarz. Minimal model semantics for nonmonotonic modal logics. In *Proceedings of LICS-92*, IEEE Computer Society Press, Washington, DC, 1992.

K. Segerberg. An essay in classical modal logic. *Filosofiska Studier. Uppsala: Filosofiska Foreningen och Filosofiska Institutionen vid Uppsala Universitet*, 13, 1971.

R. Solovay. Provability interpretation of modal logic. *Isreal J. Math.*, 25:287–304, 1976.

A. Tarski. Der aussagenkalkul und die topologie. *Fund. Math.*, 31:103–134, 1939.

M. Truszczyński. The modal logic S4F, the default logic, and the logic here-and-there. In *In Proceedings of the 22nd National Conference on Artificial Intelligence (AAAI 2007)*. AAAI Press, 2007.

D. van Dalen. *Logic and Structure*. Springer, 4th edition, 2004.

N. N. Vorob'ev. A constructive propositional calculus with strong negation (in russian). *Doklady Akademii Nauk SSR*, 85:465–468, 1952a.

N. N. Vorob'ev. The problem of deducibility in constructive propositional calculus with strong negation (in russian). *Doklady Akademii Nauk SSR*, 85:689–692, 1952b.